Beyond *SPUTNIK*

Beyond *SPUTNIK*

U.S. Science Policy in the Twenty-First Century

*Homer A. Neal, Tobin L. Smith, and
Jennifer B. McCormick*

THE UNIVERSITY OF MICHIGAN PRESS | ANN ARBOR

Copyright © by the University of Michigan 2008
All rights reserved
Published in the United States of America by
The University of Michigan Press
Manufactured in the United States of America
⊛ Printed on acid-free paper

2011   2010   2009   2008   4   3   2   1

*A CIP catalog record for this book is available from the
British Library.*

Library of Congress Cataloging-in-Publication Data

Neal, Homer A.
  Beyond Sputnik : U.S. science policy in the twenty-first century /
    Homer A. Neal, Tobin L. Smith, and Jennifer B. McCormick.
        p.   cm.
    Includes index.
    ISBN-13: 978-0-472-11441-2 (cloth : alk. paper)
    ISBN-10: 0-472-11441-7 (cloth : alk. paper)
    ISBN-13: 978-0-472-03306-5 (pbk. : alk. paper)
    ISBN-10: 0-472-03306-9 (pbk. : alk. paper)
    1. Science and state—United States. 2. Technology and
state—United States. 3. Research—Government policy—United
States. I. Smith, Tobin L., 1966– II. McCormick, Jennifer B.,
1969– III. Title.

Q127.U6N398   2008
338.973'06—dc22   2008007854

ISBN13 978-0-472-02745-3 (electronic)

*Dedicated to those individuals
who have mentored us in both
science and policy*

&

*to our families
who have stood by us
in our endeavors . . .*

. . . including the writing of this book.

Contents

Preface

Over the last half century, many societal advances have been driven by progress in science and technology. Devastating diseases have been conquered, our quality of life and national security have been enhanced, and new economic and intellectual frontiers have been opened. Yet the public generally sees this progress as good fortune, not recognizing that it is largely the result of a sustained commitment by the nation to support science through policies enacted immediately after World War II and in response to the Soviet launch of *Sputnik*. This book is an attempt to increase awareness of the importance of science and "policy for science" among scientists, policymakers, and the public, in the hope that the lessons from the past half-century can help guide choices for the future.

In 1999, I began a national science policy course at the University of Michigan that attracted undergraduate and graduate students from physics, engineering, medicine, education, and other disciplines. In searching for a suitable text for the course I was very surprised to find that none existed. I proceeded by drawing upon articles on science policy and other resources such as the National Science Foundation's *Science and Engineering Indicators,* and the AAAS "Science Policy Yearbook" and its "Annual Report on the Federal Budget."

In the spring of 2002, I had lunch with Tobin Smith, who was the University of Michigan's director of federal relations for research. In that capacity Toby had been making presentations on the importance of working with Congress, offering them to science and engineering faculty and students in both Ann Arbor and Washington, DC. We lamented the absence of a book that outlined the basic elements of national science policy. Given the growing interest within academic circles and the increasing number of university courses being offered to examine how public policy can support, regulate, and guide the conduct of scientific research, we were both surprised by our shared conclusion that no such book existed. The day after our lunch, I invited Toby to join me in writing it.

To test what we believed to be true, we conducted a review of existing books on science policy. We found none that provided a comprehensive account of science policy and related issues. Indeed, we were hard pressed to find one that contained a basic definition of science policy. While some books addressed aspects of the topic, we found that the existing literature (1) focused narrowly on specific issues; (2) was oriented toward the politics of science and how science fits in a democracy; (3) provided a historical perspective on science policy and the relationship between the government and the scientific community; or (4) addressed only how science and scientists influenced broader public policies, such as those relating to the environment or health care, not how policy influenced science itself.

It was to fill this void that we wrote this book. Early in the process, knowing how busy our schedules were, we sought the help of Jennifer McCormick. Jen was a graduate student in public policy who had taken a keen interest in science policy and had been one of the outstanding students in my science policy class.

In this text we focus on how policy is formulated to guide and influence the conduct of science. Our hope is to provide a basic text that can be used in an introductory undergraduate or graduate course on science policy. At the same time, the information we provide may be of value to a scientist, policymaker, or layperson who wants to learn about science policy and the context in which decisions about it are made.

From their individual backgrounds the authors have brought eclectic perspectives on science policy. Toby is currently the associate vice president for federal relations for the Association of American Universities (AAU) and represents the research interests of the nation's most prestigious collection of research universities, a post that requires daily interactions with science policymakers. He previously worked on Capitol Hill as a legislative assistant to former representative Bob Traxler (D-MI). Jen holds a doctorate in biology and a master's degree in public policy, has studied national science policy extensively, and is a postdoctoral fellow at the Stanford Center for Biomedical Ethics. My own experience has come as a high-energy physicist involved in large-scale research projects, and as an administrator responsible for academic and research programs at several of our nation's major research universities. I have served on the National Science Board, the Board of the Center for International and Strategic Studies, and on the boards of several national laboratories and a corporation with major R&D investment. I have also provided testimony in numerous congressional hearings on subjects ranging from the support of national laboratories to international collaboration and undergraduate science education. This diversity in the authors' backgrounds has led to many a lively discussion and sustained a healthy analysis of the topics in the text. We hope the result is a fair and comprehensive representation of the state of U.S. science policy and the prospects that lie ahead.

We have tried to make our text both substantive in content and interesting to read. We want to help readers—even those with limited knowledge of science or policy—to look deeply into issues of scientific exploration. Science flourishes in an open environment, yet we live in a world that imposes growing restrictions on it. Ethical conflicts in science are likely to grow and will require public involvement. We hope this book can help inform the public of emerging science policy issues. Moreover, we hope that the issues of science policy we discuss, the stories we tell, and the questions we raise will capture the interest of any concerned citizen.

To accomplish this goal, we review the historical role of national science policy in addressing the health, welfare, and security needs of the nation. We provide an organizational map to help the reader better understand how the federal government develops and executes its science policy and why it funds science. We explain how universities, national laboratories, and industry partner with the federal government to carry out scientific research, and why states are developing their own scientific and technological support structures.

Another goal is to provide the reader with a better understanding of the context in which science policy is made. We review the major issues, the interactions between the scientific community and policymakers, and the grand challenges that face science and society, including environmental preservation, transportation, power generation, and prevention and cure of diseases. We show that meeting these grand scientific and societal challenges depends on, and can in turn impact, public policy.

This book is published shortly after the fiftieth anniversary of the launch of *Sputnik*. Perhaps more than any other event in history, that launch caused the nation to realize the dangers of losing technological superiority. It changed national policies both scientific and social, giving the United States the boost it needed to create the space program, invest in science education, and encourage industrial R&D. Many of the technological benefits we now enjoy are the result of the commitments made by the nation in the wake of *Sputnik*.

Some observers have wistfully suggested that, in the face of contemporary dangers, we need another *Sputnik*. While we would not welcome another such threat to our national existence, we do argue that support for science and technology is essential if an advanced nation is to provide for the health, prosperity, and security of its citizens. We call upon scientists and policymakers to learn about each other's world, so that we can better work together to address the distinct challenges our nation faces in the twenty-first century.

—Homer A. Neal
Ann Arbor, Michigan

Acknowledgments

We want to extend our special thanks to the contributions of a number of individuals to this effort. They include David Abshire, Mariel Bailey, Jonathan Baugh, Audrey Benner, William Blanpied, Carol Blum, Jennifer Bond, Joanne Carney, Katharyn Cochrane, Dan Cohn, Dave Coverly, Benjamin Croner, Cinda Sue Davis, Tony DeCrappeo, Steve Dorman, Beth Demkowski, Jameson Dyal, Christine Feigal, Genene Fisher, Howard Gobstein, Robert Hardy, Lee Harle, Peter Harsha, Mary Hashman, Jeremy Herr, Erik Hoffmann, John E. Jankowski, Julia Jester, Dianne Jones, Gary Jones, Kei Koizumi, Ashley Martin, Natalia Melcer, Nicholas Moustakas, Bradley Nolen, Judy Nowack, Christian Osmeña, Marvin Parnes, Steven Pierson, Craig Rader, Carol Sickman-Garner, Sean Smith, Chelsea Stevens, Michael Telson, Bill Valdez, Jason Van Wey, Mike Waring, and Charles Wessner. There are others not mentioned who have reviewed parts of the text and contributed in various ways to this book. These include colleagues at the University of Michigan, Stanford University, and the Association of American Universities as well as at other associations and federal agencies. Their contributions also deserve recognition.

We wish to extend a special thanks to Jim Reische, our editor at the University of Michigan Press for most of this project, who, while leaving the Press before the book's final publication, played a major role in helping it come to fruition.

Finally, we hope the final publication of this work will help explain to our families why we had to spend the hours each week in extended Skype sessions as we labored in separate time zones to complete this project. We would like to acknowledge their support.

A Note on the Text

The text is organized into four sections, of five chapters each. The first section gives the reader a general overview of what science policy is, how U.S. science policy originated, the processes and players involved in its formation, the rationale for U.S. science policy, and the means used by the government to evaluate the effectiveness of its scientific programs. Complementing this overview, the second section aims to identify the partners that work with the federal government to conduct science and therefore who have a major stake in the development of policy on science. These include universities, federal laboratories, industry, the states, and the general public. In particular, we intend this section to address the respective roles of these various interests, explaining their unique roles both in the conduct of science and in science policy.

The third section takes on some of the major science policy issues that have been a part of the fabric of science policy, that still face the nation today, and that will continue to be of importance as we move forward. Finally, the fourth section looks toward the future and highlights some of the issues that will be relevant in an era of increasing globalization. The chapters in the second half of the book point to areas in which we believe the nation's scientists and policymakers must be engaged to insure the continued advancement of scientific research.

In each chapter, the reader will find at least one policy discussion box. The point of these is to not only to provide information on a particular issue, but also to raise questions and stimulate discussion. Policy is often debated and implemented in a context of differences of opinion and conflicting views. The questions raised by each side are often grounded in the values and perspectives they bring to the discussion. Policy outcomes many times result from compromises that balance multiple views. The discussion boxes highlight major questions in science policy and conflicts that have been and continue to be central to science policy. With the questions discussed in these boxes we hope to push the reader to think about the decisions that policymakers are forced to make on science policy.

Finally, to the extent that the text needs further explanation, we have included additional information in the notes for each chapter. In these notes we also provide a comprehensive listing of references for readers who are interested in further discussion of an issue we touch upon in the text.

Overview of
U.S. Science Policy

CHAPTER 1

Science Policy Defined

What Drives U.S. Science Policy?

On October 4, 1957, the Soviet launch of *Sputnik I* sent shock waves around the world—shock waves felt most strongly in the United States, where the news of the launch of the world's first artificial satellite indicated that the country's Cold War rival had beaten the United States into space. The result was widespread panic among the American people, a fear that the nation had lost its scientific and technological superiority.

At the time of *Sputnik,* the struggle between the United States and the Soviet Union was more than a chess game, an ideological struggle with science and technology as surrogates for the issues involved. At stake for the United States was a potential nuclear attack and takeover by a Communist nation. Coming on the heels of the McCarthy era, *Sputnik* produced a climate of near-hysteria, fueled by a sense that there was now an eye in the sky capable of looking down on the United States at will. Perhaps bombs could eventually be released from outer space—weapons against which the country had neither the scientific nor the technological ability to defend itself.

More than any other event in U.S. history, the *Sputnik* crisis focused the attention of the American people and policymakers on the importance of creating government policies in support of science and of education, with the aim of maintaining U.S. scientific, technological, and military superiority over the rest of the world.

The year 1958 was a milestone in the history of science policy, as the United States undertook a series of major actions that cemented the foundation for more than half a century of national science policy. A little over a month after the launch, President Eisenhower appointed James R. Killian, president of MIT, to be the first special assistant to the president for science and technology. Killian's appointment was a sign of the ascension of science to a new position of importance: as his memoir notes, "Only when Jefferson was his own science advisor and Vannevar Bush was advising Franklin Roosevelt during World War II was science so influential in top government councils."[1]

Sputnik also led to passage of the Space Act of 1958, which created the National Aeronautics and Space Administration (NASA). NASA was charged with carrying out the space program and developing long-term aerospace research for civilian and military purposes. That same year, Congress also enacted the National Defense Education Act, which was designed to encourage a new generation of students to pursue degrees in science and engineering.

Finally, 1958 saw Eisenhower's creation of the Advanced Research Projects Agency (ARPA)—now known to many as "DARPA"—within the Department of Defense (DOD). ARPA was charged with preventing technological surprises like *Sputnik* and with developing innovative, high-risk research ideas that held the potential for significant technological payoffs.[2]

Funding for existing science agencies also increased dramatically during the years immediately following *Sputnik*. In 1959, Congress increased funding for the National Science Foundation (NSF) to $134 million, from a figure of just $34 million the year before. This explosive growth was characteristic of the entire post-Sputnik era. The NSF's budget grew from just $3.5 million in its first full year (FY1952) to total funding of $500 million by 1968.[3]

At the same time, further activities and policies sparked the development of a new university and national laboratory system, which would eventually nurture unparalleled scientific growth. Perhaps the first major building block in this structure was a report delivered a dozen years before the *Sputnik* crisis: *Science—the Endless Frontier,* prepared by Vannevar Bush. It was requested by President Franklin D. Roosevelt and submitted to President Harry Truman in July 1945.[4] This document was the foundation for modern American science policy, and provided the impetus for Truman's signature on the legislation that created the National Science Foundation.

Other major research agencies had been emerging from the late 1940s onward. The Office of Naval Research and the Atomic Energy Commission—the precursor of today's Department of Energy—were both created in 1946 to channel government sponsorship of major research. The army and the air force created their own research offices in 1951 and 1952, respectively. New health institutes had been created in the late 1940s, including the National Institute of Mental Health, the National Heart Institute, and the National Dental Institute; in 1948, Congress passed legislation aggregating them under the new name of the National Institutes of Health (NIH).

The building blocks put in place at the end of World War II and in response to *Sputnik* established the general structure in which science is conducted in the United States today. Major policy decisions established universities as the primary vehicle through which government-sponsored basic research would be conducted, created our system of national laboratories for the purpose of advancing science in support of national security and other needs, and inspired a generation of students to pursue degrees in science and engineering. Immediately following World War II and during the early years of the Cold War, government support of science grew, and the scientific enterprise flourished. During the 1960s science was at the heart of one of the nation's goals—namely, sending a person to the moon. The 1970s and 1980s brought energy shortages and crises, and citizens once again turned to science to provide a solution. Scientists and engineers were instrumental in meeting the nation's defense needs as the Cold War progressed. With the end of the Cold War in the early 1990s, the U.S. scientific enterprise turned its attention to meeting the demands of an aging population and finding cures for major diseases. This shift translated into significant funding increases beginning in the late 1990s for health research at the NIH.

Today, some scientists and others fear that public enthusiasm for government support of science may be waning, even as science and innovation grow more important to our economy and national security. Indeed, concerns have emerged that America's global leadership in science may be in danger from neglect and inattention. These range from fears that have emerged post September 11, 2001 that U.S. policy makers have failed to understand the importance to science of openness and the free movement of ideas and people, to worries about the politicization of science, to laments about governmental neglect of science and math education.[5]

Further and quicker scientific advances will be necessary if the United States is to outrun its competitors in the new, knowledge-driven global economy. This will require a renewed commitment to science and to the government policies that support it. Concern over the commitment of the government and the public to science has led some, including Representative Vernon Ehlers R-MI, the first PhD physicist ever elected to Congress, to ask, "Where is Sputnik when we need it?"[6] Others, such as Microsoft chairman Bill Gates, have echoed this wish for a *Sputnik*-like event, a so-called Sputnik moment, that would once again lead to farsighted government science policies.[7]

The serious tone of a 2005 National Academy of Sciences report, Rising above the Gathering Storm, provides good reason for focusing on and reevaluating our existing national policies for science.[8] Unlike *Sputnik,* the next crisis may be difficult to detect at first, with no advance warnings to capture the attention of the American public and leading policymakers. In fact, some individuals, such as Shirley Jackson, former president of the American Association for the Advancement of Science and president of Rensselaer Polytechnic Institute, have suggested that we already face a "quiet crisis."[9] During a National Press Club event in February 2005 at which business and academic leaders outlined their concerns about growing competition from emerging Asian nations such as China and India—competition that threatens U.S. scientific and technological superiority—Intel CEO Craig Barrett remarked: "It's a creeping crisis, and it's not something the American psyche responds to well. It's not a Sputnik shot, it's not a tsunami."[10] Award-winning *New York Times* columnist Thomas L. Friedman claims that this crisis "involves the steady erosion of America's scientific and engineering base, which has always been the source of American innovation and our rising standard of living."[11]

There are many reasons why we should be concerned. Our students perform poorly on international science and math tests. Many of our industries are losing their traditional leadership roles to companies from abroad. More and more U.S. jobs are being outsourced to foreign nations. Major research advances are being made outside the United States, as other countries increase their commitments to their own scientific research efforts. More and more of the brightest students from around the world decide against enrolling at American universities, choosing foreign institutions that are beginning to rival our own in the quality of research and instruction. Meanwhile, the United States faces major scientific and technological hurdles to national challenges such as reducing our dependence upon foreign oil, addressing the potential for a global pandemic, and defending against the threat of biological attack.

While these problems are not likely to generate the kind of popular alarm provoked by the Soviet launch of *Sputnik,* they should not be ignored. After all, these challenges, and many others like them, beg for increased attention, and thus for public awareness of how our government guides our country's scientific and technological advancement. It would be unfortunate if the United States were to become a follower, as opposed to a world leader, in science and technology. At the same time, policies must be in place to ensure that science continues to be conducted, and its results used, in an ethical and socially acceptable manner. Determining what policies meet these conditions is a major challenge, as will be made clear in the remainder of this book.

Beyond Sputnik focuses on governmental policies that affect the conduct of science. We explore areas where government regulation of the areas of inquiry and the practices followed by scientists is clearly required, such as the use and protection of human research subjects. We explain how the government has devised structures and policies intended to advance science and technology, and discuss governmental and nongovernmental approaches to supporting research and development (R&D). Moreover, we try to address as fully as possible the questions of what science policy entails, and what policymakers must do to derive maximum advantage from scientific and technological advances. We also address how science policies sometimes emerge as a result of larger societal needs and goals.

Before we start to do all this, however, we should first define what we mean by science and public policy.

Science Defined

What exactly is science? The word itself derives from the Latin *scientia,* meaning "knowledge." The term *science* can be used to describe both a process and an outcome—the process of obtaining knowledge, and the knowledge that is obtained. Thomas Kuhn, a physicist and historian of science, hints at this duality when he notes that science is "the constellation of facts, theories, and methods collected in current texts," while "scientists are the men [and women] who, successfully or not, have striven to contribute one or another element to that particular constellation."[12]

Carl Sagan, the late astronomer, popularizer, and dogged critic of pseudoscience, highlighted this same tension. Even though he was not certain what science *meant* when he was young, he was rapt with the splendor of the stars and the expansive night sky. He wanted, he said, to take part in the activities leading to both the discovery of new things (process) and the understanding of what it all meant (outcome). "Science," Sagan wrote, "is more than a body of knowledge; it is a way of thinking."[13]

Science is ultimately about both the search for "truth" and new knowledge. Central to its pursuit is the conviction that truth must be obtained in an objective and systematic manner, by incorporating standard models and methods, statistical analyses, controlled experiments, and replication. Its goal is to better understand the world in which we exist, and to create rational and probable models that explain occurrences within it. Science is ideally impersonal and value free. In fact, one important test of scientific validity is the susceptibility of findings from one scientist or group of scientists to duplication by others.

While there is no single, exact scientific protocol, researchers employ what is referred to as the scientific method. The steps involved include observation and characterization of a phenomenon or group of phenomena; the development of hypotheses and theoretical explanations; the use of evidence to predict phenomena or observations; and the use of experiments to test those hypotheses and predictions. The scientific method is used cooperatively over time by all scientists (including the social scientists) to describe aspects of the world and its inhabitants in a reliable, nonarbitrary manner.

After World War II, when the U.S. government began to formally provide major support for scientific research, there was much debate about whether to include the social sciences. Those favoring exclusion won out in the first round, and when the National Science Foundation was established in 1950, its mandate did not include the social sciences. Only four decades later, in 1991, did the NSF establish its Directorate of Social, Behavioral, and Economic Sciences. Today, the NSF is not the only agency funding the social sciences. The National Institutes of Health and the Department of Defense are also heavy sponsors of social science research: the mission of the NIH includes not only biomedical but also social and behavioral research, while the DOD is increasingly interested in these areas as they pertain to national defense. The departments of Agriculture and Commerce have also traditionally devoted funding to work in economics.

However, there is a tendency to consider disciplines of the social sciences—political science, economics, sociology, and psychology—as different from physics, chemistry, biology, and geology. The former have often been referred to as the "soft" sciences, while the latter are known as the "hard" sciences. This distinction refers not to the level of intellectual talent or skill required from practitioners, nor to the use of the scientific method, but rather to the ease with which a discipline's observations can be predicted and reproduced by future experiments, or whether a phenomenon can be explained from its component parts. Advances in computation and mathematical modeling now enable social scientists to make far more reliable predictions than ever before, and differences between the methodologies used by the various science disciplines is narrowing. Nevertheless, the social sciences are still periodically attacked by policymakers as unworthy of government support.

Science versus Technology, Research versus Development, and Science versus Engineering

The general public, policymakers, and even scientists are sometimes confused about differences between *science and technology,* or S&T, on the one hand, and *research and development,* or R&D, on the other. Confusion about these terms is understandable, inasmuch as they are frequently used synonymously: one often sees *S&T* and *R&D* used to refer to the same activity, for example. For our purposes, however, it is important to define science and technology, and to distinguish them from research and development, even though in some cases the lines between them are blurred.[14]

As we explained earlier, science may be thought of as the objective pursuit of knowledge and understanding through the scientific method. The understanding produced by science is articulated through concepts, words, theories, and equations. Science may also be viewed as the world's store of knowledge about the natural universe and those who inhabit it. Technology, in contrast, derives from a conscious attempt to draw upon existing scientific or engineering knowledge for the purpose of achieving a specific material result.[15] The use of both science and technology can significantly affect our lives, in positive or negative ways.[16]

Research may be thought of as the process through which scientific principles are developed and tested. The NSF defines research as systematic study directed toward fuller knowledge or understanding of the subject studied. In contrast, it defines development as systematic use of the knowledge or understanding gained from research, directed toward the production of useful materials, devices, systems, or methods, including design and development of prototypes and processes.[17] Quality control, routine product testing, and production are all excluded from this definition.

Research is often classified by federal agencies as either "basic" or "applied," depending upon the objective of the sponsoring agency. *Basic research* is aimed at gaining more comprehensive knowledge or understanding of the subject under study without specific applications or products in mind. *Applied research* is aimed at gaining new knowledge or understanding to meet a specific, recognized need. It focuses on the creation of knowledge that has a specific application or commercial objective relating to products, processes, or services. In contrast, development can be thought of as the use of knowledge gained from research to produce useful materials, devices, systems, and methods. Development includes designing and developing prototypes and related processes.[18]

The term *R&D* comprises basic research, applied research, and development. When we refer to funding for R&D, this includes funds spent for R&D personnel, program supervision, and administrative support directly associated with R&D activities. Expendable or movable equipment needed to conduct R&D—for example, microscopes or spectrometers—is also included.

In examining *science policy,* we often use the component terms of *science* and *engineering,* or *S&E,* interchangeably. We would note, however, that science and engineering are quite different, with engineering most often referring to the practical application of science to specific problems. Just as there are several different disciplines of science, there are several different disciplines of engineering, including aerospace, electrical, chemical, civil, mechanical, environmental, and computer science.

Pasteur's Quadrant

One reason for the frequent confusion between *science and technology* and *research and development* is that it is not always clear where one ends and the other begins. It is important to keep in mind that many of the policies formulated since World War II operate under the assumption that research and development exist on a sequential continuum, starting with basic research, then going to applied research, then to development and the creation and deployment of new technology, with each stage building upon the one preceding it. This understanding of the transformation of scientific knowledge into technology is known as the *linear model* (see fig. 1.1). The linear model has been used by both scientists and policymakers as the primary paradigm for interpreting the nature of research since World War II, and it continues to be used even today.[19]

Questions are now being raised about the relevance of the linear model to twenty-first-century research.[20] Much of the skepticism has been precipitated by Donald E. Stokes's book *Pasteur's Quadrant,* published in 1997.[21] Stokes argues that research falls into one of four quadrants (see fig. 1.2). The first represents what is traditionally viewed as pure, "basic," or largely theoretical research, in which researchers have no interest in seeking potential uses for their findings, but are working solely to advance knowledge. This quadrant is exemplified in Niels Bohr's research on the structure of the atom. The second quadrant represents strictly "applied" research, or research with a practical end in mind, as represented by Thomas Edison's quest to create an effective lightbulb. The third quadrant includes work that is neither basic nor applied. Examples include taxonomic or classificatory research, which, while important, is not conducted with the creation of new knowledge or the development of practical solutions in mind.

The fourth and final quadrant represents what Stokes defines as "use-inspired basic science." He labels this "Pasteur's quadrant," after Louis Pasteur, whose work had significant theoretical *and* practical applications. In this quadrant, the researcher works to advance scientific knowledge, but remains acutely aware of the potential practical applications for his or her findings.

Stokes's model illustrates that the relationship between science and technology is more complex and dynamic than the linear model suggests.[22] Under this new paradigm, knowledge can be initiated in any of the quadrants and may ultimately have an impact on all of the other quadrants. Indeed, Stokes's work suggests the path by which scientific knowledge is applied and used is not necessarily linear or sequential; that use and desired applications can and often do pull both science and technology. One might therefore think of science and technology as being

Research is inspired by:

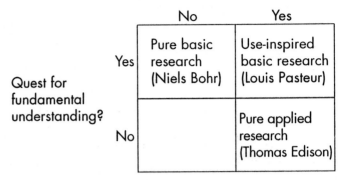

FIG. 1.2 Pasteur's quadrant. (Donald E. Stokes, *Pasteur's Quadrant: Basic Science and Technological Innovation* [Washington, DC: Brookings Institution Press, 1997], figs. 3–5.)

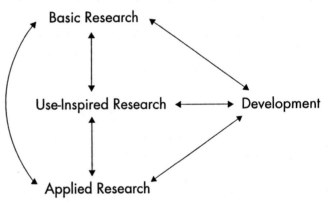

FIG. 1.3 A dynamic and parallel model of research and innovation

parallel tracks of cumulative knowledge that continually interact with each other and that have many connections and interdependencies (fig. 1.3).[23]

Stokes's revised dynamic model better describes the interaction and feedback that occur between knowledge generation, the application of knowledge, and technological innovation. As Stokes points out, adopting a more complete understanding of the relationship between scientific advancement and technological innovation is important in order to craft effective science policies and to renew the post–World War II compact that emerged between science and the government.[24]

Scope of the Scientific Enterprise

Scientific investigation in the United States is carried out in a large number of venues, ranging from national labo-

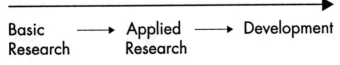

FIG. 1.1 The linear model

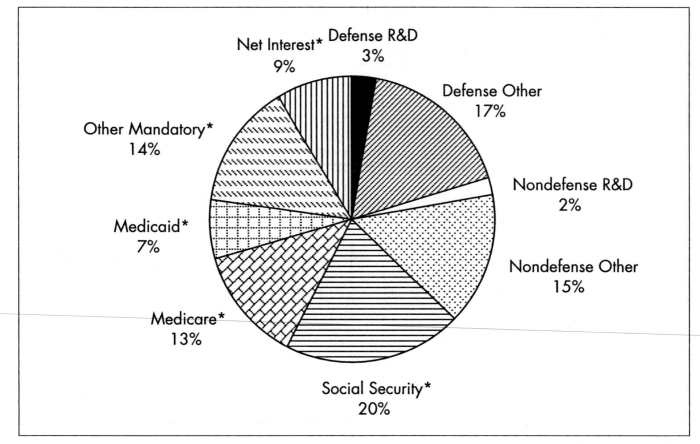

FIG. 1.4   Composition of the FY2006 federal budget. * = nondiscretionary spending. (Office of Management and Budget, *Budget of the U.S. Government FY2008*, as taken from AAAS, table I-2, "Distribution of the FY2008 Budget [outlays in billions of dollars]," http://www. aaas.org/spp/rd/08ptbi2.pdf and AAAS, table 1, "R&D in the FY2008 Budget by Agency [budget authority in millions of dollars]," http://www.aaas.org/spp/rd/prev08tb.htm.)

ratories to universities, research institutes, and industry. The researchers who conduct science are commonly referred to as the *scientific community*. In 2003, more than four million workers in the United States were employed in formal science and engineering occupations in the United States, and there were more than fifteen million S&E degree holders in the workforce.[25]

The federal government provided approximately 30 percent ($93 billion in FY2004) of all private and public funds expended for research and development in the United States ($312 billion in FY2003).[26] In FY2005, R&D expenditures (defense and nondefense combined) represented 5.4 percent of the overall $2.4 trillion federal budget and 16.1 percent of the proportion available for domestic discretionary (nonmandatory) spending.[27] While overall defense and nondefense R&D spending represents a relatively modest share of the total U.S. budget (see fig. 1.4), the federal government's R&D investment plays an essential role in supporting the nation's science and technology enterprise. Indeed, a majority of the support provided to basic research comes from the federal government.

Public Policy Defined

Notable political scientists have asserted that the search for a specific definition of public policy can quickly "degenerate into a word game."[28] Indeed, many definitions of public policy have been developed, and agreement on a single definition does not exist.[29] Even the Supreme Court has been unable to render a precise definition of the term, stating that "no fixed rule can be given by which to determine what is public policy": it is, the Court notes, "a very uncertain thing" and "impossible to define with accuracy."[30]

For our purposes, public policy may be thought of as the processes and players involved in making governmental decisions, the factors that influence their decisions, and the manner in which those decisions are carried out. This definition encompasses the roles of the various players in the process, the factors that motivate them, and the effectiveness of their actions. Concisely defined, *public policy* refers to the outcome produced when public officials arrive at some decision regarding the best course of action for addressing an issue of public concern. The policy itself is typically expressed in the form of a law, an executive directive, an agency policy, a rule, or a regulation. The outcome may be reached through legislative, executive, or judicial action or through a public referendum.

Like *science, policy* refers to both a process and a product.[31] We will spend a great deal of time examining the process of public policy, exploring the hows, whos, and whats of the process as it relates to the conduct of science. Who are the players in making science policy? What mechanisms and tools do they use? At what levels are key science policy decisions made? Who implements and enforces them? How? What motivates individuals involved in science policy to act as they do? What role do external actors and forces, including practitioners of science, play in influencing decisions on policy?

Once we've answered these fundamental questions, we will turn to outputs, examining specific science policy issues, and describing how policymakers have labored to address them.

Science Policy Defined

National science policy refers to the federal rules, regulations, methods, practices, and guidelines under which scientific research is conducted. It also refers to the dynamic, complex, and interactive processes and procedures—both inside and outside government—that affect how these rules, regulations, methods, practices, and guidelines are devised and implemented. In a sense, national science policy is nothing more than public policy governing matters of science (and, to some extent, technology), including research, development, regulation, and overall support of the national scientific community.[32] Therefore, it should be consistent with the general tenets of "public policy" and "public policy–making." Science policy involves, at some level, all three branches of government and is the result of a continuing dialogue between policymakers and the scientific community. As is the case for all public policy, science policy is often neither rational nor comprehensive. It is incremental and somewhat disjointed by nature, and may seem especially so to those who do not fully understand how it is created.[33]

Ideally, science policy should support the needs of citizens, and intrude on the conduct of science only when such intrusion enhances the public good. Moreover, it should not interfere when doing so would limit the progress of science without a concomitant reduction of public risk. In reality, however, specific science policies often do not live up to this high standard. Furthermore, the incremental nature of the policy-making process means that unforeseen problems often have to be fixed "on the fly," in response to outcries from those in the scientific community who are most greatly affected. Finally, most of those making science policy are nonscientists with no clear concept of the scientific method, much less the expertise and research experience needed to evaluate a particular line of research. Thus, the creation of science policy, like all public policy, is far from exact. As aptly described by political scientist Charles Lindblom in his discussion of the incremental and evolutionary—as opposed to revolutionary—nature of the policy-making process, science policy is, itself, a "science of muddling through."[34]

The Difference between Science and Science Policy

As the analyst Phillip Griffiths has noted, science policy is very different from the conduct of science itself. While science is ideally value free and objective, as we have noted, Griffiths describes science policy as "concerned with the incentives and the environment for discovery and innovation; more mundanely, science policy deals with the effect of science and technology on society and considers how they can best serve the public. As such, it is highly visible, value-laden, and open to public debate."[35]

Because of the subjective nature of science policy, whether a specific policy is "right" or "wrong" is often impossible to prove. Moreover, the evaluation of science policy outcomes is often driven by ideology, as opposed to provable facts. This has led many in the scientific community to shy away from engagement in the policy process. Ironically, the scientific voice has often been absent from debates over major policies affecting the scientific community and its work.

In a democracy, government-funded research must be conducted within the rules, regulations, and laws governing the society in which the work is embedded. For example, researchers may not violate the constitutional separation of church and state by using federal funds to promote or question the beliefs of specific religions. Scientists must also be accountable for the expenditure of public funds used in carrying out their projects. As with all sound public policies, those regulating publicly funded research should always have the welfare of society as their preeminent consideration.

These lofty principles, however, pose many dilemmas for scientists and policymakers seeking to balance short-term and long-term risks. Is it more efficient, for example, to invest in activities that will produce incremental but certain advances? Or is it wiser to invest in radical ideas that, though unlikely to produce immediate results, might make remarkable advances in the future? Consider Christopher Columbus sailing west from Europe into uncharted waters, in search of a shorter passage to India. As we know, his voyages and exploits took several years and did not produce the desired end, but it did result in his finding the New World, one of the greatest discoveries in history. A more recent example is the laser, which grew out of government-sponsored research. One of the most influential technological achievements of the twentieth century, at the time of their invention lasers were dubbed "a solution looking for a problem," for their specific applications and societal benefits were yet unknown.[36] Today lasers are essential to the operation of compact discs, bar code readers, computer printers, long-distance telephone communications over fiber optic cable, LASIK eye surgery, and many other applications.

It would be helpful, of course, to articulate clear principles that would govern the conduct of all publicly funded research. Unfortunately, this will never occur: scientific research represents the search for knowledge and understanding, and for the foreseeable future, nature's full truth will not be known. Nevertheless, policies embraced by scientists and public alike offer the best chance for the steady advancement of both science and society. Such a common understanding can be difficult to achieve. Many of the most challenging issues for twenty-first-century science are encountered right where the scientific community's views diverge from those of the broader public.

"Policy for Science" versus "Science for Policy"

Science can both influence and be influenced by government policy. Science can and does affect decisions in many areas of societal importance (e.g., health, energy, environment), and this interaction can in turn affect science policy. Harvey Brooks, a pioneer in the study of science and public policy, is credited with first making the distinction between *policy for science* and *science for policy*. Brooks characterized *policy for science* as decision making about how to fund or structure the systematic pursuit of knowl-

POLICY DISCUSSION BOX 1.1

Who Knows Best: Scientists or Society?

As one of the primary tools permitting society to extend the frontiers of understanding, science is often caught between existing societal beliefs and the attempt to redefine them. The fate of Galileo, condemned as a heretic for proclaiming that the sun—not the earth—was the center of the known universe, exemplifies how the search for truth can place a scientist at odds with the cherished beliefs of society.

Publicly funded research may also lead to findings that challenge deeply held societal and religious beliefs, a tension alive in current debates over evolution versus creationism and intelligent design. Similar arguments are raging over the use of human fetal tissue and embryonic stem cells in scientific research and the creation of genetically modified crops. On the one hand, these research methods and the scientific advances derived from them may conflict with religious, moral, and ethical beliefs. On the other hand, human embryonic stem cell research may ultimately lead to cures and treatments for spinal cord injuries, heart disease, and Parkinson's disease, while genetically modified crops may be more nutritious and disease resistant, helping to feed the hungry in poorer nations around the world.

Conflict between science and religious, moral, and ethical beliefs lies at the heart of some of today's most contentious debates over policy. In the end policymakers are left to grapple with daunting questions: Who knows best, scientists or society? Whose wishes should predominate in instances of conflict? Do moral and religious concerns trump the need for new knowledge that might lead to scientific advances and beneficial technologies? And finally, at what point does depending upon what society believes to be true inhibit the progress of the very science that might ultimately contradict those beliefs?

edge, while *science for policy* concerns the use of knowledge to assist or improve decision making.[37]

Because of their complementary nature, the distinction between *policy for science* and *science for policy* is often murky.[38] While the two are easily confused, the former is a direct application of policy to regulate and oversee the conduct of science, or policy for science. It must be differentiated from policy that is informed by science, that is, where science is used specifically to inform the policy-making process, or science for policy. For example, policies that rely on science—such as laws, regulations, and standards pertaining to air and water quality, pesticide usage, food processing and handling, drug usage and building codes, to name a few—are not policies governing science. Rather, they are largely policies that have been informed by science. While such policies are discussed in this book, the reader will find that much of the focus is on government policies designed specifically to shape, guide, and regulate science and its conduct.[39]

If a better understanding could be reached between scientists and policymakers about the two very different worlds in which they operate, perhaps we could more fully realize science's potential to inform policy-making. Such increased understanding would also go a long way toward ensuring that policymakers do not unintentionally or needlessly constrain scientific research.

Why Do We Need a National Science Policy?

It may be surprising to some that the United States has a national policy for science. Aren't many current science policies just by-products of efforts to achieve other national objectives? Certainly, there is some truth to this assumption; in fact, prior to World War II, there was no well-defined U.S. strategy for the support of science.[40] However, the war itself led to recognition of the value in knowledge for its own sake. New knowledge, it was agreed, was critical to progress in the war against disease; to the creation of new products, industries, and jobs; and to the development and improvement of weapons for national security. Such new knowledge could be obtained only through basic research. Vannevar Bush's *Endless Frontier* report put the case succinctly: "Science, by itself, provides no panacea for individual, social, and economic ills. It can be effective in the national welfare only as a member of the team, whether the conditions be peace or war. But without scientific progress no amount of achievement in other directions can insure our health, prosperity, and security as a nation in the modern world."[41]

The level of science conducted in the United States is largely determined by the amount of federal funding made available to researchers in universities, national laboratories, government agencies, and industries. National science policy provides the structure for determining how much funding will be allocated to the federal agencies that support research, how they are to use and distribute funds, and the policies and regulations that will govern research conducted with federal dollars. U.S. science policy also provides mechanisms for promoting science education and *technology transfer,* activities that fuel economic growth, where scientific results or technology developed by one entity is transferred to another entity, oftentimes for development for commercial use.[42] Essentially, national science policy is meant to ensure that science is conducted in a way that enhances the public good; that the nation's research enterprise is supported and advanced; and that uniform guidelines exist for conducting science within the United States.

The scientific community, like most other organized groups, operates most effectively when it shares norms and guidelines. That said, most members of the scientific community prefer that these norms and guidelines be created by the community itself, rather than imposed upon it from outside. An example of such self-regulation is the case where molecular biologists decided to put a moratorium on the use of recombinant DNA (rDNA) after the initial development of the technique. These scientists recognized the concern the rDNA might raise among policymakers and the public and believed that it would be better to establish guidelines and policy recommendations themselves while more was being learned about the potential impact of the new discoveries. In such cases, involvement of the public in deliberations is very important, both to make sure that its concerns are addressed and to reassure policymakers that community-developed policies and guidelines are adequate.

The Importance of National Science Policy

To appreciate the importance of science policy, one must recognize the enormous impact science has on modern society. Science and technology have underlain nearly every major advance in quality of life over the past century. Scientific discoveries have enabled us to fight polio, smallpox, tetanus, cholera, and many other debilitating diseases. Noninvasive scanning devices employing x-ray, PET (positron emission tomography), and MRI (magnetic resonance imaging) technologies have led to innumerable medical advances. Famine, at least for the present, has become an issue largely of food distribution, not supply—a remarkable stride from just a few decades ago. Radio, television, cellular phones, and high-speed travel all tie individuals together in ways once unimaginable.

In addition to these spectacular advances, more routine technological innovations have changed the way we experience our world. Ancient mariners considered themselves lucky if they were able to determine their position to within hundreds of miles; now, with a small, twenty-dollar Global Positioning System (GPS), weekend hikers or boaters can instantly know their position to within a few feet. Today's average car has more onboard computer power than the original lunar lander module. Handheld devices have many times the computing power of the early, room-sized supercomputers. We have seen stunning technological advances over recent decades, and the pace of scientific advance is quickening.

Economists attribute as much as half of our economic growth over the last fifty years to scientific advances and technological innovation.[43] Technological advances resulting from past government support of scientific research have contributed immeasurably to our nation's economy, spawning whole new industries and market niches. For example, federally supported research in fiber optics and lasers helped create the telecommunications revolution that brought about unprecedented American economic expansion and job creation during the late 1990s, with the telecommunications and information technology now comprising one-seventh of the U.S. economy and representing almost $1 trillion in annual revenues.[44] Meanwhile, the inception of research in molecular biology during the 1940s and 1950s, and of recombinant DNA research during the 1960s and 1970s, opened the door for today's multibillion-dollar biotechnology industry.[45] And the Internet, with its vast impact on our daily lives, was spawned by government-sponsored research conducted by the Department of Defense and the National Science Foundation.[46]

Besides enhancing the overall economy and quality of life, scientific and technological advances enable us to better understand the universe in which we live. Because of science, we know more than ever about the structure of the atom, the rules under which atoms form molecules, how molecules array themselves to make proteins, and the composition of the basic building block of life itself—DNA. With new insights into how the universe may have formed and evolved, including such novel discoveries as black holes, we have gained an added appreciation of how our solar system developed. Such strides address some of the oldest questions of humankind—our deep desire to understand the universe, how and when it was formed, and the relationship between the earth and the other parts of the solar system and the universe.

Where did we come from? What are we made of? How are the characteristics of one generation passed to the next? Such questions exemplify the inquisitiveness of human beings—our ability to question, to dream, to wonder about the world and universe that envelop us. The quest for answers leads to the discovery of new ways to enrich life itself, often spawning stunning scientific, humanistic, and artistic advances. A vibrant, inquiring society that is knowledgeable about the present and intrigued about the future is best prepared to advance technologically, and to improve quality of life for all its people.

We can now look toward a horizon of new and exciting breakthroughs: the colonization of space, the use of nanomachines that can be sent into the body to repair specific cells, the creation of computers the size of dust particles that can control our personal environment, the growth of transplantable organs from single cells. The world ahead will no doubt be as awe-inspiring to us as the current one is to our grandparents. The advances necessary to bring us safely to this new world will require robust science policies, to guide and govern its evolution—to serve as a bridge to our future.

Science in Support of Public Policy

As we have already mentioned, Vannevar Bush, an engineer and inventor, served as science advisor to President Roosevelt during World War II. Before the end of the war, Roosevelt requested recommendations on the role that science should play in the nation's future. Bush's response came in the form of his aforementioned report, *Science—the Endless Frontier,* which was received by President Harry Truman on July 25, 1945. In it Bush suggested that government support for research be directed toward improvements in three major areas: national security, health, and the economy.[47] American science policy has been focused on these three areas ever since.

Nowhere in his report did Bush recognize the role that science and federal support of science might play in informing and guiding the formulation of public policies themselves. Because it generates new knowledge, science can in fact help to create better public policy. This is true, for example, in the formulation of environmental laws and regulations. As science has allowed us to better understand the impact of human actions on ecosystems, and as the public's environmental consciousness has grown, new standards have been developed and laws enacted to protect our wilderness, better manage our natural resources, reduce the pollution of our air and water, and preserve our planet for future generations. The increasing use of science in addressing environmental issues and formulating public policy led Jim McGroddy, former vice president for research at IBM, to conclude that "no serious student of history . . . would today substantively revise [Vannevar] Bush's rationale or conclusions in any major way, other than perhaps to add a fourth area of impact, the improvement and management of our environment."[48]

Despite the power of science to influence policy, there is an inherent tension between science and policy-making. The norms and processes that drive science are profoundly different from the politics of democratic institutions. This has led some to conclude that science and democracy (or perhaps the political processes within democracy) are "like marriage partners who get along best when they respect each other's differences."[49]

Such tensions result, in part, from the deductive process through which new scientific knowledge is generated. Scientific knowledge often raises more questions than it answers. Policy formation, on the other hand, is inherently inductive and aimed at concrete solutions. This difference in purposes often leaves policymakers frustrated with the inability of science to provide clear answers to political questions. The Honorable Sherwood Boehlert (R-NY), former member of the House of Representatives and chairman of the House Science Committee, summed up the role of science in policy-making: "Science is a necessary, but not sufficient basis for policy. It must inform all of our policy decisions, but it cannot be the determinate of any of them. . . . And asking scientists to answer what are essentially political questions can only distort the science and muddle the policy debate."[50]

The Positive and Negative Potential of Science

Any treatment of science policy should acknowledge that scientific advances can have negative effects. Knowledge may be used to inflict harm on others; or incomplete knowledge may be used to produce short-term gains at the expense of long-term good. Indeed, such considerations emphasize the need for the development of sound science policies.

Science has, for instance, enabled us to create nuclear weapons, deploy biological agents, and pollute the environment at a prodigious rate. Its dual-edged sword is also evident in the debates surrounding cloning and the use of human embryonic stem cells. While most people agree that "reproductive" cloning—the creation of an exact replica of a human being—is an immoral use of science and should be illegal, much disagreement exists over the morality of therapeutic, or research, cloning for the treatment of cystic fibrosis, Huntington's disease, or diabetes.[51] These ethical and moral debates will be discussed in more detail in chapter 14.

An important distinction must be drawn between the creation and the use of knowledge. For the purposes of our discussion, we assume that knowledge is inherently good and that the creation of knowledge should be supported. At the same time, we realize that science can be used for both good and ill, and that actions enabled by science can have adverse impacts. Sound public policy therefore must be developed to guide the use of knowledge. We do not, however, subscribe to the notion that new knowledge should not be sought out of fear that someone might misuse the resulting knowledge.

Given the importance of science and technology to civilization's progress, one is led to ask if our national policies are optimally structured to foster advances. In spite of remarkable advances, we continue to face huge scientific challenges. Along with these challenges come policy issues of mammoth proportions that may not be easily addressed by science policy.

Policy Challenges and Questions

Each of the subsequent chapters includes a section that raises specific issues of policy. In this introductory chapter we call upon the reader to contemplate the scope of the questions that must be addressed in the decade ahead if the United States is to increase its standard of living and protect its national security. Examples are these: How will we educate our children to be globally competitive in science and technology? How will we decide the scope of national investments in research? How should universities, national laboratories, and industry partner with the government to best meet national objectives? What is the role of the states and the public in setting science policy? How will we decide what avenues of research should have priority?

To illustrate the complexity of just one of these issues, consider the challenge of determining what fraction of a research budget should be provided to one field or another. In spite of the structures that have been developed for review of research proposals and to assess the relative needs of various disciplines, no system has been perfected to determine how much funding should be provided to each scientific discipline.

Over the years, the nation has shifted its priorities to achieve specific public and political goals. President John F. Kennedy's goal of placing a man on the moon, for example, required the dedication of millions of R&D dollars to the space program. During the oil embargo and energy crisis of the early 1980s, the federal government made significant investments in research on energy.

Our country has recently been involved in emphasizing research in the life sciences. To many, this seems an appropriate priority: the baby boomers are approaching old age, and the life sciences may generate major medical advances. Others note, however, that an overconcentration of resources in one area may inhibit progress in others. According to this argument, the unpredictable nature of fundamental research means that targeted funding may not lead to desired advances. In addition, we should remember that breakthroughs can come from unexpected directions: without the discovery of DNA by Francis Crick (a physicist), James Watson (a biologist), Rosalind Franklin (a physical chemist), and Maurice Wilkins (a physicist), today's tremendous opportunities in the life sciences would not exist. It is also worth noting that many of the tools used by life scientists—including mass spectrometers, electron microscopes, and automated sequencers—are the results of advances in the physical and engineering sciences. Thus, focus on one discipline at the expense of other disciplines may be ill-advised.

Such complex issues can only be resolved through deliberations involving all relevant players. Given the government's substantial support of research, government officials will obviously play a major role in determining how these funds are applied for, awarded, and expended. At the same time, members of the scientific community must play a significant role in developing policies that affect their research on behalf of the nation. Moreover, the broader public also needs to be engaged, since these policies can affect their lives in many ways.

NOTES

1. James R. Killian Jr., *Sputnik, Scientists, and Eisenhower: A Memoir of the First Special Assistant to the President for Science and Technology* (Cambridge: MIT Press, 1977), xv.

2. Defense Advanced Research Projects Agency, *Powered by Ideas* (Arlington, VA: Defense Advanced Research Projects Agency, February 2005), 1. See also Mark Greenia, "ARPA and the ARPANET (Early History of the Internet)," in *The History of Computing: An Encyclopedia of the People and Machines That Made Computer History* (Antelope, CA: Lexikon Services, 2003), http://www.computermuseum.li/Testpage/99HISTORYCD-ARPA-History.HTM (accessed April 30, 2007).

3. National Science Foundation, "50th Anniversary Web Site: An Overview of the First 50 Years," http://www.nsf.gov/about/history/overview-50.jsp (accessed April 30, 2007).

4. Vannevar Bush, *Science—the Endless Frontier*, July 5, 1945, NSF 40th Anniversary Edition, NSF 90-8 (Arlington, VA: National Science Foundation, 1990).

5. See Harry Kreisler, "Conversations with History Blog: Neglecting Science Policy," interviews with Zhores Alferov, Charles M. Vest, David King, and Clyde Prestowitz, http://conversationswithhistory.typepad.com/conversations_with_histor/2005/10/neglecting_scie.html (accessed April 30, 2007).

6. Rep. Vern Ehlers (R-MI) at Yale University's Faculty of Engineering Sesquicentennial Forum, "Challenges to Innovation in the 21st Century," May 3, 2002, http://entity.eng.yale.edu/150yrs/ehlers.ppt (accessed April 30, 2007). See also Steve Lilienthal, "Where Is Sputnik When We Need It?" *Accuracy in the Media*, April 6, 2006, http://www.aim.org/guest_column/A4476_0_6_0_C/ (accessed April 30, 2007).

7. Todd Bishop, "Gates: Computer Science Education May Need Its Own 'Sputnik Moment,'" *Seattle Post-Intelligencer Reporter*, August 3, 2004.

8. Committee on Science, Engineering, and Public Policy, *Rising above the Gathering Storm: Energizing and Employing America for a Brighter Economic Future* (Washington, DC: National Academies Press, 2007).

9. Shirley A. Jackson, "The Quiet Crisis and the Future of American Competitiveness," Presidential Symposium, American Chemical Society Fall National Meeting, Washington, DC, August 29, 2005.

10. Andy Sullivan, "Lawmakers, Businesses Call for More R&D Funding," Reuters, February 16, 2005.

11. Thomas L. Friedman, *The World Is Flat: A Brief History of the Twenty-First Century* (New York: Farrar, Straus and Giroux, 2005), 253.

12. Thomas S. Kuhn, *The Structure of Scientific Revolutions*, 3rd ed. (Chicago: University of Chicago Press, 1996), 1. Other philosophers and historians of science have also attempted to explain what science is (or is not). See the works of Karl Popper, Imre Lakatos, or Paul R. Thagard, for example. See also Martin Curd and J. A. Cover, eds., *Philosophy of Science: The Central Issues* (New York: W. W. Norton, 1998).

13. Carl Sagan, *The Demon-Haunted World: Science as a Candle in the Dark* (New York: Ballantine, 1996), 25.

14. For a comprehensive discussion concerning the confusion that often takes place in distinguishing science from technology, see Joseph Agassi, "The Confusion between Science and Technology in the Standard Philosophies of Science," *Technology and Culture* 7, no. 3 (1966): 348–66.

15. Alan J. Friedman, "Science vs. Technology," *Directors Column*, New York Hall of Science Press Room, http://www.nyhallsci.org/nyhs-pressroom/nyhs-directorscolumn/dc-sciencevstech.html (accessed April 30, 2007).

16. Qijie Zhou, Zhihong Chu, and Xiaowei Wang, "The Changes of Modern Science and Technology Notion and New Economic and Social Mode," *Nature and Science* 2, no. 3 (2004): 66–69.

17. Definitions are taken from the National Science Foundation, *Survey of Research and Development Expenditure at Universities and Colleges, FY 2004*, NSF Form 411 (08-06) (Arlington VA: National Science Foundation, 2004).

18. See National Science Board, *Science and Engineering Indicators 2006*, NSB-06-01 (Arlington, VA: National Science Foundation, 2006), 4.8. *Science and Engineering Indicators* is published biennially in two volumes. In volume 2 are the appendix tables; all other material is in volume 1.

19. Donald E. Stokes, *Pasteur's Quadrant: Basic Science and Technological Innovation* (Washington, DC: Brookings Institution Press, 1997), 10. See also Dudley S. Childress, "Working in Pasteur's Quadrant," *Journal of Rehabilitation Research and Development* 36, no. 1 (1999): xi–xii.

20. For example, see National Academies Committee on Department of Defense Basic Research, *Assessment of Department of Defense Basic Research* (Washington, DC: National Academies Press, 2005), 8–14.

21. Stokes, *Pasteur's Quadrant*.

22. Stokes refers to his model as "a revised dynamic model" (*Pasteur's Quadrant*, 88).

23. The notion of science and technology as being "parallel streams of cumulative knowledge" is attributable to Harvey Brooks. Brooks described science and technology as "two strands of DNA which can exist independently, but cannot be truly functional until they are paired." See Harvey Brooks, "The Relationship between Science and Technology," *Research Policy* 23 (September 1994): 479.

24. Stokes, *Pasteur's Quadrant*, 89 and chaps. 4 and 5.

25. National Science Board, *Science and Engineering Indicators 2006*, chap. 3.

26. National Science Foundation, Division of Science Resources Statistics, *National Patterns of R&D Resources: 2004 Data Update*, NSF 06-327 (Arlington, VA: National Science Foundation, 2006).

27. Kei Koizumi, "Federal R&D in the FY 2007 Budget: An Introduction," in *AAAS Intersociety Working Group Report XXXI: Research Development FY 2007* (Washington, DC: American Association for the Advancement of Science, 2006), 11.

28. In stating that there is a lack of consensus on the definition of public policy, Birkland refers to noted political scientist Thomas Dye's argument that trying to find a precise definition of public policy can "degenerate into a word game." See Thomas A. Birkland, *An Introduction to the Policy Process* (New York: M. E. Sharpe, 2001), 19.

29. Different definitions of public policy include the following: "Public policy is 'a goal-directed or purposive course of action followed by an actor or a set of actors in an attempt to deal with a public problem'" (James E. Anderson); "Public policy consists of political decisions for implementing programs to achieve societal goals" (Charles L. Cochran and Eloise F. Malone); and "Stated most simply, public policy is the sum of government activities, whether acting directly or through agents, as it has an influence on the life of citizens" (B. Guy

Peters). These definitions are taken from Birkland, *Introduction to Policy Process,* table 1.3.

30. See Richard H. W. Maloy, "Public Policy—Who Should Make It in America's Oligarchy?" *Detroit College of Law at Michigan State University Law Review* 4 (1998).

31. Richard Barke, *Science, Technology, and Public Policy* (Washington, DC: CQ Press, 1986), 8.

32. Barke defines science and technology policy as "a governmental course of action intended to support, apply, or regulate scientific knowledge or technological innovation" and specifically notes that it is "often impossible to separate 'science policy' and 'technology policy'" (ibid., 11–12).

33. John W. Kingdon, *Agendas, Alternatives, and Public Policies,* 2nd ed. (New York: Addison-Wesley Educational Publishers, 1995), 78–79.

34. Charles E. Lindblom, "The Science of Muddling Through," *Public Administration Review* 14 (Spring 1959): 79–88.

35. Phillip A. Griffiths, "Science and the Public Interest," *The Bridge* 23, no. 3 (1993): 4.

36. Laura Garwin and Tim Lincoln, eds., *A Century of Nature: Twenty-One Discoveries That Changed Science and the World* (Chicago: University of Chicago Press, 2003), 107–12.

37. Harvey Brooks, "The Scientific Advisor," in *Scientists and National Policymaking,* ed. Robert Gilpin and Christopher Wright (New York: Columbia University Press, 1964), 76–77.

38. Atul Wad, "Science and Technology Policy," in *The Uncertain Quest: Science, Technology, and Development,* ed. Jean-Jacques Salomon (Tokyo: United Nations University Press, 1994), chap. 10, http://www.unu.edu/unupress/unupbooks/uu09ue/uu09ue00.htm (accessed April 30, 2007).

39. Richard Barke has noted that in some instances, science—just like economic data, political pressures, or threats abroad—can be used as an input to shape government policies in several areas, for example, health, energy, or environmental policy. This represents *science for policy.* In other instances, science is the output of government policies, or *policy for science,* for example, funding for research into pesticide accumulation in the environment or regulations on animal use in scientific research. See Barke, *Science, Technology and Public Policy,* 4.

40. In his *Endless Frontier* report, Vannevar Bush stated, "We have no national science policy for science. The Government has only begun to utilize science in the Nation's welfare. There is no body within the Government charged with formulating or executing a national science policy. There are no standing committees of the Congress devoted to this important subject. Science has been in the wings. It should now be brought to the center of the stage—for in it lies much of our hope for the future" (12).

41. Ibid., 11.

42. The Office of Management and Budget defines technology transfer as "efforts and activities intended to result in the application or commercialization of Federal laboratory-developed innovations by the private sector, State and local governments, and other domestic users. These activities may include, but are not limited to: technical/cooperative interactions (direct technical assistance to private sector users and developers; personnel exchanges; resource sharing; and cooperative research

and development agreements); commercialization activities (patenting and licensing of innovations and identifying markets and users); and information exchange (dissemination to potential technology users of technical information; papers, articles, reports, seminars, etc)." See Office of Management and Budget, Circular A-11, 1994.

43. This statement has it origins in work originally done by economist Robert Solow. See Robert Solow, "Technical Change and the Aggregate Production Function," *Review of Economics and Statistics* 39, no. 3 (1957): 312–20; Edwin Mansfield, "Academic Research and Industrial Innovation," *Research Policy* 20, no. 1 (1991): 1–12; Gregory Tassey, *R&D Trends in the U.S. Economy: Strategies and Policy Implications,* 99-2 Planning Report (Washington, DC: National Institute of Standards and Technology, U.S. Department of Commerce, April 1999); and Al Gore, The White House, Office of the Vice President, remarks, Microsoft CEO Summit, Seattle, May 8, 1997.

44. Telecommunications Industry Association, *Telecommunications Market Review and Forecast* (2006), 3–8; Reed E. Hundt, Chairman, Federal Communications Commission, "From Buenos Aires to Geneva and Beyond," speech before the World Affairs Councils, Philadelphia, October 22, 1997, http://www.fcc.gov/Speeches/Hundt/spreh759.html (accessed April 30, 2007).

45. See, for example, "MIT, the Federal Government, and the Biotechnology Industry: A Successful Partnership," Massachusetts Institute of Technology, December 1995; Biotechnology Industry Organization, *BIO 2005–2006 Guide to Biotechnology* (Washington, DC, June 2005), http://www.bio.org/speeches/pubs/er/ (accessed April 30, 2007); European Biopharmaceutical Enterprises, *History of Healthcare Biotechnology: Milestones of Discovery Leading to Concrete Benefits for Society,* http://www.ebe-biopharma.org/index.php?option=com_publications&task=section&Itemid=115 (accessed May 2, 2007).

46. For background information on the history of the Internet, see Barry M. Leiner, Vinton G. Cerf, David D. Clark, Robert E. Kahn, Leonard Kleinrock, Daniel C. Lynch, Jon Postel, Larry G. Roberts, and Stephen Wolff, *A Brief History of the Internet,* found at Internet Society, "All about the Internet, Histories of the Internet," http://www.isoc.org/internet/history/brief.shtml (accessed April 30, 2007); Richard T. Griffiths, "History of the Internet, Internet for Historians (and just about everyone else)," Leiden University, History Department, http://www.let.leidenuniv.nl/history/ivh/frame_theorie.html (accessed April 30, 2007); and Computer History Museum, "Exhibits, Internet History 1962–1992," http://www.computerhistory.org/exhibits/internet_history/ (accessed April 30, 2007).

47. Bush, *Science—the Endless Frontier;* see also House Committee on Science, *Unlocking Our Future: Towards a New National Science Policy,* 105th Cong., September 1998, Committee Print 105-B, 9, http://www.access.gpo.gov/congress/house/science/cp105-b/science105b.pdf (accessed May 5, 2007).

48. James C. McGroddy, before the House Committee on Science, "National Science Policy Part II: Defining Successful Partnerships and Collaborations in Scientific Research," 105th Congress, 2nd sess., March 11, 1998.

49. Bruce L. R. Smith, *American Science Policy since World War II* (Washington, DC: Brookings Institution Press, 1990), 12; see also Don K. Price, *The Scientific Estate* (Cambridge: Belknap Press of Harvard University Press, 1965).

50. Sherwood Boehlert, "Science: A View from the Hill," keynote address, EPA Science Forum 2002: Meeting the Challenges, May 1, 2002.

51. Therapeutic cloning is also called somatic cell nuclear transfer, research cloning, and cell nuclear replacement.

CHAPTER 2

U.S. Science Policy before and after *Sputnik*

The U.S. Research Enterprise: A Unique System for the Support of Science

The U.S. research enterprise, in which taxpayer dollars are appropriated for government-sponsored scientific research and awarded on the basis of scientific excellence, is envied around the world, and many countries are now trying to replicate it. Its tremendous success is due to an emphasis on basic research conducted and guided within a structure of interlocking relationships between government agencies, universities, national laboratories, and industry, all under the umbrella of post–World War II national science policy.

Whereas the governments of other countries, such as Japan, historically devoted large portions of their support of R&D to specific industrial objectives, the United States favored a system aimed at generating new fundamental knowledge and ideas in support of the core missions of federal agencies and broader societal needs. Moreover, whereas other countries centralized support for science under one central ministry or department of science, the United States chose a decentralized approach, funding research within agencies with an eye toward advancing the agency's broader mission. Indeed, the multiagency—or "pluralistic"—U.S. system that grew up after the war, while at times seeming complex and cumbersome, has attracted and helped retain the best scientific talent from around the world for over a half-century. We believe it has also been a major driver of American innovation and economic growth.

Other characteristics also distinguish the U.S. research system. First, projects are supported based on merit in the competitive marketplace of ideas. Second, the federal in-vestment in university-based research does double duty, paying for research while helping to educate the next generation of scientists, engineers, and scholars. And third, the government shares in the costs of research—both direct costs such as salaries, equipment, and supplies, and indirect costs such as facilities and administrative (F&A) expenses incurred during the research.

In this chapter, we provide historical insights into this unique system and address questions such as these: What are the system's constitutional underpinnings? How did World War II and the Soviet launch of *Sputnik* shape the policies that serve as the foundation for today's scientific enterprise in the United States? And finally, how have major events such as the fall of the Soviet Union and the events of September 11 affected U.S. science and the policies that drive it?

Constitutional Underpinnings of the Federal Role in Science

Throughout U.S. history a healthy tension has existed between those who wish to see government play a strong role in its citizens' activities, and those who wish to see minimal government involvement. In the arena of science and technology this debate takes shape as a dispute between those who believe the government's control of the research agenda, through its ability to set funding priorities and regulate research conduct, is so intrusive that it limits the progress of science itself, and others who believe that the important and far-reaching effects of science and technology demand greater governmental oversight. What is the government's appropriate role in regulating scientific research done and how it is conducted? Should the government merely provide funding, without assuming any concomitant oversight of the resulting research?

While the Constitution does not directly specify a governmental role in the support of science and technology, it does empower Congress "to promote the progress of science and useful arts, by securing for limited times to authors and inventors the exclusive right to their respective writings and discoveries."[1] It is on the basis of this clause that the entire U.S. patent and trademark framework is founded. The founding fathers clearly recognized the importance of invention, as we would expect from individuals such as Benjamin Franklin and Thomas Jefferson. As we shall see, however, this constitutional statement about the importance of invention falls far short of specifying what role government should play in the management of research.

The government's role in modern scientific research is the product of many incremental decisions, most of which are the result of various congressional and presidential decision-making processes. Because a notable portion of the research conducted in the United States is funded with taxpayer dollars, those in charge of setting and approving budgets (the president and Congress, respectively) have an interest in what kind of research is being conducted. In addition, the Constitution gives Congress and the president responsibility for creating agencies they deem necessary to determine and pursue specific national goals. Our contemporary governmental structure includes several agencies involved in scientific research, through both internal *(intramural)* and nongovernmental *(extramural)* research programs. This multiform structure has promoted the development of links between America's government and its national labs, universities, and industries.

Toward a National Science Policy

Who was involved in the creation of our current research system? What drove its creators to focus so heavily on science's potential to contribute to the national welfare? Certainly President Roosevelt's early recognition of the role of science and technology fueled the establishment of our complex system of research universities, national labs, and industries. But Roosevelt was notably joined in this endeavor by the influential and opinionated Vannevar Bush, and by Senator Harley M. Kilgore (D-WV).

Beginning in 1939 and continuing through the war years, Bush had rendered part-time service as a consultant to the government. A respected engineer and science administrator, he had the ear and the attention of President Roosevelt.[2] Harley Kilgore, despite his admitted ignorance of science and technology, took a significant interest in the management of military research.[3]

In spite of their apparent differences, the two men shared a deep interest in ensuring a peacetime role for science. Among other things, they pointed to the success of wartime initiatives like those in medicine. During the war, research supported by the Office of Scientific Research and Development (OSRD) led to the production and widespread use of penicillin; an effective and standardized means to fight against malaria, including the development of DDT; new blood substitutes, such as serum albumin; and the use of immune globulin to fight infection. These medical advances saved thousands of lives during the war.[4]

These wartime results suggested that similarly organized efforts to attack peacetime challenges might be at least as productive. But while Bush and Kilgore shared a vision of ongoing federal support for research, especially on university campuses, they held very different opinions about the government's role in directing the research it was funding.[5]

For his part, Senator Kilgore introduced a series of bills in Congress in 1942, 1943, and 1945, all aimed at creating a new agency to oversee, promote, and fund scientific and technological advancement.[6] His 1942 bill, the Technology Mobilization Act (S. 2721), would have created an independent Office of Technological Mobilization (OTM) to centralize the deployment of wartime technology resources, by incorporating all of the government's technically oriented bureaus—military and civilian—into one agency.[7] The director of the OTM would be appointed by the president. Kilgore, who had arrived in Washington at the beginning of 1941, before the United States had formally entered the war, was even then concerned about the nation's ability to quickly mobilize resources, including technical resources, in case of war. Part of the impetus for his 1942 legislation was his outrage at what he found in the capital: a high concentration of government research contracts among a very limited number of corporations, and the granting of rights to intellectual property derived from government-supported research to these same corporate contractors. He and his like-minded colleagues attributed the ongoing shortage of strategic materials, including synthetic rubber, to these industrial patent cartels, and to agreements the government had made with foreign nations prior to the war. The Technology Mobilization Act was meant to ensure that the nation's technological operations would be available for the war effort, and that the flow of knowledge from federally funded research projects would be unrestricted.[8]

Kilgore's OTM was able to draft outside technical personnel and facilities for its own use, finance research projects contributing to victory, and force the licensing of technology for war uses. It would also have made grants for the development of "selected" technologies. The bill, however, ran into opposition from Vannevar Bush, industry leaders, the army and navy, and some in the scientific community who went so far as to call the act totalitarian.[9] Industrialists were particularly alarmed at the tightening of controls on intellectual property resulting from government-funded research, while military leaders were aghast at the prospect of losing control over their research operations. Stalled by multifaceted opposition, the bill died in committee.

In 1943, Kilgore redrafted his legislation, attempting to address the concerns of the first bill's opponents in the new Science Mobilization Act (S. 702). This second effort would have established an Office of Science and Technological Mobilization (OSTM), which would have less power than the OTM but would still be able to award grants and loans for scientific and technical education and both pure and applied research. The office, like its proposed predecessor, would be overseen by a presidentially appointed director, advised by a broadly representative board (including not just practicing scientists, but also educators and members of industry). In spite of Kilgore's changes, however, the new bill still faced opposition on several fronts—from trade associations like the National Association of Manufacturers (NAM), from the army and navy again, and from the scientific community itself.[10] In fact, the American Association for the Advancement of Science (AAAS) went so far as to issue a resolution denouncing the legislation, out of concern that control of science would be taken away from scientists.[11] Because of the lack of support—especially, some would contend, from Bush and the OSRD—this bill also failed.

While Kilgore's first two efforts focused on mobilizing scientific and technical resources for the war, his 1944 bill focused on the postwar promotion of science. This time, Kilgore invited Bush to collaborate in drafting the legislation, apparently out of a genuine belief that the two could successfully work together. This third piece of legislation would have established an independent agency—the National Science Foundation (NSF)—that would focus primarily on promoting peacetime basic and applied research, along with scientific training and education programs. The director would be appointed by the president, and the board composed not just of scientists

and technical experts, but also members of the public. The new bill still included provisions for government ownership of intellectual property developed with federal dollars, and stipulated that, in funding the social sciences, which at the time were not necessarily considered "real sciences" by some members of the scientific community, monies would be distributed on a geographical basis, not according to merit.

While Bush admired Kilgore's intentions, he did not agree with many of the bill's premises. Driven by this difference of opinions, Bush opposed Kilgore's third effort. Roosevelt had already asked Bush for a formal report, and the recommendations Bush submitted in response were designed to preempt Kilgore's 1944 bill and perhaps get Bush's own ideas into legislation.

In the spring of 1945, under the guise of cooperation, Bush asked Kilgore to delay the introduction of his latest legislation. Believing the two camps were finally working together, Kilgore agreed, only to be greatly angered in July of that year when Senator Warren Magnuson (D-WA) introduced legislation (S. 1285) based on Bush's report on the very day it was released. Clearly, Bush had arranged this move.[12] Feeling double-crossed, Kilgore immediately introduced his bill (S. 1297). A compromise bill (S. 1720) was drafted but died late in 1945; a similar bill (S. 1850) introduced early in the new 1946 session also failed. In 1947, Senator H. Alexander Smith (R-NJ), a majority member in the newly Republican-controlled Senate, consulted with Bush and the OSRD on a bill (S. 526) that accommodated Bush's idea that appropriate administrative oversight should come from scientists, and not the public at large. While this bill was eventually passed in both chambers, President Truman vetoed it before the end of the term, citing concerns about leaving the administrative controls totally in the hands of scientists with no public accountability.

Vannevar Bush's 1945 report engendered considerable debate and controversy, but it also bolstered Harley Kilgore's campaign for the creation of a postwar science agency—an idea that would eventually be realized as the NSF.[13] The report also laid the foundation for a permanent U.S. system of federal funding for scientific research and training. To this day, *Science—the Endless Frontier* is often referred to as "the basis for the Nation's existing science policy."[14]

The Differences between Bush and Kilgore

Despite their shared vision of government support for science, Bush and Kilgore had significant philosophical differences over the details of providing this support, among other issues of science policy.[15]

Perhaps the most controversial difference was over who should control the new science agency. Believing that the federal government should support only the best science, Bush wanted administration of the agency to be in the hands of a part-time independent board of scientists, who would appoint their own administrative director. Others, including Kilgore, President Truman, and key members of his administration, were reluctant to give wide-ranging authority to a board composed wholly of scientists. Instead they favored government control in the form of a presidentially appointed director and an advisory board.

This controversy over administrative control was so heated that it prevented the consensus needed to pass legislation on the structure of support for research. Indeed, a historical analysis reveals that controversy delayed the creation of the NSF. For example, President Truman vetoed Senator Smith's 1947 legislation because, in keeping with Vannevar Bush's ideas, the bill granted scientists complete discretion over the agency's spending and management. In his veto message, the president noted that the Smith's bill would allow the agency to become "divorced from control by the people to an extent that it implies a distinct lack of faith in democratic processes."[16]

A little over a year after Bush issued his report, Truman issued an executive order creating the President's Scientific Research Board (PSRB). The board's first chairman was John R. Steelman, assistant to the president, who was charged with reviewing current and proposed research and development programs both inside and outside the federal government.[17] The PSRB's goal was to create a rational system for managing research programs and coordinating them across government, industry, universities and colleges, and other institutions. The board's final report, issued in August 1947, was entitled *Science and Public Policy: A Program for the Nation*.[18] It is not surprising that a report written by New Deal supporters was generally in line with Kilgore's preference for governmental control. The Steelman report went a bit further

POLICY DISCUSSION BOX 2.1

A Conflict Then, a Conflict Now

President Harry Truman once noted, "There is nothing new in the world except the history you do not know." This statement can be applied to science policy. To a new observer, today's debates over science policy may seem unprecedented, yet in fact they replay the debates between Vannevar Bush and Senator Harley Kilgore, a New Deal Democrat, as they struggled over philosophical differences concerning the creation of the NSF.

For example, intense argument rages in Congress today over who should maintain intellectual property rights when federally funded research leads to commercially profitable technology and when the government should step in to ensure such technologies should be licensed to companies for further development and public gains. This debate blossomed in the Bush-Kilgore days and is still in flower across all federal research agencies. Another example is whether taxpayer dollars should be used to fund R&D in industry; many Republicans think this is "corporate welfare," while many Democrats think it is democracy at its finest. Opinions also continue to differ on the appropriate balance between fundamental and strategic or applied research, and the extent to which funds should be distributed on the basis of geography rather than the quality of the proposed research.

A final example involves funding of the social sciences. Early on, Bush and some others viewed the social sciences as not worthy of federal support, and argued against supporting these disciplines as a part of the mission of the NSF. Others, such as Kilgore, believed the social sciences could provide insight into human behavior and society that could be valuable to policymakers in addressing social issues. While today the social sciences are a part of the NSF, it took some forty years for this to happen. Even today, however, social science research funded by agencies such as NSF and NIH comes under attack by some people as having questionable value and being a waste of taxpayer dollars.

Issues such as these extend far beyond the role of the NSF; indeed they engulf the entire science policy arena. Which side of these debates do you support? At the inception of NSF, who was right, Bush or Kilgore? How do their disagreements reflect partisan politics today? Has diversity of agencies been helpful to the nation's R&D, or hobbled what we might have achieved? Why is Vannevar Bush heralded as the father of modern U.S. science policy, while Kilgore is barely mentioned—though many of his ideas have had a significant impact on the current NSF?

than Kilgore's bill, however, in recognizing the importance of research to America's economic competitiveness and international economic development. Indeed, the report suggested that the United States should aid European and Asian nations in rebuilding their scientific establishments—that is, that science should become part of the Marshall Plan.

The sharp contrast between Bush's ideas and Steelman's meant that the Steelman report's recommendations would remain strictly theoretical. Once again summoning the full measure of his influence and authority, Bush opposed the Steelman report on grounds similar to those he had employed against Kilgore. With the 1946 elections handing control of both the House and Senate to the Republicans, Bush's conservative leanings suddenly became an asset. Truman and the Democrats soon found themselves stymied by an opposition Congress.

Finally, in 1950, Congress passed and President Truman signed into law legislation creating the NSF. Truman called the new foundation "a landmark in the history of science in the United States." It was a hollow victory for Vannevar Bush, however, as many of his principles had been sacrificed to get the legislation passed and signed. Moreover, during the debate preceding the vote, several agencies, including the Atomic Energy Commission (predecessor to the Department of Energy), the NIH, and the army and navy all claimed responsibility for civilian scientific research—a prerogative that Bush had originally hoped would be granted exclusively to the NSF.[19]

Many of the controversies surrounding the creation of the NSF have persisted into recent years. Who should hold rights to patents for work done using federal funds? Should the social sciences be placed on the same footing as the natural sciences? Should federal research funds be distributed geographically, or solely on the basis of merit? What is the role of the federal government in supporting applied research? Along with these concrete problems comes an assortment of more abstract but still controversial issues: What are the roles of scientists and policymakers in determining, distributing, and administering funding? Who is in the best position to determine priorities? These topics will be covered more fully in the chapters that follow.

The Cold War

From the end of World War II until about 1985, the United States was embroiled in an intense rivalry with the Soviet Union. No direct confrontation between the armies of these superpowers ever took place. Instead, the Cold War took the shape of a competition for the ability to inflict the most harm on the other side. It was a race based largely on technological capacity. Who could make the most powerful bombs? Whose missiles had the longest range and could deliver the largest payload? How many times over could one side destroy the other? A single miscalculation could have destroyed civilization as we knew it. As one might imagine, both rivals poured resources into science and technology.

For a short period immediately after World War II, the United States had allowed itself to grow complacent about the advancement of science and the development of new technologies. Even the technology-reliant military showed diminished interest in science during the mid-1950s. This lull was short-lived, however. For most Americans, the wake-up call came in October 1957, with the Soviets' successful launch of *Sputnik*.[20] *Sputnik,* and the subsequent launch of the Russian *Soyuz* spacecraft, created great alarm in the United States, suggesting a possibly catastrophic shift in the Cold War balance of power and risk. American investment in space research skyrocketed.

The ensuing revival of patriotism and scientific interest lured a whole generation of young people to pursue careers in science and engineering. Although many of them are still in the workforce today, most are nearing retirement age, and no *Sputnik*-like event has come along in the ensuing years that would encourage a new generation of young people to choose scientific and engineering careers.

During the Cold War years, though, American researchers enjoyed substantial government support, with defense concerns dominating federal funding decisions. At the time, space research was considered a crucial part of national defense, a means to military superiority. Such considerations played a major role in setting the funding priorities and national science policy of the period (see fig. 2.1). By the late 1960s, funding for space research was receiving more federal support than all other nondefense research combined. This funding level was not matched until the late 1990s, when the government began to invest heavily in the health and life sciences. Indeed, while the late 1970s and early 1980s saw slight increases in research spending, largely as the result of high oil prices and the energy shortage, it was not until 2000 that nondefense research funding levels finally surpassed those of the late 1960s.

Funding for Cold War university research was predicated on a fairly simple, unwritten agreement: universities would educate a cadre of well-trained science and engineer-

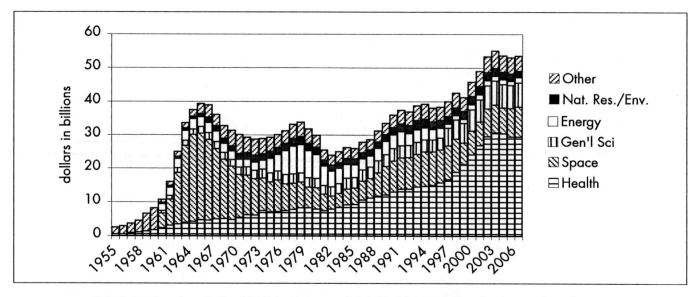

FIG. 2.1   Trends in Federal R&D by Function, FY1955–2007. Outlays for the conduct of R&D in billions of constant FY2007 dollars. Some energy programs shifted to general science beginning in FY1998. Constant dollar conversions based on GDP deflators. (National Science Foundation, Federal R&D by Budget Function Historical Tables, 2007, from AAAS table, "Trends in Federal R&D by Function, FY 1955–2008," http://www.aaas.org/spp/rd/histda08tb.pdf.)

ing students; their faculty would advance research in areas relevant to national security; and in exchange, the federal government would provide funding not only for this type of research but also for more basic research of special interest to universities. Maintaining or exceeding technological parity with the Soviet Union thus became a major factor in defining national science policy during the Cold War.

Post–Cold War Issues

With the demise of the Soviet Union, U.S. science policy adapted to a new, more complicated environment. Many in Congress and elsewhere asked why significant federal outlays were needed to support research when there was no threat as obvious or powerful as the USSR had been during the Cold War. These questions translated into smaller budgets for research in the physical sciences and engineering and raised doubts about the future of the country's national laboratories. Ironically, their success in developing weapons of extraordinary capability had helped end the Cold War, yet now this success threatened the existence of the laboratories themselves. American R&D had been closely aligned with national defense for over thirty years: now the need for a strong defense seemed less urgent.

To some extent, the decreased interest in defense technologies was compensated for by efforts to retool the scientific infrastructure to focus on improving the country's economic competitiveness and national health. By the late 1980s our manufacturing processes were far less efficient than those of several other nations, and our trade deficit was enormous. Under the circumstances, it was natural to wonder whether scientists and engineers, newly freed from their defense work, could help bolster the nation's economy. Many national laboratories and universities, for example, created incubator facilities to help business start-ups. Such partnerships fostered new cooperative research and development agreements (CRADAs) through which national labs provided technical expertise to industry.

The Post–September 11 Environment

As noted earlier, following World War II, the mission of American science shifted from defeating the Axis powers to adapting wartime innovations for peacetime purposes. Attention then turned toward winning the Cold War, and finally, during the 1990s, to addressing the nation's economic problems and to improving the health and longevity of the nation's people.

In the aftermath of September 11, 2001, and the onset of a new type of war—the so-called global war on terrorism—America is now asking science and technology to adapt once again by contributing to the improvement of homeland security.

The terrorist attacks of September 11 raised the issue of how an extended, open, democratic society can protect itself from those wishing to inflict harm on its citizens. Technology has a role to play in this effort, by identifying weapons-grade materials, tracking individuals' movements, enhancing travel security, and intercepting hostile communications, to mention just a few examples. While many of these efforts can be managed by security agencies and national laboratories, some academic involvement is required: most of the country's experts on biological agents, for example, are based at universities.

Yet these efforts to use science and technology for security-related purposes force scientists and policymakers to confront difficult issues. We have, for example, entered a new era of biological advancement in which the same breakthroughs that can advance health and cure disease may also be used to commit deadly acts of terror. New laws have been passed and regulations developed that restrict certain academic scientists from gaining access to select biological agents and related materials. Such laws, along with increasing restrictions on the publication of scientific results, call into question the fundamental scientific principles of openness, the free flow of information, and the exchange of knowledge—principles that have underpinned many of the last century's greatest scientific advancements.

In one sense, the science and policy challenges that the United States faces after September 11 are simply new manifestations of issues that have been debated since World War II. What role can and should universities play in national security? How can we encourage and train American students to pursue careers in science and math? Does our reliance on foreign nationals to fill research positions leave us vulnerable, or is the internationalization of scientific talent critical to the global advancement of science? Finally, is it possible to isolate knowledge within national borders in the era of "globalized" science?

Policy Challenges and Questions

This chapter has described the evolution of U.S. science policy in the period surrounding the launch of the Soviet spacecraft *Sputnik*. It portrays the shock experienced by America at the prospect that it was not technologically supreme and describes how it responded. In the ensuing years many significant developments occurred. The long Cold War with the Soviet Union effectively ended as the latter disintegrated. Efforts were then made to marshal the national laboratories to help the nation improve its economic competitiveness. We now face the challenge of using the nation's scientific and technological prowess to reduce the threat of terrorism. These twists and turns have required major shifts in national science policy, and have raised issues that will require careful analysis in the decades to come. How can domestic security be maintained without snuffing out the spark of creativity, traditionally championed in the United States as it welcomed people from all parts of the world and encouraged the sharing of pathbreaking information? How can basic research support, with long-term goals, be justified when the public and the government seek short-term results?

While individuals such as Vannevar Bush and Harley Kilgore helped prepare a national science policy for the Cold War, new visions are required to meet the challenges of our era, when science and technology will play a significant role in ensuring national health, in securing the homeland, and in meeting the economic and global challenges of the twenty-first century.

NOTES

1. U.S. Constitution, art. 1, sec. 8.
2. Under Bush's leadership, the Office of Scientific Research and Development (OSRD) was established in June 1941 to mobilize and coordinate the nation's science and research activities for application in the war effort and nation's defense. The OSRD supported thousands of civilian scientists and oversaw programs that developed some of the powerful weapons used in World War II including the Manhattan Project, which resulted in the development of the atomic bomb. A full account of Bush and the OSRD can be found in G. Pascal Zachary, *Endless Frontier: Vannevar Bush, Engineer of the American Century* (Cambridge: MIT Press, 1999).
3. Zachary, *Endless Frontier*, 232–34.
4. Bush, *Science—the Endless Frontier*, appendix 2, "Report of the Medical Advisory Committee." See also Pauline Maier, Merritt Roe Smith, Alexander Keyssar, and Daniel J. Kevles, *Inventing America: A History of the United States*, vol. 2 (New York: W. W. Norton, 2003), 826–28.
5. The differences between Bush and Kilgore are detailed in Daniel J. Kevles, "The National Science Foundation and the Debate over Postwar Research Policy: 1942–1945: A Political Interpretation of Science—the Endless Frontier," *Isis* 68, no. 1 (1977): 4–26; Daniel L. Kleinman, *Politics on the Endless Frontier: Postwar Research Policy in the United States* (Durham, NC: Duke University Press, 1995), chap. 4; and Smith, *American Science Policy*, chap. 3.
6. Kevles, "National Science Foundation."

7. Kleinman, *Politics on the Endless Frontier*, 78–83.

8. See Kevles, "National Science Foundation," 6, 8; Kleinman, *Politics on the Endless Frontier*, chap. 4.

9. See Kevles, "National Science Foundation," 9.

10. In both instances, many in the scientific community opposed Kilgore's bills, for several reasons including that federal support of academic research threatened academic freedom. This is ironic given the present day when scientific societies often call upon Congress to provide more funding for their specific research disciplines.

11. Kleinman, *Politics on the Endless Frontier*, 86–87.

12. See ibid., 119–41; J. Merton England, "Dr. Bush Writes a Report: 'Science the Endless Frontier,'" *Science* 191, no. 4222 (January 9, 1976): 41–47.

13. In its own historical account of the events that led to its creation in 1950, the National Science Foundation describes Bush's *Science—the Endless Frontier* as "an antidote to the Kilgore suggestions." See National Science Foundation, *The National Science Foundation: A Brief History*, NSF 88-16, (Arlington, VA: National Science Foundation, 1988), 2. July 15, 1994, http://www.nsf.gov/about/history/nsf50/nsf8816.jsp (accessed June 3, 2007), chap. 1.

14. House Committee on Science, *Unlocking Our Future*, 8.

15. For additional background, see Kevles, "National Science Foundation."

16. Quoted in National Science Foundation, *National Science Foundation: A Brief History*, 3.

17. Executive Order 9791, "Providing for a Study of Scientific Research and Development Activities and Establishing the President's Scientific Research Board," *Code of Federal Regulations*, title 3 (1943–48), 578. For additional background, see William A. Blanpied, "Inventing US Science Policy," *Physics Today*, February 1998, 34.

18. John R. Steelman, *Science and Public Policy: A Program for the Nation* (Washington, DC: U.S. Government Printing Office, 1949).

19. The significant role already assumed for research by other agencies is illustrated by the fact that by 1955 almost 80 percent of the federal government's expenditure on R&D was devoted to the Department of Defense, 10 percent to the Atomic Energy Commission, and 6.6 percent to the Department of Health, Education, and Welfare (of which the National Institutes of Health was a part). By this time, the National Science Foundation's share was only 0.1 percent. See David Dickson, *The Politics of Science* (New York: Pantheon, 1984), 27.

20. For more information on *Sputnik* and its impact see Paul Dickson, *Sputnik: The Shock of the Century* (New York: Walker, 2001); Robert A. Divine, *The Sputnik Challenge: Eisenhower's Response to the Soviet Satellite* (New York: Oxford University Press, 1993); and Killian, *Sputnik, Scientists, and Eisenhower*.

CHAPTER 3

The Players in Science Policy

What Is the Federal Government's Role in Making Science Policy?

The creation of federal policy has been compared to sausage making so often that it has become a cliché.[1] Perhaps nowhere, though, is this aphorism better illustrated than in the formation of science policy.

Science policy-making is a messy process, engaging innumerable federal agencies and congressional committees, more than almost any other type of federal policy-making. While plurality has its benefits, it also makes the creation and comprehension of science policy more difficult than comparable work in transportation, housing, or education. Adding to this complexity is the difficulty—often impossibility—of defining exactly where science policy ends and broader policy with an impact on science begins. These two issues—how policy affects science and how science affects policy—seem at times hopelessly tangled.

One of the primary reasons for this difficulty is that science policy, or policy that affects the conduct of science, often results from broader public policy discussions, and from concerns that stretch far beyond the realm of science. Such discussions might be aimed, for example, at developing standards for national defense, homeland security, health, education, energy, or the environment, with science simply the means to desired ends. Science may also provide the foundation upon which specific policies in other areas, such as clean air and water laws and regulations, are based.

Such is the impact of science on other policy areas. As to the reverse—the influence of broader policy debates on science—policymakers may be subject to moral considerations, political constraints, or budgetary priorities.

Science policy is not—cannot be, in our system—always based solely on considerations of what is good for science. This fact is a common source of frustration and confusion in the scientific community, whose members may not fully understand the federal policy-making process.

Given the complexity of the U.S. policy system, it is amazing that the country ever accomplishes anything at all. But it does. In fact, the science policies of the last fifty years have generally provided sound guidance, helping the United States to maintain and preserve its global scientific and technological leadership.

Before providing the details on *how* science policy is made in the United States, however, it is important to explain *who* makes it.

Who Makes Science Policy?

At the federal level, direct responsibility for science policy-making is shared by the executive and legislative branches. This collaborative approach is fundamental to American policy-making overall. It is, in fact, a consequence of the very design of our constitutional system. Political scientist Walter Oleszek explains, "The Constitution . . . creates a system not of separate institutions performing separate functions but of separate institutions sharing functions (and even competing for predominate influence in exercising them). The overlap of powers is fundamental to national decision making."[2]

At times the judiciary branch also joins in shaping science policy. This occurs when laws regulating science and scientific conduct are subject to competing interpretations, or appear to conflict with the Constitution. The courts, for example, have recently played a significant role in science policy issues relating to intellectual property and the teaching of intelligent design in public schools. The judiciary's role in science policy-making, though limited at present, is likely to expand significantly as science's intersection with policy and law grows more complex.

Federal policy-making is generally characterized by multiple decision points, any of which can produce delays or blockages. In some situations, the impact of a given policy on science is ignored or considered only as an afterthought. For instance, the Health Insurance Portability and Accountability Act (HIPAA) of 1996, which actually took effect in the spring of 2003, benefited patient privacy—its intended purpose—but it simultaneously increased requirements for research projects using human subjects. By imposing new limits on access to patient information, HIPAA inadvertently limited opportunities to recruit participants for new clinical trials. In some instances, it has become more difficult to develop and test new medications and protocols.[3]

Policy-making engages a wide range of participants, including elected officials, appointees, and career civil servants. Each group has its own interests and views, all of which must be brought to bear during the decision-making process. The president and key members of Congress occupy the top tier. But career civil servants or congressional staff are often the real drivers of science policy. A great deal of interpretation of policy is done at the lowest levels of government, empowering the people who carry out federal programs to interpret and enforce policies as they see fit.

Almost every federal agency relies upon science and technology to carry out its mission, or is involved in the implementation of science policy. Some agencies are much more involved than others. Those that directly support and fund research do so either through grants and contracts with organizations outside the federal government (*extramural* research) or by supporting research within the federal agencies themselves or the national laboratories (*intramural* research).

Congressional treatment of science funding, like so many legislative activities, is characterized by fragmentation and decentralization. In a 1998 hearing held before the House Science Committee, for example, Admiral James Watkins, then president of the Consortium for Oceanographic Research and Education and former secretary of energy, noted that nine federal agencies and forty-seven congressional committees or subcommittees claimed jurisdiction over oceanographic research alone.[4]

Complaints about duplication in the system have led to calls for the creation of a "Department of Science" that could centralize and better coordinate science policy. This debate dates back to Vannevar Bush's original vision of a unified federal agency with responsibility for the conduct of civilian science.[5] Since then, the proposal to consolidate all federal science agencies into one government-wide department of science has been considered and rejected on several occasions.[6]

The centralized model for government support of science is used by several other countries, including Japan. Others, such as China, place some responsibility for the conduct of scientific research within their Ministry of Education, where an explicit function of the ministry is involvement in scientific research at institutions of higher education.[7] This is in contrast to the United States, where the Department of Education has little if any connection with scientific research. While the advantages and disadvantages of a centralized approach can be debated, bureaucratic structures already in place in the United States are not easily reorganized and are indeed resistant to change. There is little likelihood that one single department of science will be created in this country in the near future.

At present, more than thirty cabinet-level departments and federal agencies provide funding for, or have a role in, science and engineering research. Because the responsibility for science is spread among so many federal agencies, multiple congressional *authorization* committees are involved in establishing science policy for the agencies over which they have specific jurisdiction and oversight responsibility. Meanwhile, over half of the congressional funding subcommittees, or *appropriations* subcommittees, currently have responsibility for allocating funds to key federal agencies for the conduct of science.

The complexity of the system makes it hard to understand, even for those charged with its oversight. So a scientist unfamiliar with public policy-making may conclude that the process by which decisions are made, including research funding, is chaotic. For junior scientists, the task of seeking funding for a new idea in this decentralized system appears daunting, if not insurmountable. For example, a biologist who wants to study the effects of radioactivity on cells might apply to the National Science Foundation, the National Institutes of Health, the Department of Energy, the Department of Defense, or the Environmental Protection Agency, to list just a few possibilities. How would such an individual know where to apply?

Our purpose here is to briefly describe the executive branch offices, federal agencies, legislative branch committees, and other governmental, quasi-governmental, and nongovernmental organizations that shape science

POLICY DISCUSSION BOX 3.1

Do We Need a Department of Science?

The debate over how best to coordinate science policy within and among federal agencies and congressional committees is an old one. Vannevar Bush's vision of a single federal agency coordinating national science policy was never realized. Instead, during the five years of debate about how to structure and organize this proposed agency—what ultimately became the National Science Foundation—existing and new agencies rushed into the breach, starting research programs aimed at fulfilling their individual missions.

Those in favor of a single department of science suggest that if there were only one agency responsible for science, duplication of effort would be reduced, efficiency increased, waste reduced, and funding streamlined. Under such a system, a researcher with an idea to be explored would apply for support only to the department of science. The department would decide how the project, if worthy, was to be funded. Standards for judging ideas could be made common, rates of funding and practices could be uniform, and so on.

At the same time, having only one department of science would greatly reduce the number of congressional committees involved in establishing science policy. It would also mean that only two funding subcommittees (one in each chamber) would be instrumental in providing the funds for government-sponsored research, as opposed to the current system, in which over half the appropriations subcommittees have some responsibility for funding scientific research. So what is wrong with

this idea, and why has no bill ever been passed to bring about this major reorganization in government?

Opponents of one single department of science argue that each respective agency involved in policy looks at scientific problems with a slightly different approach. That is, each agency considers a research question with its own mission in mind, making grants and soliciting proposals to advance its specific goals. While there is some duplication of scientific effort among the agencies, many observers argue that this is a strength of the U.S. system, not a weakness. They maintain that the funding framework has vitality because many agencies have a vested interest in supporting research as a component of achieving their broader missions. Thus, the Department of Energy, for example, has long been invested in what the effects of radioactivity are on living organisms, and it is willing to divert some percentage of its substantial budget to see that these effects are studied. DOE would not want to see any of its budget diverted to another agency (the department of science) it could not control. A single department of science might mean that study of the effects of radioactivity on human health would decline.

How do you think the U.S. government should be organized to meet national R&D needs? Does the current system work effectively? Under our decentralized system, is there excess duplication in the efforts of federal agencies with research responsibilities? What value might such duplication have? Are there better ways to coordinate research efforts among federal agencies or to consolidate congressional oversight of science? How would you change the system to make it more effective?

---

policy. In the pages that follow, we describe the location of these organizations within or outside of government, their origins, staffing, basic functions, and purpose. We also assess the different cultures that drive these organizations, their respective interests in science policy, and how these cultures interact with other governmental and nongovernmental structures in the creation of science policy.

The Executive Branch

The President

The Constitution assigns to the president a panoply of roles, including chief executive, commander in chief of

the armed forces, chief diplomat, and head of state. The president also assumes informal roles, acting as the national agenda setter and the leader of his or her party and even the world.

While the responsibility for making science policy is not explicitly assigned to the president by the Constitution—and, in fact, is hardly touched upon by that document in any context—the increasing importance of science has required that presidents take an active role in key science policy decisions. Some have chosen to become engaged more actively than others. President Nixon, for example, had little interest in science policy, and even expressed distrust of the scientific community. Other presidents, such as Gerald Ford, were much more willing to embrace the scientific community and grant it a role in their administrations.[8]

While World War II and the Russian launch of *Sputnik* were crucial spurs to policy, presidential appreciation for the significance of science to national policy dates back to Thomas Jefferson. Not only did Jefferson himself take a strong interest in science, but he also recognized the president's power to promote the exchange of scientific ideas. In opposing tariffs on learned treaties, Jefferson affirmed, "Science is more important in a republic than in any other government."[9]

Despite some presidents' personal interest in science and science policy, presidents have come to rely on advisors, particularly the science advisor, to shape policy and engage with Congress. In our own era, the president directly involves himself in science policy decisions primarily (1) when international considerations and cooperation are required, as with the International Space Station or the Kyoto Protocol; (2) as a means to improve national security, such as President Reagan's push for the Strategic Defense Initiative (SDI); (3) for political, moral, or ethical reasons, as when President George W. Bush chose to limit researchers' use of human embryonic stem cells; (4) for nationalistic reasons, such as when President Kennedy captured the nation's imagination with the race to put a man on the moon; and (5) as a means to broadly demonstrate their vision for the future.

The President's Science Advisor

The position of presidential science advisor dates back to Vannevar Bush, who served as Franklin Roosevelt's unofficial advisor during World War II. Despite recommendations that the position be formalized in the executive branch, President Truman created a science advisor within the Office of Defense Management, backed by an advisory committee. Eisenhower was comfortable with this arrangement when he first assumed office. Everything changed with the success of the Soviet *Sputnik* mission, however, after which Eisenhower moved the advisor's office to the White House, and appointed then-MIT president James Killian for the job. Eisenhower announced Killian's appointment on November, 7, 1957, via a radio broadcast to the nation:

> I have made sure that the very best thought and advice that the scientific community can supply, heretofore provided to me on an informal basis, will now be fully organized and formalized so that no gap can occur. The purpose is to make it possible for me, personally, whenever there

appears to be any unnecessary delay in our development system, to act promptly and decisively.

> To that end, I have created a new office called the office of Special Assistant to the President for Science and Technology. This man, who will be aided by a staff of scientists and a strong Advisory Group of outstanding experts reporting to him and to me, will have the active responsibility of helping me follow through on the program of scientific improvement and of our defenses. . . .

> In conclusion, although I am now stressing the influence of science on defense, I am not forgetting that there is much more to science than its function in strengthening our defense, and much more to our defense than the part played by science. The peaceful contributions of science—to healing, to enriching life, to freeing the spirit—these are the most important products of the conquest of nature's secrets.[10]

The advisor's influence has ebbed and flowed ever since, depending largely on the individual president's own interest in science, and the extent to which he has wished to incorporate related issues into his political and legislative agendas.

Even under presidents with an appreciation for science, the science advisor has not always enjoyed tremendous influence, especially since the advisor has no direct authority over budgets or policies. Instead, the advisor's primary tool is often the power of persuasion.[11] As one former staff member noted, "How effective [the advisor] is in convincing agencies to shape or modify their R&D budgets depends largely on the strength of his personal relationship with inner circles of the White House."[12]

The relationship between many presidents and their science advisors has been rocky at best. For example, President Johnson, angered by the scientific community's lack of support for the Vietnam War, banned his science advisor and other members of the President's Science Advisory Committee (PSAC) from the White House dining hall. President Nixon so disliked scientists that in 1973, at the beginning of his second term, he pushed his science advisor to resign, together with members of the PSAC, effectively eliminating the job. Whatever advisory responsibilities remained were shifted to the director of the National Science Foundation, H. Guyford Stever.[13]

When Gerald Ford took over the presidency in 1974, he reinstated the science advisor and, in spite of considerable opposition, worked with Congress to codify the position. In the spring of 1976, Ford signed the National Science and Technology Policy, Organization and Priorities Act (P.L. 94-282), which created the Office of Science and Technology Policy (OSTP) within the White House, with the presidential science advisor as its director.[14] Ford then asked Stever to step down from the directorship of the NSF to serve as his new science advisor.

It comes as no surprise that science advisors have had an uneven relationship with their bosses. First, unlike so many presidential advisors, who have been chosen specifically because their political views are similar to the president's, science advisors are often chosen for their credentials and prominence within the scientific community. Most do not have the same personal relationships with the president as other top-level advisors.

The science advisor's influence is also limited by their tendency to consider science more important than politics and policy. From the president's perspective, however, what is good for the gander is not always good for the goose. When they do consider political and other factors, advisors are often criticized by the scientific community for failing to take the community's best interests into account.[15] It is essentially a no-win situation.

Even when the science advisor does have a good relationship with the president (e.g., Jerome Wiesner's deep friendship with John F. Kennedy), his or her focus on science may conflict with the views of other White House staff.[16] Wiesner's troubles, for example, reportedly began when he shared information with Russian colleagues during the Kennedy-Khrushchev summit of April 1961.[17] Wiesner eventually realized that he was being quietly excluded from discussions of national security issues, in part because of the increasing influence of the National Security Council. Nor was Wiesner's belief in disarmament and arms control shared by the Pentagon or other Kennedy aides.[18]

The Office of Science and Technology Policy

It was the 1976 National Science and Technology Policy, Organization, and Priorities Act that formalized the position of presidential science advisor and created the Office of Science and Technology Policy (OSTP). The predecessor to the OSTP, the Office of Science and Technology (OST), had only existed informally until President Kennedy, prompted by Jerome Wiesner, institutionalized it in 1962. Reorganization Plan Number 2, prepared by Kennedy and submitted to Congress in March 1962, called for "strengthening of the staff and consultant resources . . . available to the President in respect of scientific and technical factors affecting executive branch policies and . . . also [to] facilitate communication with the Congress."[19] Its successor, the OSTP, is now the leading office within the executive branch for the formulation and development of presidential S&T policies.

The OSTP's responsibilities include advising the president on the formulation of policies that directly involve or affect S&T; working with the president and the Office of Management and Budget (OMB) to formulate S&T budgets and cross-cutting S&T initiatives for federal agencies; communicating the president's S&T policies and programs; working to foster strong cooperation on S&T matters among federal, state, and local governments; and maintaining the partnership between the federal government and the scientific community at large.

The office typically comprises more than forty full-time staff members, including the director and two associate directors, one for science and one for technology.[20] The director and associate directors are presidential appointees and are confirmed by the Senate. Under each of the associate directors are several assistant directors with responsibility for narrower areas of science and technology. In George W. Bush's administration, for example, the assistant directors focus on the health and life sciences; the environment; the physical sciences and engineering; education and the social sciences; space and aeronautics; telecom and information technology; technology; national security; and homeland security. Assistant directors serve at the will of the director and are not political appointees, so some may span multiple administrations. They are not, however, traditional civil servants, and a new director has the authority to bring in his or her own team of assistant directors when the office changes hands. The director may also reorganize the office more broadly, including changing the number and focus of the assistant directors, whose positions are not dictated by law. The current incarnation of the OSTP also includes a chief of staff, chief legislative counsel, a staff overseeing legislative and public affairs, and a budget staff.[21]

The President's Council of Advisors on Science and Technology

The President's Council of Advisors on Science and Technology (PCAST) was created by President George H. W. Bush via a 1990 executive order, to provide "critical links to industry and academia."[22] PCAST serves as the highest-level science advisory group to the president that is *not* composed of federal government officials. Its immediate predecessor was the President's Science Advisory Committee (PSAC), which was disbanded by President Nixon as an outgrowth of Nixon's aforementioned animosity toward the scientific community.

An executive order, signed in 2005, dictates that PCAST will consist of a maximum of forty-five members, one of whom shall be the president's science advisor (also OSTP director) and serve as one of the group's cochairs.[23] The other members are to be top scientists, engineers, and scholars from outside the government. Like the members of many government advisory committees, PCAST members serve voluntarily and without pay.

PCAST is designed to bring the views and opinions of distinguished scientists, engineers, and other nongovernmental experts to bear on national science policy. Because of its independence, the panel is one of the best opportunities for lay individuals to influence the development and direction of science policies. It also helps the president gather expert information on specific S&T topics. PCAST is thus a critical link between the White House and the scientific and technical communities.

The National Science and Technology Council

While PCAST dates back to the Eisenhower and Truman administrations, President Bill Clinton's 1993 establishment of the National Science and Technology Council (NSTC) marked a more recent effort to coordinate science, space, and technology advice.[24] Created on November 23, 1993, under Executive Order 12881, the NSTC is charged with (1) coordinating the science and technology policy-making process; (2) ensuring that science and technology policy decisions and programs are consistent with the president's stated goals; (3) helping integrate the president's S&T policy agenda across the federal government; (4) ensuring that science and technology are considered in the development and implementation of federal policies and programs; and (5) furthering international S&T cooperation.[25] Unlike PCAST, the NSTC is composed entirely of government officials, with the primary purpose of coordinating science policies among government agencies.

Before the NSTC, other, less formal mechanisms were used to achieve this coordination, most notably the Federal Coordination Council for Science Engineering and Technology (FCCSET). FCCSET was created by the National Science and Technology Policy, Organization, and Priorities Act of 1976, the same legislation that created the OSTP. It was founded in order to "consider problems and developments in fields of science, engineering, and technology and related activities affecting more than one federal agency, and to recommend policies designed to provide more effective planning and administration of federal scientific, engineering, and technological programs."[26] The council's activities attracted significant attention during the tenure of D. Allan Bromley, President George H. W. Bush's science advisor, for helping to coordinate multiagency initiatives in areas including global climate change, high-performance computing, and math and science education.[27]

Based upon the successful model established by the National Security Council (NSC)—which convenes cabinet secretaries and other top-level advisors on matters of national security and foreign policy—the NSTC was created to involve high-level executive branch officials in science and technology policy. As he does on the NSC, the president acts as chair of the NSTC, which also includes the vice president, the president's chief science advisor, and cabinet secretaries and agency heads with significant S&T responsibilities.[28] Other top-level presidential officials on the NSTC include the president's national security, economic policy, and domestic policy advisors. Despite being modeled after the NSC, the NSTC has yet to obtain the same stature within the White House, and many, even within the scientific community, are unaware of its existence.

Still, the NSTC has tremendous potential. Since its creation, it has played an important role in coordinating cross-agency initiatives in nanotechnology, information technology, and climate change. On a more limited basis, it has also attempted to ensure consistency in science policies across federal agencies, although its effectiveness has been limited, in part because of the difficulty of overcoming agency-oriented cultures and resistance to a different way of doing business.

The Office of Management and Budget

The Office of Management and Budget (OMB) is the executive branch office with primary responsibility for overseeing preparation of the presidential budget request to Congress, and for the negotiations with Congress that follow. OMB is perhaps the single most influential executive branch office in the establishment of scientific priorities. This is because the OMB determines the funding levels contained in the presidential budget request for all federal research agencies and science programs.

In carrying out its duties, the OMB establishes funding priorities for all federal agencies and balances competing demands. The OMB also has primary oversight and administrative responsibility for the budgetary functions of all executive branch agencies and ensures that all agency reports, rules, and congressional testimony are consistent with the administration's policies and the president's budget.

In addition, the OMB has assumed increasing responsibility for evaluating programs' effectiveness and efficiency, with the results used in budget preparation. The specific programs used to conduct these evaluations include the Government Performance and Results Act (GPRA), the President's Management Agenda (PMA), and the Program Assessment Rating Tool (PART). Working cooperatively with grant-making agencies and the grantee community, the OMB leads the effort to assure that grants are managed properly and that federal dollars are spent in accordance with applicable laws and regulations.[29]

The OMB has four resource management offices overseeing budgeting in natural resource, human resource, national security, and general government programs. A program associate director (PAD) appointed by the president oversees each office, which is further broken down into divisions.[30] The OMB's Energy, Science, and Water Division exerts particular influence on the scientific community through its oversight of the budgets of the NSF, NASA, and the DOE. The OMB's Health Division oversees the NIH, while responsibility for funding of defense research falls to the National Security Division PAD.

Working under each PAD are career civil servants who oversee specific federal agencies and programs, as well as the policies and initiatives for those agencies. Perhaps the most important of these people from the standpoint of science policy is the OMB budget examiner. Each budget examiner in OMB is assigned to work with top-level officials in the agency on baseline budgetary analyses and policy alternatives. They ensure that programs are consistent with government-wide management and administrative initiatives. Each key science agency (e.g., NSF, NASA, NIH, the departments of Energy, Defense, Agriculture, etc.) has at least one budget examiner working on initial budget recommendations. The budget examiner reports to an OMB branch chief and a division director, who in turn report to the PAD. Because the PAD is a political appointee, he or she plays a major role in reconciling the president's political agenda and the administration's budget priorities with recommendations from the OMB budget examiners, branch chiefs, and division directors.

The importance of OMB's budget examiners to science funding is not widely appreciated, and they are often overlooked by members of the scientific community in trying to influence S&T budget decisions. However, a budget examiner can be an agency's primary advocate within the OMB or its worst enemy. It is critically important that each budget examiner fully understand the returns that the scientific community and society at large can expect from the programs whose budgets they oversee. It is every agency's job to make sure that specific programs and goals are properly explained to the examiners. If an agency does not convey the importance of its programs, the examiner will likely reduce the funding for that program. Other OMB offices, such as the Office for Information and Regulatory Affairs (OIRA), play an important role through their enforcement of federal regulations and information requirements, and the development of policies to improve the government's management of statistics and information.

As a regular part of the office's annual preparation of the budget, the OMB evaluates the effectiveness of programs, policies, and procedures, using performance measures that are established in collaboration with the agencies themselves. All of the key science agencies, including the DOE, the NSF, and the NIH, have developed standards for this process in concert with their respective scientific constituencies. It is not always easy, however, for science programs—particularly those supporting basic research, which may have no clear near-term benefits—to develop such indices.[31]

Federal Departments and Agencies

Multiple federal cabinet-level departments, as well as independent agencies and commissions, share responsibility for supporting scientific research and guiding science policy (see table 3.1). As part of the executive branch, all of these agencies report to the president, and are charged with carrying out his or her agenda. Their staffs are composed of both career civil servants and political appointees, some of whom require Senate confirmation.[32]

This group includes the cabinet-level Department of Defense and Department of Energy, as well as independent agencies including the NSF, NASA, and the EPA. Some are headed by presidential political appointees (e.g., the EPA), others (such as the NSF and NASA) by someone with a term appointment, meaning that he or she may span multiple administrations.

Some of the most important actors in federal science policy are offices or institutes within cabinet-level departments, such as the National Institutes of Health (NIH), part of the Department of Health and Human Services; the National Institute of Standards and Technology (NIST), and the National Oceanographic and Atmospheric Administration (NOAA), both part of the Department of Commerce; the Defense Advanced Research Projects Agency (DARPA), in the Department of Defense; and the Department of Energy's Office of Science.

TABLE 3.1   Federal Departments and Agencies with Science and Technology Responsibilities

Departments	
Department of Agriculture	Department of Commerce
Department of Defense	Department of Education
Department of Energy	Department of Health and Human Services
Department of Homeland Security	Department of Interior
Department of Justice	Department of State
Department of Transportation	Department of Veterans Affairs

Agencies and Commissions	
Defense Advanced Research Projects Agency	Federal Bureau of Investigation
Federal Aviation Administration	National Aeronautics and Space Administration
Federal Communications Commission	National Institutes of Health
National Institute of Standards and Technology	National Science Foundation
National Oceanic and Atmospheric Administration	National Security Agency
National Technology Transfer Center	National Transportation Safety Board
National Telecommunications Information Administration	Patent and Trademark Office
Smithsonian Institution	U.S. Geological Survey
Environmental Protection Agency	

Source: OSTP listing of Science and Technology Federal Departments and Federal Agencies/Commissions at http://www.ostp.gov/html/_federal departments.html and http://www.ostp.gov/html/_federalagencies.html (accessed on July 15, 2007).

Agencies approach scientific fields and disciplines very differently, depending on their missions. The NSF's goal of facilitating basic research and science education leads it to support scientists and engineers across a wide variety of disciplines, while NIH's focus on biomedical and behavioral research naturally leads it to support work in the biological and life sciences. The Department of Energy is a major funder of the physical sciences because of its historical role in the conduct of nuclear and energy-related science. And NASA provides significant levels of support to several engineering subdisciplines, as do the Department of Defense and the Department of Energy.

Some agencies such as the Environmental Protection Agency and the Food and Drug Administration (FDA) that one might expect to provide significant support for extramural scientific research do not, in reality, provide it. These agencies are focused upon promulgating and enforcing regulations pertaining to the environment and to food and drugs, respectively, and thus, much of the research they conduct is targeted to fulfilling their regulatory responsibilities.

The National Science Foundation

The NSF, which receives less than 4 percent of the total federal R&D budget, is the only federal agency with responsibility for basic research and education across all areas of science and technology. Research supported by the agency has resulted in many scientific breakthroughs. For example, it supported the development of magnetic resonance imaging (MRI, a now-familiar medical diagnostic tool), the Internet, speech recognition technology, Web browsers, bar codes, Doppler radar, fiber optics, computer-aided design, improved tests for artificial heart valves, and nanotechnology and micro-electro-mechanical systems (MEMS).[33] The Web search engine Google has its origins in research conducted by two graduate students—one an NSF graduate fellow—and sponsored by the NSF-led Digital Library Initiative.[34]

To accomplish its basic research mission, the NSF makes a wide range of research grants from relatively modest individual investigator awards to multimillion-dollar research and engineering centers. In recent years, the NSF has been playing a greater role in funding very large research facilities. These projects are supported through funding provided to NSF's Major Research Equipment and Facilities Construction (MREFC) account.

The NSF is also responsible for encouraging young people to seek careers in science, math, and engineering, which it does by supporting elementary- and secondary-level science and math education programs. It supports efforts to encourage undergraduate students to engage in research and provides for graduate and postdoctoral fellowships. The agency has also increasingly been requesting, and in some instances even requiring, that major research proposals include educational or public outreach components.

The NSF has a staff of approximately thirteen hundred people overseen by a director and a deputy director, both presidential appointees. The director serves a six-year term. More than any other science agency, the NSF relies upon members of the research community to help meet its personnel needs. Much of the NSF's staff consists of individuals, commonly referred to as *rotators,* who have taken temporary leave (usually two to three years) from academic positions to work for the agency.

The NSF is divided into seven directorates. Six are directly responsible for funding discipline-oriented basic and applied research: Biological Sciences (BIO); Computer and Information Science and Engineering (CISE); Engineering (ENG); Geosciences (GEO); Mathematical and Physical Sciences (MPS); and Social, Behavioral, and Economic Sciences (SBE). The remaining directorate oversees the NSF's Education and Human Resources (EHR) activity. The NSF also uses management offices to coordinate and oversee key agencies' activities. These include the Office of Cyberinfrastructure; Office of Integrative Activities; Office of International Science and Engineering; Office of Polar Programs; Office of Information and Resource Management; and Office of Budget, Finance and Award Management.

NSF policy is guided by the National Science Board (NSB), an independent science policy body established by Congress in 1950. The NSB's twenty-four members are top academic and industrial researchers from across the S&T spectrum. The NSF director acts as an ex officio member of the board, whose members are appointed by the president and confirmed by the Senate.[35] The NSB also advises the president and Congress on S&T policy issues identified by the president, Congress, or the board itself.

A comprehensive review of federal agency management carried out by the OMB under President George W. Bush gave the NSF higher marks than almost any other federal government agency for the quality of its management. In the FY2006 budget, for example, the NSF received three "green lights" on the Executive Branch Management Scorecard, and was the only agency to receive the highest rating for every program that underwent the OMB's Program Assessment Rating Tool (PART) evaluation. Less than 5 percent of the NSF's budget is spent on administration and management, with the other 95 percent going directly to research and education in the form of grants, contracts, and cooperative agreements.[36]

During a November 2001, speech at the National Press Club, then OMB director Mitch Daniels called the NSF one of the

true centers of excellence in this government . . . where more than 95 percent of the funds you provide as taxpayers go out on a competitive basis directly to researchers pursuing the frontiers of science, at very low overhead cost. It has supported eight of the 12 most recent Nobel Prize awards earned by Americans at some point in their careers. . . . Programs like these, and there are many, many others that perform well, that are accountable to . . . taxpayers for reaching for real results and measuring and attaining those results, deserve to be singled out, deserve to be fortified and strengthened.[37]

The Department of Health and Human Services

The Department of Health and Human Services (DHHS) houses more than ten agencies devoted to the protection of human health. Several, including the NIH, the Centers for Disease Control and Prevention (CDC), the Health Resources and Services Administration (HRSA), and the FDA play a role in science policy. The NIH, which funds university and college biomedical research, as well as its own intramural research program, has the largest science-policy footprint of this group, and thus will be the focus of our discussion in this section.

The NIH's official mission is "science in the pursuit of fundamental knowledge about the nature and behavior of living systems and the application of knowledge to extend healthy life and reduce the burdens of illness and disability."[38] Its goals include (1) discovery, innovation, and application to improve human health; (2) human resource maintenance and production in health research and related fields; (3) expansion of the knowledge base in the medical and associated sciences, which, in turn, stimulates private development in health-related fields and economic growth; and (4) ensuring scientific integrity, public accountability, and social responsibility in the conduct of health-related research.[39]

The Hygienic Laboratory, established in 1887, was renamed the National Institute of Health (singular) by the passage of the Ransdell Act (P.L. 71-251) in 1930. Private philanthropic support had long been sought to fund an institute that would pursue fundamental knowledge in medicine. When this quest proved futile, the Ransdell Act was passed to establish federal sponsorship of the Hygienic Laboratory.

The approval of the Ransdell Act marked a shift in attitudes toward public funding of biomedical research. The National Cancer Institute (NCI), created with bipartisan support by the National Cancer Act of 1937 (P.L. 75-244), was the first NIH topic-focused institute to be created.[40] Its approval reflected growing concern about the deadly nature of various forms of the disease and set a precedent for the NIH's categorical, disease-specific structure.[41] As of 2006, the NIH comprised twenty-seven institutes and centers.[42] Other institutes with a topical focus include the National Human Genome Research Institute, the National Institute on Aging, the National Institute of Mental Health, and the National Library of Medicine. Other NIH institutes and centers are dedicated to understanding, and eventually curing, diseases and chronic conditions such as diabetes, deafness and blindness, drug abuse, and heart, lung, and blood diseases.[43]

The NIH is second only to the Department of Defense in the total amount of federal R&D it sponsors. It is the number one supporter of non-defense-related R&D and the largest sponsor of both basic and applied research at colleges and universities.[44] Approximately 85 percent of the agency's budget is distributed to extramural performers, with a majority devoted to college and university research. Another 10 percent goes to the agency's own intramural research, much of which is conducted at facilities in Bethesda, Maryland. The remaining 5 percent is spent on management, administration, and facilities.[45]

The director of the NIH is a political appointee who requires Senate confirmation. Each institute is overseen by an institute director, all of whom are appointed by the Secretary of Health and Human Services except the director of the NCI, who, as a result of the National Cancer Act of 1971 (P.L. 92-218), passed as an outgrowth of President Nixon's "war on cancer," is a political appointee but does not require Senate confirmation.[46] The NIH, unlike the NSF, does not have a rotator system, although many of its senior administrative positions are filled by individuals with doctoral training in science. Also unlike the NSF, the NIH has a fairly large, permanent research staff—scientists who conduct research in many labs across the different institutes. Most of the other agencies discussed here also have intramural, or on-site, research labs and thus a staff of in-house scientists.

The National Aeronautics and Space Administration

Congress and the president established NASA with the National Aeronautics and Space Act of 1958 (P.L. 85-

568), and charged it to "provide for research into problems of flight within and outside the earth's atmosphere, and for other purposes."[47] NASA leads the civilian space R&D effort through its work in space science, earth science, astrophysics, aero- and astronautics, environmental science, and bioscience, which it conducts at both universities and its own centers. The NASA administrator and deputy administrator are appointed by the president and confirmed by the Senate.

The agency was created in direct response to the launch of *Sputnik,* and concern over the threat it posed to national defense. NASA is thus a very good example of how Congress and the president can use the creation of a new agency to focus national attention on a real or perceived national crisis.[48]

While best known for its achievements in human space flight, NASA has also had a long history of advances in aeronautics, space science, earth science, and space applications. Its investments in unmanned probes have enabled us to explore the moon, the planets, and other parts of the solar system. NASA's R&D funding and the technologies produced from it have played a significant role in the development and competitiveness of the U.S. aerospace industry.

The Department of Defense

The DOD is the largest single sponsor of R&D in the federal government, accounting for approximately half of all research and development. Much of this funding is devoted to the "D," rather than the "R," particularly to the development of new and improved weapons systems.[49] These range from missile defense systems to improved weapons for use by ground troops; they include protection against biological and chemical warfare and the development and advancement of military communication systems.

Even with this heavy emphasis on development, the DOD still plays a key role in science policy. Unlike, for instance, the EPA and the FDA, whose research programs are largely devoted to fulfilling their regulatory missions, the DOD supports a significant amount of basic and applied research. It also maintains a relatively strong extramural research program. In fact, DOD is the largest sponsor of engineering research conducted on university and college campuses. In FY2003, DOD sponsored 37 percent of total federal funding for engineering research at academic institutions (compared to NSF, which sponsored 35 percent).[50]

These expenditures have often resulted in noticeable civilian sector spin-offs. For example, the Defense Advance Research Projects Agency's (DARPA) ARPANET, a communications network designed to withstand a nuclear attack, eventually became the Internet. The military also played a role in the development and deployment of the Global Positioning System (GPS), which now enables average citizens to identify their location on the surface of the earth to within a few feet. In recent years, Congress has been augmenting medical research funding at NIH with additional funding for medical research at the DOD. In FY2002, for example, Congress devoted a total of $464 million in defense spending for medical research, of which $146 million was devoted to breast cancer research. Another $83 million went to prostate cancer research, $10 million for ovarian cancer, $62 million to general medical research, and $41 million for new prion (mad cow disease) research program.[51]

The secretary of defense is one of the key members of the president's cabinet. Secretaries of defense have often been willing to support investments in science education, with the view that it is important to ensure the availability of a pool of well-trained scientists and engineers, both within the armed forces and in the defense industry and supplier base that provide them with services and supplies.

The Department of Energy

The DOE is the fourth-largest sponsor of R&D in the federal government, behind the DOD, the NIH, and NASA. It is the leading sponsor of research in the physical sciences and a significant supporter of mathematics and computer science research. The DOE also ranks first among federal agencies in its support for R&D facilities.[52] DOE supports a considerable amount of basic research as well as R&D in support of its weapon, environmental management, and energy missions. Its portfolio of science-related responsibilities includes the DOE national laboratories, among them the three that supervise America's nuclear weapons stockpile, often referred to as the *weapons laboratories.*

The agency's science funding flows mainly through the Office of Science, which supports basic research in such areas as nuclear and high-energy physics, fusion and plasma science, biological and environmental research, basic energy research, and high-performance computing. Applied research in nuclear energy; fossil energy; solar, wind, and renewable energy sources; and energy efficiency is carried out by other DOE offices. The DOE also provides significant support for defense-related research related to the nuclear stockpile through the National

Nuclear Security Administration (NNSA), a semiautonomous agency within the department.

The Office of Science supports several experimental fusion devices, specialized nuclear linear accelerators, neutron reactors, synchrotron light sources, and supercomputers. Scientists and engineers from universities, industry, and other federal agencies use many of these facilities, which are therefore referred to as *user facilities.*

DOE's user facilities have recently started to play a major role in supporting multidisciplinary research. For example, biologists use the synchrotron light sources—instruments developed by physicists to study particles and matter at high speeds—to determine the structure of biological molecules and examine molecular interactions and cellular anatomical localization. These DOE facilities have been used, for example, to study the formation of the plaques and tangles in the brain that characterize Alzheimer's disease.

The Office of Science is overseen and directed by the DOE under secretary for science, a position appointed by the president and confirmed by the Senate. The under secretary for science also has responsibility for providing advice to the secretary of energy on science policy and for coordinating scientific activities across the entire department and national laboratories and technology centers. The under secretary position was created by the Energy Policy Act of 2005 (P.L. 109-58).

The Department of Commerce

The DOC fosters national industry and commerce. Its interest in R&D stems from the importance of research and technology to the creation of jobs and the balance of trade. While the department does not have a large external funding program, it does support work in specific areas and promotes cooperation among industry, national laboratories, and universities through such programs as the Manufacturing Extension Program (MEP) and the Advanced Technology Program (ATP), recently renamed the Technology Innovation Program (TIP). It also supports standards-related work in science and technology, much of which is done by the National Institute of Standards and Technology (NIST) and the National Oceanic and Atmospheric Administration (NOAA).

The secretary of commerce often plays a significant role in decisions on national science policy. Indeed, encouragement from the commerce secretary for particular initiatives can often marshal administration support.

NIST was founded in 1901 as the nation's first federal physical science research laboratory. It provides standards of measurement for commerce and industry, and its R&D investments have enabled major advances in lighting and electric power, materials testing, and temperature measurement. The institute and the scientists that it has supported have made significant contributions to the development of image processing, DNA diagnostic "chips," smoke detectors, automated error-correcting software for machine tools, atomic clocks, x-ray standards for mammography, scanning tunneling microscopy, pollution-control technology, and high-speed dental drills.[53]

NOAA was created on October 3, 1970, to serve a national need "for better protection of life and property from natural hazards . . . for a better understanding of the total environment . . . [and] for exploration and development leading to the intelligent use of our marine resources."[54] NOAA conducts research and gathers data on the oceans, atmosphere, space, and sun, applying this knowledge in ways that touch the lives of all Americans. The agency studies climate change, manages and protects fisheries and marine ecosystems, and promotes environmentally sound transportation and commerce. The agency's work is often mistakenly ascribed to the EPA, which, as noted earlier, is more of a regulatory enforcement agency.

The Department of Homeland Security

The events of September 11, 2001, prompted a new focus on homeland security and a sweeping reorganization of the U.S. government. While many existing agencies were already performing functions related to homeland security, lawmakers pointed to a lack of interagency coordination as a factor contributing to the surprise the attackers achieved.

The Department of Homeland Security (DHS) was created by the Homeland Security Act of 2002 (P.L. 107-296) in an effort to consolidate the many agencies that shared responsibility for homeland security. This consolidation was intended to eliminate unnecessary duplication of effort, reduce turf wars between agencies, and enable better information sharing and utilization of resources. The creation of the DHS represents the most significant reorganization of government since the Department of Defense was created in 1949.

Within the DHS is the Science and Technology (S&T) Directorate, which has primary responsibility for research sponsored by DHS. Objectives of the directorate include developing technologies to prevent and mitigate the consequences of biological, chemical, explosive, nuclear, and radiological attacks, and detecting materials used in such attacks. Testing and assessing vulnerabilities and existing and emerging threats is another area of focus. The S&T Directorate also oversees training of the next generation of homeland security researchers.[55] Modeled after DOD's DARPA and operating under the S&T Directorate is the Homeland Security Advanced Research Projects Agency (HSARPA). HSARPA works with industry, academia, government, and other sectors to engage them in innovative research and development, rapid prototyping, and technology transfer to meet DHS's needs.[56]

Additionally, a DHS Scholarship and Fellowship program was established in 2003 as a means of encouraging undergraduate and graduate students across all disciplines of science, including the social sciences, to pursue basic research contributing to the overall DHS mission.[57] The DHS has also supported major centers at universities that conduct basic research in support of homeland security and the DHS's basic research needs.[58]

The secretary of DHS is a full member of the presidential cabinet. The department's R&D role is still being clarified, but could eventually be significant, as it marshals industry, university, and national laboratory R&D for such goals as the detection of radioactive material and the war against cyberterrorism and bioterrorism.

Some, however, have questioned the effectiveness of DHS with regard to science and technology research. Since being established in 2003, the department, and specifically the S&T directorate, has had to deal with leadership turnover, staffing reorganizations, and budget cuts.[59] Although the DHS was created to consolidate activities, research activities related to homeland security are still conducted by several of the traditional research agencies. Among them is the National Institutes of Health, whose budget for homeland security exceeds that of the DHS. Funding for homeland security going to the NIH is largely devoted to combating biological terrorism.[60]

The Department of Agriculture

The U.S. Department of Agriculture (USDA) is the sixth-largest federal agency in terms of research and development funding, supporting research on everything from farming to food to forestry.[61] Its overall mission is to "provide leadership on food, agriculture, natural resources, and related issues based on sound public policy, the best available science, and efficient management."[62] In keeping with this responsibility, the USDA plays a significant role in making policy on a range of scientific and societal topics, including hunger, nutrition, global farming, and American agriculture.

The history of government-supported university research can be traced back to the USDA.[63] In fact, the agency was the primary supporter of research and dissemination through the early twentieth century, in large part because agriculture still dominated the U.S. economy during that period. In 1910, for example, one-third of the nation's workforce was involved in farming, and farm products made up slightly more than half of the nation's total exports.[64] The USDA's research and dissemination work was also encouraged by several major legislative proposals passed during the late nineteenth and early twentieth centuries.

The Morrill Act of 1862 set aside federal land for the creation of state colleges and universities "for the benefit of agriculture and the mechanic arts."[65] Many state universities were established as a direct result of this act, some of which are still referred to today as "land grant" or "A&M" schools because of their original designation as "Agricultural and Mechanical" institutions (e.g., Texas A&M).

Fifteen years later, in 1887, the Hatch Act created a nationwide system of state agricultural experimental stations to be administered by the land grant colleges. A second Morrill Act, passed in 1890, extended the original act to include the historically black schools already in existence in southern states and allowed for the establishment of new African American colleges in southern states where they did not yet exist. Finally, the Smith-Lever Act of 1914 (P.L. 63-95) designated funding "to provide for cooperative agricultural extension work between the [state] agricultural colleges . . . and the United States Department of Agriculture."[66] Taken together, these new laws helped define the USDA's preeminence in agricultural research.

The bulk of USDA R&D funding supports research on the environment, natural resource management, food safety, crop yields and production, global competitiveness and emerging markets, and plant and animal diseases. It carries out these research activities through four major agencies.

The Agricultural Research Service (ARS). The ARS is the USDA's in-house agricultural research agency. With one hundred research facilities throughout the United States and the world, the ARS currently supports over two thousand scientists and six thousand other employees through an annual budget of approximately one billion dollars.[67] The ARS also manages the National Arboretum in Washington, DC. The arboretum, which was established by Congress in 1927, conducts horticultural research and educational activities, and maintains a wide range of gardens for conserving and showcasing plants.

The Cooperative State Research, Education, and Extension Service (CSREES). The CSREES coordinates the USDA's relationships with external partners including state agricultural experimental stations, state cooperative extension systems, and state universities and land grant colleges. A significant proportion of CSREES research funding is provided to these partners in the form of *formula funds,* which are distributed according to a statutory formula, as opposed to an open, competitive grant solicitation. The CSREES also provides funding through *competitive research grants* and *special research grants.* CSREES special research grants are often designated, or "earmarked," by Congress for special funding. The use of formula funding and the degree of congressional involvement in USDA funding decisions make the USDA's research support system very different from the peer-review-based systems at agencies such as the NSF and the NIH.[68]

The Forest Service. The Forest Service is "the largest forestry research organization in the world and the national and international leader in forest conservation."[69] The largest agency within the USDA, it promotes ecologically sound management of the 192 million acres encompassed by U.S. National Forests and Grasslands. The Forest Service is separate from both the ARS and the CSREES, reporting to a different assistant secretary. And because the Forest Service's dedication to managing the nation's forests and open grasslands is so similar to that of the Department of Interior, its budget is approved by the Interior Appropriations Subcommittee, rather than the Agriculture Appropriations Subcommittee.

The Economic Research Service (ERS). The Economic Research Service's mission is to "inform and enhance public and private decision making on economic and policy issues related to agriculture, food, natural resources, and rural development."[70] To this end, the ERS employs 450 people, many of whom are highly trained economists and social scientists. The ERS receives a budget of over $80 million annually.

The Environmental Protection Agency

As noted earlier, the EPA conducts very little environmental research, instead focusing on the development and enforcement of environmental regulations. The EPA does, however, provide some funding for R&D. The Superfund, for example, is a trust fund administered by the EPA and other agencies to pay for the cleanup of hazardous waste sites. Under Superfund, the EPA supports efforts to develop innovative treatment technologies, as well as monitoring and measurement devices that can assist in the cleanup of Superfund sites.

The agency also has more than a dozen labs scattered across the country whose mission-directed research programs are supposed to help the EPA be more effective in its regulatory role. EPA rules often have a powerful, if indirect, effect on national R&D; for example, stricter emissions rules can, in principle, lead automakers to direct their research toward the design and production of lower-emissions vehicles. Finally, the EPA's National Center for Environmental Research runs a peer-reviewed extramural research program, the Science to Achieve Results (STAR) program. STAR awards grants and fellowships for research in environmental science and environmental engineering.

The Department of State

The Department of State is not a major participant in the direct support of science and technology activities. It does, however, play a significant role in science policy through its influence over international policy, a role that is likely to expand as scientific cooperation increases around the globe.

In 2000, the Department of State created the position of science and technology advisor to the secretary (STAS). The advisor serves as the department's principal liaison with the national and international scientific communities, marking recognition of the department's burgeoning involvement in science and technology. As the State Department notes on its STAS Web site: "science and technology are ubiquitous to the functioning of the modern world and the framing and execution of domestic policies and international relations. Science and technology—the engines of modern industrial economies—are seminal to international cooperation and are the 'bricks and mortar' of the three pillars of national security—intelligence, diplomacy and military readiness."[71]

The State Department's role in U.S. science policy has recently been demonstrated by its participation along with the DOE in multilateral negotiations on the International Thermonuclear Experimental Reactor (ITER). ITER is an ambitious international scientific construction project aimed at harnessing the power of fusion energy; it involves Japan, Russia, China, India, South Korea, and the European Union in addition to the United States.

Among its other roles, the Department of State oversees international access to U.S. research and educational institutions through the visa process. As a part of this process, Department of State consular affairs officers

screen foreign students and scholars coming to study or to conduct research on sensitive technical subjects in the United States, and have the ability to approve or deny their requests for a visa. The Department of State also regulates technology licensing agreements with other nations and has authority for implementation and enforcement of the International Traffic in Arms Regulations (ITAR), export control rules relating to the transfer of weapons and munitions or technical information about such items, to foreign nationals. The manner in which the department carries out its visa and export control responsibilities has implications for U.S. scientific research and how it is conducted.

The views of the secretary of state, a key member of the cabinet, on science policy can be critical in terms of both individual initiatives and the government's overall tone. Every country in the world with which the United States maintains formal diplomatic relations is represented by a country office, often referred to as country "desk," in the State Department. These desks, in tandem with the diplomatic corps, can determine how the United States will treat international and multinational scientific initiatives and partnerships.

Other Federal Agencies with
S&T Responsibilities

The Department of Transportation (DOT), the Department of the Interior (DOI), the Department of Veterans' Affairs (VA), and the Department of Education (DOEd) all provide small amounts of funding for intramural and extramural research. The Smithsonian Institution, established by Congress in 1846, also conducts both scientific and collections-based research with partial federal support.

The U.S. Department of Transportation. The DOT conducts much of its research under the auspices of the DOT Research and Special Programs Administration (RSPA). Among its other objectives, the RSPA works to advance intermodal transportation research and technology, to protect the public from dangers inherent in the transportation of hazardous materials, and to improve and enhance transportation safety. The RSPA also oversees the John A. Volpe National Transportation Systems Center in Cambridge, Massachusetts, an internationally recognized center of transportation and logistics research and expertise. While the Volpe Center is part of the RSPA, it differs from most federal organizations and research centers in that it receives no direct congressional appropriations, but is instead completely funded through a fee-for-service structure, in which all of the center's costs are covered by sponsored project work. The RSPA's Office of Innovation, Research, and Education also supports more than twenty university transportation centers around the country.

The Department of the Interior. The primary science agency within the DOI is the U.S. Geological Survey (USGS), created by a legislative act of Congress in 1879 as an "independent fact-finding agency that collects, monitors, analyzes, and provides understanding about natural resource conditions, issues, and problems."[72] The USGS maintains extensive expertise in the natural sciences, and supports a vast array of extramural research in the earth sciences and biological sciences. Its programs provide scientific information aimed at preserving important ecosystems and managing natural resources, including water, ocean, and mineral resources. It also supports a number of research programs targeted at anticipating and minimizing damage from natural disasters such as earthquakes, floods, hurricanes, and tsunamis.

The Department of Veterans' Affairs. The Department of Veterans' Affairs (VA), primarily known for its role in providing benefits and health care to the nation's veterans, also conducts some of its own research. Most VA research is performed within the department's own medical facilities. However, these facilities are in some instances affiliated with a university medical complex, so that VA researchers may hold joint or adjunct appointments on university campuses. The work supported by the VA naturally focuses on issues relevant to veterans, such as aging, chronic disease, environmental exposure, mental illness, and prosthetic research.

The Department of Education. The DOEd funds several small research programs and centers dedicated to improving education and our understanding of learning at all educational levels. Its Office of Postsecondary Education Programs includes the Fund for the Improvement of Postsecondary Education (FIPSE), which supports efforts to improve postsecondary education. The department's National Institute on Disability and Rehabilitation Research seeks a better understanding of the educational needs of individuals with disabilities. Other department programs, such as those supported by the Institute of Education Sciences and the Office of Educational Research and Improvement, are dedicated to enhancing teaching and learning at the primary and secondary levels. The department is also a funding source for education at both the K–12 and college levels.

The Smithsonian Institution. When many people think of the Smithsonian Institution, they immediately picture its sixteen museums, including the museums of Natural History, American History, and Air and Space in Washington, DC. What may surprise some is that the Smithsonian also conducts both scientific and collections-based research. The Smithsonian oversees nine research institutes, including the National Zoo's Conservation and Research Center in Virginia, the Smithsonian Astrophysical Observatory in Massachusetts, the Smithsonian Tropical Research Institute in Panama, and the Smithsonian Environmental Research Center near Maryland's Chesapeake Bay. The Smithsonian is an independent trust overseen for the U.S. government by a board of regents; approximately 70 percent of its funding comes from direct federal appropriations. The rest comes from trust funds, including contributions from private donors and the Smithsonian Business Ventures operation.

The Legislative Branch

Congress

Members of Congress, like the rest of us, have become increasingly dependent on science and technology to assist them in the decision-making process: computers, e-mail, and the Internet have fundamentally changed the way business is conducted on Capitol Hill. Likewise, C-SPAN's television coverage of the House and Senate means that legislators and their staff can follow key floor discussions even when they are not in the chamber itself.

The use of scientific information also helps members of Congress to make better-informed decisions and pass laws in the public interest. Knowledge of the scientific facts is important to the passage of effective clean air and water legislation, the formulation of sound health and educational policies, and the clear assessment of international trade and security agreements.

Of course, Congress also makes laws that govern the conduct of science itself, provides the federal funds to support scientific research, and oversees the federal departments and agencies that are responsible for promoting S&T work. One would be hard pressed to find a modern congressional committee that is not in some way influenced by or involved in science and technology.[73]

Congress can be a difficult body to understand, with its organization and rules of operation born incrementally, out of tradition and history—out of process, power, and politics, rather than logic. When studying the workings of Congress, therefore, it is generally best to ask not why but, rather, what is.

What Motivates a Member of Congress

To be effective, members of Congress need to attend to both their constituents back home and their colleagues in Washington. This dual allegiance is best known from the work of political scientist Richard Fenno, who characterizes the members' concern with the needs and expectations of their constituents as their *home style*. Fenno notes, however, that there is an inherent tension between that home style and the route to power and sound public policy in Washington.[74]

Another political scientist, David Mayhew, sees reelection as members' overriding objective: "Whether they are safe or marginal, cautious or audacious, congressmen must constantly engage in activities related to reelection."[75] In this interpretation, the desire for reelection drives members of Congress to pay attention to the needs and interests of their districts. To simplify, in Fenno's scheme, home style often conflicts with what might best advance sound public policy on the national level, while in Mayhew's analysis, home style wins out because it is what gets and keeps members of Congress elected.

A key part of a legislator's job in Washington relates to the specific committees on which he or she serves. All members of Congress serve on committees, each of which comprises smaller working groups, or subcommittees. Committee assignments can be based on several factors, from seniority (based upon the number of years of service in Congress) to personal interest. Once members are assigned to a committee, they typically retain that assignment for as long as they hold their seat in Congress or until they choose to switch to another committee for which there is a vacancy.

In order to ensure reelection, members strive to appear as responsive as possible to the people who live in their home districts. At the same time, they try to stockpile campaign funds to ensure that they can communicate the value of their activities in Washington to voters back home. This has led members of Congress to focus ever more intensely on campaign fund-raising. It has also led them to increase the size of their personal staffs. For example, in 1956 members of the House of Representatives were limited to nine full-time staff members, to be divided between their Washington, DC, and home district offices. Today that number has doubled. This is not to mention the additional committee staff that help members to carry out their legislative responsibilities and committee-related work.[76]

Legislators' need to justify their actions to constituents back home creates a problem for members who are ac-

tive in science policy. Unlike other issues such as health care, education, national defense, and veterans' affairs, scientific activities may not immediately benefit voters. If they do, those benefits may be unclear to the average voter—or, for that matter, to the member of Congress, many of whom have a cursory understanding of science. As a result, members are far less likely to champion science policy than concerns of greater interest to their constituents. The fact that most members of Congress themselves do not come from scientific backgrounds also limits their interest in science policy.

A 1992 survey on congressional decision-making found that 75 percent of the legislative offices surveyed said they paid a "great deal" or "quite a bit" of attention to communications from constituents when deciding their positions and rating the relative importance of specific issues—outstripping any other source of information.[77] However, while senators and representatives receive thousands of letters and e-mails every month on the cost of prescription drugs, the right to own guns, abortion, or medical care for veterans, they typically receive very few communications from their constituents about science and technology.[78]

This lack of communication from constituents about the importance of science, combined with the scientific community's historically rather modest support for individual candidates, means that while many members of Congress appreciate the importance of science, they tend to assign a higher priority to other issues.[79] In Congress, science is most always the bridesmaid, and rarely the bride.

The Responsibilities of Congress

Congressional responsibilities for creating, authorizing, funding, and overseeing agencies and programs extends into all areas of policy, including science. In carrying out these responsibilities, Congress helps set priorities within science, as well as balancing it with other domestic and international needs. More indirectly, Congress is responsible for generating revenues to fund science programs it believes are necessary. These responsibilities are largely managed through legislation.

William Wells, noted S&T policy expert and former OSTP chief of staff, has remarked that "congressional policy-making can often be overly driven by localism; afflicted with partial, piecemeal solutions when the problem calls for an overarching, national solutions; saddled with ceremonial 'make people feel good' solutions that solve nothing; and frozen in a state of near paralysis pending the arrival of a crisis."[80] Wells goes on to note, however, that the same traits that produce these inefficiencies have also contributed to the system's tremendous effectiveness, especially in times of crisis. Says Wells, "When crisis finally arrives . . . Congress can respond quickly and, most of the time appropriately. Moreover, Congress provides the place and the circumstances to hammer out consensus on controversial issues—taking months, years or decades, if necessary."[81]

This effectiveness, however, is not always recognized, in part because many people incorrectly assume that legislators' sole job is to *pass* laws, and that, if laws are not passed, then Congress is not working. This misperception overlooks the fact Congress is also responsible for preventing bad laws from being passed and impeding the adoption of policies on which there is no public consensus. As Wells notes: "The public and the media may be clamoring for action that Congress cannot produce. But the fact is that Congress is often doing a good job of representing and reflecting the fact that no consensus exists on an issue among the American people."[82]

Congressional Support Agencies

Congressional support agencies are legislative branch agencies that support the work of the Congress and are paid for with legislative branch appropriations, but which are not part of the members' personal or committee offices. These organizations conduct studies, develop policy papers, provide budgetary analysis, and help craft legislation at the request of congressional members and committees. Some of the most important congressional support agencies affecting science policy are listed here.

The Congressional Research Service (CRS). The Congressional Research Service is "the public policy research arm of the United States Congress," exploring legislative topics for members of Congress.[83] As an agency within the Library of Congress, CRS works exclusively and directly for members of Congress and their personal and committee staffs on a confidential, nonpartisan basis. One might think of CRS as a congressional encyclopedia, with the ability to provide useful information to Congress at a moment's notice. It is often the first place to which members of Congress and their staffs turn when they need information quickly.

CRS was originally created in 1914 as the Legislative Reference Service. It was renamed the Congressional Research Service as a part of the Legislative Reorganization Act of 1970 (P.L. 91-510). While CRS serves as a nonpar-

tisan source of information and analysis for the Congress and its members, CRS does not make policy recommendations.

The CRS staff comprises nationally recognized experts in a range of issues and disciplines and is organized into five interdisciplinary research divisions: American Law; Domestic Social Policy; Foreign Affairs, Defense and Trade; Government and Finance; and Resources, Science and Industry. CRS produces policy and issue briefs on a wide range of topics, and conducts comprehensive legislative comparisons and analysis. It also handles requests for information from individual legislators and staffers via educational seminars and workshops, phone conversations, in-person briefings, written analytical reports, and confidential memoranda.

The Government Accountability Office (GAO). While the CRS is comparable to a congressional encyclopedia, the Government Accountability Office—formerly known as the General Accounting Office—may be thought of as a congressional detective agency investigating and evaluating existing government programs.

The GAO's primary functions include: (1) evaluation of government policies and programs; (2) auditing of agency operations to ensure that funds are appropriately and efficiently used; (3) investigating allegations of illegal or improper government activities; and (4) issuing legal decisions and opinions.[84] Some observers have referred to it as the "government's accountability watchdog."[85] However, the office does not have the authority to reprimand or to make rules. Instead, its power is indirect: the detailed studies, evaluations, and recommendations that the GAO provides to Congress often lead to new legislation and regulations.

The GAO has certainly had an impact on science policy through its assessments of such major science projects as the Superconducting Super Collider (SSC) and the International Space Station. It has also conducted reviews and issued reports on issues such as technology transfer; university indirect costs; agency peer review; export controls and university research; federal science and math education programs; and the effect of changing U.S. immigration policies on foreign student enrollment at U.S. universities, particularly in science-related fields.

The Office of Legislative Counsel. The Office of Legislative Counsel provides Congress with the legal expertise needed to draft legislation. While members of Congress and their staffs, individual and committee, usually write the first draft of new legislation, most bills introduced in Congress have been reviewed and revised by the House or Senate Office of Legislative Counsel.

POLICY DISCUSSION BOX 3.2

Does Congress Need More Technical Support?

A congressional support agency that used to exist specifically for science policy issues was the Office of Technology Assessment (OTA). Created by an act of Congress that was signed into law by President Nixon in October 1972, the OTA was designed to provide "congressional members and committees with independent, objective and authoritative analysis of the complex scientific and technical issues." The OTA was given wide latitude so that it could fulfill its role to provide in-depth analyses of policy challenges confronting Congress in science and technology. The OTA was more than just a gatherer of information. It was also able to work proactively, exploring uses for science and technology in crafting effective policies across a wide range of issues.

So what happened? The 104th Congress eliminated the OTA in 1995 as a cost-cutting measure, part of the Republican leadership's Contract with America.

Proponents of the cut argued that the OTA and CRS were duplicative. They also claimed that the OTA took too long to conduct its analyses, which were consequently of little or no value to congressional decision makers.

Some members of Congress, however, believe that a revived OTA is needed now more than ever. In fact, Congressman Rush Holt (D-NJ) has introduced legislation that would bring the OTA back. Many in the scientific community naturally believe that doing so would greatly enhance congressional ability to use science in formulating effective policy.

What do you think? Did the goals of OTA duplicate those of the CRS or the GAO? Do members of Congress need "a focal point for the timely analysis of science and technology-related legislative issues, to develop strategies to address matters that cut across multiple committees, and to facilitate the acquisition and dissemination of information on S&T-related activities"?[86] Should the OTA be revived? Is there a role in the current science policy-making process for the OTA?

The Congressional Budget Office (CBO). The mission of the Congressional Budget Office is to "provide the Congress with objective, nonpartisan, and timely analyses to aid in economic and budgetary decisions on the wide array of programs covered by the federal budget and [with] the information and estimates required for the Congressional budget process."[87] Created by passage of the Congressional Budget and Impoundment Act of 1974 (P.L. 93-344), the CBO acts in a much more focused and narrow manner than other congressional support agencies, such as the GAO and the CRS. The CBO does not make policy recommendations, but instead formulates budgetary recommendations and forecasts. Although its influence over the budget does give it some indirect influence over science policy, among other domestic and national security issues, the CBO wields considerably less power over science policy than its sister office in the executive branch, the OMB.

The Judicial Branch

While not directly involved at the front end of the science policy-making process, the judiciary's power to interpret laws and rule on their constitutionality gives it a not-insignificant role in shaping the future of science and technology. As Supreme Court justice Stephen G. Breyer has noted, "the practice of science depends on sound law—law that at a minimum supports science by offering the scientist breathing space, within which he or she may search freely for the truth on which all knowledge depends."[88] Since court verdicts are used as precedents in future decisions, today's rulings will have a dramatic and sometimes unanticipated impact on future policies. Like Congress, the courts are also the beneficiaries of new scientific knowledge, with one of the best-known examples being the use of DNA testing to establish innocence or guilt.

As the judiciary struggles to balance the demands of ethics, morality, and science, the interpretation of specific science policies is only becoming more complicated. We are in the midst of an explosion of court cases involving issues such as intellectual property, privacy, informed consent, and the relationship between morality and research. A few recent federal court decisions have particular potential to influence the future of science and technology.

In *Felten v. RIAA* (2001), the Recording Industry Association of America (RIAA) challenged the right of Edward W. Felten, a Princeton professor, and his research team to conduct and publish the results of their work on encryption and anticircumvention. Felten and his team filed a complaint against the RIAA, claiming infringement of their First Amendment right to free speech. Their case was eventually dismissed by a federal district court.[89]

In the area of intellectual property, *Madey v. Duke University* raised questions about whether researchers and the institutions they work for are exempt from normal intellectual property requirements when using patented materials in research or for experimental purposes. Whereas in the past it was believed that there was, in fact, an "experimental use" exemption protecting those involved in such activities from accusations of infringement, the Court of Appeals for the Federal Circuit ruled that this is not necessarily the case. Specifically, the court held that there is only a "very narrow and strictly limited experimental use defense," which does not apply if the use is "in furtherance of the alleged infringer's legitimate business," regardless of the user's "profit or nonprofit" status.[90]

In the area of biotechnology, *Diamond v. Chakrabarty* (1980) is a landmark case in which the U.S. Supreme Court ruled that life created in the research laboratory can be patented. This case involved Ananda Chakrabarty, a microbiologist, who when working for General Electric genetically engineered a *Pseudomonas* bacterium to break down crude oil and other toxic materials. Chakrabarty's patent application had originally been denied on the basis that living things are not patentable. Majority opinion in the final case found Chakrabarty's genetically engineered bacterium to be "a nonnaturally occurring manufacture or composition of matter—a product of human ingenuity" and thus patentable.[91]

Courts have also recently been asked to weigh in, for example, on the validity of evolution versus intelligent design. In *Kitzmiller v. Dover* (2005), the court was specifically asked to rule whether inclusion of intelligent design in biology curriculum is constitutional. In 2004, the Dover, Pennsylvania, school district implemented a policy incorporating the teaching of intelligent design along with the theory of evolution and natural selection in science classes. A group of parents and students filed suit against the school board on the grounds that intelligent design is not science but rather is simply creationism with a different name. The U.S. district court ruled in favor of the students and parents, declaring that intelligent design is not science, and that intelligent design "cannot uncouple itself from its creationist, and thus religious, antecedents."[92] Because such cases determine what can and cannot be taught in public school science courses, their outcomes are of great interest to the scientific community.

We also note that science is playing an increasingly important role as a resource for the nation's judicial system. More and more cases are being decided on the basis of scientific evidence (e.g., DNA matches). As scientific technologies and data become more commonly used to inform court decisions, either as evidence or as tools for obtaining evidence, judges will increasingly be asked to distinguish between legitimate and junk science—yet another reason for the scientific community to increase its interactions with the judicial branch.[93]

In the case of *Daubert v. Merrell Dow Pharmaceuticals*, the Supreme Court attempted to define the role of scientific "experts," providing a list of guidelines that judges use to assess the admissibility of scientific testimony.[94] In this gatekeeping role, judges are presumed to be keeping out unsupported evidence while allowing testimony that meets tests of plausibility. One of the concerns raised by *Daubert*, however, was trial court judges' lack of scientific understanding, which leaves them ill-equipped to evaluate the merits of scientific evidence. Indeed, the results of a survey published in 2002 suggested that out of four hundred state trial court judges, only a few questioned the quality of evidence allegedly based on science, and only a fraction had a clear understanding of error rate and falsifiability. This suggests that most trial judges are not fully prepared to appreciate the diversity and complexity of the scientific facts presented before them, and that "many judges [do] not recognize their lack of understanding."[95] Like their colleagues in the legislative and executive branches, many judges are poorly informed about science and its conduct. Nor does the judicial process allow as much input from interested individuals and organizations as do the other two branches of government.

Recognizing this, the Supreme Court went further in *Daubert*, authorizing judges to request that recognized scientists assist them in evaluating scientific evidence when deciding whether to allow it. This is just one case in which the scientific community has been called upon to serve the public interest, and may signal the beginning of increased cooperation between science and the judiciary. It does also, however, raise questions about the proper role of science and scientists in informing judicial decisions: Who is qualified as an expert witness on scientific matters? What criteria should be used to make such determinations? Is it possible to ensure that court decisions are based upon a consensus of scientific opinion, as opposed to one individual's interpretation of scientific facts? How do we foster an exchange among judges, lawyers, and scientists so that all are better informed about how judicial decisions can affect the conduct of science, and how science can play an important role in judicial decisions?[96]

Daubert and the other cases we have discussed highlight why it is vital that members of the scientific community not only pay attention to Congress and the executive branch, but also learn how to better interact with the judiciary. Scientists have an obligation to enhance our courts' abilities however possible, and supplement judicial knowledge with their own expertise.

Nongovernmental Entities Involved in Decisions on Science Policy

Quasi-governmental and nongovernmental organizations also have tremendous influence on science policy. A discussion of some of the most influential of these organizations follows.

The National Academies

The National Academies consist of three member organizations: the National Academy of Sciences, the National Academy of Engineering, and the Institute of Medicine. The National Research Council is also a part of the National Academies, but is not a membership organization. Rather it serves as the functional arm of the National Academies that helps fulfill and respond to congressional and government needs for scientific and technical information. The National Academies are generally viewed as impartial advisors on science policy, and both Congress and the executive branch often turn to them for advice on scientific issues and projects. As the judicial branch becomes more involved in debates over science policy, it, too, may come to rely on the National Academies.

The first of the three member organizations to be established was the National Academy of Sciences (NAS), a "private non-profit self-perpetuating society of distinguished scholars engaged in scientific and emergency research, dedicated to the furtherance of science and technology and to their use for general welfare."[97] The NAS was chartered through an act of incorporation passed by Congress and signed into law by President Lincoln in 1863. Since its creation, the organization has provided essential assistance to Congress and the executive branch's efforts to formulate sound science and technology policy.

As scientific issues became increasingly complex, the NAS found it necessary to expand the scope of its services to the government and the scientific and engineering communities. This led to the formation of new organizations under the NAS umbrella, beginning with the National Research Council (NRC) in 1916, which was created largely in response to the government's increasing need for scientific advice during World War I. The creation of the NRC was followed by the National Academy of Engineering (NAE) in 1964 and the Institute of Medicine (IOM) in 1970.

The NAE is subject to the same charter as the NAS, but its mission focuses on "marshalling the knowledge and insights of eminent members of the engineering profession" to "promote the technological welfare of the nation," while the Institute of Medicine's charge is to "serve as adviser to the nation to improve health . . . [and] provide unbiased, evidence-based, and authoritative information and advice concerning health and science policy to policy makers, professionals, leaders in every sector of society, and the public at large."[98] The IOM has its own charter and bylaws, approved by the NAS.

The NRC is now the primary mechanism by which the NAS, NAE, and IOM provide service to the government, the public, and the scientific and engineering communities. The NRC is overseen by the NAS, NAE, and IOM. The broad range of advisory services it provides to government officials and the nation, largely through study panels and reports, differentiates it from the scientific academies of other nations. The NRC's reports and studies are often oriented to describing the scientific, technical, and policy challenges facing the scientific community, the private sector, the United States, and foreign governments. These reports are widely used by industry, government policymakers, and the international community.

The NRC is comprised of six units under which many of the National Academies' projects and programs fall: the Division of Behavioral and Social Sciences and Education, the Division on Earth and Life Studies, the Division on Engineering and Physical Sciences, the Institute of Medicine Programs, the Policy on Global Affairs Division, and the Transportation Research Board. Although each has its own distinct focus, all work toward the same overarching goals. The Committee on Science, Engineering, and Public Policy (COSEPUP), which falls under the Policy on Global Affairs Division, is one of the primary National Academies' committees conducting studies on cross-cutting issues in science and technology policy, monitoring key developments in science policy, and addressing "the concerns and requests" of the key science policymakers in the federal government.[99] COSEPUP also serves as an information conduit, putting the National Academies in the forefront of American science and technology policy. Other National Academies' units heavily involved in science policy include the Committee on Science, Technology, and Law (STL) and the Board on Science, Technology, and Economic Policy (STEP).

The NAS and NAE charters mandate that they "investigate, examine, experiment, and report upon any subject of science or art" when called upon to do so by any governmental department.[100] Likewise, the IOM charter requires that it "respond to requests from the federal government and other public and private agencies for studies and advice on matters related to health and medicine."[101] In recent years, and especially since the demise of the OTA, Congress has increasingly relied upon the National Academies to review science projects or agency policies and practices affecting the conduct of science. While Congress requests and funds these studies, the academies themselves determine the process and procedures by which they will be carried out. National Academies' studies are also carried out with private support.

The academies are all honorary societies, whose members and foreign associates are chosen annually from a pool of candidates nominated by current members for their "distinguished and continuing achievements in original research."[102] Election is one of the most prestigious honors that can be bestowed upon a scientist or engineer.

Scientific Societies and Higher-Education Associations

The professional societies and associations representing the scientific and research communities—including the American Physical Society, the American Chemical Society, the Federation of American Societies for Experimental Biology, and the American Council of Education—are also influential players in the policy-making process. These groups keep their members apprised of pending legislation and encourage them to actively engage their congressional delegation. These societies also have public or government affairs committees and staff who keep members informed, arrange visits to members of Congress, and organize other advocacy activities. The societies thus present a united scientific voice on key topics. Each year, for example, when the appropriations subcommittees are busy working on their appropriations bills, the societies encourage their members to contact members of Congress in support of funding for the relevant federal agencies.

The American Association for the Advancement of Science (AAAS) is a large general professional society representing all disciplines of science and engineering. Because it has the largest membership of any professional science society, the AAAS plays an important role in representing the scientific community to the government, and the government to its members. The detailed analysis AAAS publishes each year on the president's proposed budget is referred to by many advocates in science policy, and the AAAS's Web site offers continually updated budget analyses as appropriations bills are drafted and brought to the floor for vote. Unlike many scientific societies, however, the AAAS does not engage in grassroots lobbying or take official positions on specific pieces of legislation. It does, however, issue position statements on important science policy subjects.

Several higher-education associations, including the Association of American Universities (AAU), which represents sixty public and private U.S. research universities; the National Association of State Universities and Land Grant Colleges, which represents 211 public state and land grant institutions; and the Council on Governmental Relations, which represents 150 research-intensive universities, represent the interests and concerns of the academic research community as a whole. These associations represent their member universities in Washington on a broad range of issues, including research funding, graduate education, and research policy matters. Increasingly, the efforts of these groups are being complemented by the growing number of universities that maintain their own Washington-based offices.

A further group of professional organizations and trade associations have sprung up to represent specific categories of university officials, such as the Association of International Educators (also known as NAFSA, an acronym based upon the Association's former name, the National Association of Foreign Student Advisors), the Association of University Technology Managers, the National Council of University Research Administrators, and the National Association of College and University Attorneys. These organizations also take positions on policy matters of specific interest to universities.

While all of these associations' members are important, their staffs—most based in Washington, DC—manage day-to-day activities and are responsible for regular contact with members of Congress, executive branch offices, and other societies and associations.

Coalitions and Ad Hoc Advocacy Groups

Scientific societies, associations, universities, and businesses have also begun banding together into informal coalitions or ad hoc groups. Examples include the Ad Hoc Group for Medical Research, Research!America, the Coalition for National Science Funding, The Science Coalition, the Energy Sciences Coalition, the Coalition for National Security Research, and the Joint Steering Committee for Public Policy. This pooling of resources allows groups with similar interests to avoid unnecessary duplication. For instance, the American Society for Cell Biology, the Genetics Society of America, and the Society for Neuroscience share an interest in many issues, such as the use of animals in research. By partnering in the Joint Steering Committee for Public Policy, they coordinate their efforts and reach out to members of Congress with a unified message.

The science community has also formed coalitions with public-citizen and advocacy groups. The biomedical community, for example, has found that uniting with patient-advocacy groups behind a particular issue (e.g., human embryonic stem cell research, increased funding for mental illness research or breast or colon cancer research) gives them an effective way to appeal to members of Congress. This reliance on personal appeals from disease groups is particularly effective with the many legislators who have themselves had personal experience with such diseases (e.g., many have had family and/or friends who have had or even died from such diseases or they themselves may have suffered from them).

Think Tanks and Policy Support
Organizations

Think tanks and other organizations in Washington provide policy support to Congress, the executive branch, and the federal agencies. Some favor a specific ideological position, while others are more issue-oriented.

In times of especially high tension between the major political parties, it is difficult to get a hearing for new ideas on science policy. If a proposal arises on one side of the aisle, the other side will almost reflexively shout it down, or label it a Trojan horse. Well-meaning representatives who cross the aisle to collaborate on important topics may be tagged as traitors by their parties.

Broadly based think tanks composed of scholars with recognized expertise can help overcome the stalemates that naturally arise in a two-party system. Other, more partisan think tanks such as the Heritage Foundation, the Brookings Institution, the American Enterprise Institute, the Cato Institute, and the Progressive Policy Institute provide proponents of a particular idea or position with the data and arguments they need to make their case. These organizations are powered by a cadre of highly trained scholars who bring their intellectual powers to bear on the issues most important to their party or political constituency.

The Center for Strategic and International Studies (CSIS) is one example of a more broadly based, nonpartisan think tank. The CSIS, located in downtown Washington within walking distance of the White House, employs about 190 research scholars and support staff. Highly valued by both legislators and the executive branch, its board includes several former government leaders, including David Abshire, Henry Kissinger, William Cohen, Zbigniew Brzezinski, James Schlesinger, and Brent Scowcroft. CSIS, like many think tanks, conducts analyses on a wide variety of issues, including some important to science policy. For example, CSIS has conducted a major review of the Department of Energy laboratory security systems; assessed ways in which the national laboratories could contribute to U.S. economic competitiveness; examined the impact of globalization on science and technology; made recommendations concerning new R&D in response to terrorism; and been actively involved reviewing the impact of post–September 11 visa and export control policies on the conduct of science.

One well-recognized CSIS effort was the Strengthening of America Commission chaired by Senators Sam Nunn (D-GA) and Pete Domenici (R-NM), which in 1992 issued an action plan to put the U.S. fiscal house in order by reducing the federal budget deficit, replacing the current tax code with a consumption-based income tax, and establishing an "endowment for the future" to support investment in R&D, technology, and education.[103] The tax proposal contained in this plan later evolved into the USA (Unlimited Savings Allowance) Tax. Although in this instance, CSIS was not itself an advocate for any specific policy change, the commission's report provided the educational and intellectual foundations to make the case for broad and comprehensive budget and tax reform and served as the basis for tax legislation (S. 722) that was introduced by Senators Domenici and Nunn in 1995.[104]

The RAND Corporation, the National Bureau of Economic Research, and the Hastings Center all occupy similar positions of nonpartisan influence. These entities powerfully influence the shape of laws affecting research and development. To the extent that both ideological and nonpartisan think tanks provide expert analysis on topics of intense national interest, their role in the formation of policy must be regarded as positive.

Industry

As a—if not the—primary beneficiary of federal R&D investment, industry has an obvious interest in federal science policy. However, because the benefits to industry of federal research investment are often indirect, industrial leaders do not always appreciate the importance of science policy to their work. Instead, corporate leaders tend to focus their advocacy on more directly relevant issues like tax or trade policy.

Still, when industry does speak up on broader federal policies, it can have a significant impact. This impact, evident in such outcomes as the many extensions of the R&D tax credit and the American Competitiveness Initiative (ACI) announced by President George W. Bush in January 2006, has been achieved through sector-wide groups like the Industrial Research Institute, the National Association of Manufacturers (NAM), the American Electronic Association, Tech Net, and the Technology CEO Council. Industry has also begun to cooperate with higher-education associations and scientific societies to make the case for increased federal funding of basic research. Their major collaborations include the Council on Competitiveness, the Alliance for Science and Technology Research in America (ASTRA), and the Task Force on the Future of American Innovation. Other business groups such as the Business Roundtable have taken a special interest in improving science and math education.

Policy Challenges and Questions

In this chapter, we have provided a comprehensive overview of the major federal and nonfederal actors responsible for making and influencing science policy. Were there only one major federal science agency, this review would be easy; however, this is not the case in the United States. The U.S. system for support of science is, in fact, pluralistic, with several federal departments and independent agencies maintaining responsibility for the conduct of science that corresponds with the agencies' overarching missions.

While this pluralistic system is not easy to understand, it has helped make the United States a worldwide leader in many fields of science. Still, in an attempt to enhance the policies that drive science in the United States, it is likely that new suggestions will be put forward to better coordinate scientific activities and to improve upon the current infrastructure for science policy. Policymakers will likely continue to challenge the existing system and to ask if, in fact, our pluralistic system for government support of science will be effective in meeting our future scientific challenges.

Many issues that have been raised in the past are likely to continue to be the topic of discussions in the future. For example, should the science advisor be given more authority and access to the president? Would consolidation be a better way to manage existing science programs among government agencies? What should be the role of the OSTP in coordinating science across federal agencies? Should OSTP play a greater role in establishing government-wide interagency priorities for scientific research? Does Congress need better scientific advice? If so, would it make sense to bring back the OTA? And finally, are there enough people employed in the federal government with scientific training and adequate technical backgrounds (e.g., in Congress, the courts, and key federal agencies such as the State Department) to ensure that science is adequately considered and understood, especially in making decisions that will affect the very conduct of science itself? Some of these questions and others will be further examined in the chapters to follow.

NOTES

1. The metaphor comparing lawmaking to sausage making is widely attributed to German chancellor Otto von Bismarck (1815–98) who stated, "If you like laws and sausages, you should never watch either being made." See Suzy Platt, ed., *Respectfully Quoted: A Dictionary of Quotations Requested from the Congressional Research Service* (Washington, DC: Library of Congress, 1989), 190. For a discussion of whether this metaphor is still applicable to the modern-day legislative policy-making process, see Alan Rosenthal, "The Legislature as Sausage Factory," *State Legislatures,* September 2001, 12–15. For a discussion of the complexities that make policy-making a messy process, see Elizabeth Moore and Michele Corey, "Beyond Civics: A Primer on Political Realities and the Forces That Shape Public Policy," *MSU Connect Magazine,* Fall 2003, http://www.fact.msu.edu/Connect/policy.htm (accessed January 19, 2007).

2. Walter J. Oleszek, *Congressional Procedures and the Policy Process,* 5th ed. (Washington, DC: CQ Press, 2001), 4.

3. See Susan Ehringhaus, statement before the National Committee on Vital Health Statistics, U.S. Department of Health and Human Services, Meeting of the Subcommittee on Privacy and Confidentiality, Silver Spring, Maryland, November 19–20, 2003; Gerald W. Woods, *Impact of the HIPAA Privacy Rule on Academic Research* (Washington, DC: American Council on Education, November 2002); and Jennifer Horner and Michael Wheeler, "HIPAA: Impact on Research Practices," *ASHA Leader,* November 8, 2005, 8–9, 26–27.

4. Admiral James Watkins, before the House Committee on Science, "National Science Policy Study, Part III: International Science," 105th Cong., 2nd sess., March 25, 1998, http://commdocs.house.gov/committees/science/hsy084000.000/hsy084000_0f.htm (accessed May 5, 2007); see also House Committee on Science, *Unlocking Our Future,* 56.

5. See Smith, *American Science Policy,* 139, 159–66; Raymond L. Orbach, at the Arnold O. Beckman Lecture in Science and Innovation, University of Illinois at Urbana-Champaign, "Pluralism as a Foundation for U.S. Scientific Preeminence: Advancement of Knowledge and the Freedom and Happiness of Man," May 5, 2004, http://www.er.doe.gov/sub/speeches/speeches/Beckman_Lecture.htm (accessed May 5, 2007).

6. William C. Boesman, "A Department of Science and Technology: A Recurring Theme," *Congressional Research Service Report on Congress,* February 3, 1995, 95-235, http://digital.library.unt.edu/govdocs/crs//data/1995/upl-meta-crs-197/95-35_1995Feb03.html?PHPSESSID=6210b91796f76e247228762510d48843 (accessed May 5, 2007).

7. Explicit functions of the Chinese Ministry of Education include "to plan and direct the research work of higher education institutions in natural science, philosophy and social science; to undertake macro-instruction on higher education institutions in the application, research and popularization of new hi-technology, in the transform of scientific and research achievements and in the efforts of combining industry, teaching and research; to coordinate and supervise higher education institutions in the implementation of state key scientific and research programs and national-defense scientific-tackling programs; to direct the development and construction of national key laboratories and research centers for national projects" (http://www.moe.edu.cn/english/ministry_f.htm, accessed May 5, 2007). The Chinese Ministry of Science and Technology, http://www.most.gov.cn/eng/organization/Mission/index.htm (accessed May 5, 2007), also has significant responsibility of scientific research conducted as does the National Natural Science Foundation of China, http://www.nsfc.gov.cn/e_nsfc/2006/01au/01mr.htm (accessed May 5, 2007), an entity similar to the U.S. NSF.

8. The late congressman George Brown, former chair of the House Science Committee, noted that "President Gerald Ford helped redefine the Federal role in science policy with the signing of the Science Policy Act of 1976, a major work of the House Science and Technology Committee. While never fully implemented, this Act led to the further definition of the Federal role in technology transfer and advanced technology development in the 1988 Trade Bill signed by President Reagan." See George E. Brown, "Unlocking Our Future," *APS News,* January 1, 1999, http://apsweb.aps.org/publications/apsnews/199901/future.cfm. Some in the scientific community have been critical of President George W. Bush claiming that he has not used the scientific community to assist him in making key scientific decisions and on the grounds that under his administration major scientific review panels have become increasingly politicized. See "Restoring Scientific Integrity in Policy Making," statement issued by over sixty-two leading scientists, Union of Concerned Scientists, Washington, DC, February 18, 2004, http://www.ucsusa.org/scientific_integrity/interference/scientists-signon-statement.html (accessed May 5, 2007).

9. Steven Goldberg, *Culture Clash: Law and Science in America* (New York: New York University Press, 1994), 28. The quotation from Jefferson comes from a letter he wrote to Hugh Williamson. See Paul L. Ford, ed., *The Works of Thomas Jefferson* (New York: Knickerbocker Press, 1904), 458–59.

10. President Dwight D. Eisenhower, "Radio and Television Address to the American People on Science in National Security" November 7, 1957. The original text can be accessed at the American Presidency Project, http://www.presidency.ucsb.edu/ws/index.php?pid=10946&st=&st1 (accessed May 5, 2007).

11. Presidential expert Richard E. Neustadt has suggested that one of the primary powers at the disposal of the president is the "power to persuade." According to Neustadt, presidents who are the most effective are those that are exceptional at politics and the art of persuasion. See Richard E. Neustadt, *Presidential Power and Modern Presidents: The Politics of Leadership from Roosevelt to Reagan* (New York: Free Press, 1991). The same can be said of presidential science advisors; however, the person whom they most need to persuade is the president, not members of Congress or the general public.

12. Office of Technology Assessment, *Federally Funded Research: Decisions for a Decade,* OTA-SET-490 (Washington, DC: U.S. Government Printing Office, May 1991), 76.

13. For additional background, see Bruce L. R. Smith, *The Advisors: Scientists in the Policy Process* (Washington, DC: Brookings Institute, 1992), 163–79.

14. Ibid.; William Golden, ed., *Science Advice to the President,* 2nd ed. (Washington, DC: AAAS Press, 1993), 120–24.

15. Golden, *Science Advice,* 93.

16. Ibid.

17. Referring to Soviet leader Nikita Khrushchev.

18. Smith, *The Advisors,* 166–68.

19. John F. Kennedy, "Presidential Message to Accompany Reorganization Plan No. 2 of 1962," March 29, 1962. The full text of the plan along with the accompanying message from President Kennedy is available at www.access.gpo.gov/uscode/title5a/5a_4_71_2_.html (accessed May 5, 2007).

20. Under the Clinton administration, the OSTP had four associate directors. In addition to the associate director for science and the associate director for technology, the OSTP also had an associate director for the environment and an associate director for national security. Shortly after President George W. Bush took office in January 2001, a decision was made to restructure the OSTP and to consolidate the number of Senate-confirmed OSTP associate directors from four to two.

21. An organization chart for OSTP is available at the White House Office of Science and Technology Policy, http://www.ostp.gov/images/orgchart2.jpg (accessed May 5, 2007).

22. Executive Order 12700, January 19, 1990, "President's Council of Advisors on Science and Technology," *Federal Register* 55 (January 23, 1990): 2219.

23. Executive Order 13385 expanded the number of PCAST members from twenty-five to forty-five and assigned the role and responsibilities formerly performed by the President's Information Technology Advisory Committee, authorized by Congress under the High-Performance Computing Act of 1991 (P.L. 102-194) and the Next Generation Internet Act of 1998 (P.L. 105-305) as a Federal Advisory Committee, to the President's Council of Advisors on Science and Technology (PCAST). See Executive Order 13385, September 29, 2005, "Executive Order: Continuance of Certain Federal Advisory Committees and Amendments to and Revocation of Other Executive Orders," *Federal Register* 70 (October 4, 2005): 57989. Executive

Order 13385 extended and amended Executive Order 13226, September 30, 2001, "President's Council of Advisors on Science and Technology," *Federal Register* 66 (October 3, 2001): 50523.

24. Executive Order 12881, November 23, 1993, "Establishment of the National Science and Technology Council," *Federal Register* 58 (November 26, 1993): 62491.

25. Ibid., sec. 4.

26. The National Science and Technology Policy, Organization and Priorities Act of 1976, P.L. 94-282 (May 11, 1976), Title IV.

27. Office of Technology Assessment, *Federally Funded Research*, 77–78.

28. President Clinton established the NSTC on November 23, 1993, by Executive Order 12881. According to this executive order, membership on the council shall consist of the president of the United States, vice president, secretary of commerce, secretary of defense, secretary of energy, secretary of health and human services, secretary of state, secretary of the interior, administrator of NASA, director of the National Science Foundation, director of the Office of Management and Budget, administrator of the EPA, assistant to the president for science and technology, national security advisor, assistant to the president for economic policy, assistant to the president for domestic policy, and such officials of executive departments and agencies as the president may, from time to time, designate.

29. See the Office of Management and Budget, "Grants Management," http://www.whitehouse.gov/omb/grants/ (accessed May 5, 2007).

30. An organization chart for OMB is available at the White House Office of Management and Budget Web site, http://www.whitehouse.gov/omb/omb_org_chart.pdf (accessed May 5, 2007).

31. For further discussion, see Committee on Science, Engineering, and Public Policy, *Evaluating Federal Research Programs: Research and the Government Performance and Results Act* (Washington, DC: National Academies Press, 1999); Committee on Science, Engineering, and Public Policy, *Implementing the Government Performance and Results Act for Research: A Status Report* (Washington, DC: National Academies Press, 2001); and William Schulz, "Assessing Federal R&D Efforts," *Chemical and Engineering News*, March 4, 2002, 10.

32. Office of Personnel Management, Office of Workforce Information, Central Personnel Data File, "Political Appointments by Type and Work Schedule," September 2001, http://www.opm.gov/feddata/html/POL0901.asp (accessed May 5, 2007).

33. These and other advances, which have been generated in part as the result of NSF's investment in basic research, can be found in the NSF's *Nifty50*, http://www.nsf.gov/od/lpa/nsf50/nsfoutreach/htm/home.htm (accessed February 25, 2007). It should be noted that the science contributing to some of these advances can also be rightfully claimed by other federal agencies, including the Department of Energy and the Department of Defense. See *National Science Foundation, Celebrating 50 Years: Resource Guide 2000*, NSF 00-87 (Arlington, VA: National Science Foundation, 2000), http://www.nsf.gov/od/lpa/nsf50/nsfoutreach/htm/home.htm (accessed May 5, 2007).

34. David Hart, "The Origins of Google," *NSF Discoveries*, National Science Foundation, http://www.nsf.gov/discoveries/disc_summ.jsp?cntn_id=100660 (accessed May 5, 2007).

35. National Science Foundation, National Science Board, http://www.nsf.gov/nsb/ (accessed May 5, 2007).

36. National Science Foundation, "About NSF," "Orientation to the NSF," "NSF Goals," http://www.nsf.gov/about/career_opps/orientation/goals.jsp (accessed May 5, 2007).

37. Mitchell Daniels, Director, OMB, remarks to the National Press Club, November 28, 2001, http://www.whitehouse.gov/omb/speeches/natl_press_club.html (accessed May 5, 2007).

38. See National Institutes of Health, "About NIH," http://www.nih.gov/about/ (accessed May 5, 2007).

39. See ibid.

40. National Cancer Act of 1937, P.L. 75-244 (August 5, 1937), 50 Stat. L. 559.

41. Victoria A. Harden, "A Short History of the National Institutes of Health," National Institutes of Health, Office of NIH History, http://history.nih.gov/exhibits/history/docs/page_04.html (accessed May 5, 2007).

42. The National Institutes of Health Reform Act of 2006 (P.L. 109-482) limits the number of NIH institutes and centers to twenty-seven.

43. An in-depth discussion of the different NIH institutes, "Institutes, Centers, and Offices," can be found on the NIH Web site, http://www.nih.gov/icd/ (accessed May 5, 2007).

44. Kei Koizumi, "National Institutes of Health in the FY 2007 Budget," in *AAAS Intersociety Working Group Report XXXI*, chap. 8.

45. Ibid.

46. When the National Cancer Act was originally passed in 1971, its intent was to remove the NCI from the NIH and have the NCI director report directly to the president. After concerns were expressed by the scientific community about the adverse impacts of such a move, a compromise was reached that gave the NCI special privileges, including the presidential appointment of the NCI director and the ability to develop a bypass budget, which would be delivered directly to the president, circumventing approval by the NIH director. To assure, however, that the NIH director had a higher status than the NCI director, it was agreed that the NIH director would be a presidential appointee requiring Senate confirmation, while the NCI director would not require Senate confirmation. See Alan Rabson, "Report of the Acting Director, NCI," Board of Scientific Advisors, National Cancer Institute, Meeting Minutes, Bethesda, Maryland, November 13–14, 2001, http://deainfo.nci.nih.gov/Advisory/bsa/bsa1101/bsa1101min.htm (accessed May 5, 2007).

47. See National Aeronautics and Space Act of 1958, P.L. 85-568, http://www.hq.nasa.gov/office/pao/History/spaceact.html (accessed May 5, 2007).

48. See National Aeronautic and Space Administration, History Division, "A Brief History of NASA," http://history.nasa.gov/brief.html (accessed May 5, 2007).

49. Kei Koizumi, "R&D in the FY 2007 Department of Defense Budget," in *AAAS Intersociety Working Group Report XXXI*, chap. 6.

50. National Science Board, *Science and Engineering Indicators 2006*, chap. 5 and appendix table 5.9.

51. Koizumi, "R&D in the FY 2007 Department of Defense Budget."

52. Michael S. Lubell, "The Department of Energy in the FY 2007 Budget," in *AAAS Intersociety Working Group Report XXXI*, chap. 9.

53. See National Institute of Standards and Technology, "About NIST," http://www.nist.gov/public_affairs/nandyou.htm (accessed May 5, 2007).

54. See National Oceanic and Atmospheric Administration, "Background Information," http://www.publicaffairs.noaa.gov/back.html (accessed May 5, 2007).

55. The Department of Homeland Security, "Directorate of Science and Technology," http://www.dhs.gov/xabout/structure/editorial_0530.shtm (accessed May 5, 2007).

56. The Department of Homeland Security, "Research," http://www.dhs.gov/xres/ (accessed May 5, 2007).

57. The Oak Ridge Association of Universities, "DHS Security Scholarship and Fellowship Program," http://www.orau.gov/dhsed/ (accessed May 5, 2007).

58. Department of Homeland Security, "Fact Sheet: Homeland Security Centers of Excellence: Partnering with the Nation's Universities," January 10, 2005, http://www.dhs.gov/xnews/releases/press_release_0586.shtm (accessed May 5, 2007).

59. Spencer S. Hsu, "DHS Terror Research Agency Struggling," *Washington Post*, August 20, 2006, A08.

60. Genevieve J. Knezo, *Homeland Security Research and Development Funding, Organization and Oversight*, RS21279 (Washington, DC: Congressional Research Service, Library of Congress, August 22, 2006), 1–2.

61. National Science Board, *Science and Engineering Indicators 2006*, figs. 4.4–4.13.

62. Department of Agriculture, *Strategic Plan for FY 2002–2007*, September 2002, 2.

63. See Office of Technology Assessment, *Federally Funded Research*, 113–16.

64. Wayne D. Rasmussen, *Taking the University Back to the People: Seventy-five Years of Cooperative Extension* (Ames: Iowa State University Press, 1989), 41.

65. For further information on the Morrill Act see Allen Nevins, *The State Universities and Democracy* (Urbana: University of Illinois Press, 1962); Fred F. Harderoad and Allan W. Ostar, *Colleges and Universities for Change* (Washington, DC: University Press for the American Association of State Colleges and Universities, November 1987); and Ralph D. Christy and Lionel Williamson, eds., *A Century of Service: Land-grant Colleges and Universities, 1890–1990* (New Brunswick, NJ: Transaction, 1992).

66. Rasmussen, *Taking the University Back*, 47.

67. See Agriculture Research Service, "About ARS," http://www.ars.usda.gov/aboutus/ (accessed May 5, 2007).

68. See Cooperative State Research, Education, and Extension Service, U.S. Department of Agriculture, "Funding Mechanisms," March 2005, http://www.csrees.usda.gov/about/pdfs/funding_mechanisms.pdf (accessed May 5, 2007).

69. See Forest Service, "Research and Development," http://www.fs.fed.us/research/ (accessed May 5, 2007).

70. See Economic Research Service, "About ERS," http://www.ers.usda.gov/AboutERS/ (accessed May 5, 2007).

71. See U.S. Department of State, "About State, Organization—Bureaus and Offices A–Z, Science and Technology Advisor," http://www.state.gov/g/stas/ (accessed May 5, 2007).

72. See the U.S. Geological Survey, "About USGS," http://www.usgs.gov/aboutusgs.html (accessed May 5, 2007).

73. Carnegie Commission on Science, Technology, and Government, *Science, Technology, and Congress: Expert Advice and the Decision-Making Process* (New York: Carnegie Commission on Science, Technology, and Government, 1991), http://www.carnegie.org/sub/pubs/science_tech/cong-exp.txt. (accessed May 5, 2007).

74. Richard F. Fenno Jr., *Home Style: House Members in Their Districts* (Glenview, IL: Scott, Foresman, 1978).

75. David Mayhew, *Congress: The Electoral Connection* (New Haven: Yale University Press, 1974).

76. A historical overview of congressional staff and management can be found in the Final Report of the Joint Subcommittee on the Organization of Congress, *Organization of the Congress*, 103rd Congress, 1st sess., December 1993, http://www.rules.house.gov/Archives/jcoc2.htm (accessed May 5, 2007).

77. Burson-Marsteller, *Attention Congressional Staff Give Selected Communications and Comparative Frequency of Such Communications* (Washington, DC: Burson-Marsteller, 1992). See also Bob Smucker, *The Nonprofit Lobbying Guide*, 2nd ed. (Washington, DC: Independent Sector, 1999), 23–33, http://www.clpi.org/CLPI_Publications.aspx (accessed May 5, 2007).

78. American Institute of Physics, "Communicating with Congress—Correspondence," *FYI: The American Institute of Physics Bulletin of Science Policy News*, May 7, 1999.

79. This lack of participation in the political process by the scientific community changed for the 2004 presidential election. See chap. 20, "Science, Science Policy, and the Nation's Future," for a discussion of this subject. In addition, see Scientists and Engineers for America, http://www.sefora.org/index.php (accessed January 19, 2007).

80. William G. Well Jr., *Working with Congress: A Practical Guide for Scientists and Engineers* 2nd ed. (Washington, DC: American Association for the Advancement of Science, 1996), 23.

81. Ibid.

82. Ibid.

83. See the Congressional Research Service, "About CRS," http://www.loc.gov/crsinfo/whatscrs.html#about (accessed May 5, 2007).

84. See Government Accountability Office, "The Background of GAO," http://www.gao.gov/about/history.html (accessed May 5, 2007).

85. Ibid.

86. Carnegie Commission on Science, Technology, and Government, *Science, Technology, and Congress*, 22.

87. Congressional Budget Office, "About CBO," http://www.cbo.gov/aboutcbo/ (accessed May 5, 2007).

88. Stephen G. Breyer, "The Interdependence of Science and Law," *AAAS Science and Technology Policy Yearbook, 1999* (Washington, DC: American Association for the Advancement of Science, December 1998), chap. 9.

89. For additional information see the Electronic Frontier Foundation, *Felton et al v. RIAA et al* http://www.eff.org/IP/DMCA/Felten_v_RIAA/ (accessed January 19, 2007).

90. *Madey v. Duke University,* 307 F.3d 1351 (Fed. Cir. 2002).

91. *Diamond v. Chakrabarty,* 447 S.Ct. 303 (1980).

92. *Kitzmiller v. Dover,* 400 F.Supp.2d 707 (M.D. Pa. 2005).

93. *Daubert v. Merrell Dow Pharmaceuticals, Inc.* 509 S.Ct. 579 (1993).

94. See National Science Board, *Science and Engineering Indicators 2006,* 7.18.

95. S. A. Dobbin, S. I. Gatowski, J. T. Richardson, G. P. Ginsburg, M. L. Merlino, and V. Dahir, "Applying *Daubert:* How Well Do Judges Understand Science and Scientific Method?" *Judicature* 85, no. 5 (2002): 244–47; S. I. Gatowski, S. A. Dobbin, J. T. Richardson, G. P. Ginsburg, M. L. Merlino, and V. Dahir, "Asking the Gatekeepers: A National Survey of Judges on Judging Expert Evidence in a Post-*Daubert* World." *Journal of Law and Human Behavior* 25, no. 5 (2001): 433–58.

96. Many of these same questions were raised by an ad hoc committee of the National Academies, which met in January and March 2005 to consider the impact of *Daubert* and subsequent Supreme Court opinions and to identify questions for future study. See National Academies Committee on Science, Technology, and Law, *Discussion of the Committee on Daubert Standards: Summary of Meetings* (Washington, DC: National Academy Press, 2006).

97. National Academy of Sciences, "About the NAS," http://www.nasonline.org/site/PageServer?pagename=ABOUT_main_page (accessed May 5, 2007).

98. National Academy of Engineering, http://www.nae.edu/nae/naehome.nsf (accessed May 5, 2007); Institute of Medicine, "About IOM," http://www.iom.edu/CMS/3239.aspx (accessed May 5, 2007).

99. The National Academies, Committee on Science, Engineering, and Public Policy http://www7.nationalacademies.org/cosepup/ (accessed May 5, 2007).

100. National Academy of Sciences, "About NAS."

101. Institute of Medicine, "Charter and Bylaws," September 2002, http://www.iom.edu/Object.File/Master/16/562/charter%20and%20bylaws%202002%20-%20final%20copy-web.pdf (accessed January 6, 2007).

102. National Academy of Sciences, "About NAS."

103. Center for Strategic and International Studies, *The CSIS Strengthening of America Commission/First Report* (Washington, DC: Center for Strategic and International Studies, October 1992).

104. Sam Nunn, "Reforming the Tax System through the Unlimited Savings Allowance (USA) Tax Proposal," *Tax Features* 39, no. 9 (1995): 4–5, http://www.taxfoundation.org/files/e9208 1219e30e8abcc8fd8f13aaac58e.pdf (accessed May 5, 2007).

CHAPTER 4

The Process of Making Science Policy

How Is Science Policy Made?

The process of making science policy can appear messy to the outsider. In truth, the process is quite similar to the creation of any kind of policy—it just requires more ingredients. Indeed, science policy is the product of multiple inputs, mechanisms, and interactions. Its pluralist nature and the involvement of many different executive agencies, congressional committees, and executive branch offices make its formation more difficult to understand than many other areas of public policy.

Like other types of policy, science policy can be made in many different ways: by legislative and federal entities, or through administrative, legislative, and judicial mechanisms. The resulting decisions are also made and implemented at many levels. Implementation sometimes involves legislative or executive branch action; sometimes happens as a result of the budget process, which itself involves multiple entities; and sometimes comes about through the actions of individuals—political appointees or career civil servants—working in federal agencies. Once a policy is created, its interpretation may greatly influence its real-world impact. Complicating the matter further, the judiciary may be called on to resolve disputes over the interpretation of specific policies.

As is the case with public policy more generally, the process by which science policy is made is oftentimes counterintuitive and difficult to understand. Notes political scientist John Kingdon: "The processes by which public policies are formed are exceedingly complex. Agenda-setting, the development of alternatives, and choices among those alternatives seem to be governed by different forces. Each of them is complicated by it-self, and the relations among them add more complications. These processes are dynamic, fluid, and loosely joined."[1]

Thus, the study of policy-making is, in fact, its own science and art. A thorough understanding can never be achieved unless we study human nature in context: policymakers' motivations, and thus the policies they produce, would not necessarily appear rational when considered without reference to the political world of which they are a part. Moreover, one must remember that policies are the result of negotiated outcomes and represent compromises and trade-offs between the preferences of many different governmental and nongovernmental interests. Therefore, an achievable outcome might not always be the "best" policy outcome.

The circumlocutions involved in setting science policy often frustrate members of the scientific community who struggle to rationalize the process. When asked by the *New York Times* about scientists' and researchers' skills in presenting to Congress, the late George Brown, longtime chair of the House Science Committee, replied: "They, generally speaking, have too great a faith in the power of common sense and reason. That's not what drives most political figures, who are concerned about emotions and the way a certain event will affect their constituency."[2] Indeed, many scientists find the political process so difficult to understand that they refuse to try, despite science policy's impact on the science in which they are engaged.

In this chapter, we explain the policy-making process for science. While we cannot hope to provide a perfect understanding, we can familiarize the reader with the basics. We begin by looking at the ways in which political scientists have explained how public policy is made.

A Political Science Perspective on Policy-Making

Incrementalism, Garbage Cans, Primeval Soup, and Natural Selection

For years, political scientists have tried, through the use of metaphors and analogies, to help those not involved in the policy-making process understand the emergence, development, and implementation of policy. Scholars have proposed different models to describe "how things work" in Washington. John Kingdon, for example, refers to ideas for new policies as "seeds." Although many seeds are planted, only a few germinate and grow into full-blown policies.[3] A combination of factors must come together in order for new ideas to take root: they must be planted in fertile soil and receive plenty of nourishment. In this sense, the success of a particular idea has less to do with who originally proposed it than with the suitability of the environment.

One of the most widely cited theories of policy-making is *incrementalism*. Incrementalism has been described by Charles Lindblom as the process by which policymakers make small changes to existing policies, as opposed to enacting large-scale changes.[4] Under the incremental model, policymakers have limited time and ability to be thorough; they do not undertake comprehensive analysis and therefore avoid sweeping initiatives. Instead, they "muddle through" by making small adjustments to existing policies as needed. These incremental adjustments amount to notable change only if one compares one decade to the next.

Frank Baumgartner and Bryan Jones have borrowed the concept of *punctuated equilibrium* from evolutionary biology to describe the policy-making process.[5] In this model, policy is characterized by periods of major and rapid change, sandwiched between long periods of stability. Changes are the result of major shifts in public opinion concerning problems and in the balance of power between special interests. This notion of the process differs significantly from incrementalism. While incrementalism suggests that policy in a particular area undergoes relatively small, but constant, incremental change, in the punctuated equilibrium model, change occurs quickly and is major.

Social scientists Michael Cohen, James March, and Johan Olsen originally suggested a "garbage can" model to understand how organizations make decisions.[6] Others have modified the model to fit the federal government's policy-making process.[7] In the revised model, three types of processes occur at the federal level: (1) problem recognition, (2) creation and refinement of policy proposals, and (3) politics.[8] These processes generally take place independently. The outcomes of the three processes then come together in a "garbage can." The mix of "garbage" that results at any one time (the designation of specific problems, proposal of policy, and political advocacy and lobbying) can often be recycled as a "new" policy proposal, largely because political mood has changed. According to the modified model, these three processes are always ongoing and run in parallel but independent streams. The greatest opportunities for changing the policy agenda occur when the mixture of garbage is properly balanced and when two or more of the streams meet, creating an opportunity for policy change.

Another metaphor is the primeval soup model, which suggests that ideas for new policies float around in a primeval policy soup.[9] Just as elemental gasses once mixed to form organic molecules and create life, ideas float and congeal as proposals are drafted and bills are introduced. Some ideas "bump" into others and join into a bigger idea, while others bounce around aimlessly until they're forgotten. Just as nature selects for traits that are beneficial to an organism, so this system of natural selection favors some ideas over others.

Whirlpools, Iron Triangles, and Sloppy Messy Hexagons

Additional and more detailed models have been developed for understanding the actors and subsystems involved in U.S. policy-making. First, the *whirlpool model* refers to "whirlpools of special social problems and interests"—vortices of relationships that must be considered in order to understand how policy is made. These relationships involve the entire subset of legislators, administrators, lobbyists, and scholars—or *subgovernments*—interested in a particular problem.[10]

Another well-recognized subgovernment model is the *iron triangle*.[11] Congressional committees and subcommittees, federal agencies, and interest groups (trade and industry) form tight and impenetrable issue networks within this triangle, from which narrowly defined public policies emerge. Rudimentary as it is, this model enables political scientists to simplify the web of players who form policy. One example that has been cited in support of this model is the early and rapid expansion of the NIH.[12] Expert witnesses involved in biomedical research were called to testify in congressional hearings on possible expansion of the NIH, but because of their connection to the agency, the NIH could be certain the testimony would be favorable.

Today political scientists generally agree that the issue networks that shape public policy are often more sophisticated and complex than the cozy and impenetrable little triangles envisioned by the iron triangle model. They view the model as an incomplete definition of the complex relationship and interactions that shape policy.[13] New and more widely accepted concepts suggest that as issues gain increased public attention and grow in importance, they result in more loose and permeable policy networks. These networks might be better thought of as large and sloppy messy hexagons, as opposed to neat triangles.[14]

While not amounting to a comprehensive list, these models are all helpful in explaining the complex process of policy-making and the issues networks that shape it. As political scientist Thomas Dye points out, however, "a model is merely an abstraction or representation of political life."[15] So while these models can help one understand the process, they are not the real thing. Rather, they are intended to help students of public policy gain an approximate grasp of the process. We hope that introducing them here will make the reader more comfortable with becoming involved in making better science policy. In what follows we provide our own subgovernment framework to help the reader to better understand the networks that shape and create science policy.

The Science Policy Web

For the purposes of this book, science policy may be thought of as emerging from the center of a complex web of actors (see fig. 4.1). At the center of this web are key White House offices, such as the OSTP and OMB; federal agencies; congressional committees and subcommittees; and scientific interests, as represented by scientific societies, university associations, and nongovernmental organizations such as the National Academies. The web's center is connected to peripheral groups largely by the press. These outlying groups include universities, industry, national laboratories, members of scientific societies, state and local governments, and the general public—all of which have specific interests in science policy. The groups at the center of the web are continually influenced by their interactions with those along the outer threads.

In general, we are most likely to see public participation in science policy when issues beyond science are being discussed, for example, the dispute over teaching

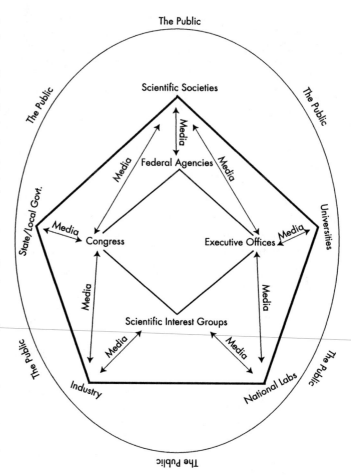

FIG. 4.1   The science policy web

creationism versus evolution. In such cases public opinion may become a key factor in decisions on science policy.

While the policy-making process often seems opaque, understanding both the official and unofficial actors involved can help us navigate through the system to solve a specific policy problem. Also important is to understand the different levels at which science policy is made and the mechanisms used to make it.

Levels at Which Key Decisions Are Made

Policy issues with the most significant potential impact, or the most controversial issues, frequently are handled at the highest level of the policy-making process. These individuals and agencies occupy themselves with such issues as reproductive cloning, strategic missile defense, climate change, health care (e.g., AIDS), projects (such as the Space Station or the Superconducting Super Collider) that involve international agreements or large sums of money, and so on. Decisions with major moral, social, economic, or political dimensions may require direct presidential involvement. Decisions on these issues ideally maintain a subtle balance between the needs of the scientific community and the opinions and expectations of other constituencies, such as religious and business groups. Decisions that require the president's direct participation are few and far between. One highly visible example, however, was George W. Bush's use of his first televised presidential address in 2001 to announce his decision to limit federal funding of human embryonic stem cell research.

Congress can also adopt broad positions on policy, sometimes in response to the president's statements. At other times, Congress is first to act, with the president becoming involved by deciding to support or oppose the congressional position. As mentioned in chapter 2, this was the case when Congress first passed legislation in July 1947 calling for the establishment of a National Science Foundation (NSF) overseen by an autonomous director and board of scientists, with no requirement that it report to the president. President Truman vetoed the legislation, later signing a bill requiring that the NSF director and board members be presidentially appointed, and that the agency itself be accountable to the White House.[16] More recently, President George W. Bush has used his veto authority to reject legislation that would have loosened federal funding restrictions on embryonic stem cell research.

At the next lower level decisions are made by the OSTP and the OMB in cooperation with federal agencies, or by key congressional committees. Such decisions usually have a sizable impact on the scientific community. They also often play a key role in setting future policy directions across agencies, as was the case with multiagency initiatives on information technology, nanotechnology, and climate change research, all developed and first funded during the Clinton administration. Policy decisions of this type might also be coordinated by the NSTC. Congress also plays a role, often through the budget, appropriations, and authorization processes. Major science authorization bills, such as the one approved for the NSF in December 2002, fall into this second category. Congress may also pass laws that regulate specific scientific practices and procedures, or require federal agencies to provide rules and guidelines for specific types of scientific research, for example, regulating the use of animals in research.

At a third level, federal agencies, Congress, and the scientific community work together to set priorities for specific fields and subfields of science. In most instances, federal agencies take the lead in this process, by having

their program officers establish goals for specialized subdisciplines and research fields.[17] Much of this work is done through the strategic distribution of funding, and by setting the requirements for individual grant programs. Such agency-level decisions sometimes need to be made with attention to international scientific agreements or joint efforts.

Finally, the fourth level is the individual. Policy at this level is perhaps best understood as a decision made by an individual who independently interprets a law, regulation, or agency policy. These individuals can be grant, contract, or program officers, or officials based in regulatory and/or enforcement agencies. Challenges to the decisions made by these civil servants are generally resolved by moving to a higher level within the agency. When a dispute over interpretation of a policy cannot be resolved within the agency itself, congressional staff may get involved, using their clout and oversight powers to settle the dispute. The case might also end up in the hands of a judge, who would be asked to rule on how a case should be interpreted in light of constitutional and case law.

Mechanisms for Making Science Policy

Congress and the federal agencies have several means at their disposal to influence science policy. (See table 4.1.) Congress can propose bills and approve laws that shape policy. Another mechanism at their disposal is committee *report language*. When congressional committees approve, or *report*, a bill for floor consideration, they usually accompany the bill with a *committee report*. The report includes language that clarifies the committee's intent and often provides guidance to government agencies and executive branch offices above and beyond what is in the bill itself. The inclusion of report language is often used by congressional committees to influence science policy.

Congress can also increase or decrease an agency's budget, or authorize the creation of a new agency. Combined with its ability to conduct hearings and investigations, Congress can use these mechanisms to exert pressure on federal agencies, directly shaping science policy. Indirectly, the House Ways and Means Committee and the Senate Finance Committee can change the tax system to generate revenue for research. Finally, the Senate can exert its right to advise and consent on key presidential appointments, including cabinet secretariats and other top-level executive offices whose occupants will play key roles in setting science policy. The roster of such posi-

TABLE 4.1 Mechanisms That Determine Science Policy

Office	Mechanisms					
President	Executive directives	Appointments	The budget	Veto authority	Agency coordination	Treaties
Executive branch offices	OMB circulars	Interagency memos				
Agencies	Agency policy	Interpretation of laws	The budget	Rule-making	Implementation	
Congress	Laws, bills, and committee report language	Creating new government entities	The budget/ appropriations bills	Oversight	Senate approval of presidential appointments	
Judiciary	Interpreting laws	Determining constitutionality				

tions includes the directors of the OSTP, the Department of Energy's Office of Science, the NIH, the NSF, and the EPA, along with the administrator of NASA.

For his part, the president can feature key science agencies and programs in the annual budget proposal submitted to Congress. The president can also issue executive orders and directives and make appointments and nominations. Executive branch agencies, such as OMB and OSTP, can influence policy by providing executive guidance in the form of memos, letters, and reports. OMB also issues *circulars,* which provide specific instructions or information to federal agencies concerning government-wide policies on budget and accounting procedures, financial management, procurement, data collection, and interactions with educational and nonprofit organizations. OMB circulars have a tremendous impact on certain aspects of science policy.

Federal agencies have a great degree of flexibility in establishing their own internal and external operating procedures and policies, as long as they follow the broad guidelines established by federal law. Their choices, particularly regarding how research dollars are dispensed, can have a major impact upon broader science policy. In the case of the NIH, for example, while the congressional Appropriations Committee designates the funding amounts for each institute, the NIH itself has considerable freedom to allocate those funds. Since agencies occupy the first step in the budget process, each agency annually develops a budget proposal that it submits to the OMB for review.

One of the most significant mechanisms agencies use to influence policy is their power to implement, interpret,

and enforce laws passed by Congress and signed by the president. This interpretation process presents the scientific community with a tremendous opportunity to influence policy—an opportunity that is, unfortunately, generally underused.

Finally, as described earlier, the judiciary influences science policy by assessing the constitutionality of specific rules and laws.

Congressional Committees

Congress uses a system of committees to carry out many of its day-to-day activities. These committees review and sometimes propose legislation. They also establish guidelines to which the president must adhere in the administration's annual budget proposal. The Senate as a whole can also influence science policy by approving or disapproving individual presidential nominees.

The intricacy of the congressional committee structure complicates the science policy picture, especially since multiple committees may share jurisdiction over a given issue. Indeed, jurisdictional ambiguity often results in turf wars between congressional committees over who will take lead on a piece of legislation.[18] When members of Congress introduce bills that contain multiple provisions, many committees are often involved since different committees have jurisdiction over different portions of the bill.

When it comes to science, almost every congressional committee makes decisions that influence or are influenced by science. Broadly, however, there are two general categories of committees: *authorization committees,* which convey statutory authority for scientific agencies and govern their operations; and *appropriations committees,* which allocate funding for science and scientific agencies.

Authorization Committees

Authorization committees are responsible for producing bills that set policy, establish or "authorize" the creation of federal agencies and programs, and recommend funding levels for these agencies and programs. Bills originating with authorization committees are generally effective over several years, and do not need to be approved annually. The House authorization committees with the most direct role in science policy include Science and Technology; Energy and Commerce; Agriculture; and Armed Services. On the Senate side, the relevant committees include Commerce, Science, and Transportation; Health, Education, Labor, and Pensions; Energy and Natural Resources; Agriculture, Nutrition and Forestry; and Armed Services. Other authorization committees, including Judiciary; International Relations; Small Business; and Government Reform, also play an important role in forming science policy on a case-by-case basis.

Increasing decentralization in Congress, with decision-making responsibility spread over many committees and subcommittees, sometimes impedes the development of comprehensive and coherent science policy. Making things even more difficult, the House and Senate each have their own independent committees rosters. For example, the House maintains a committee that exists specifically to look at science matters: the House Science and Technology Committee. The Senate has no parallel. Instead, Senate jurisdiction over science agencies is dispersed among several committees.[19]

Appropriations Committees

Appropriations committees produce the legislation that approves the expenditure of money for federal agencies and programs. Appropriations bills must be passed annually. Under the congressional schedule, appropriations bills are supposed to be signed into law before October 1, the beginning of the federal fiscal year. In recent years, however, Congress has routinely failed to meet the October 1 deadline, and has instead been forced to pass *continuing resolutions*—often referred to as CRs—which provide the statutory authority required to continue to finance government functions until the necessary appropriations bills have been passed.

The powerful House and Senate Appropriations committees arguably exert the greatest influence on science programs.[20] Indeed, the chairs of the appropriations subcommittees are often referred to as "cardinals."[21]

The Roles of Authorizing and Appropriations Committees

A simplified way to understand the difference between authorization and appropriations committees is to imagine the two as a newly married couple. One member—the authorizer—picks out a brand-new house and all furnishings the couple would like to have. The other determines which of the "authorized" items the couple can actually afford. In certain instances, the couple may need to spend less than it would like. The "appropriator" also keeps track of bills, such as the mortgage, that must be paid each month or year.

Historically, senators who sit on the Appropriations Committee also hold key positions on major authorization committees. For example, during the early 1990s, Senator Bennett Johnston (D-LA) chaired the Energy and Natural Resources Committee (the lead Senate authorization committee for energy-related matters) at the same time that he chaired the Energy and Water Appropriations Subcommittee, which sets the funding for the Department of Energy.

Membership on the House Appropriations Committee is much more exclusive, primarily because a House member with a seat on the Appropriations Committee is often not offerred any other committee assignments. For example, a member of the House Science and Technology Committee, the authorizing committee with jurisdiction over the NSF, is very unlikely to also sit on the Appropriations subcommittee that has funding authority for the NSF. This often results in greater tension between appropriators and authorizers in the House than in the Senate—especially on matters relating to science, since there is rarely an overlap of House members who sit on both the authorizing and appropriations committees responsible for science policy and funding.[22]

Congressional Laws

The power of Congress to propose and pass legislation is its primary mechanism for affecting public policy, including science policy. Legislative action on science policy ranges from the authorization of large projects like the Space Station, the Superconducting Super Collider, and the Human Genome Project, to passing appropriations

POLICY DISCUSSION BOX 4.1

A Balancing Act: Funding for Immediate Needs versus Long-Term Research Investments

The system by which Congress appropriates funding for federal research agencies often forces science to compete directly with other popular nonscience programs funded by nonresearch agencies. Members of Congress have to balance short-term immediate public needs and priorities with longer-term, yet still essential, investments in research.

For example, funding for research conducted by the NIH is contained in the same congressional appropriations bill as funding for important K–12 and higher-education programs supported by the U.S. Department of Education. The NSF and NASA compete for funding with the Community Oriented Policing (COPS) program, a very popular program at the Department of Justice aimed at local crime prevention, because all three

are in the same appropriations bill. The Department of Energy appropriations are contained in the same bill as support for the U.S. Army Corps of Engineers, so members of Congress have to choose between funding DOE science versus important infrastructure programs and water projects (e.g., dams, levees, and river dredging) that might directly benefit their home districts. Even within the Department of Defense, research must compete against the immediate need of the troops, including food, medical care, and military salaries, not to mention the money needed to build major new weapons systems and to pay for major military engagements such as the Iraq war.

With this in mind, how should Congress set priorities for appropriations when so many constituencies are involved? How does it balance immediate societal and military needs with long-term research investments? Who speaks to the importance of federal support of research, given that popular programs with which science must compete often have well-organized constituencies?

bills in line (or not in line) with the president's budget proposal.

The Constitution grants Congress the responsibility to formulate and pass laws. Major changes in science policy are likely to be the result of changes in statute or budgets, both of which depend largely upon congressional passage of laws. NASA, the National Science Foundation, the National Institutes of Health, the Department of Energy, and the Department of Defense were all created by authorization bills. The Department of Homeland Security was formed by act of Congress as recently as 2002.

The authorization of key government agencies is often allowed to lapse, however, either because authorization committees fail to act, or because the House and Senate committees cannot work out fundamental differences. This is a common experience for several agencies that fund science. For example, until recently, the only reauthorization bill ever approved for the Department of Energy was the Energy Policy Act of 1992 (P.L. 102-486). Much of the legal authority provided by this act expired in the mid-1990s, and was not renewed until 2005.[23] When authorization bills are allowed to lapse, and no new reauthorization bill is passed, agencies simply continue to adhere to the language of the most recently enacted authorization, and the original act under which they were granted statutory authority.

The Patent and Trademark Law Amendment Act of 1980, more commonly referred to as the Bayh-Dole Act (P.L. 96-517), is another example of legislation's effect on science. The act, named for its two primary sponsors, Senators Bob Dole (R-KA) and Birch Bayh (D-IN), provided a uniform framework for commercializing innovations generated by federally funded research, thereby allowing universities and other nonprofit organizations to retain title to these inventions and work with companies to bring them to market. This law is widely credited with increasing the level of technology transfer between U.S. universities and industry by creating incentives for university researchers to consider the practical applications of their discoveries, and for universities to search out commercial partners to develop these applications. Among other recently passed laws pertinent to the conduct of science are the USA PATRIOT Act and the Bioterrorism Act of 2002. Passed in response to concerns about the potential for terrorist attacks against the United States, these two laws restrict access to certain biological research agents and regulate their handling and storage.

On the appropriations side, Congress must approve agency budgets annually, before funds can be dispensed. If an appropriations bill does not pass, then legally the agency has no funding for that fiscal year. Operations may come to a halt, as, for example, in November 1995, when Congress and the White House failed to reach an agreement on the FY1996 budget. That year, when no agreement could be reached on several appropriations bills, the government's spending authority simply ran out, leading to a partial government shutdown.

Finally, it is important to note that science policy can be determined by the laws or amendments that Congress chooses *not* to approve. For example, in July 2003 Representative Pat Toomey (R-PA) put forth an amendment to the Labor, Health, and Education appropriations bill that would have rescinded the funds from five grants that had already been awarded by the NIH. The amendment was motivated by Rep. Toomey's assertion that the specific research projects being funded appeared irrelevant to curing disease and were a waste of taxpayer dollars. While Congress does have oversight authority for science agencies and how they spend money, this amendment was of concern because it overruled the NIH's authority to determine which research proposals deserved funding, and to select awardees based upon a peer-review, or merit-based, process. In the end, the House of Representatives narrowly defeated the Toomey amendment. Had it passed both the House and the Senate, the precedent would have permanently curtailed NIH and NSF reliance on the peer-review process.[24]

"The Dance of Legislation": How a Bill Becomes a Law

Members of Congress, from either the House or the Senate, are the only people who can formally introduce legislation. The individual who introduces a piece of legislation is referred to as its *sponsor*. In an effort to gain support for his or her legislation both before and after its introduction, the sponsor of a given bill typically seeks out other members of Congress to act as *cosponsors*. When a member of Congress cosponsors a bill, he or she is essentially expressing support for the legislation and promising a positive vote if the bill is considered on the House or Senate floor.

Legislative scholar Eric Redman puts it this way: "Like a college mixer, the 'dance of legislation' first becomes earnest and fascinating with the search for partners. In Congress, however, unlike a college dance hall, one is permitted as many partners for each dance as one can muster, and since every partner helps, there are rarely any wall-flowers except by choice."[25] In order to persuade a member to cosponsor legislation, the chief sponsor often sends a letter to all members of his or her chamber, or targets a subset of members who are most likely to sup-

port the bill. These letters are commonly referred to as "Dear Colleague" letters. Members use them not only to seek cosponsors to legislation, but also to encourage other members to cosign letters to committees (e.g., the Appropriations Committee) or the administration; they are also used to communicate important information and requests.

Sometimes a member of Congress generates an idea for legislation himself or herself. At other times, a bill is the product of congressional staff's response to concerns raised by interest groups, constituents, or the administration. Often a piece of legislation is written in large part by someone on the member's personal or committee staff, with significant input from advocacy organizations. While the president cannot formally introduce legislation, the administration often presents a legislative package to Congress for consideration, lining up members in the majority or minority leadership to introduce legislation on the administration's behalf.

There are four types of legislation: bills, concurrent resolutions, joint resolutions, and simple resolutions. A bill, signed by the president, is the only one of the four that becomes a law, although a joint resolution has the force of law when signed by the president. Concurrent and simple resolutions do not become laws. A concurrent resolution represents an agreement between the House and Senate but is not sent to the president for signature. The Congressional Budget Resolution, which sets the congressional spending and revenue levels, is an example of a concurrent resolution. Concurrent resolutions can also be used to express a joint opinion of the House and Senate, commonly referred to as a *sense of the Congress*. A simple resolution is passed by only one chamber and may, for example, be used to create a new House or Senate committee or to express an opinion, or sense, of the House or Senate.[26]

After a bill is introduced, it is reviewed and then referred to a standing committee, in accordance with the rules of the House or Senate. In the House, the Speaker has the authority to assign the legislation to a primary committee for review. The Speaker often delegates this task to the parliamentarian, who is not a member of Congress but advises the Speaker on details of House rules. Although the Senate also has a parliamentarian, the Rules Committee normally makes this determination in the Senate.

Once referred to a standing committee, the bill may be further assigned to a subcommittee. The committee or subcommittee then determines whether to put the bill on its agenda for further consideration or to "kill" it. Most legislation dies in committee and is never given a public hearing.

If a bill makes it onto the committee or subcommittee agenda, committee members may hold a public hearing on it. They then "mark up" the bill—that is, amend its text. If a majority of committee members agree with the final language, a formal report is produced by the committee, after which it is *reported,* or sent, to the House or Senate floor for further consideration. This report accompanies the bill when it is considered by the full chamber. It specifies the intent and scope of the bill and the relevant positions of the executive branch and any dissenting committee members. Report language often allows members of Congress to direct the activities of federal agencies, or strongly encourage certain behaviors from them, without creating new statutory requirements.

A public floor debate ensues after the committee reports a bill back to the chamber in which it originated. In the House, the terms of the debate—the time allotted, as well as the order of floor amendments—is determined by the Rules Committee. In the Senate, which has no such committee to set time limits on debate (despite the similarity of its name, the Senate's Rules Committee has different responsibilities), members may *filibuster* by delivering long speeches in an attempt to delay or block legislation. The only way a Senate debate or filibuster can be terminated without unanimous consent is through a device known as *cloture,* which requires a two-thirds majority voice vote.

Both the House and the Senate generally allow amendments to be considered during floor consideration of a piece of legislation. The House considers amendments in a much more orderly fashion than does the Senate. In the House, the Rules Committee enforces a requirement that all amendments be germane to a bill. The House rules committee can also provide for a closed rule that prohibits members of the House from offering amendments during floor consideration. In the Senate, however, members are free to attach amendments that have little to do with the original legislation. The only exceptions to this are made during consideration of general appropriations and budget measures and to matters being considered under cloture.

Once all amendments are considered during floor debate, a final vote is taken on the bill. If it passes, it is referred to the other chamber, which may either take up the bill or drop it from consideration.

If the other chamber greatly alters a bill's language or passes similar, but still different, legislation, a *conference committee* of members from both chambers is formed to iron out the differences. The majority and minority leaders of the House and Senate appoint the members of this committee, or *conferees*. These conferees often include members of the committees involved in the initial consideration of the legislation. For example, if a bill (or any particular piece of it) that is going to conference was originally voted on by the House Science and Technology Committee, it is likely that at least some of the Science and Technology Committee's members will be among the conferees. When appropriations bills are being discussed, the conferees are usually members of the House and Senate subcommittees that initiated the bills.

If differences between the House- and Senate-passed versions of the bill cannot be resolved during conference deliberations, the bill dies. If the conference committee does reach agreement on a compromise, a conference report is made to both chambers, which then vote to approve the compromise version. If the bill passes both chambers, it is sent to the president for signature. Often members of both the House and the Senate simultaneously introduce similar pieces of legislation, referred to as *companion* legislation. This sometimes speeds up the legislative process, paving the way for quick passage and a conference agreement.

The president may either sign or veto legislation. If the president signs a bill, it becomes law and is given a public-law number. If, on the other hand, the president strongly opposes the legislation, he or she may veto the bill or refuse to sign it. Congress can override a presidential veto by a two-thirds majority vote in both chambers. Even the threat of a presidential veto can influence the legislative process. Once the House or Senate has passed a bill, for example, the president often issues a *statement of administration policy* (SAP). This statement, usually in the form of a memo issued by the Office of Management and Budget, expresses the president's support or opposition. Even if the administration looks favorably upon a piece of legislation, the SAP provides an opportunity to voice concerns about specific provisions. When the president has major concerns regarding a specific piece of legislation or any of its provisions, SAP may be used to threaten a veto. If Congress wishes to avoid a veto, it may respond by trying to work with the administration on a compromise.[27]

Congressional Oversight and Review

Once Congress enacts a law, it is responsible for seeing that the legislation is carried out according to congressional intent. This is often referred to as *congressional oversight.* Unlike the president, who has direct authority over executive branch offices and agencies, Congress often uses indirect means to monitor the functions of government departments.

Congressional power to reauthorize bills enables legislators to indirectly govern agency priorities. They do so by revising the legislation that provides statutory authority for a given agency. Very few science agencies, however, are reauthorized by Congress on an annual basis. The only real exception is the Department of Defense, for which the Armed Services committees in the House and Senate traditionally pass an annual defense authorization bill. Other agencies, such as the NSF, the NIH, and NASA, are reauthorized approximately every three to five years. The reauthorization schedule is usually defined in the bill itself: bills may be one-year or multiple years in scope. The specific number of years in an authorization bill is determined by the presiding committee.

In some instances, an authorization bill expires. Ideally Congress will have crafted, introduced, and passed a new authorization so that the agency does not have to operate under old statutory authority. However, controversy can sometimes frustrate efforts toward consensus.

Congress also uses oversight hearings and letters to agencies to ensure that the government is functioning in accordance with its wishes. Personal contacts between committee chairs and high-level officials also play a role. In the end, if Congress is not happy with an agency's functioning, it can enact legislation addressing its concerns.

Finally, the funding reports submitted to key appropriations subcommittees can shape policy by requesting that studies be conducted to examine the operations of an agency. Because appropriations committees control purse strings, agency officials are often responsive to concerns raised in report language. If agency administrators are unresponsive, they risk being sanctioned by the appropriations committee and losing their funding.

How Funds Are Awarded

A significant amount of federal research funding is dispensed through mission-oriented agencies such as the Department of Energy, the Department of Defense, the Department of Health and Human Services, and NASA, so called because they conduct research in support of a broader mission. Each institute within the NIH, for example, issues requests for proposals (RFPs) that embody its mission. Likewise, NASA research is aimed at supporting space exploration, while DOD research is aimed at enhancing national security.

The NSF, unlike these mission agencies, supports a broad range of research for its own sake. In deciding which projects to fund, the NSF evaluates the status and needs of various science and engineering fields, and coordinates its programs with other federal and nonfederal agencies. The NSF's grants and contracts, then, enhance scientific and engineering research and education at all levels, while also supporting specific activities related to international cooperation, national security, and the nation's general well-being.

Peer Review

Regardless of their focus, most of these agencies use some form of organized, competitive peer-review process for awarding research funds.[28] This system grants funds based on merit, as determined by the evaluations of specialists in the relevant field. Each member of a grant-review panel typically reviews, provides feedback on, and scores proposals independent of their colleagues. Panel members then meet to confer and make final decisions, with the goal of funding as many of the top-scoring grants as possible. The panel is aided in its work by program officers, who review panel critiques and make final recommendations to the agency or institute director. Reviewers' anonymity is preserved throughout the process.

Competitive peer review is designed to ensure that the highest-quality and most promising research receives federal support. The system is also designed to guarantee federal support of leading-edge science; help agencies develop research priorities by revealing research trends and gaps; provide expert feedback that can help scientists improve their project proposals; and ensure that tax dollars are spent wisely. The peer-review system, then, is important in keeping the United States at the forefront of scientific discovery. Even this system is not perfect, though. The peer-review system has been criticized for being too conservative and failing to fund high-risk projects that do not coincide with mainstream scientific views. It also tends to favor senior faculty better known and established in the field, making it difficult for junior faculty to be successful in competing for funding.[29]

There is no uniform federal policy for peer review, but all federal agencies funding science are encouraged by the OSTP to use peer review to determine what research projects are funded. Further, OSTP encourages agencies to tailor peer review policies toward the agency's mission and the type of research funded by the agency. Generally, each proposal is scored on its own merit. Program officers use the scores (given by the panelists) to rank the proposals. Based on the amount of funding available, proposal awards are made to the top-scoring proposals. How many are funded relies largely on the amount of funding available. In some cases, proposals with good scores are not awarded funding because of the lack of money. The degree to which programs are able to fund meretorious proposals, commonly referred to as the *success rate,* is often used as a measure of the programs' overall health.

Each agency has its own criteria it asks panel reviewers to use in scoring proposals. For example, some of the criteria both NSF and NIH use include significance (does the study address an important problem? how will scientific knowledge be advanced?); approach (are the design and methods well developed? are problem areas addressed?); innovation (are there novel concepts or approaches? are the aims original and innovative?); investigator (is the investigator appropriately trained?); and environment (does the scientific environment contribute to the probability of success? are there unique features of the scientific environment?). NSF has several additional questions it asks panel reviewers to use: How well does the activity advance discovery and understanding while promoting teaching, training, and learning? How well does the proposed activity broaden the participation of underrepresented groups (e.g., gender, ethnicity, disability, geographic, etc.)? To what extent will it enhance the infrastructure for research and education, such as facilities, instrumentation, networks, and partnerships?[30]

Overall, the peer-review system has served the scientific community well. It has become a hallmark of science, differentiating good science from poor science and serving as a mechanism by which scientists evaluate, and provide feedback about, each other's work. There are flaws within the system, but most would agree that it is much better for the scientific community to improve the system than abandon it.

Earmarking and Pork

In some cases Congress directs research funds to go to specific recipients. The process whereby a member of Congress includes a provision in an appropriations bill specifically designating funds to a specific project (research or otherwise), often based in the member's home district or state, is commonly referred to as *earmarking,* or *pork barreling.* Items introduced in this way are called *pork* or *earmarks.* In budgetary speak, earmarks include *add-ons* and *carve-outs* appropriated by Congress that place specific restrictions on who and where the funds can go.[31] The merits and quality of congressional earmarked projects are often called into question, especially in the case of science, when funds targeted for a research project or facility have not been subjected to the same level of scientific scrutiny and peer review as research projects awarded by a federal agency.

When a congressperson votes, for example, in favor of a hydroelectric project, it is clear all along that the project will be based in a particular geographical location. But a bill that institutes a new initiative (e.g., agroterrorism research), and which has been written in a way that the research must be done in a specific congressional district, contradicts laws requiring that federal funding be awarded to the most meritorious projects. In the past, earmarking by congressional appropriators has been the subject of heated congressional debate and this practice often has been criticized by administration officials as being wasteful of taxpayer dollars. Earmarking of research projects has also been the subject of much discussion and controversy within the academic and scientific community.[32]

Presidential Directives and Executive Orders

While the president does not have direct constitutional authority to enact laws without congressional approval, presidential directives give the president power over many policy areas, including science. The Congressional Research Service notes, "From the earliest days of the federal government, Presidents, exercising magisterial or executive power not unlike that of a monarch, from time to time have issued directives establishing new policy, decreeing the commencement or cessation of some action, or ordaining that notice be given to some declaration. The instruments used by Presidents in these regards have come to be known by various names, and some have prescribed forms and purposes."[33]

The oldest and perhaps best-known kind of presidential directive is the *executive order.* The president can use an executive order to guide government operations, encouraging or prohibiting agencies from taking part in specific activities. Executive orders are official government documents and are consecutively numbered as the president issues them. They are printed daily, in their entirety, in the *Federal Register,* a legal newspaper published daily during the business week by the National Archives and Records Administration, and in the Code of Federal Regulations (CFR), an annually published volume of general and permanent rules appearing in the *Federal Register.*[34]

While orders and directives do not have the full force of law, presidents frequently use them to clarify administration policies: the President's Council of Advisors on Science and Technology (PCAST) and the National Science and Technology Council (NSTC), for example, were both created by executive order. Some directives and executive orders are explicitly directed at scientific research, such as President Eisenhower's Executive Order 10668, which further defined the membership and powers of the National Academy of Sciences' National Research Council and amended Executive Order 2859; or President Clinton's directive of March 4, 1997, which banned the use of federal funds for human cloning.[35]

Finally there is a distinction between presidential directives and legislation. The latter are public laws and are codified. To change a law requires congressional action in the form of additional legislation. While FCCSET, OSTP, and the OSTP director position were created by legislation passed by Congress and signed into law by the president, presidential executive orders established both the NSTC and PCAST. As a result, they have no permanent status and may be eliminated by the president at any time without congressional action.

Other Presidential Powers

In addition to powers to make annual budget proposals, draft legislation, and issue executive orders and directives, the president also has constitutional authority to make international treaties (with the advice and consent of the Senate); appoint ambassadors and other top-level executive and judicial officials, including Supreme Court justices and federal judges (again with Senate approval); and to make recommendations to Congress.

The president's influence on science policy is often exerted through the power to nominate key officials, including the directors of the OSTP, NSF, and the EPA; the administrator of NASA; and the secretaries of Health and Human Services, Defense, Energy, State, and Commerce. Nominations put forward by the president are subject to Senate approval.

Advisory Reports and Statements

As we noted earlier, several formally established panels, including the OSTP, the NSTC, PCAST, and the National Bioethics Advisory Commission, counsel the government on matters of science and science policy, typically by means of panel studies and written reports. It is then up to the president, Congress, the agencies involved, and the scientific community to heed their suggestions. Publication of the reports can increase awareness of an issue, in which case the media also plays a role by bringing public attention to the reports and the issues (or opting not to do so).

The OMB also advises the president. Although its suggestions are primarily focused on the annual budget, the importance of government funding to American science means that the OMB wields significant influence over scientific research. The OMB generally sets limits on the funding available for key science accounts, rather than determining major policy directions. Policy directions and cross-agency science initiatives are more likely to be established by the OSTP, with guidance from the NSTC and PCAST.

One OMB budget examiner dealing with a key science account remarked that "if OSTP is not willing or able to make the case for increasing funding in a particular area or for making a fundamental budgetary change that would affect science policy, don't expect that OMB will do so. We take our lead from them." In reality, however, the OMB does not always wait for OSTP's lead before taking actions that will affect science policy. Moreover, when the OMB and the OSTP disagree on funding levels for an agency, the OMB's opinions often win out. The OSTP's authority within the executive branch has historically been relatively weak in comparison to that exerted by the OMB.

The Budget Process

John Marburger, science advisor to President George W. Bush, has stated that "Science policy entails more than setting budgets, but that is the bottom line of the policy process."[36] Obviously, the budget process plays a critical role in the formation of science policy: even though a new program or policy has been created, its final enactment depends upon the availability of funds. The budget process might best be thought of as a competition for available funds, in which science programs compete both with each other and with nonscience programs.

The budget process often forces members of Congress to determine (and reveal) their true priorities in policy. Before the critical budget agreement is reached, members of Congress may lend their support to many different ideas for policy. Even the congressional authorization process, which only sets funding limits, does not force members to choose among the many different existing policy ideas: programs are often created or authorized by Congress but never fully funded at the authorized levels (this is often done for symbolic reasons).

One former member of Congress who sat on the House Science Committee noted, "In the final analysis, after science and technology decisions have been subject to the judgment of conflicting objectives . . . they are then subject to the reality of the Federal budget process. First research and development programs must compete with other Federal programs for the availability of limited Federal dollars . . . for there will always be more programs and projects than there will be funds to implement them. Thus another set of choices in how to allocate the funds to gain the greatest benefits must be faced."[37]

In modern-day politics, the budget process never really ends. (See "A Timeline for the Annual Federal Budget Process" text box. See also fig. 4.2.) By the time Congress completes one fiscal year's budget, the next is being developed by federal agencies with input from the OMB and OSTP. Further complicating the process, budget discussions begin two years in advance: for example, in FY2005 (which began October 1, 2004), Congress was already debating the FY2006 appropriations bills, and agencies were planning their FY2007 budgets.

Federal agencies usually submit an initial budget request to the OMB for review in mid-September, more than a year in advance of when the fiscal year budget being planned for is set to begin. The OMB surveys the initial numbers and suggests revisions. Agency heads are prohibited from engaging in formal lobbying with the OMB; however, agency officials are often engaged in clarifying and explaining their budget request to OMB staff. The budget request is an agency's one opportunity to make its scientific priorities known to the budget examiner and program associate director (PAD).[38]

In Washington circles, the point at which the OMB returns revised budget numbers to agencies is known as the *pass back,* and it usually occurs just before Thanksgiving. After the pass back, agencies may appeal the OMB's suggested funding levels. Agency heads choose their battles carefully, however, knowing that the OMB will consider only a few appeals. Agencies usually appeal the OMB's funding recommendations only when they want to clearly express a high funding priority, or when they are under significant political pressure to increase funding levels.

In December, the OMB reviews appeals and places the finishing touches on the president's budget proposal for the upcoming fiscal year. All major decisions have usually been made by Christmas Eve. The only changes likely to be made after this time are on budget initiatives that the president wishes to stress in the coming year—perhaps items that will be mentioned in the annual State of the Union address in late January. After this final round of political decisions has been made and the president has approved the budget, the proposal is sent to the Government Printing Office. The president's budget for the upcoming fiscal year (which begins on October 1) is usually released during the first week in February, at which time the federal agencies hold a series of specific briefings to explain its details to the public and the press.

With regard to the federal budget process, it is often said that "the president proposes, and the Congress disposes." In other words, the president proposes a budget, which Congress either accepts or rejects. Congressional consideration of the president's budget request begins with the House and Senate Budget committees.

The House and Senate Budget committees were established in 1974 by the Congressional Budget and Impoundment Control Act. The primary role of the Budget Committee is to draft an annual budget plan, or *budget resolution,* which responds to the president's budget proposal and sets an overall target for the desired size and composition of the federal budget. The budget resolution outlines budget totals looking at planned costs of federal programs, expected revenues, and the size of the surplus or deficit. To help to determine the overall limits of the budget, the budget resolution broadly defines funding priorities by focusing resources among broad budget categories, or *budget functions,* such as education; health; defense; and science, space, and technology.

The function of congressional budget committees is to determine the overall size of the budget "pie" and the proper mix of ingredients that will make up the pie, consisting of expected revenues, discretionary spending, and required spending for entitlement programs.

Assuming that an agreement can be reached between House and the Senate on a budget resolution—which is not always the case—it then becomes the responsibility of the appropriations committees to divide the pie and to determine the funding levels that will be allocated to specific government programs. If an agreement on a budget resolution cannot be reached by mid-May, then the Appropriations Committee will move forward based upon guidance provided by the chamber's own Budget Committee. This guidance concerning the overall size of the budget pie is known as a *302(a) allocation.*

Upon receipt of the 302(a) allocation, which usually occurs not later than mid-May, the Appropriations committees in the House and the Senate set about dividing the budget pie. The individual slices of the pie that are allocated to each subcommittee are referred to as *302(b) allocations.* Once the 302(b) allocation has been made by the Appropriations chairman in each chamber, the House and Senate appropriations subcommittees begin their work of further dividing their slice of the budget pie into specific agency and program funding levels.

All appropriations bills originate in the House; however, it is not unusual for the House and Senate versions of the appropriations bills to be very different. In recent years, the difference between many of the House and Senate versions of the appropriations bills has been so great

A Timeline for the Annual Federal Budget Process

Early Fall

Agencies send initial budget requests to the OMB.

November

The OMB reviews and modifies requests and sends them back to agencies.

December

Agencies make final appeals to the OMB.

January

The OMB resolves appeals and assembles the final budget request.

February

The president's budget proposal is released to Congress, traditionally on the first Tuesday in February.

On February 15 the Congressional Budget Office submits its report on projected spending for the forthcoming year.

February–March

Administration and agency officials testify in support of the budget request.

Appropriations subcommittees (House and Senate) hold hearings with agency heads and outside public witnesses.

March

Congressional committees submit their "views and estimates" to the Budget Committee.

House and Senate Budget committees develop respective budget resolutions. House votes on its budget resolution.

April

Senate votes on its budget resolution.

House-Senate conferees develop conference report on budget resolution and each chamber votes on the resolution conference report. The final budget resolution determines the 302(a) allocation that determines the overall level of discretionary spending the Congress can appropriate.

April–May

Authorizing committees develop reconciliation legislation (if necessary) and report legislation to the Budget committees. Budget committees package reconciliation language and report to respective chambers. After passage in each chamber, House-Senate conferees develop conference report on reconciliation and bring to floors of both chambers.

May

The Appropriations committees (House and Senate) make 302(b) allocations. The 302(b) allocations represent the amount of funding provided by each respective chamber's appropriations committees to be spent by its respective subcommittees.*

June

House appropriations subcommittees prepare appropriations bills; Senate appropriations subcommittees revise the House-passed bills.

July–August

House passes spending bills; Senate passes revised bills.

September

House-Senate conference committees resolve differences and agree on final versions of spending bills.

The president signs or vetoes final bills.

October 1

Fiscal year begins. Congress passes continuing resolutions to maintain funding for any agencies affected by appropriations bills that have not been passed and signed by the beginning of the fiscal year.

*The designations *302(a)* and *302(b)* refer to the sections of the law governing these procedures.

Adapted from the American Mathematical Society (http://www.ams.org/government/budget-timeline.html) and the House Republic Office of the Committee on the Budget, Budget Calendar (http://www.house.gov/budget_republicans/budgcalendar.pdf).

that individual approval of the bills has not occurred, with all appropriations bills in the end being rolled into one major piece of legislation known as an *omnibus appropriations* bill.

It should be noted here that the pluralism of the science policy-making process is reflected not only in the manner in which science laws are passed, but also in the participation of more than half of the twelve federal appropriations subcommittees in funding science. Since there is, in fact, no single budget for science, this makes understanding science funding even more difficult (see figure 4.2 for a diagram illustrating the complexities of the budget process).

Agency Policies

Federal agencies have a great deal of flexibility in their operations, and are often in a position to shape science policy, both formally and informally. For example, inter-

nal accounting practices and priority-setting processes affect final policy, as do rules guiding internal peer review. Sometimes these policies are created and implemented in response to laws passed by Congress; if so, the relevant policies are usually announced in the *Federal Register* for public comment. At other times, an agency might influence policy by specifying the requirements in a request for proposals (RFP), by dictating, for example, that proposals must have an educational outreach component, or that only a certain number of proposals will be accepted from any given institution. Similarly, agency policies might restrict the percentage of an award that can be used for "indirect costs" for facilities and administrative work.

Regulations and the Rule-Making Process

The formal policies established to implement the law are commonly referred to as *regulations* or *rules*. Specific

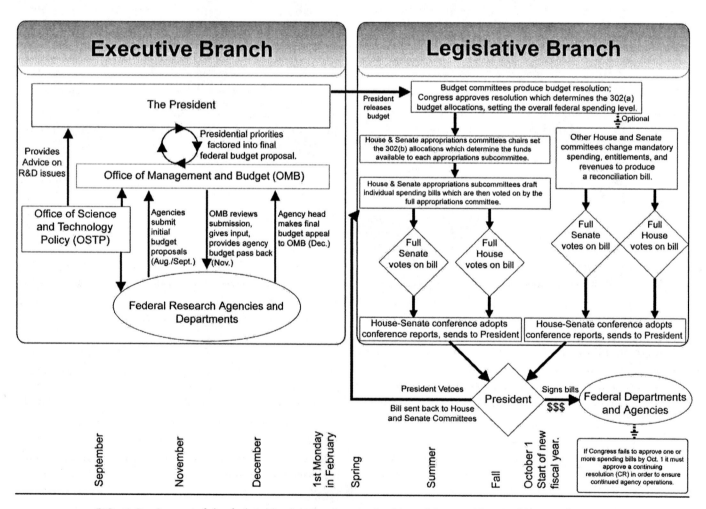

FIG. 4.2 Stages of the federal budget process in the United States. (Created by Bradley Nolen, Association of American Universities.)

agencies are responsible for overseeing the promulgation of certain rules. For example, under the Animal Welfare Act, the USDA's Animal Plant Health Inspection Service (APHIS) regulates the use of animals in research. Although the agencies' rule-making powers are often overlooked by members of the scientific community, knowledge of regulatory policy and politics is essential to a full understanding of the development of science policy. As political scientists Richard Harris and Sidney Milkis have noted, "Regulatory politics—the struggle for control over the administrative levers of power and the policy shaped within government agencies—is central to government activity in the United States. Thus, the study of regulation sheds light on the defining elements of contemporary American politics."[39]

Of course, the pertinent laws establish parameters for what is allowable. But these laws often need significant interpretation, which is provided through regulations developed by federal agencies. As part of the executive branch, these same agencies are also responsible for enforcement duties. An agency's choice of how to interpret a law can thus have a tremendous effect on the conduct of science.

On July 12, 2002, for example, President George W. Bush signed into law the Bioterrorism Act. The act, which had been proposed in response to the events of September 11, 2001, and the subsequent anthrax attacks on Capitol Hill, is formally known as the Public Health Security and Bioterrorism Preparedness and Response Act of 2002 (P.L. 107-188). It created new reporting requirements, and imposed screening requirements on individuals with access to select biological agents, as called for in the USA PATRIOT Act (the full name is the Uniting and Strengthening America by Providing Appropriate Tools Required to Intercept and Obstruct Terrorism Act).

While the Bioterrorism Act set general expectations regarding the screening of individuals and laboratories where research on these agents was being conducted, it left regulatory and enforcement details up to the secretary of Health and Human Services.[40] Thus, to implement the provisions of the Bioterrorism Act, the Centers for Disease Control and Prevention issued an interim final rule titled "Possession, Use, and Transfer of Select Agents and Toxins" in the *Federal Register* on December 13, 2002.[41] After a comment period, the rule was modified, with the final rule published in the *Federal Register* on March 18, 2005, and enacted on April 18, 2005.[42]

Often laws leading to agency rule-making require scientists and scientific institutions to behave in a certain fashion. One cannot read the relevant laws in isolation, however, since many merely set goals and mandate that certain federal agencies establish more specific rules to attain them.

Federal rule-making can be a drawn-out process. When a law is passed, the federal agency tasked with its enforcement develops a notice of proposed rule-making (NPRM). Before an NPRM can be released for public review, it must be vetted within the federal government. This usually involves scrutiny from top-level departmental advisors (e.g., for rules developed by the NIH, the reviewers will most likely be high-level DHHS officials) and the OMB.

Once a rule has been developed, it must be printed in the *Federal Register*. Publication of a proposed rule gives the public an opportunity to comment and suggest changes before implementation. All of the comments on a proposed rule are reviewed and analyzed by agency staff, who are responsible for drafting the final rule. This final draft is again published in the *Federal Register*. Public input is vital to the process, since a final rule, once issued, has the force of federal law.

How Policy Is Evaluated

Any discussion of the mechanics of science policy must leave open the question of how a particular policy is identified as "good" or "bad." Similarly, one might wonder whether it is possible to effectively evaluate a policy before it has been implemented.

Of course, the measure of a policy's effectiveness depends on where one sits; while everyone would like policy to be objective, our democratic system leaves a great deal of room for subjective politics. Thus, the policy goals of the scientific community—for example, the authorization of human embryonic stem cell research and therapeutic cloning or continued funding for research-oriented university nuclear reactors—may not be shared by a majority of the public. In reality, the best policy solutions often balance concerns and interests, rather than completely satisfying any one interest group.

Recent attempts to provide better up-front analysis of public policies has taken a quantitative, scientific approach to the search for solutions to society's problems. Utilizing techniques such as operations research, systems analysis, and cost-benefit and cost-effectiveness analyses, this new generation of policy analysts is "searching for feasible courses of action, generating information and marshaling evidence of the benefits and other conse-

Steps Involved in the Federal Rule-Making Process

1. *Proposed rule is drafted.* The federal agency or office within an agency responsible for implementing a specific law, or specific portions of a law, develops a notice of proposed rule-making (NPRM).
2. *Internal government review of proposed rule.* The NPRM is reviewed and approved by top-level agency officials and the OMB.
3. *Publication of proposed rule.* The NPRM is published in the *Federal Register.*
4. *Comment period.* Each NPRM is followed by a period set aside for public comment. Comments are accepted electronically through a Web site and by postal mail for a specific period following publication (e.g., sixty days). The purpose of the comment period is to provide an opportunity for interested and affected parties to influence the outcome by raising issues and questions that can be addressed before the regulation is finalized.
5. *Public inspection of comments.* Comments received are made available for public inspection. Sometimes these comments are made available for viewing as they are received. They are almost always available for public viewing, often on the agency or office Web site, after the comment period has ended.
6. *Analysis of comments.* Comments are analyzed and summarized, and responses are prepared by the agency or office staff directly responsible for the rules content.
7. *Publication of final rule.* The final rule is published in the *Federal Register.* The final rule often includes a summary of the comments and responses to the comments, including any changes that were made to the proposed regulation as a result of comments.

quences that would follow their adoption and implementation in order to help the policymaker choose the most advantageous action."[43]

Policy analysis attempts to go beyond idealized examinations of policy outcomes in isolation; it attempts, in other words, to make evaluation objective, by accounting for organizational and political difficulties that may arise from particular decisions and their implementation.[44] Unlike natural science, policy analysis does not identify a true or false answer, but rather identifies better and worse policy solutions. "Better" might be defined in terms of cost, timing, or palatability to a particular target group. As always, a solution that seems better to one person will seem worse to someone else. Therefore, policy analysis, despite its quest for objectivity, may be much more value-laden than natural science. E. S. Quade, a policy analyst at the RAND Corporation, observes, "We have not been and never shall be able to make policy analysis a purely rational, coldly objective, scientific aid to decision making that will neatly lay bare the solution to every problem for which it is applied."[45] In fact, Quade suggests that policy analysis has actually done more to reveal the complexities of societal problems than to determine their correct solutions. However, as Quade also notes, "this alone can be of tremendous help."[46]

Setting these qualifications aside, policy analysis is otherwise conducted in a manner similar to other types of scientific research. First, analysts define the problem to be addressed. Then they gather observable data. Finally, the data are analyzed, resulting in an evaluation of options. As one scholar and former policymaker notes, however, "it is never easy to forecast the long-term effect of policy choices, both positive and negative. In the end, those who would provide policy advice must also do so recognizing that they are taking a leap of faith. Such advice should only be offered with humility, recognizing that any policy advisor has a responsibility to watch how a policy evolves, to evaluate it, to listen to those whose lives are affected by it, and—if necessary—to willingly change one's mind when the evidence suggests that other policy approaches are necessary."[47]

Policy Challenges and Questions

In this chapter we have laid out the basics of how science policy is made in the United States. We have discussed models for public policy and provided our own model to help the reader better understand the complex network of actors who play a role in the formation of science policy.

We have also discussed the levels and mechanisms involved in making science policy. While laws are thought of as the primary means through which policy is formed, we have discussed many other mechanisms that can influence science policy. The mechanisms range from regulations, presidential directives, budget and appropriations decisions to presidential appointments to key agencies and to scientific and technical advisory panels. Furthermore, persons other than elected politicians can influence science policy. Civil servants—the men and women who work for federal agencies—also have a certain amount of power. They, after all, are the individuals that interpret and enforce laws, regulations, and executive branch policies.

Of course, to understand the congressional role in science policy, it is essential that one understand the legislative process and the key committees involved in making science policy. In particular, it is important to understand the difference between the roles played by authorization and appropriations committees. It is also essential to understand that while Congress makes the laws, the job of interpreting, implementing, and enforcing the laws is usually left to specific federal agencies. It is also important to understand the role that both the president and Congress play in the budget process and the impact that the budget can have on policy.

Finally, evaluating a policy is an important but challenging part of the policy-making process. While policy analysis is not an exact science, we do know that individuals and groups can influence policy outcomes. Since the scientific community is often—although not always—capable of agreeing on its policy preferences, scientists can sway policy in the direction of their preferred results.

Despite the opportunity that exists for scientists to shape science policy, the scientific community has often shied away from the policy-making process. Indeed, some find the political process so difficult to understand that they refuse to try, despite the deleterious impact of this decision.

An issue, then, is engaging scientists in the science policy process. This is essential, some think (including us), because of the number of science policy questions that face policymakers, who could be greatly informed by those whom these policies will most affect. For example, how do we ensure that high-risk projects are funded under the system of peer review? What mechanisms can be put in place to promote the funding of young investigators? What can the scientific community do to protect peer review and to avoid earmarking? What are the proper funding levels for various areas of science and for agencies critical to advancing scientific research? Are existing policies working for science, whether they are the result of laws, regulations, or other executive or agency decisions? If they are not working, how should they be changed, and who should change them?

We must find ways to encourage more dialogue between scientists and policymakers. As John Edward Porter, former chair of the House Labor, Health and Human Services, and Education Appropriations Subcommittee and a strong advocate for NIH research, noted: "While I realize that scientists by nature often feel uncomfortable with advocacy, if we all stayed within our comfort zones, little would be accomplished. Though perhaps they are not well understood, scientists are highly respected in our society. They are also highly credible. When they speak with a unified voice, the people listen."[48]

For the casual observer, including many scientists, the key to understanding the policy-making process lies in accepting that one may never comprehend every detail of this world but can nevertheless learn how to navigate through it. Acceptance of this complexity will make it much easier for us to appreciate the process for what it is, and work within it for positive changes. We hope the information we have provided on the making of policy will assist in this navigational process.

NOTES

1. Kingdon, *Agendas, Alternatives, and Public Policies*, 230.

2. Claudia Dreifus, "A Conversation with George E. Brown, Jr.: The Congressman Who Loved Science," *New York Times*, March 9, 1999, F3.

3. Kingdon, *Agendas, Alternatives, and Public Policies*, 76–77.

4. Lindblom, "Science of Muddling Through."

5. Frank Baumgartner and Bryan D. Jones, *Agendas and Instability in American Politics* (Chicago: University of Chicago Press, 1993).

6. Michael D. Cohen, James G. March, and Johan P. Olsen, "A Garbage Can Model of Organizational Choice," *Administrative Science Quarterly* 17 (1972): 1–25.

7. Kingdon, *Agendas, Alternatives, and Public Policies*, 84–89; Jonathan Bendor, Terry M. Moe, and Kenneth W. Shotts, "Recycling the Garbage Can: An Assessment of the Research Program," *American Political Science Review* 95, no. 1 (2001): 169–90.

8. Kingdon, *Agendas, Alternatives, and Public Policies*, 87.

9. This notion of a policy primeval soup is based on the use of biological natural selection in modeling the cooperative interactions among individuals. See Robert Axelrod and William Hamilton, "The Evolution of Cooperation," *Science* 211 (March 27, 1981): 1390–96; and in modeling economic change Richard Nelson and Sidney Winter, *An Evolutionary Theory of Economic Change* (Cambridge: Belknap Press of Harvard University Press, 1982). For a detailed discussion see Kingdon, *Agendas, Alternatives, and Public Policies*, 116–44.

10. Ernest S. Griffith, *Impasse of Democracy* (New York: Harrison-Hilton, 1939).

11. Douglas Cater, *Power in Washington* (New York: Random House, 1964).

12. Kingdon, *Agendas, Alternatives, and Public Policies*, 33–34.

13. Political scientist Hugh Heclo has noted that the "iron triangle concept was not so much wrong as it is disastrously incomplete." See Hugh Heclo, "Issue Networks and the Executive Establishment," in *The New American Political System,* ed. Anthony King (Washington, DC: American Enterprise Institute, 1978), 88.

14. Charles O. Jones, *The United States Congress: People, Place, and Policy* (Homewood, IL: Dorsey Press, 1982), 358–65.

15. Thomas R. Dye, *Understanding Public Policy,* 7th ed. (Englewood Cliffs, NJ: Prentice Hall, 1992), 44.

16. See Smith, *American Science Policy,* 52.

17. Office of Technology Assessment, *Federally Funded Research,* 71.

18. David C. King, *Turf Wars: How Congressional Committees Claim Jurisdiction* (Chicago: University of Chicago Press, 1997).

19. For a fuller discussion of some of the complications for the development of science policy that arise from the existing congressional committee structure, see Carnegie Commission on Science, Technology, and Government, *Science, Technology, and Congress: Organization and Procedural Reforms* (New York: Carnegie Commission on Science, Technology, and Government, 1994).

20. Audrey T. Leath, "Congressional Committees with Jurisdiction over R&D, Education," *FYI: The American Institute of Physics Bulletin of Science Policy News,* April 8, 1999.

21. See Dick Munson, *The Cardinals of Capital Hill: The Men and Women Who Control Government Spending* (New York: Grove Press, 1993).

22. Congressional Quarterly, *How Congress Works* (Washington, DC: CQ Press, 1983), 86.

23. Many Department of Energy programs were finally reauthorized with the passage of the Energy Policy Act of 2005, P.L.109-58, which was signed into law by President George W. Bush on August 8, 2005.

24. See Ted Agres, "Politicizing Research or Responsible Oversight?" *GenomeBiology.com,* July 14, 2003, http://genome biology.com/researchnews/default.asp?arx_id=gb-spotlight -20030714-01 (accessed May 5, 2007); American Psychological Association Policy Alert, July 14, 2003, "NIH Funding: Sexual Behavior Research at NIH Still Threatened by Possible Senate Amendment," http://www.lgbthealth.net/archive03/weekly updates03_7.html#TWO-FIFTYNINE (accessed May 5, 2007).

25. Eric Redman, *The Dance of Legislation* (Seattle: University of Washington Press, 2001), 77.

26. Monty Rainey, "Types of Legislation," the Junto Society, 2002, http://www.juntosociety.com/government/legislation. htm (accessed January 9, 2008).

27. A useful figure outlining the process by which a bill becomes law can be found in David McKay's *Essentials of American Government* (Boulder, CO: Westview Press, 2000), fig. 8.1.

28. Peer review has generally been defined as "a process that includes an independent assessment of the technical, scientific merit of research by peers who are scientists with knowledge and expertise equal to that of the researchers whose work they review." See General Accounting Office, *Federal Research: Peer Review Practices at Federal Science Agencies Vary,* GAO/RCED-99-99 (Washington, DC: U.S. Government Accountability Office, March 1999), 2.

29. Daryl E. Chubin and Edward J. Hackett, *Peerless Science: Peer Review and U.S. Science Policy* (Albany: State University of New York Press, 1990).

30. *Report of the National Science Board on the National Science Foundation's Merit Review System,* NSB-05-119, September 30, 2005, 5, http://www.nsf.gov/nsb/documents/2005/0930/ merit_review.pdf (accessed January 5, 2008).

31. The OMB defines an earmark as funding that is "provided by the Congress for projects or programs where the congressional direction (in a bill or report language) circumvents the merit-based or competitive allocation process, or specifies the location or recipient." Further, if the administration's ability to control particular aspects of the allocation of those funds is limited, then OMB also considers those funds to be earmarks. See Director Rob Portman, OMB, in a Memorandum for the Heads of Departments and Agencies (M-07-09), "Collection of Information on Earmarks," January 25, 2007, 1. This memorandum was written as the result of President George W. Bush's call on Congress to put in place reforms on the practice of earmarking, reducing the number and dollar amount of earmarks by one half. In order to establish a transparent baseline from which to measure the goal, OMB set in place the data collection process outlined in this memo.

32. For an extended discussion of academic earmarking and the controversy that has accompanied it, see James D. Savage, *Funding Science in America: Congress, Universities, and the Politics of Academic Pork Barrel* (New York: Cambridge University Press, 1999).

33. Harold C. Relyea, *Presidential Directives: Background and Overview,* 98-611GOV (Washington, DC: Congressional Research Service, Library of Congress, January 7, 2005).

34. See the National Archives, Executive Orders FAQs, http://www.archives.gov/federal-register/executive-orders/about .html; Federation of American Scientists, Intelligence Resource Program, Official Intelligence-Related Documents, Executive Orders, http://www.fas.org/irp/offdocs/eo/index.html (both accessed May 5, 2007).

35. Executive Order 10668, May 10, 1956, "Amendment of Executive Order No. 2859, of May 11, 1918, Relating to the National Research Council," *Federal Register* 21 (May 12, 1956): 3155; the text of President Clinton's cloning directive can be found at http://grants1.nih.gov/grants/policy/cloning _directive.htm (accessed May 5, 2007).

36. John H. Marburger III, Director, Office of Science and Technology Policy, April 22, 2004, at the American Association for the Advancement of Science 29th Forum on Science and Technology Policy.

37. Don Fuqua, "Science Policy: The Evolution of Anticipation," *Technology in Society* 2 (1980): 372, quoted in Office of Technology Assessment, *Federally Funded Research,* 71.

38. This process occurs for all agencies, not just those involved with science.

39. Richard A. Harris and Sidney M. Milkis, *The Politics of Regulatory Change: A Tale of Two Agencies* (New York: Oxford University Press, 1989), viii.

40. See American Society for Microbiology, http://www .asm.org/Policy/index.asp?bid=8654 (accessed May 5, 2007).

41. See *Federal Register* 240, no. 67 (December 13, 2002): 76886–905.

42. See *Federal Register* 70, no. 52 (March 18, 2005): 13241–92.

43. E. S. Quade, *Analysis for Public Decisions: A RAND Corporation Research Study,* 3rd ed., rev. ed. by Grace M. Carter (New York: Elsevier Science, 1989), 4–6.

44. Ibid.

45. Ibid., 11.

46. Ibid.

47. Rebecca M. Blank, "Poverty, Policy and Ethics: Can an Economist Be Both Critical and Caring?" *Review of Social Economy* 61, no. 4 (2003): 447–69.

48. Kenneth D. Campbell, "Former Congressman Gives Faculty Lesson on Science Lobbying," *MIT Tech Talk,* May 9, 2001.

CHAPTER 5

Federal Funding for Research: Rationale, Impact, and Trends

Why Does the U.S. Government Fund Research?

Democratic governments seek to administer to the needs of their people, particularly in areas where centrally coordinated efforts are more effective than the initiatives of individual citizens. Whereas the choice of evening meals is best left to the individual family, national defense is most effectively managed by the government. Our question here is whether research should be a local concern, or one that should be centrally coordinated. We think the proper answer is "both." Research is vital to the national good, but derives from the creativity of individuals who freely pursue their dreams and visions.

Chapter 2 cited several examples of how the nation has benefited from federally coordinated research. For instance, the Lewis and Clark expedition might not have taken place without Thomas Jefferson's backing. And World War II may have ended very differently were it not for the Manhattan Project.

But why should the federal government support research today? Each era presents its own challenges. Science has historically aided the nation, helping to remedy the effects of war, famine, and disease. Scientific research led to significantly increased crop yields in the early twentieth century, helped to spawn the industrial revolution, and was critical in ending World War II. Since World War II, scientific advances have been central to U.S. military and economic preeminence. The challenges we now face include providing jobs, fighting old and new diseases, and guarding against new forms of aggression. These goals can best be achieved through a coordinated effort, in which federal government partners with industry, national laboratories, universities, and state governments. Nor should we fail to include the public in this list: they provide the tax dollars and, through advocacy and the ballot box, can express their support—or disdain—for science and for federal support of research.

This chapter addresses the issue of why the federal government invests in research. It also examines how the impact of past investments can be assessed and how this assessment can affect future funding and policy decisions. Finally, the chapter describes historical and current trends in federal support for scientific research.

Measuring the Value of Research Investments

While many in the research community advocate for federal funding of scientific research, proving the value of such investment can be a challenge. Just as a private investor seeks to ensure a positive gain when making financial investments, so too does the federal government. Vannevar Bush's report *Science—the Endless Frontier* noted that the government should invest in scientific research because of its responsibilities: to look out for the public's well-being, to promote a strong economy, to assist the battles against disease and ill-health, and to defend the homeland and its citizens. The technologies used in achieving these goals are all derived from advances in scientific research, many of which are funded by federal dollars. But how do you put a numeric value on these benefits? How do we measure whether these investments are worthwhile? And how do we know the correct amount for the nation to spend on research?

Economic Analyses

Most nations aspire to have a booming economy. Citizens with jobs that allow them to easily purchase the basic necessities of life, educate their children, enjoy cultural activities, receive medical care, and to be cared for in their twilight years, have achieved much of what a society can make possible in life. Regrettably, this dream is not within reach of a significant percentage of the world population. The effort to improve the quality of life relies in part upon science and technology.

Economists have long argued that technology and technical progress contribute to the growth of a nation's economic output, or gross domestic product (GDP).[1] In 1957, economist Robert Solow, applying economic theory and mathematical analyses to data from 1909 through 1949, looked at changes in GDP with respect to changes in the number of workers and amount of capital input into the economy. He identified a third factor, in addition to labor and capital, that contributed to economic growth, and speculated that it could be attributed to technological advances.[2] Solow calculated that approximately 87 percent of U.S. economic growth between 1909 and 1949 was due to such advances.[3] In 1987, he won the Nobel Prize in Economics for his efforts.

Solow's work strongly suggests that technology fuels economic growth. Because technology is derived from scientific research, government funding of scientific research is often justified on the basis of its contribution to growing the U.S. economy. And a growing economy, as we have noted, is associated with higher living standards and the overall well-being. As Solow noted in his Nobel acceptance:

> Technological progress, very broadly defined to include improvements in the human factor, was necessary to allow long-run growth in real wages and the standard of living. . . . Gross output per hour of work in the U.S. economy doubled between 1909 and 1949; and some seven-eighths of that increase could be attributed to "technological change in the broadest sense" and only the remaining eighth could be attributed to conventional increase in capital intensity. . . . The broad conclusion has held up surprisingly well in the thirty years since then. . . . Education per worker accounts for 30 percent of the increase in output per worker and the advance of knowledge accounts for 64 percent.[4]

Since science and technology are components of economic growth, investment in research benefits society.[5] The quantifiable value of science is thus a question of interest to economists and investors alike.[6] With regard to government support of research, calculating what is referred to as "the social rate of return" is key, the "private" return being the return to the original investor, and the "social" return the payoff to society as a whole.[7] Its generation of new knowledge makes science an enterprise in which the social gains may be just as great as, or even greater than, the private returns. Science—or rather the new knowledge it produces—can thus be referred to as a "public good," meaning that another person can use the knowledge generated without incurring any cost and that the knowledge can be used by multiple persons simultaneously.[8]

This consideration is particularly relevant for U.S. research, much of which is funded by public tax dollars. Members of Congress and the public at large are particularly keen to ensure that there is a social return on public dollars invested in funding research. In a study published in 1991, economist Edwin Mansfield evaluated the role of university research in product and process development, by focusing on seventy-six major U.S. firms.[9] His findings suggest that about 10 percent of the new products and processes commercialized between 1975 and 1985 in seven different R&D-intense industries would not have been developed, or would have been substantially delayed, without access to recent academic research. Mansfield's tentative estimate of the social rate of return from investment in academic research during the time period was approximately 28 percent. In 1992, Mansfield reestimated the social return at an even more impressive 40 percent.[10] Based on data from a second survey, Mansfield estimated that 15 percent of new products and 11 percent of new processes would not have been developed without the benefit of academic research.[11]

In response to a 1993 request from Representative George Brown, the Congressional Budget Office issued a technical review of Mansfield's 1991 study.[12] The CBO report confirmed Mansfield's findings and noted that while the methodology would not be appropriate for calculating the optimal level of federal research investment, the "return from academic research, despite measurement problems, is sufficiently high to justify overall federal investments in this area."[13] For this reason, the notion demonstrated by the Mansfield study that investment in academic research yields substantial benefits has been used by many policymakers, including the George H. W. Bush administration, to strengthen their case for federal support of academic research.[14]

One need only consider the complexity of valuing the more abstract products of research, such as training, the creation of research tools, and the development of formal and informal researcher networks, to grasp the complexity of this calculation. While all of these products are socially valuable, they were explicitly left out of Mansfield's calculation because of the difficulty in quantifying them. It is particularly difficult to quantify the value of knowledge—the primary product of scientific research. The calculation is further complicated by the difficulty of tracing a given technological innovation back to a single moment of creation. New inventions are almost invariably the products of multiple discoveries and innovations: the lightbulb, for example, could not have happened without breakthroughs in physics, electrical engineering, and material sciences. Even if one could chart the genealogy of ideas behind such an invention, it would be almost impossible to place a value on its intangible effects, such as improvements in quality of life. While the calculation of private rates of return is a bit more straightforward, even here the result is often undervalued.

Some economists have tried to quantify the benefits of research investment by using patents as a proxy for innovation. Bibliometric methods, which look at publication patterns by analyzing citation rates, are used in these types of studies.[15] Tracing the references cited in U.S. patents, for example, bibliometrician Francis Narin and colleagues showed in a 1997 study that patents refer more often to publicly funded research than to industry research publications. Specifically, they found that in the more than 300,000 U.S. patents issued during the periods 1987–88 and 1993–94, approximately 73 percent of references were to publicly funded science papers (from universities, government, and other public institutions in the United States and overseas), whereas only 27 percent were to industry papers.[16]

Research on patents has also been used to determine the positive geographical spillovers from university research. Work in this area suggests that university research encourages industrial innovation, thereby implying a strong link between publicly funded research and public goods.[17]

Using another approach, Narin and colleagues have also tested the assumption that companies that optimize their use of technology outperform those who do not.[18] Their ranking of companies in certain high-technology industries, based on three indices ("citation intensity" of patents, "science link," and "technology cycle time"), has been proven a reliable predictor of future stock price.[19] The portfolios that Narin and colleagues have created using this model have significantly outperformed the Standard & Poor's, leading to the conclusion that federally funded research propels commercial innovation, creating jobs and increasing shareholder value.

Anecdotal Evidence

In the absence of decisive economic analyses, those who advocate for federal science investment often depend on anecdotes and historical examples when making their case. They point out to policymakers and the public that technological breakthroughs are the result of the knowledge from many different sources being obtained and pieced together over many years. Examples include the expansion of communication networks and transportation systems that investments in scientific research have promoted, the vaccines that prevent widespread epidemics, and new, more efficacious medicines that allow us to "bounce back" when we are ill. Their arguments also refer to technologies that have clearly had a significant impact on the economy: for instance, the Internet and the World Wide Web; fiber optics; GPS; the development of the artificial heart; the discovery of statins (a class of pharmaceuticals that effectively lower blood cholesterol); and the development of new classes of antidepressants, with fewer side effects.[20] A contemporary example with high prevalence of a technology resulting from knowledge from multiple sources is the MP3 player.[21] (See fig. 5.1.) These are all things we take for granted in our daily lives, improving our quality of life and general well-being. Despite these anecdotal accounts, because the fundamental research underpinnings of such major technological advances frequently have many different origins, it is often difficult to trace these technologies directly back to federal investments in research.

In spite of these difficulties, some researchers have tried to link individual technologies to their scientific origins, with the idea that the links would offer solid justification for public research funding. Perhaps the first such effort was "Traces," a study conducted in 1969 by Narin long before his 1997 bibliometric study of patents and technology. The Traces study, which was supported by NSF, asked experts in select technologies to trace the genealogy of those technologies, identifying the factors that led to their eventual widespread use. Respondents were asked to classify the stages in this evolution as non-mission-oriented research, mission-oriented research, and development and application. The Traces researchers then tried to measure the relative proportions of these various stages.[22] Similar efforts have been made as part of the NAS's "Beyond Discovery" series, the NSF's "Nifty Fifty," and the DOD's "Project HINDSIGHT."[23]

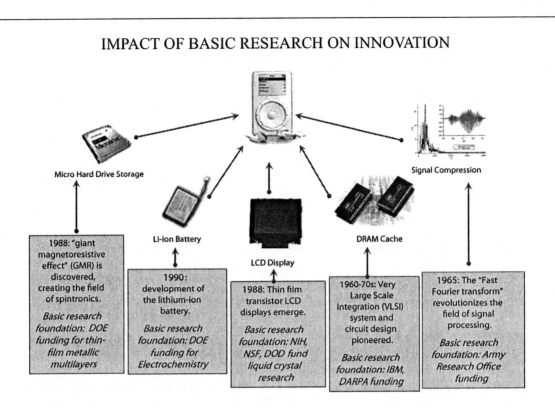

IMPACT OF BASIC RESEARCH ON INNOVATION

Micro Hard Drive Storage

Signal Compression

Li-ion Battery

DRAM Cache

LCD Display

1988: "giant magnetoresistive effect" (GMR) is discovered, creating the field of spintronics.

Basic research foundation: DOE funding for thin-film metallic multilayers

1990: development of the lithium-ion battery.

Basic research foundation: DOE funding for Electrochemistry

1988: Thin film transistor LCD displays emerge.

Basic research foundation: NIH, NSF, DOD fund liquid crystal research

1960-70s: Very Large Scale Integration (VLSI) system and circuit design pioneered.

Basic research foundation: IBM, DARPA funding

1965: The "Fast Fourier transform" revolutionizes the field of signal processing.

Basic research foundation: Army Research Office funding

The development of MP3 technologies illustrates the unexpected benefits of basic research. In 1965, a hand-sized storage and playback device that would hold 15,000 recorded songs was the stuff of science fiction. Even simple hand-held calculators were rare and expensive at that time. Research funded by the Department of Defense, the National Science Foundation, the National Institutes of Health, the Department of Energy, and the National Institute of Standards and Technology contributed to the breakthrough technologies of magnetic storage drives, lithium-ion batteries, and the liquid crystal display, which came together in the development of MP3 devices. The device itself is innovative, but it built upon a broad platform of component technologies, each derived from fundamental studies in physical science, mathematics, and engineering.

FIG. 5.1 Tracing the development of the MP3 player. (*American Competitiveness Initiative: Leading the World in Innovation*, Domestic Policy Council, OSTP, February 2006, 8.)

The value of government-funded science was, for most of its history, an article of faith, based on a limited number of economic analyses, on history and anecdotes. Belief held that there was inherent value in the generation of new knowledge, and that any country investing in knowledge development would derive significant benefits, economic and otherwise.[24] New methods of analysis are now providing a firmer basis for such assumptions. To support the development of these methods, the NSF, with its FY2007 budget request, launched an initiative tentatively called "science for science policy." The hope is this initiative will lead to about a half-dozen research centers across the United States as well as many individual investigator grants. While those from the NSF's Social, Behavioral, and Economic Sciences (SBE) division are cautioning against expectations of fine-tuned precision, the director of SBE believes that what can be expected is the development of "a more evidence-based understanding of what happens to our R&D investments."[25]

How the Government Evaluates the Effectiveness of Its Investment in Science

As we noted earlier, it is profoundly difficult, if not impossible, to calculate the rates of return on public investment in research. What makes a research program effective? How does one gauge the impact of a program's funding awards? What determines whether a set of research experiments was successful?

Over the years, the OMB has relied on several processes to annually determine the effectiveness of programs and agencies across the federal government. However, it is extremely difficult to quantify knowledge, as we have already seen. In addition, recall that there is disagreement over how we should define success in scientific research. It is especially tricky to make such evaluations on an annual basis, as the OMB is expected to do, since the payoff for most research occurs many years after the original investment: the Global Positioning System, for example, is in part a product of subatomic basic research conducted during World War II. What mechanisms, then, does the OMB use to determine the effectiveness of science-related programs?

The Government Performance and Results Act of 1993

The aim of the Government Performance and Results Act of 1993 (GPRA) was to hold federal agencies accountable for results, rather than process. The GPRA introduced more emphasis on performance metrics and program *outcomes* (rather than program *plans*); its ultimate goal was to move the federal government to a higher level of operational effectiveness and efficiency.

The act was the brainchild of John Mercer, former mayor of Sunnyvale, California (a city where performance-based management and budgeting metrics had been in use since 1973) and at the time governmental affairs staff advisor to Senator William V. Roth (R-DE).[26] In early 1990, Roth authorized Mercer to write legislation based on the Sunnyvale system. Roth introduced the resulting bill in October 1990, and again in 1991.[27] It was finally passed and signed into law in early 1993.

Under the GPRA, federal agencies were, for the first time, mandated by federal law to develop long-term strategic plans and publish annual performance reports defining their goals, delineating a road map for reaching these, and documenting their progress toward their goals. The required annual report must include a comparison of actual goals met versus those originally established by the agency in its strategic plan. The GPRA also requires the plans to include performance metrics that can be used to measure how well the agency is meeting its objectives.

The GPRA was an attempt to eliminate the "waste and inefficiency of federal programs" and increase the government's ability to "address adequately vital public needs."[28] Under the law, an agency's strategic plan is expected to reflect the views of entities (individuals and groups) who are potentially affected by a particular program. While there are no procedural requirements guiding the solicitation of these views, each agency is supposed to be open to comments from interested parties. Therefore, science agencies such as the NSF and NIH usually solicit public comments on a draft of their strategic plan before finalizing it. All plans and annual reports are, of course, a matter of public record. Finally, and significantly, the GPRA does not limit congressional ability to "establish, amend, suspend, or annul a performance goal."[29]

The version of the GPRA legislation that was passed in 1993 also featured a pilot project. The idea was that no fewer than ten agencies would participate in the GPRA during each of the next three years, starting in 1994. This phase-in was designed to gradually bring all federal agencies into compliance with the GPRA. The initial agencies would test the effectiveness of the process, by developing a strategic plan and publishing progress reports. The GAO was to report back to Congress on its cost-benefit analysis of the GPRA's success, after which the legislation could be amended as necessary. The phase-in plan called for government-wide compliance with GPRA by FY1998.

Congress saw GPRA as a means to gather performance information to guide resource allocations.[30] A *Science* article observed that the GPRA's effects on science could reach far beyond the intended goals of improved accountability and resource allocations.[31] The law's emphasis on annual performance outcomes and metrics encourages a focus on the short term. But science, and particularly basic research, is fundamentally oriented toward the long term. Of concern for many at the time of its implementation was that GPRA and its attention to quantitative indicators and accountability would shift the focus of government-funded research away from basic science to applied research, where results come more quickly and are easier to identify and quantify.

The National Academies have issued at least two reports on the application of the GPRA to federal research programs. The first, published in 1999, reached several conclusions:

> Both applied research and basic research programs supported by the federal government can be evaluated meaningfully on a regular basis. . . . The most effective means of evaluating federally funded research programs is expert review. Expert review—which includes quality review, relevance review, and benchmarking—should be used to assess both basic research and applied research programs. . . . The development of effective methods for evaluating and reporting performance requires the participation of the scientific and engineering community, whose members will necessarily be involved in expert review.[32]

The report also included six recommendations, among them that effectiveness of applied research programs may be measured by progress toward practical outcomes. Effectiveness of basic research programs, however, should be measured by indicators of quality, relevance, leadership (worldwide in the particular area), and practical outcomes measured over a longer period of time. The report also noted the hazards of applying inappropriate metrics, which could "lead to strongly negative results" and recommended that agencies "use expert review to assess the quality of research they support, the relevance of that research to their mission, and the leadership of the research."[33]

The National Academies published a follow-up report in 2001, based on workshops and focus groups with the five federal agencies that receive the majority of the government's research funding: the NSF, NIH, DOD, DOE, and NASA.[34] It concluded that the agencies were making a good-faith effort to meet GPRA requirements, but that some were encountering difficulties, including increased costs for reporting staff and resources, and poor communications with oversight bodies. The report also noted that agencies did not always understand how the oversight bodies were using GPRA results in their programmatic decision making.

The Academies committee again concluded that expert review is the most effective means of evaluating research programs, and recommended that agencies continue their efforts to improve compliance and communications with oversight bodies.

The President's Management Agenda

In August 2001, President George W. Bush announced an aggressive plan for improving management of the federal government. The President's Management Agenda (PMA) was designed to "address the most apparent deficiencies where the opportunity to improve performance is the greatest."[35] Five government-wide initiatives were established: (1) strategic management of human capital; (2) competitive sourcing; (3) improved financial performance; (4) expanded electronic government; and (5) budget and performance integration. Across-the-board standards were instituted to ensure agency compliance.

The PMA also included something called the Executive Branch Management Scorecard. The scorecard has been used by the OMB since June 2002 to track department and major agency adherence to the five government-wide management initiatives. Each federal agency's performance in all five areas is graded on a quarterly basis. These performance ratings are used by the OMB in its policy and funding decisions, with progress assessed on a case-by-case basis against the agency's timelines and deliverables. There are three possible scores:

> *Green:* implementation is proceeding according to plans agreed upon with the agency.

> *Yellow:* some slippage or other issues exist requiring adjustment by the agency in order to achieve the initiative objectives.

> *Red:* the initiative is in serious jeopardy and unlikely to realize the agreed-upon goals absent significant management intervention.

The PMA focuses solely on federal agencies per se, unlike the GPRA, which also looks at specific programs within each agency. Under the PMA, agencies are evaluated on their operational and functional performance, particularly on financial and budgetary matters. The administration's rationale for its institution of the PMA was that the existing GPRA framework was not working well, and needed to be supplemented with additional performance measures. Because the GPRA is a matter of law, it could not easily be revoked, and so the PMA was conceived as supplemental to, rather than a replacement for, the GPRA.

Incidentally, the NSF was the only agency to be awarded a rating of "green" when the first scorecards were issued in June 2002, and continues to be among the highest-scoring agencies assessed by the PMA. Despite the NSF's high scores, however, it was not granted any significant additional support from OMB and the administration in response to these high ratings, even as agencies with lower PMA scores, such as NASA, received substantial budget increases. This has led to questions about how the PMA figures into final budgeting decisions, and suggests that performance and effectiveness are only two of many variables that the president and the OMB use to establish their funding priorities.

Program Assessment Rating Tool

To reinforce the PMA, the OMB developed the Program Assessment Rating Tool (PART), which links the GPRA to the budget process. The OMB has used PART since the FY2004 budget process to evaluate the performance of each agency and determine which scores are received.

PART focuses on the specific programs administered by each federal agency or department. The goal is to rate the effectiveness of these programs, so that the highest-ranking will be more likely to benefit from budget increases. Again, however, some programs—particularly those involving scientific research—are not open to simple "success versus failure" evaluations. It is understandable, in our political system, why a policymaker or politician would want the investment of resources in a program to yield clear results, in the form of solutions to problems. For scientists, however, the generation of new knowledge is the sole marker of success. That is, if a set of experiments yields new knowledge, allowing for the asking of

new questions, then the experiments are a success. Consequently, members of the scientific community fear that no scientific programs will fare well under PART.

The questions asked by PART about each federal program fall into four broad categories: (1) program purpose and design; (2) strategic planning; (3) program management; and (4) program results and accountability. The questionnaire is intended to provide consistent and systematic criteria for rating programs across the federal government. Programs can receive one of five different ratings: effective, moderately effective, adequate, ineffective, and results not demonstrated.

Our discussion of the concerns about PART is not meant to dismiss the importance of efficiency and effectiveness—especially in the distribution of extramural research awards. But given the difficulty of evaluating scientific outputs and the creation of knowledge—whose benefits are often not fully realized until years after initial discovery—scientists are rightfully concerned about how well current evaluation schemes will work for re-

POLICY DISCUSSION BOX 5.1

The Tricky Business of Forecasting What Basic Science Will Lead To

We have talked a lot about the difficulty of putting a value on investments made in research and the difficulty in tracing a given technology back to the fundamental scientific findings that led to its development. We can frame the issue another way by asking, how simple is it to predict what will be developed from a given focused area of research?

Hyperlinks are a good example. Today hyperlinks allow almost anyone to easily navigate the World Wide Web and to use many service-based Web sites, for example, Amazon.com, eBay, and CheapTickets.com, without having to worry about entering complex network addressing or protocol information. Put simply, hyperlinks abstract the complexity of computer networking in a way that make it simple to understand, lowering barriers to use by those who are not computer experts. Without hyperlinks, the World Wide Web and e-commerce as we know it today would not exist.

Yet the original conceptual research for hyperlinks began in the mid-1960s, even before work began on what would become the Internet. At the time that DARPA first funded researchers to study the concept of creating

a "web" of hyperlinks, no one envisioned that it would eventually become the foundation upon which the World Wide Web and a multibillion-dollar e-commerce and service based industry would eventually be built. Hyperlinks instead were simply seen as an easy way to link the world's information together in an efficient and effective way, a means to create a sort of a universal information library.

In 1991, Tim Berners-Lee at CERN publicized his "World Wide Web" concept, which employed hyperlinks to bridge information and servers on the academic Internet. Shortly thereafter, in 1994 the first Internet "shopping malls" appeared. And the rest is history . . .

If the early research on hyperlinks had been evaluated under GPRA in the 1960s, how would this research program have fared? Would the agency funding this research have been given a high rating under the PMA? How could anyone have predicted the long-term outcome and economic impacts that would result from these research investments?

This raises broader questions: What makes a scientific research program successful? Who should be involved in evaluating scientific programs? Should scientists working with policymakers determine what programs should receive a high performance rating under GPRA or the PMA?

search agencies such as the NIH, NSF, and DOE's Office of Science.

Historical and Current Funding Trends

Having reviewed why the federal government invests in scientific research and how it measures progress made through this funding, we should also understand some of the general trends of federal research funding. These trends include what percentage of the nation's research is supported by industry, the link between national priorities and investments in science, spending on research as a percentage of the nation's GDP, and the notion of balance in funding across disciplines. We will also look briefly at agency funding trends.

Increasing the Role of Industry in Support of R&D

During the 1990s, overall spending on R&D from all sources, including government and industry, rose significantly.[36] The increase was, however, driven not by in-

creased government spending, but rather by a rise in commercial investment, much of which was in development, not basic research. When the figures are adjusted for inflation, it turns out that federal R&D investment during this period actually declined.[37] Until the mid-1970s, the government acted as the primary source of U.S. R&D spending. In 1964, for example, the federal government was providing over 66 percent of R&D funds. Through the 1970s and 1980s, however, federal dollars for R&D steadily declined, falling to just over 25 percent of the total in 2000, the lowest level since these data have been tracked (see fig. 5.2).

In 2004, approximately 64 percent of the total R&D conducted in the United States in 2004 was funded by industry—a significant increase from the approximately 45 percent industry spent in 1975.[38] The growth in private investment has been driven primarily by the commercial sector's growing dependence on technology, and the explosion of high-tech industries resulting from the biotechnology and information technology revolutions.

The reversal of roles is cause for concern, as suggested in a 2002 report from the President's Council of Advisors on Science and Technology (PCAST). The PCAST report

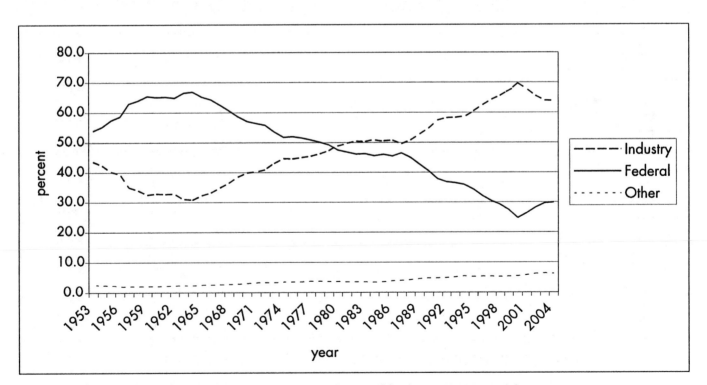

FIG. 5.2 National R&D expenditures, by source of funds: 1953–2004. Other = universities and colleges, nonprofits, state and localities, and other nonfederal sources. (Data from National Science Foundation, Division of Science Resources Statistics, National Patterns of R&D Resources annual series; and National Science Foundation, *Science and Engineering Indicators 2006*, chap. 4, fig. 3; and app. table 4.5.)

pointed to two problems. First, industry R&D funding follows the business cycle: if profits decline, R&D investment is generally cut back. Second, industrial R&D understandably focuses heavily on the development of proprietary technologies that will boost company profits and produce private returns on industrial investments. Investment in basic research, which drives fundamental knowledge creation and has broader societal value, is largely left to the federal government and conducted by universities—which tend to do very little development work—and to the federal laboratories. A net decline in federal research funding may ultimately have an adverse impact on the level of basic research that can be conducted, unless we see more money flowing from industry to universities in the form of grants, which has not been the case.[39] And if industry were to provide significantly more funding for campus research, it would likely mean a reorientation of university research priorities. Concerns have been raised that increased industrial sponsorship of university research tends to conflict with university's traditional education and scholarship missions and can impede the strongly held principle of free exchange of research results in an academic setting.[40]

National Priorities and Funding Research

The proportion of funding for R&D coming from industry versus the government is not the only area in which science policy trends have changes over time. Priorities have also shifted dramatically over the years in the focus of the research that the government supports. These changes are for the most part in step with changes in national priorities. We saw spikes in funding during the space race and the Cold War during the early 1960s, and during the energy crisis of the 1970s. In fact, it is interesting to note that the amount that the United States has been willing to spend on energy-related R&D corresponds with rises and falls in annual oil prices. (See fig. 5.3.)

Further surges in R&D spending occurred in the 1980s as a result of President Reagan's emphasis on defense (e.g., the Strategic Defense Initiative, derisively labeled "Star Wars") and then in the 1990s driven by growing interest in combating human disease and in improving health care to address the needs of the aging baby boomers.[41]

Since September 11 the nation has turned its attention to biodefense and the fight against terrorism—shifts that were reflected in a mammoth FY2002 increase to the Na-

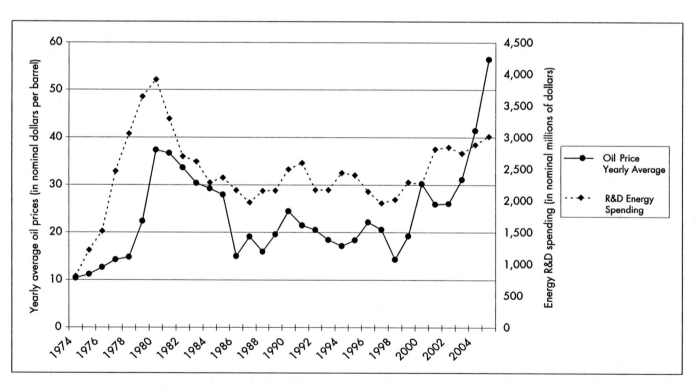

FIG. 5.3 Oil prices compared to energy R&D spending. (Oil prices are from the Federal Reserve Bank of St. Louis, based upon the Dow Jones and Company data, http://research. stlouisfed.org/fred2/data/oilprice.txt. Energy R&D spending data are from the International Energy Agency, http://www.iea.org/Textbase/stats/rd.asp.)

tional Institutes of Health budget, largely for biodefense research at the National Institute for Allergy and Infectious Diseases and at the Centers for Disease Control and Prevention.[42] Details aside, it is clear that the trends in research support over the last fifty years have most often been driven by a real or perceived sense of crisis or a major shift in national policy focus (e.g., the space race during the 1960s, the energy crisis during the 1970s, the concerns about health and the aging baby boomers during the 1990s, and the increased awareness of homeland security, national security, and bioterrorism during the first decade of the 2000s; see also fig. 2.1, chapter 2).

Funding for Defense R&D

Science funding is often analyzed in terms of defense versus nondefense spending.[43] In these terms, President Clinton's FY2000 budget proposal represented the first time since the Carter administration that funding of nondefense R&D surpassed its defense counterpart. This shift away from defense was due largely to the end of the Cold War and to the boom in nondefense health and biomedical research, which began in the late 1990s. Since the terrorist acts of September 11, 2001, and the start of the Afghan and Iraqi wars, attention has again turned toward the development of major weapons and defense systems, which has been reflected by specific growth in the defense development funding.[44] Despite this recent growth in Defense R&D funding, funding specifically for defense research has remained essentially flat in recent years with no significant increases.[45]

The end of the Cold War created an opportunity to reorient American science and technology.[46] "The Cold War is over, Japan and Germany won," Massachusetts senator Paul Tsongas said in 1992, neatly expressing the sense that America's new challenges were economic, rather than military.[47] Science and technology programs that had been created for national defense were now being scrutinized for their potential to enhance the country's global economic competitiveness. Efforts were even made to deemphasize the military origins of these programs: the Defense Advanced Research Projects Agency (DARPA) was temporarily renamed the "Advanced Research Projects Agencies." Meanwhile, longtime DOD programs such as ManTech (short for Manufacturing Technology), a program designed to support the maturation and validation of manufacturing technologies related to weapons production, were being retooled to improve civilian manufacturing technologies.[48] In fact, ManTech deserves credit for several key innovations, including development of numerically controlled machine tools, some of the basic processes and tools for microelectronics manufacturing, and integrated computer-aided design (CAD).[49]

Indeed, there is a strong case to be made that the recent spikes in R&D spending for defense and homeland security have been far too oriented to the development of large and expensive weapons and security systems without adequate attention to funding for basic and applied research. Defense research has played a historically significant role in both ensuring our national security and leading to scientific and technological advances that have resulted in significant economic growth. The strong linkage that exists between national and economic security have led some to call for significant increases in the federal budget devoted to defense research.[50]

Research Funding as a Percentage of Gross Domestic Product

Funding for R&D can also be evaluated as a percentage of the total U.S. gross domestic product (GDP).[51] Total U.S. R&D funding (industry and government sponsored) as a percentage of GDP declined steadily through the 1970s, but began to increase in the early 1980s, reaching a high by the middle of the decade. It has remained relatively stable since it peaked in the mid-1980s. It is worth noting that the historical high-water mark of nearly 3 percent of GDP has not been achieved since 1964.[52] Within this measure, industry-funded R&D as a percentage of GDP has been increasing since 1953 (although the last several years have seen a slight decrease). Federally funded R&D as a percentage of GDP has decreased from 1.25 percent in 1985 to 0.68 percent in 2000.[53] Through 2004, there was a notable increase; however, that trend has reversed, and recent years once again show a decline.[54]

As mentioned earlier, economic growth is linked to progress in technology. And, technological developments are the result of research. Some have suggested that the nation's economy is better off when increases in national R&D investments are directly tied to growth in GDP and that the decline in the proportion of the U.S. share of GDP devoted to R&D is troubling.[55]

Balance in Funding across Disciplines

One of the most persistent questions of science policy concerns the balance in funding among academic disciplines. Should federal funding be increased evenly across the board, for all agencies that support research? What is the correct level of funding for the life sciences, physical sciences, and engineering and for agencies with very different missions, such as the DOD, DOE, and DHHS? At what point does underfunding in one area adversely affect other areas?

The NIH has recently been the beneficiary of a generally steady increase in funding, particularly since the late 1990s, although this trend began to change in 2004. The increase during the 1990s and first few years of the new millennium was, in part, the result of the widening search for new medical treatments and cures, driven by the aging of the baby boomers. However, during this time of increased funding for NIH, funding for other disciplines remained flat. As eloquently stated by Harold Varmus, president of Memorial Sloan-Kettering Cancer Center and former director of the NIH: "[The] disparity in treatment [across agencies] undermines the balance of the sciences that is essential to progress in all spheres, including medicine. . . . Scientists can wage an effective war on disease only if we—as a nation and as a scientific community—harness the energies of many disciplines, not just biology and medicine."[56]

Indeed, PCAST's 2002 assessment of the nation's R&D investment also recommended that funding for the physical sciences and engineering be brought to parity with support of the medical and biological sciences. In FY1970, federal support for three main areas of research—physical and environmental sciences, life sciences, and engineering—was relatively proportional. In contrast, the life sciences received about 48 percent of federal R&D support in FY2002, while the physical sciences received 11 percent and engineering 15 percent.[57] As the panel noted, "Continuation of present patterns will lead to an inability to sustain our nation's technical and scientific leadership." Further, "long-term breakthroughs in biological and life sciences will also rely on strengthening the physical sciences and engineering as well."[58]

Following its doubling period (1998–2003), the NIH began to experience flat or declining budgets, while science funding at other agencies (e.g., NSF, DOE science) experienced marginal increases. With the National Academies' publication of *Rising above the Gathering Storm* in 2005 and President Bush's American Competitiveness Initiative (ACI) announced in 2006, there was an increased recognition that the nation could no longer neglect the physical and other nonbiomedical sciences. This shift in funding emphasis is not necessarily the most effective answer, since sustaining what was created during the NIH budget doubling requires NIH budgets that at least keep up with inflation. If anything should be learned from the dramatic imbalance during the late 1990s it ought to be that gradual balanced growth for research is the better

federal funding strategy. Only time will tell whether and how balance in funding across disciplines will be again achieved.

Funding Trends by Agency

There are several agencies central to U.S. R&D, including the Department of Defense, the Department of Energy, the National Institutes of Health, the National Science Foundation, and the National Aeronautics and Space Administration. Many additional agencies have more modest R&D budgets. The American Association for the Advancement of Science annually publishes a comprehensive analysis of the forthcoming R&D budget, followed by timely updates throughout the budget process.[59] The following pages describe key trends in each agency or department's funding.

The National Institutes of Health

The NIH's budget is often constructed with an eye toward how the funds will be distributed across the agency's twenty-odd institutes and centers. Between FY1993 and FY2001, federal funding for the NIH grew by 60 percent in constant dollars.[60] Much of this growth was the result of a commitment made in 1998 by key members of Congress, including House and Senate Labor-HHS-Education Appropriations Subcommittee chairs, Representative John Porter (D-IL) and Senator Arlen Specter (R-PA), to double funding for the NIH over a period of five years. Because of these efforts, between FY1998 and FY2003, the NIH saw annual increases of 14 to 16 percent.[61] Even after this period of doubling was over, many NIH institutes continued to see small annual increases. However, in fiscal year 2006 this changed with institutes' budgets remaining flat—decreasing when inflation is taken into consideration.[62] Of the agency's funds, 97 percent are dedicated to R&D, and the remaining 3 percent to training and facilities.

Why did funding for the NIH increase so dramatically from FY1998 to FY2003? One frequently cited reason is a very effective advocacy campaign coordinated by NIH-sponsored researchers and dozens of patient advocacy groups. These groups, including the Juvenile Diabetes Research Foundation, the American Cancer Society, and the American Heart Association, have been able to directly mobilize grassroots support for research cures for the diseases they fight.[63]

Concern over the spread of AIDS/HIV and the growing threat of bioterrorism has also benefited the NIH, particularly the National Institute for Allergies and Infectious Diseases (NIAID). NIAID's budget increased in the aftermath of the fall 2001 anthrax attacks such that by 2003 it moved past the National Heart, Lung, and Blood Institute to become the second-largest institute (next to the National Cancer Institute). In the aftermath of September 11, the NIAID budget jumped by about 22 percent in FY2002, and grew for several years following that.[64]

Some NIH-watchers have expressed concerns about whether the agency can cope with the drastic budget increases that occurred between 1998 and 2003. The agency's prosperity during that period allowed it to make more awards than usual, many of them multiyear; but increases in the national deficit and changing national priorities will likely mean much smaller or even no annual increases in the years to come, raising the question of how the agency should balance payouts on existing multiyear projects against the need to increase the number and size of new awards. Indeed, already the NIH is having difficulty sustaining the momentum that was built up during the doubling period, and as a result is having to decrease the total number of researchers it is able to support.

The National Science Foundation

Since its founding in 1950, the National Science Foundation has had an enormous impact on U.S. science and discovery. Although small in size, the NSF is the only agency with broad responsibility for the well-being of all science and engineering disciplines. Moreover, the NSF plays a crucial role in supporting research at universities and colleges.[65] Indeed, while the NSF is among the smaller federal agencies supporting science, it is second only to the NIH in its funding of university-based research. The NSF also provides significant support to the nation's educational programs, from K–12 programs to postdoctoral fellowships. It also grants significant support to researchers in the environmental sciences, computer science, the physical sciences, and the social sciences.[66]

The NSF has experienced budget increases during every year since 1990, with the exceptions of FY1996 and FY2005. Between FY1999 and FY2003, federal support for the agency rose to a higher level than ever before, although the increases were nowhere near as large as those seen at the NIH. The NSF Authorization Act of 2002, a bill aimed at putting the NSF on track to double its budget over five years, was signed into law by President Bush in December 2002.[67] In spite of the passage of this authorization legislation, however, the NSF did not receive the appropriations required to come close to achieving this doubling goal.

There is great hope the significant funding increases might now be in NSF's future as the result of the bipartisan support for major competitiveness initiatives such as President Bush's ACI, the House Democrats Innovation Agenda announced in November 2005, and the National Academies *Rising above the Gathering Storm* report.

The Department of Defense

Over the years, DOD funding has increased in times of perceived conflict and decreased during periods of relative stability. For example, DOD's overall R&D budget increased notably at the height of the late Cold War, in the 1980s. It then decreased after the Cold War ended in the early 1990s.[68] It has seen another spike since September 11, 2001, and the anthrax attacks that followed, and in fact has reached all-time highs in recent years.[69] This increase, however, is driven by increased investments in weapon systems development and is not always accompanied by increases in basic and applied research and technology development.[70]

While DOD funding always accounts for a significant portion of federal R&D spending, it is important to remember that most of this money goes to weapons development, testing, and evaluation. DOD's basic and applied research is a small portion of the total budget for DOD, in recent years accounting for less than 2 percent of the total budget.[71]

The Defense Department funds a large amount of basic and applied engineering and computer science research, and when support for DOD's S&T programs remains stagnant or declines, the level of available support for engineering and computer science, especially at universities, is negatively affected.[72] Since the early 1990s, the DOD has also provided funding for medical research through "congressionally directed medical research" (CDMR) programs. Federal support for these programs has increased slightly since their inception.

The Department of Energy

The R&D work funded by the Department of Energy falls into one of three categories: general science, energy, or defense. Within the department, the Office of Science manages most of the agency's basic research programs in energy, the biological and environmental sciences, and computational science, and provides more than 40 percent of the total federal funding for research in high-energy physics, nuclear physics, and fusion-energy sciences.[73]

Ranked fourth among federal agencies for its overall support of R&D, the DOE is the number one federal agency in support of the physical sciences, of R&D facilities, and of mathematics and computer science combined. It is also a significant funder of engineering research.[74] In addition, the DOE oversees and is responsible for research conducted at the more than twenty DOE national laboratories and technology centers. Thus, any decrease or leveling off of DOE support directly affects the availability of research funding in the physical sciences, math, and computer science.

Despite the importance of the DOE's role, the department did not receive significant funding increases for over a decade, either overall or specifically for its Office of Science. Between 1990 and 2006, the DOE's federal support peaked in 1992, and then decreased or remained stagnant for every year through 2006. However, recently, the department and Office of Science have seen increases in its research budgets, and additional increases are called for under President Bush's American Competitiveness Initiative.[75]

One common explanation for DOE's research funding problems is that it has had difficulty building a strong constituency for the work it performs and its advocates have not been as cohesive and vocal in their message as those for other research agencies such as the NIH, which has in the past received significant support from disease-oriented advocacy organizations. This problem was aggravated by the fact that until recently, the director of the Office of Science, in spite of managing a significant portion of the DOE's R&D budget, was a fourth-level administrator in the DOE organization chart, whereas the director of the NIH reports directly to the secretary of health and human services. This changed with the creation of a new under secretary of science position at DOE with the passage by Congress of the Energy Policy Act of 2005. The under secretary of science, who reports directly to the secretary of energy, has been assigned direct responsibility not only for overseeing the DOE Office of Science but for coordinating scientific activities across the entire DOE.

Still, however, researchers funded by the DOE must contend with budgetary competition within the agency itself, where research support is balanced against other major DOE missions such as energy production, environmental cleanup and management, and national security. The independence of the NIH and NSF budgets insulates research funding at these agencies against such internal budgetary trade-offs.

The National Aeronautics and Space Administration

The National Aeronautics and Space Administration was established in 1958, with its R&D program focused on innovation and engineering for space flight, although it does fund a range of disciplines.[76] The agency has also worked on the development of new satellite applications, as well as instrumentation for probing the solar system and the universe. NASA's Hubble Space Telescope has been used by astronomers and astrophysicists around the world since 1990, providing, for example, convincing proof of the existence of black holes, and data on the origins of gamma rays.

The 1960s, when the nation (and the world) was focused on sending a person to the moon, were a time of prosperity for NASA. The agency's budget has not been as large since, hitting an all-time low during the Reagan administration.[77] Its budget rebounded again after 1990, but has remained almost stagnant since 1995, even decreasing a bit at the turn of the century. President Bush's stated goals of continuing our moon exploration and landing American astronauts on Mars have not translated into budget increases for NASA. So far, the budgets that have been provided by the Bush administration for NASA have been met with significant criticism from the scientific community on the grounds that the funding being provided by the administration is not adequate to support the space exploration vision outlined by President Bush while at the same time continuing to support space science and aeronautics research at adequate levels.[78]

Other Agencies

Other federal departments and agencies, including the Department of Commerce (DOC), the Department of Agriculture, and the relatively new Department of Homeland Security, maintain more modest R&D budgets.

The DOC's mission, for example, includes "promoting innovation, entrepreneurship, [and] competitiveness."[79] As we noted at the beginning of this chapter, science, technological advancement, and innovation are key to the nation's economic growth and competitiveness. Therefore, the DOC is specifically devoted to R&D that furthers America's manufacturing technologies. The department also oversees the National Oceanic and Atmospheric Administration (NOAA), and the National Institute for Standards and Technology (NIST), including its Advanced Technology Program (ATP). The Bush administration and some Republican members of Congress, especially in the House, have repeatedly targeted ATP for elimination on the grounds that it subsidizes certain companies in ways that give them an unfair competitive advantage. These attacks on ATP on the grounds that it provides so-called corporate welfare have at times put the future of the program in doubt.[80] In an effort to buffer the program against these attacks, its focus has been reoriented and its name changed to the Technology Innovation Program (TIP), with the enactment of the America COMPETES Act of 2007. The department's R&D budget increased significantly between 1992 and 1995, but subsequently decreased again in the latter 1990s. It has since been in a state of flux, most recently experiencing annual decreases. In terms of constant dollars, however, Commerce's R&D budget has remained fairly consistent over the years.[81]

The USDA has a somewhat clearer role in the nation's R&D enterprise. This department has an entire unit devoted to research on the viability of U.S. agriculture, along with nutritional and food safety issues. The USDA is also involved in supporting social science research on the socioeconomics of farming productivity and food distribution. The R&D budget for the USDA has, like that of the DOC, been relatively stable over the years, with minor ups and downs. Between 1998 and 2001, the department experienced a steady but small increase in support for its R&D activities; however, the period since 2003 has seen a series of annual decreases. In spite of the importance of the USDA's work on food supply and nutrition, it has one of the smaller federal R&D budgets. There has been a push to try to increase funding for USDA research by growing support for more competitive research awards to be made, to supplement and perhaps eventually replace existing USDA research programs where funds are awarded largely on a formula (as opposed to competitive) basis.

The DHS was formed in 2002 as part of the National Strategy for Homeland Security and the Homeland Security Act. A complex department, the DHS includes, among its many divisions, a Science and Technology Directorate. The agency's R&D budget also includes the Border and Transportation Security Unit and the Coast Guard Unit. After its inception in 2002, the budget for DHS R&D grew very quickly, exceeding one billion dollars. Among other things, these DHS research programs were to focus on nuclear, radiological, biological, chemical, and explosive countermeasures; cybersecurity; shoulder-held weapons; critical infrastructure; and border and transportation

security. As the nation has gotten further away from the events of September 11, 2001, however, and as members of Congress have grown increasingly concerned about how the DHS has been managing the research funds it has been provided, funding for DHS research has begun to recede.[82] It remains to be seen how much of a continued commitment will be made to homeland security R&D in future years.

Other agencies with R&D budgets include the Department of Transportation, the Environmental Protection Agency, and the Department of the Interior. Most R&D funding for these agencies is intramural and mission-oriented.

Policy Challenges and Questions

Government support is essential to our country's scientific research. A vibrant economy, improved quality of life, and national and domestic security all are compelling justifications for governmental involvement.

The disbursement of federal funds always entails responsibility: the government must monitor how effectively those funds are being expended by their recipients. Initiatives recently established to assist federal monitoring, however, routinely run up against the difficulty of measuring the success of science, and of the government agencies and programs that support it, in comparison to many other federal programs and their objectives. Progress in science is often made even when the outcome of a study bears little resemblance to what was expected at its initiation. A null finding may be the most exciting outcome of all, even though generically outcome-based measures interpret this result as failure. Would context-sensitive tools for policymakers to identify progress while taking into account the special nature of scientific research help to avoid this dilemma? Or should policymakers just leave it in the hands of the scientific community to determine whether a research program is successful or not?

Though we know that R&D contributes to the national economy, it is difficult to calculate just how it does so, and how much this contribution amounts to. Yet in the absence of any better metric, we use such contributions to justify continued federal support. As economic modeling becomes more sophisticated, we hope to see progress toward understanding the relationship between funding and research outcomes. What could be used to measure the success of a scientific research program? How should success be defined for these programs? Should policymakers continually push for, and fund, the development of these new models? Would the findings generated from such models improve decision making at the federal level, as well as in private industry?

Trends in research funding convey important messages about America's priorities. For example, the ratio of defense to nondefense R&D spending speaks multitudes about the importance this nation places on national security. R&D spending can also be understood as a percentage of GDP, or in terms of the amount of funding given to individual federal agencies, or academic disciplines. What information can policymakers ascertain from carefully monitoring research funding trends? How should current national priorities be used to determine scientific funding levels and to establish related science policies? How should policymakers decide levels of funding for different disciplines? What mechanisms can be used to ensure balanced funding across disciplines?

Through all debate on funding, however, we should remember that the relationship between science and science policy will always be characterized, to some extent, by uncertainty: this is simply because of the beautifully unpredictable nature of scientific research itself. Science policy should never be the result of inertia in the federal system, nor should we maintain the funding status quo after its rationale has faded from memory.

NOTES

1. Solomon Fabricant, "Economic Progress and Economic Change," in *34th Annual Report of the National Bureau of Economic Research* (New York, 1954); Milton Abramowitz, "Resources and Output Trends in the United States since 1870," *American Economic Review* 46, no. 2 (1956): 5–23; Edward F. Denison, *The Sources of Economic Growth in the United States and the Alternatives before Us* (New York: Committee for Economic Development, 1962), supplementary paper no. 13, 272.

2. Edward Denison furthered Solow's work using growth accounting, an empirical tool he is given credit for advancing. Growth accounting is a method used to decompose economic growth into the components contributing to it. Growth accounting has proven to be useful in economic analyses, and studies utilizing this methodology have estimated that technology accounts for more than one-half of economic (GDP) growth in all OECD countries except Canada. See Matthias Doepke, University of California at Los Angeles, lecture notes on growth accounting for Economics of Growth, an undergraduate course (Econ C32) taught University College London, fall 2003, http://www.econ.ucla.edu/doepke/teaching/c32/sec1.pdf (accessed May 5, 2007); Michael Boskin and Lawrence Lau, "Generalized Solow-Neutral Technical Progress and Postwar Economic Growth," NBER Working Paper no. W8023 (December 2000).

3. Solow, "Technical Change."

4. Economist Robert M. Solow, Nobel Laureate, in his Nobel Prize lecture December 8, 1987, http://nobelprize.org/economics/laureates/1987/solow-lecture.html (accessed May 5, 2007).

5. The rate of return from R&D has been shown to be about four times that from physical capital, which suggests that investments in R&D should be increased by a factor of four. Charles I. Jones and John C. Williams, "Measuring the Social Returns to R&D," *Quarterly Journal of Economics* 113, no. 4 (1998): 1119–35; Charles I. Jones and John C. Williams, "Too Much of a Good Thing? The Economics of Investment in R&D," *Journal of Economic Growth* 5, no. 1 (2000): 65–85.

6. Some studies of return on investment in R&D have shown a social rate of return between 20 and 50 percent. Research has indicated significant variation across industries in the impact of R&D. However, no evidence has been produced to indicate diminishing returns from increased R&D across the range of R&D intensities found in manufacturing industries. Thus, no support exists for the argument that some industries need less R&D than others. In fact, available data support the proposition that low and moderate R&D-intensive industries (technology adsorption strategy) are increasingly in competitive trouble, as other economies boost R&D spending (Gregory Tassey, senior economist for the National Institute of Standards and Technology, personal communication). See also Gavin Cameron, "R&D and Growth at the Industry Level," Nuffield College Working Paper no. 2000-W4 (January 2000); Gregory Tassey, *R&D and Long-Term Competitiveness: Manufacturing's Central Role in a Knowledge-Based Economy*, National Institute of Standards and Technology, Department of Commerce, February 2002, http://www.nist.gov/director/prog-ofc/report02-2.pdf (accessed December 15, 2007), 22.

7. In general, the rate of return is the amount of output or gain relative to the amount of expenditure, over time. Both private rates of return and social rates of return can be calculated. The private rate of return refers to the return on an investment an individual or company or other institution gets. The return is only to the entity making the investment. Social rate of return, on the other hand, is the return to society. The investment may be made by a government (public funds) or an individual, company, or other institution. The point is that more than just those who made the investment receive some benefit or gain. Because knowledge generated from science is available to anyone, any investment in scientific research has a social rate of return (unless of course there is some way to keep that knowledge out of the public domain and *only* available to the investors).

8. These two characteristics of a public good are called nonrival consumption and nonexclusive consumption. When a good has both of them it is referred to as a *pure public good*. Not all public goods are both nonrival and nonexclusive. For additional discussion see Harvey S. Rosen, *Public Finance*, 6th ed. (New York: McGraw-Hill Higher Education, 2002), chap. 4 and pp. 79–81.

9. Mansfield, "Academic Research." The seventy-six firms were chosen from a *U.S. News and World Report* top 100 (see Mansfield's paper for details). While Mansfield did not include as part of his study differentiating all the sources from which universities receive funding for research, one can infer that the

tentative social rate of return he calculated is one for government investment in university research. Mansfield was one of the first economists to have estimated the social rate of return from investment in academic research. Interestingly, when innovations were weighted by revenue, only 3 percent of total revenue from the new innovations could not have been achieved without substantial delay, in the absence of academic research.

10. Edwin Mansfield, "Academic Research and Industrial Innovation: A Further Note," *Research Policy* 21, no. 3 (1992): 295–96.

11. Edwin Mansfield, "Academic Research and Industrial Innovation: An Update of Empirical Findings," *Research Policy* 26, no. 7–8 (1998): 773–76.

12. Representative George Brown (D-CA) was one of the biggest advocates of science and research. He chaired the House Science Committee from 1991 through 1994. His untimely death in July 1999 was a great loss to the scientific community.

13. U.S. Congress, Congressional Budget Office, *A Review of Edwin Mansfield's Estimate of the Rate of Return from Academic Research and Its Relevance to the Federal Budget Process,* CBO Staff Memorandum (Washington, DC: U.S. Government Printing Office, April 1993); both the CBO and Mansfield acknowledge that the percentage estimates are point estimates, meaning a single number rather than a range.

14. Ammon J. Salter and Ben R. Martin, "Economic Benefits of Publicly Funded Basic Research: A Critical Review," *Research Policy* 30, no. 3 (2001): 509–32, see especially p. 516.

15. Bibliometrics is a field in which information and content publication patterns are quantified and studied statistically. According to the OECD, *bibliometrics* is a generic term for data about publications.

16. Francis Narin, Kimberly S. Hamilton, and Dominic Olivastro, "The Increasing Linkage between US Technology and Public Science," *Research Policy* 26, no. 3 (1997): 317–30.

17. Adam Jaffe, "Real Effects of Academic Research," *American Economic Review* 79, no. 5 (1989): 957.

18. Zhen Deng, Baruch Lev, and Francis Narin, "Science & Technology as Predictors of Stock Performance," *Financial Analysts Journal* 55, no. 3 (1999): 20–32.

19. *Citation intensity* is a measure of how important a patent is. A surrogate of citation intensity is how many times the patent is referenced in other patents. The second index, *science link,* accounts for the relationship between the patent and scientific papers. In particular, Narin and colleagues count the number of times a given patent is linked to an underlying scientific paper. The assumption is that the greater the number of links a patent has to original research published in a reviewed journal, the stronger the patent is. The final index, *technology cycle time,* is a measure of the average age of the patents referenced in the company's recent patents. The assumption is that short-cycle patents suggest that the company is involved in rapidly changing, cutting-edge innovation, a good portent of strong future company performance.

20. Another example of the complexity of tracing the origin of inventions can be seen with the World Wide Web. A convergence of technologies led to its development, spanning a period of some forty years. See National Academies Computer Science and Telecommunications Board, *Innovation in Information Technology* (Washington, DC: National Academy Press, 2003), 6; and National Academies Computer Science and Telecom-

munications Board, *Evolving the High Performance Computing and Communications Initiative to Support the Nation's Information Infrastructure* (Washington, DC: National Academy Press, 1995), 20.

21. Domestic Policy Council, Office of Science and Technology Policy, "American Competitiveness Initiative: Leading the World in Innovation," February 2006, http://www.whitehouse.gov/stateoftheunion/2006/aci/aci06-booklet.pdf (accessed May 26, 2007), 8.

22. Francis Narin, "Patents and Publicly Funded Research," in National Academies Board on Chemical Sciences and Technology, *Assessing the Value of Research in the Chemical Sciences* (Washington, DC: National Academies Press 1998), chap. 6.

23. National Academies and National Academies of Science, *Beyond Discovery: The Path from Research to Human Benefit,* a series of articles tracing the origins of important technological and medical advances, http://www.beyonddiscovery.org/ (accessed May 5, 2007); National Science Foundation, *Nifty50: Where Discoveries Begin,* interactive Web site showcasing fifty NSF-funded inventions, innovations, and discoveries that have become commonplace, http://www.nsf.gov/od/lpa/nsf50/nsfoutreach/htm/home.htm (accessed February 25, 2007); Raymond S. Isenson, "Project HINDSIGHT," Accession number AD0495905 (Washington, DC: Office of the Director of Defense Research and Engineering, 1969).

24. Solow, Nobel Prize lecture.

25. Jeffrey Mervis, "NSF Begins a Push to Measure Societal Impacts of Research," *Science* 312, no. 5772 (April 21, 2006): 347.

26. See U.S. Senate Committee on Governmental Affairs, *Government Performance and Results Act of 1993 Report* (June 16, 1993), section IV, http://www.whitehouse.gov/omb/mgmt-gpra/gprptm.html (accessed May 5, 2007).

27. For background information about the inception of GPRA see John Mercer, "GPRA and Performance Management," http://www.john-mercer.com/gpra.htm (accessed May 5, 2007). Mercer is known as the "Father of GPRA."

28. The Government Performance and Results Act of 1993, P.L. 103-62 (January 5, 1993), section 2a, http://www.whitehouse.gov/omb/mgmt-gpra/gplaw2m.html (accessed May 5, 2007).

29. For additional information see Government Performance and Results Act of 1993.

30. Charles A. Bowsher, comptroller general of the United States, before the Senate Committee on Governmental Affairs and the House Committee on Government Reform and Oversight, 104th Congress, 2nd sess., March 6, 1996.

31. Ronald N. Kostoff, "Peer Review: The Appropriate GPRA Metric for Research," *Science* 277, no. 5326 (August 1, 1997): 651.

32. National Academies Committee on Science, Engineering, and Public Policy, *Evaluating Federal Research Programs,* 4–8.

33. Ibid., 8–9; see the executive summary for the full list of conclusions and recommendations.

34. Committee on Science, Engineering, and Public Policy, *Implementing the Government Performance and Results Act;* see the executive summary for all ten conclusions and four recommendations.

35. Office of Management and Budget, *The President's Management Agenda Fiscal Year 2002* (Washington, DC: U.S. Government Printing Office, August 2001), 1, "The President's Message."

36. In this part of the chapter, we will talk about R&D budgets. Data are taken from the National Science Foundation and from the American Association for the Advancement of Science, both of which group research and development together. For additional background see National Science Board, *Science and Engineering Indicators 2006* and AAAS, R&D Budget & Policy Program, http://www.aaas.org/spp/rd/ (accessed March 29, 2008).

37. National Science Board, *Science and Engineering Indicators 2006*, 4.11–4.12.

38. While industry *funded* 64 percent of total R&D in FY2004, it *performed* approximately 70 percent of all R&D. Some of the R&D performed by industry was funded by the federal government and other funding sources. See National Science Board, *Science and Engineering Indicators 2006*, 4.8–4.10. See also "U.S. R&D Funding by Performer, 1953–2004," http://www.aaas.org/spp/rd/usp04.pdf and U.S. and "R&D Funding by Source, 1953–2004," http://www.aaas.org/spp/rd/usr04.pdf (both URLs accessed May 5, 2007).

39. See President's Council of Advisors on Science and Technology, *Assessing the US R&D Investment* (Washington, DC: U.S. Printing Office, October 16, 2002) for discussion of this trend and recommendations.

40. Derek Bok, *Universities in the Marketplace* (Princeton, NJ: Princeton University Press, 2003).

41. Genevieve J. Knezo, *Federal Research and Development: Budgeting and Priority-Setting Issues, 108th Congress*, IB10088 (Washington, DC: Congressional Research Service, Library of Congress, July 16, 2003), 1.

42. "Federal R&D Climbs to Record High of $103.7 Billion; DOD, NIH, and Counter-terrorism R&D Make Big Gains," *AAAS R&D Funding Update*, December 28, 2001, http://www.aaas.org/spp/rd/caprev02.pdf (accessed May 5, 2007); "FY 2003 Omnibus Bill Completes the NIH Doubling Plan; Large Increases for Bioterrorism R&D and Facilities," *AAAS R&D Funding Update*, February 25, 2003, http://www.aaas.org/spp/rd/nih03f.pdf (accessed May 5, 2007).

43. See AAAS, R&D Budget and Policy Program, "Federal Spending on Defense and Nondefense R&D," http://www.aaas.org/spp/rd/histde08.pdf (accessed May 5, 2007).

44. "Congress Finalizes Record $76.8 Billion DOD Budget, Boosts Basic Research," *AAAS R&D Funding Update*, September 26, 2006, http://www.aaas.org/spp/rd/dod07c.pdf (accessed May 5, 2007).

45. Task Force on the Future of American Innovation, *Measuring the Moment: Innovation, National Security, and Economic Competitiveness*, November 2006, http://futureofinnovation.org/PDF/BII-FINAL-HighRes-11-14-06_nocover.pdf (accessed December 15, 2007).

46. Kevin P. Phillips, "U.S. Industrial Policy: Inevitable and Ineffective," *Harvard Business Review*, July–August 1992, 107.

47. Ibid., 108.

48. The Department of Defense's ManTech (Manufacturing Technologies) Program has been in existence since the 1950s. For additional background on the program, see Director of Defense and Engineering, *Department of Defense Manufacturing Technology Program: Manufacturing Technology for the Warfighter* (Washington, DC: U.S. Department of Defense, 2006), https://www.dodmantech.com/pubs/2006_Warfighter_Brochure.pdf (accessed May 5, 2007).

49. Department of Defense Joint Defense Manufacturing Technology Panel, *Strategy 2002: Improving the Affordability of America's Weapon Systems* (Washington, DC: U.S. Department of Defense, 2002), https://www.dodmantech.com/pubs/2002JDMTPStrategyBrochure.pdf (accessed February 19, 2007).

50. Newt Gingrich and Bart Gordon, "Invest More Now," *Washington Times*, January 17, 2007.

51. Kei Koizumi, "Federal Research and Development Funding in the 2008 Budget," presentation, Annual Meeting of the American Association for the Advancement of Science, February 15, 2007, http://www.aaas.org/spp/rd/praaas207.pdf (accessed May 5, 2007), slides 18 and 19.

52. See AAAS, R&D Budget and Policy Program, "U.S. R&D as Percent of Gross Domestic Product," http://www.aaas.org/spp/rd/usg04.pdf, and "Historical Table U.S. R&D as Percent of Gross Domestic Product, 1953–2004," http://www.aaas.org/spp/rd/usg04tb.pdf (both URLs accessed May 5, 2007).

53. "Research Funding Falls in 2008 Budget Despite ACI Gains; Development Hits New Highs," *AAAS Preliminary Analysis of R&D in the FY 2008 Budget*, February 7, 2007, http://www.aaas.org/spp/rd/prel08p.htm (accessed May 5, 2007), fig. 4.

54. Ibid., section titled "The FY2008 R&D Budget in Historical Context: Another Year of Decline."

55. The President's Council of Advisors on Science and Technology *Assessing the U.S. R&D Investment: Findings and Proposed Actions* (Washington, DC: Government Printing Office, October 2002), 16–17; Task Force on the Future of American Innovation, *Measuring the Moment*, 8–9.

56. Harold Varmus, "Squeeze on Science," *Washington Post*, October 4, 2000, A33.

57. President's Council of Advisors on Science and Technology, *Assessing U.S. R&D Investment*, 9.

58. Ibid., 3–4.

59. Data and analyses are available at AAAS, R&D Budget and Policy Program, http://www.aaas.org/spp/rd/ (accessed May 5, 2007).

60. Knezo, *Federal Research and Development*, 1.

61. Jonathan Fishburn, "National Institutes of Health in the FY 2005 Budget," in *AAAS Intersociety Working Group Report XXIX: R&D FY2005* (Washington, DC: American Association for the Advancement of Science, 2004), chap. 8, p. 90.

62. "Congress Caps Another Disappointing Year for R&D Funding in 2006," *AAAS R&D Funding Update*, January 4, 2006, http://www.aaas.org/spp/rd/upd1205.htm (accessed May 5, 2007).

63. Discussion of the role that disease groups have played in increasing funding for the NIH can be found in David Malakoff, "Can ASTRA Restore a Glow to the Physical Sciences?" *Science* 292, no. 5518 (May 4, 2001): 832.

64. "NIH Budget Climbs $3.2 Billion or 15.7 Percent," *AAAS R&D Funding Update*, January 4, 2002, http://www.aaas.org/spp/rd/nih02f.pdf (accessed May 5, 2007).

65. For additional background see T. L. Smith and K. B. Mathae, "National Science Foundation in FY 2005 Budget," in *AAAS Intersociety Working Group Report XXIX*, chap. 7.

66. Kei Koizumi, "Federal Funding of R&D and the Federal Budget Process," presentation, OPM Science, Technology and Public Policy Seminar, October 27, 2006, http://www.aaas.org/spp/rd/propm1006.pdf, slides number 25–26 (accessed May 5, 2007).

67. National Science Foundation Authorization Act of 2002, P.L. 107-368 (December 19, 2002).

68. AAAS, R&D Budget and Policy Program, "Trends in Defense R&D, FY 1976–2008," http://www.aaas.org/spp/rd/trdef08p.pdf (accessed May 5, 2007).

69. "Congress Finalizes Record Budget."

70. Kei Koizumi, "R&D in the FY2008 Department of Defense Budget," in *AAAS Intersociety Working Group Report XXXII: R&D FY2008* (Washington, DC: AAAS, 2007), 63–64.

71. "Actual FY2007 Obligations for DOD Amount to $507 Billion," http://www.defenselink.mil/comptroller/defbudget/fy2008/fy2008_greenbook.pdf; FY2007 total DOD basic and applied research as contained in congressional appropriations bills equals $6.88 Billion, http://www.aau.edu/budget/08DODTable.pdf (both URLs accessed May 5, 2007).

72. Koizumi, "R&D in the FY2008 Department of Defense Budget," 68–69.

73. U.S. Department of Energy, Office of Science, "About the OS," http://www.sc.doe.gov/about/index.htm (accessed May 5, 2007).

74. Michael S. Lubell, "The Department of Energy in the FY2005," in *AAAS Intersociety Working Group Report XXIX*, 96.

75. AAAS, R&D Budget and Policy Program, "Trends in Federal R&D, FY1995–FY2008," http://www.aaas.org/spp/rd/cht9508a.pdf (accessed May 5, 2007).

76. AAAS, "NASA Rebounds in 2008 with $1.1 Billion Increase," *AAAS Funding Update* (March 21, 2007), http://www.aaas.org/spp/rd/nasa08p.pdf (accessed May 5, 2007), fig. 3.

77. See AAAS, R&D Budget and Policy Program, "Historical Data on Federal R&D, FY1976–2008, By Agency (Constant FY2007 Dollars)," http://www.aaas.org/spp/rd/hist08p2.pdf (accessed May 5, 2007).

78. Maggie McKee, "Criticism of NASA Science Budget Grows," *NewScientist.com*, May 4, 2006, http://space.newscientist.com/article.ns?id=dn9110&feedId=astrobiology_rss20; Wade Roush, "NASA's 'Bizarre' Cuts," *TechnologyReview.com*, February 8, 2006, http://www.technologyreview.com/BizTech-Manufacturing/wtr_16302,309,p1.html (both URLs accessed May 5, 2007).

79. U.S. Department of Commerce, *Strategic Plan FY2004–FY2009: American Jobs, American Values*, http://www.osec.doc.gov/bmi/budget/DOCSTPLAN.htm (accessed May 5, 2007), 3.

80. See AAAS, "Congress Saves ATP, Boosts NOAA R&D," *AAAS R&D Funding Update* (December 7, 2005), http://www.aaas.org/spp/rd/doc05c.htm (accessed April 6, 2007); "New Congress Wraps Up 2007 Budget with Increases for Key R&D Programs: Overall Research Funding Flat," *AAAS R&D Funding Update* (February 1, 2007), http://www.aaas.org/spp/rd/upd107.htm (accessed May 5, 2007).

81. See AAAS R&D Budget and Policy Program, "Historical Data on Federal R&D, FY1976–2008, By Agency (Constant FY2007 Dollars)," http://www.aaas.org/spp/rd/hist08p2.pdf; "Trends in Federal R&D, FY 1995–2008," http://www.aaas.org/spp/rd/cht9508b.pdf and http://www.aaas.org/spp/rd/cht9508a.pdf; and "Trends in Federal R&D, FY 1990–2005," http://www.aaas.org/spp/rd/cht9005b.pdf (all URLs accessed April 6, 2007).

82. Winter Casey, "Lawmaker Unhappy with Homeland Security Research Agency," *National Journals Technology Daily*, March 8, 2007.

Federal Partners in the Conduct of Science

CHAPTER 6

Universities

What Are the Roots of the University?

From antiquity, the most successful organized cultures have generally provided an environment in which individuals could congregate, advance and debate new ideas, and pass knowledge from one generation to the next. One of the best-known sites of such activity was the Academy created by Plato in Athens, in 387 BC.[1] Such early universities focused on the arts, humanities, and philosophy. There being no telescopes, microscopes, and certainly no Earth-based particle accelerators, scholars could only imagine the inner workings of nature, struggling to relate abstract concepts to large-scale observations.[2] Indeed, for centuries after the creation of the Academy, scientific thinkers were largely occupied with such abstract debates. The concepts derived during that period continue to exert a mighty impact on our understanding of nature, even to this day, and on the ethics, values, and logic that we now embrace.

The invention of measurement tools, such as Galileo's telescope in 1609, opened new pathways for research on the natural world. New, more specific theories could be tested by measurement. If the tests proved successful, the theories would be tested at an even higher degree of precision when improved tools became available. In this way human societies refined their understanding of the principles that governed the world around them.

Plato's Academy laid the groundwork for modern higher education.[3] It placed a high premium on training talented individuals for public life, and helped define what it meant to be educated. It stressed the value of proof and hypothesis, and the importance of mathematics as a language for conveying complex ideas. It placed a high premium on discourse among scholars, and between scholars and their students.

The Academy was closed in 529 AD by authorities who regarded it as a center of paganism.[4] Its nine hundred years of service make it one of the longest-standing institutions of higher education in human history. But we should also bear in mind that it disappeared because it promoted the discussion of ideas that were not in keeping with the traditions of the day.

No research institution of similar stature emerged until the tenth and eleventh centuries, when centers arose in the Arab and European worlds, respectively.[5] Not only did these new universities embody many features of the Academy, but their curricula were built in large part upon the teachings of Plato and the other philosophers of his period. Among the earliest medieval universities was Al-Azhar University, founded in Egypt in 988, and the University of Bologna, established in Italy in 1088.[6] In 1209, English scholars wanting to get away from the hostile residents of Oxford began to congregate in a former Roman trading post nearby, where they eventually founded Cambridge University.[7] A young John Harvard entered Emmanuel College at Cambridge in 1627; he would later endow Harvard University, the first university in the United States, and provide a direct connection between the European educational tradition and its younger American cousin.[8]

One can argue that research has always been a part of the university's mission. Plato's Academy was originally designed to train individuals for public service by analyzing the outstanding issues of the day. The Academy's modern descendants foster the natural drive to convey existing knowledge, and to extend the boundaries of that knowledge. The organization, extension, and interpretation of knowledge and human creativity are, and have always been, core university activities.

In the nineteenth century, however, German universities specifically dedicated themselves to advancing scientific research; many U.S. universities embraced this change, and began to restructure themselves accordingly.[9] The new university model came at a time of major scientific advancement, and was fueled by society's growing appreciation of technology. At some universities, the original university mission of training students took a back seat to the new fascination with research. Ernest L. Boyer, former president of the Carnegie Foundation for the Advancement of Teaching, noted that in the course of "just a few decades, priorities in American higher education were significantly realigned . . . [as] the focus . . . moved from the student to the professoriate, from general to specialized education, from loyalty to the campus to loyalty to the profession."[10]

Since the late nineteenth century, America has gradually pushed its college and university system to accept greater responsibility for the nation's economy, and the welfare of its citizens. Indeed, the so-called land grant colleges made possible under the Morrill Act (commonly referred to as the Land Grant Act) of 1862 created a whole class of institutions expressly dedicated to advancing American agriculture and the mechanical arts.[11] The act awarded each state thirty thousand acres of publicly owned land for each of its senators and representatives, with the requirement that proceeds from the sale of these lands were to be invested in an endowment fund to provide support for colleges of agriculture and mechanical arts in each of the states.[12] The names of several state universities reflect their origins under the Morrill Act (e.g., Texas A&M, and Florida A&M, with A&M standing for *agriculture* and *mechanical arts*). But in any case, the young nation recognized the value of institutions that produced expertise and research in the service of societal needs.

The enactment of the Smith-Lever Act in 1914 established an official link between the U.S. Department of Agriculture and these land grant universities, via the Cooperative Extension Service, described in Section 2 of the legislation. The Cooperative Extension Service was created to promote university development of practical agricultural applications and the sharing of this knowledge with the surrounding communities through instruction and demonstration, among other means.[13] The Smith-Lever Act could also be considered a predecessor of present-day university technology-transfer programs.

Prior to World War II, the federal role in the support of university-based basic research was minimal. Indeed,

during the 1930s no coherent federal policy existed to support basic research at universities. What federal funding was provided for research at universities during this period was largely oriented toward promoting applied agricultural sciences research and other applied fields such as meteorology, geology, and conservation.[14] The primary sources of funding for basic science and the creation of knowledge came from philanthropic foundations, such as the Rockefeller and Carnegie Foundations, and large corporations such as DuPont and General Electric.[15] Universities tended to prefer such private contributions to their research efforts as they ensured their independence from governmental intrusion.[16]

As private funds began to dry up in the postdepression era, however, corporate and philanthropic support for basic research university research became increasingly constrained and more directed in its nature. In 1931 total grants provided by American foundations amounted to $52.5 million. Just three years later, this number had dipped to $34 million. While foundation giving rebounded some in subsequent years, by 1940 it still remained more than $10 million below what it had been only nine years earlier.[17] The increasing constraints on these private sources of research support forced universities to seek funding from alternative sources, including government programs designed to mobilize university-based knowledge to aid in economic recovery.[18]

A significant federal role in support of university research, however, did not really emerge until World War II. To provide for the science needed to support the war effort, Washington issued substantial government research grants and contracts to universities. As we have already seen, university scientists played a key role in the war effort. Academics' prominence in World War II powerfully signaled the importance of universities to national security and the role of advanced research in keeping the nation ahead of its potential adversaries.[19] Thus, by the mid–twentieth century, a long-term relationship had evolved in which government provided the stream of funds that sustained research at the nation's top universities. Although most American universities were not originally founded for research purposes, events have now led us to a point at which research has become one of the major functions for many, especially larger ones.

In the sections that follow, we will examine how U.S. colleges and universities are categorized, focusing specifically on what constitutes a research university. We will look at the various types of professionals who make up the research university community and key players in the successful operations of these institutions. We will briefly point out some of the key differences between American and foreign institutions and how and why U.S. research universities have become a model that other countries are now trying to emulate. We will also define basic aspects of research operations, including funding, indirect costs, cost sharing, and intellectual property. The chapter closes with a look at the partnership between academic researchers and the federal government, identifying some of the key issues and concerns for both partners.

The Categorization of U.S. Universities and Colleges

In its relatively short history, the United States has developed more than four thousand degree-granting institutions of higher education.[20] They include universities with a full array of courses and degrees, as well as small community colleges offering an associate degree after two years of study. Clearly, any meaningful analysis of American higher education has to categorize these institutions. The choice of categories has important implications: policies developed for one class of universities may have no relevance to (or a negative impact on) another class. The Carnegie Foundation for the Advancement of Teaching has provided useful guidelines for categorizing colleges and universities in the United States.[21] Since this book focuses on research in the United States, we take particular note of the characterization of those institutions referred to as research universities.

Up until recently, the Carnegie system identified two categories of research universities, Research-I and Research-II. Research-I (RI) universities were defined as universities that "offer a full range of baccalaureate programs, are committed to graduate education through the doctorate, and give high priority to research." To be classified as a Research-I university, it had to award fifty or more doctoral degrees each year and to receive $40 million or more in federal support annually. Institutions that received between $15.5 million and $40 million in federal support, but were otherwise like Research-I universities, were labeled as Research-II (RII) universities. A separate pair of classifications—Doctorate-I (DI) and Doctorate-II (DII)—defined categories of institutions that were similar to RI and RII universities in all respects, except that federal support was not a criterion.[22]

In 2005, Carnegie made significant modifications to its classification system, essentially doing away with its old Research-I and Research-II, as well as the Doctorate I and Doctorate II university categorizations.[23] In the current Carnegie classification scheme, research universities are accounted for under the broad category of *Doctoral-Granting Universities*. These universities award at least twenty doctoral degrees per year, excluding doctoral-level professional degrees such as the JD, MD, PharmD, and DPT. Also excluded from this category are special-focus institutions and tribal colleges.

Doctoral-Granting Universities are broken down into three additional categories:

RU/VH: Research Universities (very high research activity)

RU/H: Research Universities (high research activity)

DRU: Doctoral/Research Universities

These three subcategories of institutions are determined based upon an explicit measure of research activity that incorporates several improvements over the old classification system. Among other things, research funding is not the only measure; research sources beyond federal funding are taken into account; and both aggregate and per-capita measures of research activity are included in the analysis. Carnegie points out that the new categories should not be confused with, and are very different from, the old categories.[24]

Though this book focuses almost exclusively on research universities, we hasten to note that other universities and colleges make significant, while generally less direct, contributions to the nation's R&D capacity. Many graduate students at research institutions, for example, received their undergraduate degrees from nonresearch universities. Further, a significant fraction of the S&T workforce is composed of people who earned their bachelor of science degree from a nonresearch university.

That said, U.S. research universities awarded more than half of the baccalaureate degrees granted to those who earned their science and engineering PhDs between 1991 and 1995—even though research universities account for only 3 percent of American institutions of higher learning.[25] "To an overwhelming degree," one influential commission noted, research universities "have furnished the cultural, intellectual, economic, and political leadership of the nation."[26]

University Organization: Who Does What

Universities are complex organizations comprised of many people with varying responsibilities. These include top-level administrators such as the president or chancellor, the provost, and vice presidents or vice chancellors.[27] Working for these individuals are several professional and administrative staff. The university is also made up of deans, departmental chairs, faculty and research scientists, not to mention the many university postdoctoral, graduate, and undergraduate students who also play a critical role in university-based research. The following section briefly describes the university organization and outlines the respective roles of the individuals central to the conduct of research performed at universities. All these individuals working together allow for the successful operation of a university or college.

The Administration

Ideas for research almost always are originated by faculty. However, carrying out the research projects requires an administrative framework. All faculty members belong to departments or programs and must see that their projects are consistent with the internal policies and practices of their unit. Each department is overseen by a chair (or program director). Each department is in turn part of a college or school reporting to its own dean, whose office may impose broader policies, practices, and procedures. The deans report to a provost or vice president for academic affairs, who answers to the university president (with some variations when there is an executive vice president). This constitutes the "simple" system under which most universities operate.[28] Other important offices within the university system include the vice president for student affairs, the vice president for finance, the vice president for external relations, and the vice president for development, each of whom might play a role in the campus research structure. The main responsibility, however, for campus research administration rests within the office described next.

Though the president or chancellor and oversight board of any university are ultimately responsible for the school's actions, the day-to-day administration of campus research must be overseen by staff with the expertise to manage the myriad associated tasks. These duties are often carried out by the office of the vice president for research, or the associate or vice provost for research. Whatever the title, this officer and his or her staff support faculty research and ensure that all campus investigators adhere to university and federal research policies.

Research Administration

A large university may have several thousand active research grants at any given time, involving hundreds of millions of dollars of research expenditures, thousands of faculty, research staff, and students, and scores of regulatory and other committees. The successful management of such a complex enterprise requires a vast array of qualifications: in addition to an extensive knowledge of finance, public policy, and the private sector, senior research officers must generally have scholarly or scientific credentials of their own in order to earn the faculty's trust. Of course, most research offices are not one-person operations, but rely on staff with specialized expertise. Indeed, a modern research administration office may have a dozen internal divisions: some handling the details of research grants and contracts; others ensuring compliance with federal and state laws and regulations; and still others involved with technology transfer and intellectual property rights.

Before any research proposal leaves the university, it usually must be read by the departmental chair, who has to determine whether the project is worth pursuing, and what departmental space and other resources could be made available. The proposal then moves on to the dean's office, where it is reviewed to ensure that college or school policies are being observed. It is then forwarded to the office of the vice president for research, where it receives final university approval before being sent to a federal agency or other potential funding source.

What happens at this last stop? A few examples may be helpful. Strictly speaking, every research grant received by a faculty member at a specific university is actually a grant to that university, not the individual. Consequently, every proposal to a federal agency must first be approved by an agent of the university: research administration staff must read every proposal and check that its planned expenditures are appropriate and allowable; that the listed salaries are correct; that the proper indirect cost rates are being charged; and that the necessary approvals have been secured for any human subjects or animal research—to list just a few points. A university's research administrators are typically also responsible for planning new research facilities and infrastructure, managing intellectual property and technology transfers, imposing sanctions in cases of research misconduct, stimulating interdisciplinary initiatives and campus debates on science policy issues, and hosting visits by policy experts and esteemed researchers.

Student Affairs

The vice president for student affairs is concerned with matters of student recruitment, housing, disciplinary issues, cultural exposure and enrichment, and financial aid. Research grants can contribute a significant portion of a campus's available financial aid, granting students both a financial subsidy and the opportunity to learn through direct participation in cutting-edge research. Furthermore, many students select the university where they wish to study based on the research eminence of the university, believing that they will receive the best training from places working at the cutting edge of knowledge. Additionally, many students believe that their ultimate degree will be worth more if it is from an institution that is highly regarded as a research center. For these reasons the vice president for student affairs typically has a keen interest in the research standing of the campus.

Finance

Since research costs—including faculty and staff salaries, heating, lighting, security, facilities, and legal costs—make up a large fraction of the overall university budget, the vice president for finance and the university provost inevitably play a significant role in decisions about campus research. Indeed, these offices are usually involved, either directly or indirectly, in all major campus decisions concerning research.

External Relations

The importance of research to the overall campus mission makes it increasingly necessary for universities to interact with external constituents. Congresspersons, for example, may want details about campus activities when drafting new legislation. The public may have concerns about the impact of certain types of research on the local environment. The office of the vice president for external relations (sometimes called public affairs) or governmental affairs often manages these ever more significant relationships. Universities structure their governmental affairs and public relations offices differently. In some instances both functions are combined under a single vice president; in other cases they are separated into two vice presidential–level offices. In either case, the growing complexity of the federal government–university interaction on research policies is making the work of such offices increasingly critical.

Development

With some universities now raising more than one billion dollars from private sources annually, in support of their instructional and research missions, campus development officers are instrumental in communicating with potential donors about university research needs and opportunities.[29]

Medical Research

Our review would be incomplete without mention of the issues associated with medical schools and teaching/research hospitals. The medical school is a major center for biomedical and life science research activity on many campuses. Medical and health professional schools often train doctors and health care specialists through a *teaching hospital,* which also employs faculty conducting basic and clinical research. Research funds coming to the medical school and teaching hospital often amount to more than half of the total campus research volume, with a significant portion of this from research grants from the NIH.

The management of campus medical research is complicated by several factors, including the scale of medical research programs, their intertwining of research and training, their coupling of medical education with work in other disciplines (e.g., chemistry and biology), and the impact of federal patient care provisions. Indeed, medical research often requires a vice president for medical affairs, reporting directly to the university president or chancellor. Because of the complex web of interests involved, there are ample opportunities for conflict among the chief medical officer, the vice president for finance, the vice president for research, and the vice president for academic affairs. While we will not delve further into these matters, we do want to stress that universities with medical schools and medical care centers face extra challenges. Cuts or changes in the federal Medicare allocation available to a teaching hospital, for example, can precipitously end up affecting the number of medical students and faculty in a program and thus the amount of research that a given school can carry out.

The Faculty

Because of their role in conducting research and training students, faculty drive scientific research, not just at universities, but in industry and the national laboratories. They play multiple roles; they are teachers and researchers, managers and mentors, and collaborators and colleagues.

Balancing these roles can be tricky; especially when one considers that the faculty also has to deal with the tenure process. Often when a young faculty member is hired by a university or college, he or she is hired into what is referred to as a *tenure track* position. This does not guarantee the person tenure, which is long-term job security, but rather it ensures that the individual will have the opportunity to seek tenure after a certain period of time. The tenure process can be arduous and onerous, and is a period in a faculty's career when he or she is "graded." In large research universities, the "grade" is usually based on publication record, number of grant awards, professional organization membership and involvement, and university committee and teaching experiences, among other things.

Most of a university's research funding is generated by faculty members wishing to pursue a line of study that requires new resources: a graduate student research assistant, perhaps; or a $50,000 piece of equipment; or $25,000 worth of computing time. In order to obtain these resources, the investigator submits a proposal to a research agency, such as the NSF or NIH, in hopes that the research he or she would like to perform and accompanying resources that are required will receive agency support. Once the proposal has been submitted, many months may be required for its review and eventual funding or rejection.

In recent years, universities have had to struggle with the issue of faculty in disciplines such as the physical sciences, environmental sciences, or anthropology who conduct much of their research off campus. For example, a high-energy physics researcher from an American university may need to spend time at one of the national labs, or even a non-U.S. facility like CERN (the world's largest particle physics laboratory, located in Geneva, Switzerland). The problem arises because these faculty members' research responsibilities can be best fulfilled off campus, while their primary teaching duties must be carried out on campus. There is no easy solution to this conflict. Where appropriate research funding is available (and it rarely is), some faculty address this issue by spending some semesters at their home institutions in the classroom to fulfill their teaching obligations and other semesters focusing on their research at facilities off-campus. Another approach sometimes used is for the faculty member to teach a double load one semester and be released for full-time research another semester. One might expect in future years that there will be a growing use of video links, where professors at their distant laboratories will occasionally deliver lectures remotely. None of these are optimal options in terms of the impact on university instructional activities, but they represent an attempt to adapt to the evolving character of research as faculty in certain fields need to spend more time off-campus.

Graduate Students

Graduate students perform much of the day-to-day research work at American universities. This is in part because students occupy a special niche in the educational environment: they have an advanced level of mastery of their fields; they are highly motivated; they must pursue original research for their degree; research is typically both their source of employment as well as training; and they are often more open to new ideas and approaches than senior researchers, who may have become attached to established paradigms. In many fields, both the number of publications stemming from a project and the grant size are directly proportional to the number of graduate students working on the project.

Graduate students, however, do not come cheap. Indeed, in a university setting where grants are expected to pay the full cost of student support, the tab can be substantial (e.g., in the ballpark of $40,000 per student), including tuition, stipend support, and health benefits, as well as costs associated with research for facilities and administration, often referred to as *indirect costs*. Some funding agencies decline to support graduate students who are still taking courses. Funding awards also sometimes include country-of-residence stipulations. Given the opportunity to choose, many faculty would prefer to hire a postdoctoral researcher (see the discussion that follows), rather than a graduate student: once graduate student tuition is accounted for, the salary is about the same; and the postdoc has more experience and can typically devote more time to a project than a student who is simultaneously working on a dissertation. In the end, however, the long-term health of any field depends on its flow of graduate students, and any rational science policy should thus ensure that an appropriate flow exists.

Postdoctoral Researchers

Postdoctoral researchers—or *postdocs*—play a key role in university research. Having completed some significant research of their own in earning their PhD, they can offer a great deal of experience, while their relatively junior status keeps their salaries low. Like medical residents, postdoctoral researchers typically seek one- to five-year internships, depending on their discipline, at an institution other than the one where they received their doctorate. They thus experience research in a new environment, on a different research topic or using a technique not applied during their thesis research, while taking on increased responsibility. This is the time when most budding scientists launch their own careers, allowing them to find a faculty or research position at a university, national lab, or private firm.

The debate over postdoctoral compensation has heated up in recent years. While postdocs conduct much of the day-to-day work of research, and are thus responsible for much of the overall productivity in American science and engineering, they are not granted commensurate status and recognition within the scientific community.[30] In fact, postdoc salaries and benefits are well below those for faculty and regular research staff with similar training. Until very recently, those affected by this disparity raised their concerns only individually. However, the National Postdoctoral Association, founded in 2003, has done much to raise awareness of inequalities in pay, benefits, and work environment—issues that become more urgent as some fields (particularly the life sciences) require young scientists to remain in postdoc positions for an ever-greater length of time.[31]

Research Scientists

Many universities use the title *research scientist* to designate a professional who is beyond his or her postdoctoral studies and possesses a level of expertise comparable to that of faculty, but who is supported primarily by external research grants rather than university funds. Unlike faculty, research scientists often do not have major instructional or committee responsibilities, and the university bears no responsibility for supporting them in the long term if their external support disappears. These individuals have an enormous positive impact on a university's research capability, inasmuch as they can conduct advanced research without teaching or administrative duties. However, because research scientists, like most postdocs, rely on project funding for support, they exist on a treadmill: as soon as one project has been funded (for perhaps three to five years), they must almost immediately begin to seek a new one. They may have to prepare multiple proposals simultaneously, just to increase the probability that they will be funded without interruption. Since their jobs are tied wholly to their ability to obtain external research funding, they have much less job security than faculty members.

Undergraduate Students

For much of the past century undergraduate instruction and research were treated as quite separate undertakings by universities. *Research* was seen as looking for new knowledge, while undergraduate instruction was seen as teaching students what was "old" and known. The past few decades have seen this viewpoint change, and undergraduates are being increasingly brought into research.

The NSF Research Experience for Undergraduates (REU) Program is one such effort, which directly involves undergraduates in cutting-edge research. The REU program provides selected students with stipends and travel support so that they can spend all summer working with faculty on current research. The program was begun in response to decreasing student interest in science, and the observation that many bright and promising students had little concept of what scientists did, or why they did it. The REU program is one example of what can happen when the federal government, industry, and the academic community think creatively and practically about a problem, and cooperate on a mutually beneficial solution.[32]

Technical Staff

Almost every experimental study in modern science requires specialized instruments, yet scientific credentials do not make one a superior machinist, electronics or microscope technician, veterinary technician, or glassblower. Cutting-edge research, then, often requires the services of extraordinary technical support staff. A department of physics, for example, may have a specialized machine shop with one professional for every five faculty, and a similar ratio of electronics and computer specialists. Such staff are typically supported by external grants, as well as by university funds.

The Link between Research and Education

A university is a finely tuned instrument, in which research and education are inextricably intermeshed. For example, consider that most university research is carried out by graduate students, who can focus on specific research problems and are personally motivated to pursue the project's specified goals. Graduate students typically also spend several years as graduate instructors, doing the bulk of the teaching and grading in a department's introductory undergraduate courses. In return, the undergraduate students provide tuition that supports the university. There is thus a correlation between the number of undergraduates at a university and the number of graduate students and faculty that university is able to support. Of course, the amount of federal funding a university receives is also a major factor in determining the number of faculty and graduate students it is able to support. It is important to note that many undergraduate students find part-time employment in university research labs during the academic year to supplement their personal funding situation for tuition, books and supplies, living expenses, and so on. This finely tuned engine drives scientific and engineering progress at our nation's academic institutions.

Why Has the U.S. System for Funding Research Become a Model for Other Countries?

Universities in the United States have long been a magnet for the world, in terms of their attractiveness to students who wish to pursue graduate education and as a place of choice for distinguished scientists seeking employment. For faculty, this draw over the decades is due to the personal freedoms of the United States, its spirit of entrepreneurship, the opportunities for higher salaries, increased access to modern facilities, and the ability to associate with other distinguished faculty. Indeed, concerning the matter of personal freedom, we must note that the preeminence of many of the best schools in America today is due to the large migration of talented scientists from Europe as they escaped the horrors of the Nazis in World War II. That is what brought Albert Einstein, Enrico Fermi, and many other leading scientists to America in the 1940s. The legacies of these giants affect U.S. universities today, as their students and their student's students now hold many of the positions of scientific leadership in U.S. universities.

For students wishing to pursue graduate education, the United States has been particularly attractive because of the likelihood that their education would be "free" (via assistantship support that exists in many fields and that comes from multiple sources), and because of the potential for them to remain in, or later return to, the United States after completing their degree. Of course, the possibility of obtaining financial support depends on the student's qualifications, and thus the tendency is for only the very best students from a given country to be able to come to the United States. We note that having the best students from throughout the world come to U.S. universities intellectually enriches these institutions, permits them to make even more research advances, and makes them even more attractive to attend for future generations of students throughout the world.

In the United States a faculty member with a novel idea may be able to "shop" that idea to many agencies for possible funding. For example, a particular nuclear physics topic might be of interest to the Department of Energy, the National Science Foundation, the Department of Commerce, the National Institutes of Health, the Food and Drug Administration, the Department of Homeland Security, and the Environmental Protection Agency. Few countries offer such a plurality of options. Many observers believe that, while this multiplicity of options may represent a potential source of waste and inefficiency, it is in fact one of the great strengths of the U.S. system for funding scientific research.

For many years the basic structure of the U.S. system for providing research support to universities has been almost unique in the world, with its strong reliance on peer review, and the plurality of sources for funding specific research activities. The success of this system has not gone unnoticed by the rest of the world. As a result, a rather clear migration is taking place toward the U.S. system. Japan, for example, used to distribute its university research funding on an egalitarian basis that had at its heart the assumption that no investigator or university was more deserving of support than any other. With the clear defects in the performance of basic research in Japan, this system is now changing, with emphasis being placed on peer review and the merit of proposals.

Many nations have become concerned that so many of their top students are coming to the United States for training and then staying. This has prodded many to begin massive programs to improve their facilities, to hire more internationally trained faculty, and to offer incentives for students to study and stay at home. These factors, along with others associated with U.S. concerns about homeland security, may bring about a marked shift in the number of foreign students wishing to study in the United States, and the number of faculty who prefer the United States over their home countries as a place to spend their careers. Policymakers need to be aware that many of the special features that have made U.S. university graduate education the envy of the world are now undergoing change in a way that threatens future U.S. leadership in areas of science and technology.

Sources of Funding for University Research

Academic institutions get their funding from multiple sources. In 2005, academic institutions in the United States spent over $45 billion on research. Of this amount, about $29 billion was provided by the federal government, $8.3 billion by the academic institutions themselves, $2.9 billion by state and local governments, and $2.3 billion from industry.[33] More than half of the basic research conducted in the United State is carried out by academic institutions.

In spite of the massive amount of governmental R&D funding, the industrial sector's R&D investments outstrip those of the federal government.[34] (See also fig. 5.2, in chap. 5.) This fact is a source of considerable reassurance to some and of concern to others. One side holds that past federal investments are finally having their de-sired impact, encouraging industry to step forward and invest in future R&D. Others claim that industrial R&D investments are primarily focused on development of new products and that the government is failing to invest in the creation of new knowledge and front-end basic research at a rate sufficient to ensure future progress.

Figure 6.1 illustrates the growth in academic R&D expenditures between 1990 and 2003, including two very striking features. One is the significant growth of federal contributions. The other is the remarkable change in the amount of funds invested by academic institutions. It is surprising, given the significant growth in industrial R&D, that so little of this growth is reflected in support of academic research efforts. This trend has led to concerns about industry's focus on short-term results, as opposed to academia's longer-range perspective. Clearly, the issue of whether the nation has achieved a proper balance in its focus on near-term and longer-range results needs further investigation.

POLICY DISCUSSION BOX 6.1

Should Any Part of Student Tuition Pay for University Research?

University tuition has seen major increases over the past decade. At the same time, the fraction of university research and associated expenses being paid for by universities themselves has also been growing. Some argue that these two trends are related.

How could universities conceivably be able to fund their own research? One resource they can tap is their endowment funds, but these are predominantly designated for a specific purpose, such as building a new building or supporting a particular fellowship or program. Another source for public universities is their annual state government appropriations, but in recent years these have been decreasing. What remains as one of the largest sources of funds universities have to use toward research is student tuition.

Meanwhile, universities have to find funds to pay the costs of complying with a growing number of federally mandated research regulations. Another research-related expense that is likely to grow is the cost for constructing new research buildings and remodeling old ones, both necessary as campuses expand their research activities. The only way to avoid these increasing costs would be for universities to tell their faculty to reduce the volume of research they do. But what university would want to do that!

An analysis conducted by the Cornell Higher Education Research Institute suggests that universities' increasing support of research activities does result in higher student/faculty ratios, more frequent substitutions of part-time instructors for professorial-rank faculty, and higher university tuition. The report notes that the effect of these changes on undergraduate education is, however, "surprisingly small." It has been postulated, although not proven, that increased opportunities to interact with scientific researchers may offset the negative effects of university's increased support for research.[35]

Should undergraduate tuition go to support university research, rather than going exclusively to activities more directly connected to classroom training? Are the opportunities to interact with world-class scientists and engineers worth the tuition hikes students are experiencing? Who should pay the price for the university scientific research that ultimately benefits the nation by contributing to the economy and the overall health and welfare of the public? Do students recoup their investments as a result of ending up with a degree that is more valued if it comes from an institution heavily engaged in research? What do you think?

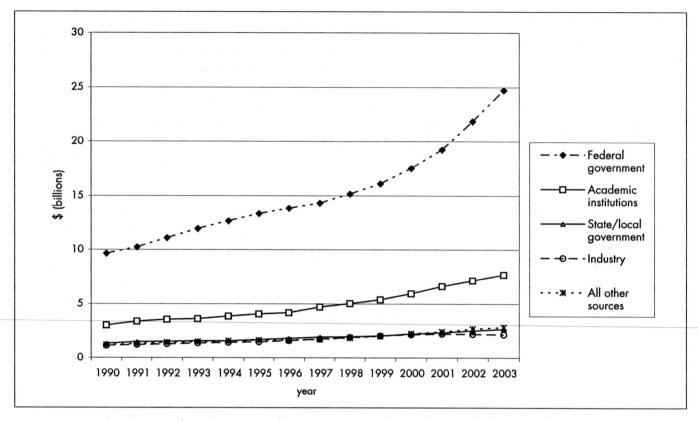

FIG. 6.1 Academic R&D expenditures, by source of funds: 1990–2003
Note: Current dollars; excludes capital expenditures. (Data from National Science Foundation, Division of Science Resources Statistics, *Academic Research and Development Expenditures: Fiscal Year 2003;* and WebCASPAR database, http://webcaspar.nsf.gov; and National Science Foundation, *Science and Engineering Indicators 2006,* fig. O-39, and app. table 5-2.)

Different federal agencies interact with university departments at different levels.[36] For example, departments in the life sciences frequently interact with the NIH, while engineering departments have stronger ties to the Department of Defense and the NSF. Such complex interplay can lead to some departments' being satisfied with federal budget decisions, while others lament these same decisions. Such disparities can, for example, lead deans and provosts to reallocate resources from one department to another, in order to maintain the desired financial equilibrium.

A source of contention between the nation's large research universities and other schools is the distribution of federal research dollars. The top ten research universities are responsible for about 17 percent of total academic R&D expenditures, while the hundreds of universities and colleges not in the top one hundred expend, in total, a comparable amount.[37] To cite one example at the core of the debate, a legislator approached one of the authors some years ago to ask why the University of Michigan should receive an amount of federal support exceeding that received by his state plus four adjoining states. The

author responded that this amount recognized the quality of Michigan's faculty, to which the legislator countered that if his state were granted similar resources, it could attract faculty of a comparable stature.

Indirect Costs

How should universities be reimbursed for the costs associated with carrying out federally sponsored research? It is relatively easy to insist that federal grants pay for researchers' salaries, equipment, and supplies. Such expenditures are often referred to as *direct costs,* the expenditures specifically required to carry out the proposed research. It is not, however, so easy to account for the *indirect costs* associated with the development and maintenance of a major research program, such as the electricity required to operate a laboratory, the janitorial services required to keep the laboratory clean, or the effort required from the university administration to manage payroll or prepare the required progress reports. Such indirect costs are typically computed as some percentage of actual direct costs. Because actual indirect costs vary from one project to the next, the government has found it advisable to calculate an average value. One unintended effect of this abstraction is that faculty often fail to appreciate that indirect costs are real costs, and that someone—the university, the federal government, or a private donor—must cover such costs.

Indirect costs are in essence expenses incurred from services and infrastructure used in the process of carrying out federally funded research. They are also referred to as facilities and administrative (F&A) costs. This terminology has been adopted to better define which expenses are being covered, and to acknowledge that the relevant resources are essential to the conduct of federally funded research.

Each university that receives federal funding negotiates an F&A rate with what is called a *cognizant federal cost-negotiating agency*. For most universities this agency is the Division of Cost Allocation of the DHHS. All others negotiate with the Office of Naval Research (ONR). In any case, the rate is based on the average F&A costs incurred, per the dollar spent on modified total direct costs (MTDC) of organized research. As implied, not all expenses in the total direct costs are used to calculate F&A rate. For example, some subcontract costs, tuition for graduate students, training stipends, plant construction and renovation, and building rental are not included in the MTDC.[38]

The negotiated rate is used by the university for most federally funded programs, although there are exceptions. For instance, the NIH caps F&A recovery at 8 percent on its training grants.[39] In some instances a lower rate may be negotiated because the proposed research will be conducted off campus and therefore may require less use of university facilities and resources.

F&A cost rates have remained relatively stable, at least through the 1990s. Because of a 1991 cap on administrative reimbursement, the administrative component of F&A costs has actually declined in the ensuing years. Facilities reimbursement, however, has increased, leading to the stability of F&A rates overall.[40] As a proportion of total costs awarded, F&A costs have even decreased slightly in recent years. An analysis of NIH award data reveals that the amount of total costs awarded (direct and F&A) increased from 1992 to 2002, while the percentage of this total designated for F&A costs decreased, from 30.1 to 28.4 percent.[41] This decrease happened even as a growing number of more costly federal rules and regulations were being put in place.

The simplest federal research grant one can imagine is one in which a single check is provided, with no strings attached. This dream has almost never been realized. Instead, when a check arrives, it is accompanied by a list of the categories in which funds may be expended. A group of individuals at the university then works with faculty to ensure that each expenditure falls within the guidelines specified in the award. For example, any students hired as project staff must be recruited following precise guidelines whose implementation must be monitored, and progress reports must be filed in a timely fashion. This oversight process often involves staff at the departmental level, the college level, and in the central administration. In order to ensure that the cost of this additional staffing is not borne by students (through increases in undergraduate tuition) or other university sources, the overall aim is to be sure that the full costs of any proposed research are borne by those sponsoring it in the first place.

This is the reason for the "A" portion of F&A costs. The reader may also wonder about the "F" portion. Why is it necessary? The answer is simple: research buildings are highly specialized facilities—wired for high-speed computer networking or high-voltage equipment, designed with special animal-care facilities, or the safe use of hazardous biological materials. They must be heated and cooled and constantly maintained and updated. These facilities are used primarily to carry out federally funded research, and so it is not unreasonable to ask the government to help pay for their construction, upkeep, and operation.

Faculty sometimes have difficulty coming to terms with F&A costs, viewing them as a university strategy for diverting funds from research. Policymakers at federal agencies also want to devote as much money as possible directly to research aimed at achieving the agency's mission. They also have to concern themselves with accountability, determining whether universities are overcharging for F&A costs.

A 2000 RAND Corporation study found that universities contribute a significant level of resources both directly and indirectly to support campus research. In addition to concluding that, as noted earlier, F&A or indirect costs have not increased overall, the report confirmed the following:

1. The federal government only pays 70–90 percent of the total negotiated amount for indirect costs. In other words, universities are not fully recovering total F&A costs incurred by federally funded research conducted on their campuses.
2. After the federal government, the largest supporter of university research is the universities themselves.
3. Typically, 75 percent of a grant award is for direct expenditures, 25 percent for supporting indirect costs.
4. Because of specific limitations on university indirect-cost reimbursement, such as the administrative cap, the actual amount awarded to universities for indirect costs is even lower than cost-structure comparisons would indicate.[42]

POLICY DISCUSSION BOX 6.2

OMB Circular A-21: How Much Should the Government Pay for the Indirect Costs Associated with University Research?

The official rules for indirect costs are contained in OMB circular A-21, first issued in 1958 and since modified multiple times.[43] This circular was adopted during a time when the United States was first establishing the university-government research partnership—in line with other efforts to promote U.S. science and technology—and acknowledged that such costs indirectly associated with, but necessary for, the conduct of research must be covered if universities were to contribute to a healthy U.S. research enterprise. The OMB's formal adoption of F&A to describe such costs, rather than indirect costs, came in 1996.[44]

One notable modification to circular A-21 came in 1991, when a 26 percent cap was established on indirect cost reimbursements for administrative expenses relating to research. Under this change, payments for clerical and secretarial salaries associated with work on federal grants were prohibited from being charged directly to the grant. This change was part of the fallout from several incidents of perceived misuse of indirect costs by universities.

One highly publicized incident at Stanford University occurred when auditing investigators learned that a portion of the operating costs of its university yacht were being claimed as an indirect cost on federal grants.[45] The university argued that if such a facility was being used in part for purposes associated with campus research, then it was proper for pertinent expenses to be part of the indirect costs pool. From the viewpoint of several legislators, however, the university was using federal funds to support a frivolous activity that had little or no relationship to federally funded projects. Public debate over this matter contributed significantly to subsequent attacks on and revisions of the entire indirect costs structure.

Today there is concern that the cap on administrative cost is hurting universities' abilities to perform research because they do not have adequate funding to support increasing administrative costs imposed in part because of an increase in federally mandated regulations associated with research. This raises a question about who should be responsible for paying for such administrative costs associated with the conduct of federally sponsored research: the university or the federal government? Should there be a cap on administrative costs? Isn't the federal government in a partnership with universities gaining some benefit from the research universities conduct? In addition, is it fair that there are no such caps on indirect costs when industry wins federal research contracts, while universities, which certainly have a lower revenue stream than most corporations, have a cap? If these funds do not come from the federal government, what other means does a university have to fund such expenses? What might the downsides be for the federal government of removing the 26 percent cap on administrative expenses?

The role of F&A costs in paying for expanded facilities and frontier research must not be underestimated. Many forward-looking universities construct major new facilities in the expectation that future federal research grants will, through depreciation accounting, defray building costs. Were it not for this expectation, the research infrastructure in the United States would be far less robust than it is today.

Several questions on science policy have arisen in regard to indirect costs. Should the federal government pay full indirect costs for all research it supports at institutions of higher education? When should universities be expected to share the costs of particular research? To what extent should universities build indirect-cost income into their budget projections? Should students be expected to pay any significant portion of the costs of campus research, given that the value of their diplomas is, in a sense, related to the research prowess of their institution? These are some of the issues that the scientific, academic, and policy communities will need to address, as federal budgets continue to come under pressure, and the costs and regulations associated with the conduct of research increase.

Cost Sharing

Cost sharing is just what the name implies—several entities "sharing" the costs of a proposed research project. Cost sharing can occur at the level of the department, the college or school, or the university (usually through the research administration or academic affairs office). Sometimes a federal agency, in issuing a request for proposals, will stipulate that submitted proposals must include cost sharing. What this means is that a line item in the proposed budget must indicate that some expenses will be paid for by an entity other than the federal agency itself. Moreover, cost sharing can sometimes be physical; for example, a department might cost-share by giving lab space to the faculty member applying for the grant—space that is above the "normal" amount given to faculty in that department, that is suited for the purposes of the proposed project, and that will be dedicated to carrying out research for the proposed project. Or the university may contribute funds toward the salaries of essential project staff.

All federal agencies also require that any cost sharing for a funded project be disclosed for auditing purposes. Because cost sharing can take the form of a faculty member's time or graduate student tuition, or a fraction of facility costs, the accounting can be complicated. Research administrative staff are usually required to manage the multiple pools of funds contributing to a single project.

Steps have been recently taken by the National Science Foundation to remove all requirements for cost-sharing on research proposals except for a statutory 1 percent contribution by the institutions. This action was taken, in part, to ensure that funds were awarded primarily upon scientific merit and not based on an institution's ability to provide the required cost share. Some had believed that a large institutional cost sharing on a project might influence reviewer's opinions on a proposal and divert attention from scientific merit as the primary basis of evaluation.[46] It was also believed that larger and better-funded research universities might more easily be able to fulfill cost-sharing requirements, disadvantaging smaller universities in competitions that required cost sharing.

Intellectual Property

Ideas are the currency in the world of research. An idea that drives the creation of a new technology, a new device, a new medicine, a new theatrical play, or a new artistic creation gains considerable value. Since universities are entities that highly prize the development of new ideas, they also need a clear definition of what a new idea is, how to specify its ownership, how to make sure that one person's idea is not stolen by another, and how the true owner of an idea or concept is properly acknowledged. This, by the way, is not to imply that the ownership of ideas—or *intellectual property,* also commonly referred to as *IP*—is significant only within the purview of uni-

versities. Indeed, entire industries are based upon the exploitation of a particular "idea," so the rules governing intellectual property are naturally of great importance to the private sector as well. But we wish specifically to highlight the role of intellectual property as a driver of creativity and to investigate how universities are structured to handle issues of intellectual property management.

In the language of the U.S. Patent and Trademark Office (USPTO), which governs much of this field, *intellectual property* encompasses "creations of the mind; creative works or ideas embodied in a form that can be shared or can enable others to recreate, emulate, or manufacture them." The USPTO identifies four ways to protect intellectual property: patents, trademarks, copyrights, and trade secrets.[47] A piece of intellectual property can be an invention, a trade secret, a piece of writing, a piece of art, a musical composition, a literary work, an architectural design, a computer program, a medical treatment protocol, a drug, or any product of intellectual creativity advanced by one or more individuals. Under this broad definition, the challenge for a university is to identify who was responsible for the generation of a particular idea, who owns it, whether the idea is marketable, how to best market it, who should share in any proceeds from the idea, and how to adjudicate cases where conflicts occur. It should be noted that the person responsible for the intellectual stimulus behind an idea is not necessarily the owner of the resulting intellectual property. Watson and Crick, for example, could not claim the intellectual property rights to the idea of genetically modified corn simply because they discovered the DNA helix used by the crop's inventor.

University policies about the treatment of intellectual property are normally promulgated by the institution's highest governing body. The following, for example, is the University of Texas's intellectual property policy, as stated by its board of regents:

> It is the objective of the Board to provide an intellectual property policy that will encourage the development of inventions and other intellectual creations for the best interest of the public, the creator, and the research sponsor, if any, and that will permit the timely protection and disclosure of such intellectual property whether by development and commercialization after securing available protection for the creation, by publication, or both. The policy is further intended to protect the respective interests of all concerned by ensuring that the benefits of such property accrue to the public, to the inventor, to the System, and to sponsors of specific research in varying degrees of protection, monetary return and recognition, as circumstances justify or require.[48]

University IP policies typically apply to everyone at the university, without exception. Thus, for example, an invention by an undergraduate employee on a sponsored research project is subject to the university's IP rules. Clearly, however, there must be some limits on the university's authority over IP created by its employees and students. For example, if someone in the mechanical engineering department conceives of a brilliant idea to cure a certain disease based on studies in an earlier biology class, should the university's policy constrain that person's private claim to the idea? Certainly not. And, indeed, university policies do take such possibilities into account.

Let us suppose that a university's procedures are sufficient to determine the ownership of a piece of intellectual property. If the owner is an individual inventor, he or she has available all the options provided by the American free-enterprise system. The property can, depending on its scope and quality, be patented, licensed, sold, or used as the basis for launching a new company. This pathway leads to specific policy questions: for example, should the inventor be allowed to establish an outside company while remaining involved in university research related to that same invention?

If the institution is the proper owner of the property, then other complex options come into play. Many universities are motivated by the desire to make the property available to society while also recovering its investments in the underlying project, or using profits from the invention to support future research. These universities have found it helpful to create an Office of Technology Transfer, or Office of Intellectual Property, to manage the process of transferring university-developed technology to the private sector. These offices generally employ licensing, business development, and legal experts, all working together to evaluate and license university-owned technologies. The funds required to maintain these offices and their staffs are, in part, recovered through the income received from start-up ventures and licensing. Interestingly, some institutions have even made technology transfer part of their mission, as the University of Michigan did in 1996.[49]

Some, however, have voiced the opinion that universities have taken too much interest in retaining intellectual property rights and generating licensing revenue. In part to respond to this criticism, a group of university research administrators published a list of nine points to consider in licensing university technology. The aim of this list is to provide guidelines for university technology transfer and to serve as a reminder that the public interest is paramount, with litigation being a last resort.[50]

What does a university do with the intellectual property it owns? If the property has commercial potential, the university will patent it through the USPTO. In such cases, the university may license the technology to an existing company, or create a small company, called a *start-up company,* based on the technology. In either case, the Office of Technology Transfer will coordinate the licensing, allowing the faculty member who originated the technology to participate as fully as possible.

The Bayh-Dole Act's Role in University Technology Transfer

The Patent and Trademark Law Amendments Act of 1980 (P.L. 96-517), more commonly known as the Bayh-Dole Act after its primary sponsors, Senators Birch Bayh (D-IN) and Robert Dole (R-KS), established the legal framework for the transfer of university-generated, federally funded inventions to the marketplace. In a nutshell, Bayh-Dole enables small businesses and nonprofit organizations, including universities, to retain the intellectual property title rights to materials and products they invent using federal funding. Bayh-Dole has been amended over the years to include licensing guidelines and expand the law's reach to include all federally funded contractors.

Prior to Bayh-Dole's passage, the products of federally funded research were owned by the federal government. If an industry wanted to license such a technology, it had to negotiate with the government agency under which the research had been funded. There was no uniform federal policy on IP transfer, and industry tended to avoid licensing federally owned technologies, rather than face the complications of the licensing process required by each federal department or agency. As a result, discoveries with a great deal of commercial potential never made it out of the university laboratory. But under Bayh-Dole, universities own the intellectual property developed by their faculty, regardless of whether they are using federal funds or not. This change provided a major impetus for universities to enhance their own technology-transfer efforts, stimulating innumerable private-public collaborations.

In 1980, when Bayh-Dole became law, the government held the titles to approximately 28,000 patents, less than 5 percent of which had been licensed to industry for commercial development.[51] Fewer than 250 patents were issued to U.S. universities each year, and discoveries were seldom commercialized for the public's benefit.[52] By contrast, in FY2005, U.S. universities received more than 3,300 new patents; signed nearly 5,000 new licensing agreements; and received royalty income from technology licensing of over $1 billion, much of which has been used to further advance university-based scientific research and education. Since 1980, more than 5,100 start-up companies have been formed on the basis of technologies discovered at academic institutions.[53] State and regional governments are becoming increasingly interested in the potential benefits of university technology transfer activities to economic development.

Bayh-Dole requires universities to obtain written agreements from faculty and research staff committing that they will assign IP ownership to the institution. This agreement obligates faculty to disclose their inventions; in return, universities are expected to share royalties with the inventors. The balance of these royalties must be applied directly toward additional research or educational endeavors—a definite benefit of the legislation. A small portion of the royalty stream also goes toward future university technology-transfer activities.

Many view Bayh-Dole as a tremendous step forward. A July 2001 NIH study notes that the Bayh-Dole Act has "yielded a dramatic return to the taxpayer through discovery of technologies that extend life and improve the quality of life and through development of products that, without the successful public-private relationship, might not be available."[54] The legislation has established a balance of rights and responsibilities that binds government, industry, and universities, and its supporters argue that any changes could upset this balance.

Others, however, have begun to question the merits of Bayh-Dole. Some critics are concerned that the Act has created a monetary incentive for universities and their researchers, which could taint objectivity and increase the likelihood of conflicts of interest: as universities and their faculty become more interested in the economic benefits of technology transfer, education, research, and scholarship may be overshadowed by the quest for profit, compromising higher education's dedication to "disinterested inquiry."[55] Moreover, when a taxpayer purchases products from a company that licenses university technology, he or she is, some claim, "paying twice."[56]

What is not acknowledged with this argument is that the technologies resulting from university research are for the most part in very early stages of development. A company licensing these technologies has to invest its own resources, over a number of years, in application and development, including expensive processes like clinical trials for pharmaceuticals. Some critics have argued that Bayh-Dole helps pharmaceutical companies use publicly funded research to develop drugs that are then significantly overpriced. Indeed, it has been suggested that the government should use Bayh-Dole "march-in" rights to take back the relevant intellectual property and offer it at a more reasonable price.[57]

Another school of criticism focuses on the practice of patenting research tools and then issuing exclusive licensing rights. Many research tools may eventually lead to therapeutic or diagnostic products, marketable to consumers. The fear is that allowing exclusive licensing unfairly limits who can and cannot use these research tools. This issue is of particular concern in genomics and proteomics, and is being heard with increasing frequency as progress is made in deciphering and understanding the data from the Human Genome Project. Critics have referred to the situation as the "tragedy of the anticommons," a reference to ecologist and microbiologist Garrett Hardin's 1968 description of the "tragedy of the commons," in which resources shared by many or all fall prey to overuse.[58] In their new spin on Hardin, the contemporary critics of Bayh-Dole have warned that the act may actually *impede* basic research in the biomedical and life sciences, because it increases the costs of gaining legal access to patent-protected materials and tools.[59] This concern about overpatenting, incidentally, extends beyond the university to the practice in general.

Finally, some industrial leaders have criticized Bayh-Dole for encouraging American universities to become more protective of the intellectual property they create, in order to increase its value. One high-level official from a major U.S. computing company told a September 2002 Senate committee hearing that "large U.S. based corporations have become so disheartened and disgusted with the situation they are now working with foreign universities, especially the elite institutions in France, Russia and China, which are more than willing to offer extremely favorable intellectual property terms."[60]

Such growing concerns have prompted some members of Congress, including Senator Ron Wyden (D-OR), to question the effectiveness of Bayh-Dole and call for reexamination of current federal technology transfer policies. Others, however, seem to think that Bayh-Dole has been successful and has done far more good than bad. A 2003 report from a PCAST subcommittee reiterated this position, stating that "existing technology transfer legislation works and should not be altered."[61]

Regardless of one's position on Bayh-Dole, however, informed observers agree on the need to monitor the interaction between research ethics and IP policy, at both the individual and institutional levels. What is the likelihood, we might ask, that a faculty member involved in both a large federal research grant and a small start-up company will use funds from the new grant to do work more appropriate for his company than for his federal grant project? What happens when similar conflicts arise for institutions? If the overlap of private and public interests is significant, the researcher may have no incentive to separate them. How should such conflicts of interest be managed?

Examples of Technologies and Products Originating from Federally Funded University Discoveries

Artificial lung surfactant for use with newborn infants, University of California

Cisplatin and carboplatin cancer therapeutics, Michigan State University

Citracal® calcium supplement, University of Texas Southwestern Medical Center

Haemophilus B conjugate vaccine, University of Rochester

Metal alkoxide process for taxol production, Florida State University

Neupogen® used in conjunction with chemotherapy, Memorial Sloan Kettering Cancer Institute

Process for inserting DNA into eukaryotic cells and producing proteinaceous materials, Columbia University

Recombinant DNA technology, central to the biotechnology industry, Stanford University and University of California

TRUSOPT® (dorzolamide) ophthalmic drop used for glaucoma, University of Florida

List taken from "The Bayh-Dole Act: A Guide to the Law and Implementing Regulations," Council on Governmental Regulations, Washington, DC, 1999.

Multidisciplinary Research

Solutions to most real-world technical problems require input from multiple disciplines. For example, the development of laser systems for the treatment of eye diseases cannot be accomplished by physicists or biologists alone. Physicists may know almost nothing about how the eye works, while biologists may have no understanding of how to focus and operate a laser. Collaboration, however, can lead to socially beneficial innovations.

Science policymakers are intimately concerned with the question of how to engender collaboration in a system that is basically designed to keep the disciplines separate. In a typical university, for example, the boundaries among the chemistry, geology, and physics departments are strictly observed. A junior faculty member in physics will be evaluated for promotion primarily on the basis of his or her contributions to the field of physics, with very little weight given to activities in other fields. There are also precious few opportunities to learn about cutting-edge research being done in another department across campus. As a result, interdisciplinary projects have traditionally been rare.

In recent years, however, university administrators have announced several initiatives to encourage collaboration. Some involve the creation of entirely new entities, bridging the disciplines where collaborative efforts might be particularly fruitful—for example, programs in applied physics, which were created to ensure a better coupling between work in physics and engineering. Or bioinformatics, where the goal is to join specialists from biology and computer science.

By creating such initiatives, universities seek to provide a space in which multidisciplinary contributions are recognized and rewarded. But the functioning of such units is often compromised because the faculty who could participate must still maintain their tenure base in an established core department. This dilemma continues to frustrate deans and vice presidents for research at many universities.

Federal agencies are also fully aware that special steps must be taken to foster interdisciplinary research. A single disciplinary program at, for example, the NSF faces the same pressures to look inward as does the corresponding department at a university. The review of a proposal with a strong interdisciplinary component would likely raise the question of why funds from one discipline should be used to help support research in another discipline when there are so many important things to accomplish within each discipline: why should funding from the biology program be used to support bioinformatics research, when the project may be equal parts computer science? High-level administrators must take a leadership role and insist that good interdisciplinary proposals be given consideration equal to that available to more traditionally discipline-bound projects.

Federal programs in support of interdisciplinary research have an uneven history. For example, the failed Research Applied to National Needs (RANN) program created in 1973 suffered from a persistent inability to develop consistent merit review standards.[62] In a more recent effort, the NSF has attempted to accommodate interdisciplinary proposals by featuring a category called *crosscutting programs:* "Crosscutting programs at the NSF include interdisciplinary programs, programs that are supported by multiple Directorates at NSF, and programs jointly supported by NSF and other Federal agencies."[63]

Science policymakers must ensure that "crosscutting" research is encouraged on campuses and appropriately reviewed and funded. Because such research differs significantly from past practice, considerable attention will be required from university administrations, federal funding agencies, and Congress itself. Such efforts, in recent years, have been increasingly supported across the government and are likely to change the way that both universities and federal agencies conduct their business.[64]

A Closer Look at a Unique Partnership

The relationship between the federal government and American universities is truly unique. Significant research funding is provided to universities in exchange for loosely defined, but extraordinarily important, anticipated returns. This partnership has served the nation well. Federal investment in university research fuels the U.S. science and engineering base and serves as the central component of our nation's advanced education and innovation system. It has generated new knowledge and discoveries that have improved human health, helped secure our nation, created jobs and economic growth, and improved our quality of life. It has also trained our generations of future leaders and entrepreneurs.

The effectiveness of this partnership is in large part due to the principles upon which it is based:

- Projects are selected based on merit in the competitive marketplace of ideas.
- Federal investment in university research does double duty: it pays for research, and helps educate the next generation of scientists, engineers, and scholars.
- The government shares in the costs associated with the conduct of research: both direct costs, including salaries, equipment, and supplies, and indirect or F&A costs, incurred in performing the research.

When we mention the *principles* of this partnership, then, it is important to keep in mind that we are referring to the underlying reasons for the system's past success. When necessary, these principles must be adjusted. A true partnership must be built upon trust, a shared understanding of expectations and roles, responsibility, and accountability. The very word *partnership* implies that both parties must be willing to work together to maximize the mutual benefits. There are, however, growing tensions in this partnership. This has led some to question whether a true partnership even exists, or whether it might more accurately be characterized as a business relationship between the federal government and universities. No matter how one evaluates the health of the current university–federal government relationship, one can easily conclude that it is time both to reassess and to renegotiate the partnership so as to ensure that universities continue to produce the best science in the world on behalf of the U.S. government and the American people for years into the future.[65]

Historical Origins of the Partnership

As described earlier, legislative initiatives such as the Morrill Act and the Smith-Lever Act are premised on an acknowledgment that academic institutions can help meet the immediate needs of society. These bills focused on specific national challenges, helping America feed its citizens by disseminating agricultural knowledge, or encouraging industrial expansion by training mechanical specialists. Each new step strengthened the partnership between universities and government. As corporate and philanthropic contributions declined during the depression, universities sought to diversify their funding sponsors for research and became more open to accepting federal funding. However, no formal mechanism or agreed-upon policy existed at the time through which universities would receive grants or contacts from the federal government.[66]

The Manhattan Project provided a clear demonstration of what could be accomplished when university scientists worked on issues of national security, and after the war, the federal government decided—largely on the basis of Vannevar Bush's recommendations—that the nation's future national security as well as many of its civilian needs might be ensured by formalizing the government-university partnership.[67] The federal research grant process was the mechanism by which the partnership was sealed. Although the precise mechanics of review and awards have undergone considerable refinement, the basic structure remains unchanged.

Key Policy Issues for Both Partners

Ensuring Balanced Federal Investments in Research across Disciplines

A strong federal research enterprise requires support for science across all scientific and engineering disciplines. But until recently, NIH funding, for example, was growing more rapidly than was funding for the physical sciences and engineering from agencies such as the NSF, DOD, DOE, and NASA. Since research developments in one field build on, and may even revolutionize, work in others, care needs to be taken to ensure that balanced support is provided for all scientific disciplines.

Across basic and applied research. While the benefits of applied research are often more immediately apparent, especially in a time of fiscal stress, it is important to remember that mission- and non-mission-based basic research is the foundation on which applied research and development are built. As agencies review their use of research in accomplishing their missions, they must keep in mind the benefits of investment in fundamental research, lest immediate needs crowd out work that may produce fundamental breakthroughs over time. This is not an idle concern: the DOD research portfolio is trending away from basic research, while the funding mix at the new Department of Homeland Security appears heavily tilted toward applied research. Such tendencies, if left unchecked, could ultimately jeopardize our system of innovation at its roots.

Across individual investigator and larger science projects. It is equally important to balance spending on investigator-driven research grants, research training, instrumentation, equipment, and facilities. Investigator-initiated grants for university-based researchers remain the wellspring of scientific advances, but funds are also needed for increasingly expensive instruments and facilities if investigators' efforts are to be fully realized.

Peer Review versus Increasing Earmarking of Funds

In recent years concerns have grown regarding the degree to which federal research funds have been earmarked by Congress for specific research projects without undergoing merit review.[68] Those who favor earmarking argue that the peer-review system is geographically biased toward wealthy or strategically located districts with existing large research universities, and that earmarking levels the playing field by supporting a diverse network of research institutions across the country (recalling Senator Harvey Kilgore's proposal for the geographic distribution of federal science dollars discussed in Chap. 2). Those opposed to earmarking argue that it undermines the peer-review system, wastes money, and represents a different type of bias. The academic community has been divided on the issue: on the one hand, individual universities or groups of schools may benefit from earmarking; on the other hand, peer review and merit-based awards are thought of as the Holy Grail of funding.

In 1983 the Association of American Universities issued a statement calling on its members and Congress to refrain from making "scientific decisions a test of political influence rather than a judgment on the quality of work to be done."[69] Less than a month later, the National Association of State Universities and Land Grant Colleges (NASULGC) issued a similar statement, which was followed by others from higher-education and science associations, including the National Academy of Sciences and the American Council of Education. These resolutions failed to stop the practice, so in 1987 the AAU went further, asking its members to join in on a moratorium on earmarking. This effort also failed.

When it comes to earmarking, the lemming mentality persists: the widespread belief that "everyone else is doing it" means that university administrators who do not seek earmarked funds are often accused of abdicating responsibility. Meanwhile, members of Congress who refuse to earmark are often similarly accused of shirking responsibility, not just to the universities in their district, but to the district as a whole.

Strains in the Partnership

The social contract on which the partnership between the federal government and universities is based has been referred to as "fragile."[70] The strains have become increasingly apparent in recent years. This has fueled a movement to reevaluate the partnership and identify and address key conflicts between the partners. A broad discussion of the issue took place at a 1996 conference at the University of Michigan, organized by one of the authors of this book, that included members of the House Science Committee, several former presidential science advisors, and leaders from industry and academia.[71] One outcome was the eventual decision by the House Science Committee to hold a series of hearings and produce a report, *Unlocking Our Future,* under the leadership of Representative Vernon Ehlers, who was also cochair of the Michigan conference.[72]

Input from the involved parties also led to action by the executive branch. The Clinton administration initiated a Presidential Review Directive, PRD 4, to examine the underlying principles of the federal government–university partnership. The end result was a 1999 report from the National Science and Technology Council, entitled *Renewing the Federal Government–University Research Partnership for the 21st Century,* and also Executive Order 13185, "Strengthening the Federal Government–University Partnership," which was signed on December 28, 2000.[73]

Some Top Issues Facing Research Universities

1. *Public funding of higher education.*[74] Particularly at the state level, reduced funding has enormous implications with respect to tuition, access, and quality. At the federal level, financial aid programs—both grants and loans—are not keeping up with costs, and direct appropriations for some key programs are being reduced.

2. *Federal funding of university research.* Large budget deficits, along with the results of other decisions made by the administration and Congress, are causing slowdowns and in some instances reductions in funding for federal programs supporting academic research, particularly basic research. Moreover, there is a disturbing trend away from fundamental research and toward more advanced applications, especially in defense. Finally, there is a continuing concern about ensuring sustained and balanced funding growth between biomedical sciences, the physical sciences and engineering, and other important scientific disciplines in social and behavioral sciences.

3. *Export controls and informal classification of information.* The imposition of restrictions on access to and publication of research on the basis of regulations such as those pertaining to export controls is undermining national security decision directive 189 (NSDD 189). NSDD 189, which was originally adopted by President Ronald Reagan in 1985, maintained that universities are exempt from restrictions on the publication and open dissemination of research results unless the research being conducted is classified for reasons of national security.

4. *Legislative threats to university autonomy.* Legislation has been proposed in connection with the reauthorization of the Higher Education Act concerning tuition, accreditation, transfer of credits, and other academic and operational issues that would greatly increase federal intervention in higher education and impose costly new regulatory and reporting burdens on universities.

5. *Impact of post–September 11 visa policies.* While taking needed steps to strengthen the security of the nation's visa system, immediately following September 11 the federal government imposed restrictive and cumbersome regulations and processes that hampered access for international students, scholars, and scientists. These visa policies sent a message that the United States was no longer a welcoming nation and came at a time of increasing competition from Europe, East and Southeast Asia, and Australia for international students, scholars, and scientists in research and graduate education. While many of these problems have now been resolved, the U.S. universities are still battling the negative perceptions that were created as a result of these policies.

6. *Intervention of nonscientific factors in awarding research grants.* The peer-review process, responsible in many ways for the extraordinary success of the U.S. research enterprise, is being threatened by congressional moves to cancel controversial grants, as well as rampant earmarking.

7. *Weak linguistic and cultural competence.* Especially in the post–September 11 environment, linguistic and cultural competencies in both the public and private sectors need to be strengthened, perhaps in the context of a new National Defense and Homeland Security Education Act that would help to create attractive career paths and courses of study in areas critical to national and homeland security, including science, math, and engineering, as well as the humanities.

8. *Threats to Bayh-Dole.* Growing concerns regarding university technology transfer and a general lack of awareness and understanding of the Bayh-Dole legislation, which has done much to promote technology transfer leading to the development of useful technologies and products, could lead to the passage of legislation that could have a serious impact on the nation's innovation capabilities.

9. *Recruitment of U.S. students into science, math, and engineering.* The United States needs to explore ways—both governmental and nongovernmental—to encourage more students to consider careers in science, math, and engineering. This could have a significant impact on long-term American competitiveness.

10. *Fair reimbursement for federally funded research.* Inadequate reimbursement for the F&A costs of performing federally funded research, including escalating compliance costs, is imposing a growing burden on universities.

11. *Fair use and open access.* There needs to be a balanced approach to the range of intellectual property issues that affect scholarly research. The ability of students and scholars to make appropriate use of copyrighted material must be protected. In addition, an appropriate balance between open access and proprietary rights in publishing is necessary to ensure wide access to research without harming the integrity of scientific publishing.

12. *Individual and institutional conflict of interest.* The conflict-of-interest problems that have afflicted the NIH—in December 2003 it was reported that scientists whose research was supported by drug companies were also paid to be spokespersons for the companies' drugs—are an important reminder that the integrity of research must be paramount at all research institutions.[75]

13. *Human subjects protection.* It is important that institutions take steps to ensure the highest standards of protection for human subjects and that a proper balance be struck between universities' traditional self-regulation and federal regulation.

During this lengthy process, the following guiding principles for the partnership were proposed:

1. Research is an investment in the future.
2. The integration of research and education is vital.
3. Excellence is promoted when investments are guided by merit review.
4. Research must be conducted with integrity.

When these principles were put forward, several university associations, including the Association of American Universities (AAU), the Council on Government Relations (COGR), and the National Association of State Universities and Land Grant Colleges (NASULGC), jointly endorsed them, stating that they were "the right principles for continuing a successful relationship while reshaping it to meet future opportunities and expectations."[76]

A University Perspective

Universities also have concerns about the partnership that are not necessarily shared by the federal government.

The Need for More Consistency across Agencies

It has become increasingly difficult to meet the requirements imposed upon grantees by federal agencies, especially because there is such variance among the agencies involved. Why, university representatives wonder, could there not be uniform federal policies and practices? While many in the university community hoped that the passage of the Federal Financial Assistance Management Act, which requires that federal agencies streamline and simplify the application, administrative, and reporting procedures for government grants, might result in a more uniform and user-friendly grants process, this has not been the case.[77] And while universities account for only one-sixth of the total amount of federal grant funds awarded, they represent one-third of the total number of awards that are made. Thus the administrative impact of nonuniform grant submission systems on universities is significant.[78]

Growing Regulatory Burdens and Unfunded Mandates

Universities comply with myriad requirements imposed through laws and regulations, including rules on the pro-

tection of human and animal research subjects, hazardous waste storage and disposal, and occupational safety and health, among others. Compliance imposes costs on university facilities and administration—costs that have increased in recent years. A July 2000 RAND Corporation study suggested that "streamlining the requirements imposed by law and regulation would enable universities to lower their costs and the federal government to reduce outlays for facilities and administrative costs."[79] Agreeing with this assessment, university representatives assert that the administration of statutory and regulatory requirements could be accomplished more efficiently, at a lower cost to taxpayers.

Sharing in the Costs of Research

Sharing in facilities costs. Historically, most research facilities have been planned and funded by universities. In committing to a major new research facility, a university takes nearly all the risks on itself, planning the building, raising the capital, and constructing the facility. Only then—and only if the institution's faculty successfully compete for research dollars—does the university recover some portion of the costs of this process via the indirect rate. This lack of parity is not healthy for the partnership and should be reviewed and reevaluated.

Removal of limitations on cost recovery. As previously discussed, in 1991 the OMB placed a 26 percent cap on the administrative costs for which universities can be reimbursed.[80] However, the government has not stopped adding regulatory burdens that increase these expenses, making it harder for universities to meet the costs of conducting high-quality research. A 2003 study by the Council on Governmental Relations concluded: "Continued increases in these substantial compliance costs cannot be borne by the universities without impairing the research enterprise. At the same time the universities recognize the need for compliance programs that address valid societal concerns. To balance these competing demands, a new, comprehensive strategy for dealing with compliance costs is necessary."[81]

Tensions between Academic Freedom and Research Sponsors

Members of the academic community should be able to advance ideas and findings openly and without restriction. Indeed, the tenets of academic freedom hold that a faculty member can freely express personal views whenever he or she wishes, without fear of retribution. Thus, restrictions imposed by government and industrial sponsors—and particularly recent requirements, fueled by homeland security concerns, that researchers obtain permission from sponsoring agencies before publishing certain types of findings—are not welcome on many campuses.

A Government Perspective

The tenor of congressional hearings has made it clear that legislators and government officials also have concerns about the state of the university-government partnership. These concerns are focused on accountability in the management of research grants, enforcement of ethical standards in research, the efficiency of scientific research, and the search for ways to measure scientific productivity and the return on federal investments. Policymakers are also worried about whether the infusion of federal research funds is causing universities to neglect their teaching responsibilities, and whether universities and national laboratories are capable of maintaining the level of security necessary for classified and sensitive research. Recently, Congress has expressed concern that terrorists may use American universities as an easy entry route to the United States, and that foreign students might use training obtained at U.S. universities to advance the technological capabilities of countries hostile to the United States.

Preserving and Improving the Effectiveness of the Partnership

Only ongoing dialogue will preserve and improve the university-government partnership. The issues involved are extremely complex, and all relevant parties must be at the table to discuss them. The federal government and the nation's universities and colleges have worked together to bring American society to its current state, and the nation's growing reliance on advanced technology suggests that this partnership should be preserved and continually strengthened.

Policy Challenges and Questions

Universities play a critical role in the nation's scientific research enterprise. They train the researchers, conduct most of the basic research, make significant contribu-

tions to applied research, and provide critical assistance to research activities in industry and national laboratories. Indeed, the university-government partnership has sustained one of the most productive periods of advancement in human history. While growing strains in this partnership need attention, and some even advocate recasting the partnership as a business relationship, doing so may compromise one of the pairing's greatest strengths.

Universities are extremely complex entities that are not well understood by those without experience in administering them. The policy-making process is just as complex, and not easily grasped by university leaders. The resulting gap in comprehension could prove dangerous if the partnership's value is underappreciated and the life-changing joint accomplishments of the past several decades are forgotten.

Numerous specific research policy issues confront universities in the decade ahead. Can there be a continuing acceptance that large regional disparities in the level of federal research support provided to universities are justified on the basis of merit? Can the urge of Congress to continue earmarking, and for certain universities to join in pursuing earmarking for selected purposes, be contained? What new structures can be put into place to ensure that young faculty with "harebrained" ideas have a chance to give those ideas a try? What can universities do to keep the trust of the public and insure that the research they conduct meets the highest ethical standards?

Universities are the backbone of R&D in the United States. Attending to issues that constrain their productivity has to be of fundamental concern to policymakers and university officials alike.

NOTES

1. See University of St. Andrews (Scotland), School of Mathematics and Statistics, History Topics Index, Ancient Greek Mathematics, Mathematicians/Philosophers, "Plato," http://www-gap.dcs.st-and.ac.uk/~history/Mathematicians/Plato.html; University of St, Andrews (Scotland), School of Mathematics and Statistics, History Topics Index, Mathematicians/Philosophers, "The Academy of Plato," http://www-gap.dcs.st-and.ac.uk/~history/Societies/Plato.html (accessed May 3, 2007).

2. See University of St. Andrews, "Plato."

3. Jill A. Singleton-Jackson and Ron Newsom, "Adapt, Adjust, Respond: Perspectives on Current Trends in American Colleges and Universities," *Higher Education Perspectives*, March 31, 2006, 192–205; Olaf Pedersen, *The First Universities: Studium Generale and the Origins of the University Education in Europe*, trans. Richard North (Cambridge: Cambridge University Press, 1997), 9.

4. David Gress, "Multiculturalism in History: Hellenic and Roman Antiquity," *Orbis* 43, no. 4 (1999): 553.

5. Al-Azhar University, English site, "About: History," http://www.alazhar.org/english/about/index.htm (accessed May 11, 2007); Islam for Today, s.v.v. "Al-Azhar," "Al-Azhar University," "Cairo" http://www.islamfortoday.com (accessed May 11, 2007); University of Bologna, English site, "Our History," http://www.eng.unibo.it/PortaleEn/University/Our+History/default.htm (accessed May 11, 2007). See also Pedersen, *The First Universities*, 1; and Herausgegeben von Walter Ruegg, *Geschichte der universitat in Europa* (Munich: C. H. Beck'sche Verlagsbuchhandlung, 1993), 24.

6. Al-Azhar University, "About: History"; Islam for Today, s.v.v. "Al-Azhar," "Al-Azhar University," "Cairo"; University of Bologna, English site, "Our History."

7. University of Cambridge, "A Brief History: Early Records," http://www.cam.ac.uk/cambuniv/pubs/history/records.html (accessed May 3, 2007).

8. Ibid; Harvard University, "The Harvard Guide: History, Lore, and More: The Early History of Harvard University," http://www.news.harvard.edu/guide/intro/index.html (accessed May 3, 2007).

9. The Boyer Commission on Educating Undergraduate in the Research University, *Reinventing Undergraduate Education: A Blueprint for America's Research Universities* (Princeton, NJ: Carnegie Foundation for the Advancement of Teaching, 1998), 6, http://naples.cc.sunysb.edu/Pres/boyer.nsf (accessed May 23, 2007).

10. Ernest L. Boyer, *Scholarship Reconsidered: Priorities of the Professoriate* (Princeton, NJ: Carnegie Foundation for the Advancement of Teaching, 1990), 13.

11. See U.S. State Department, Bureau of International Information Programs, *Basic Readings in U.S. Democracy,* http://usinfo.state.gov/usa/infousa/facts/democrac/27.htm (accessed May 31, 2007).

12. Wayne D. Rasmussen, *Taking the University to the People: Seventy-Five Years of Cooperative Extension* (Ames: Iowa State University Press, 1989), 22–26.

13. Higher Education Resource Hub, History of Higher Education, *Smith-Lever Act,* http://www.higher-ed.org/resources/smith.htm (accessed May 3, 2007).

14. Roger Geiger, who has extensively studied the history of U.S. research universities, estimates that total federal expenditure on research during the late 1930s was in excess of $100 million. However, little of these funds were going to universities, instead going to federal bureaus to support applied, as opposed to basic, research. See Roger L. Geiger, *To Advance Knowledge: The Growth of American Research Universities, 1900–1940* (New York: Oxford University Press, 1986), 255.

15. Hugh Davis Graham and Nancy Diamond, *The Rise of the American Research Universities: Elites and Challenges in the Postwar Era* (Baltimore: Johns Hopkins University Press, 1997), 27–28, 253.

16. Notes Geiger, "the very vigor of privately funded research system . . . effectively precluded government involvement in much basic research. . . . This entrenched pattern made it difficult to establish the kind of national government role in basic research that existed in other advanced countries." Geiger further attributes the lack of government support for university research during this period to a "fundamental antipathy to government involvement with science . . . rooted in the leadership of the scientific community" and the "lack of consensus on a federal interest in the advancement of science, combined with the absence of administrative mechanisms for implementing a government role" (*To Advance Knowledge,* 255).

17. Ibid., 253.

18. For a full discussion of the impact of the depression on universities and university research, and the federal government role in support of university research just prior to World War II, see ibid., chap. 6.

19. Zachary, *Endless Frontier,* prologue. See also Peter J. Westwick, *The National Labs: Science in an American System: 1947–1974* (Cambridge: Harvard University Press, 2003). Examples of other key technologies resulting from the university–federal government partnership were the proximity fuse and antiaircraft radar.

20. National Center for Educational Statistics, *Digest of Educational Statistics,* August 2005, table 243, http://nces.ed.gov/programs/digest/d05/tables/dt05_243.asp (accessed May 28, 2007).

21. The Carnegie Foundation for the Advancement of Teaching, "The Carnegie Classification of Institutions of Higher Education," http://www.carnegiefoundation.org/Classification/CIHE2000/background.htm (accessed May 3, 2007).

22. In addition to offering a full range of baccalaureate programs, the universities defined by Carnegie as Doctoral I institutions were those schools that included a commitment to graduate education through the doctorate. They had to award at least forty doctoral degrees annually in five or more disciplines. To qualify as a Doctoral II university the school had to annually award at least ten doctoral degrees in three or more disciplines—or twenty or more doctoral degrees in one or more disciplines. See National Science Board, *Science and Engineering Indicators 2006,* 2.9.

23. One of the original intentions of the Carnegie classification was to highlight differences and similarities of academic institutions for research purposes. Over the years, the classification came to mean something more, and rather than highlight heterogeneity across U.S. institutions of higher education, it appeared to cause a movement toward homogenization as more institutions strove to fit into the Research University category. For a description of the rationale for these changes, see Alexander C. McCormick and Chun-Mei Zhao, "Rethinking and Reframing the Carnegie Classification," *Change,* September–October 2005, 51–57.

24. A description of the basic Carnegie classifications as they currently exist can be found on the Carnegie Foundation for the Advancement of Teaching Web site, "Basic Classification, http://www.carnegiefoundation.org/classifications/index.asp?key=791 (accessed May 23, 2007).

25. Boyer Commission, *Reinventing Undergraduate Education,* 5.

26. Ibid.

27. In a university setting, the terms *president* and *chancellor* refer to the head of a specific university or university campus. In large university systems, such as those in California and North Carolina, the president leads the system, while chancellors are the heads of each respective university campus. It is not

the case, however, that a university president is always the head of a university system, or that the title of chancellor means one campus is a part of a university system.

28. To those not familiar with university infrastructure, operations, and administration, the "simple" system may actually seem complex.

29. John M. Cash, *Private Fund-Raising for Public Higher Education—the First Fifteen Years: A Progress Report* (Lyndhurst, NJ: Marts and Lundy, 2001), http://www.martsandlundy.com/site/pp.asp?c=ekISLeMYJvE&b=276581 (accessed May 28, 2007).

30. National Academy of Sciences, National Academy of Engineering, and Institute of Medicine, *Enhancing the Postdoctoral Experience for Scientists and Engineers: A Guide for Postdoctoral Scholars, Advisers, Institutions, Funding Organizations, and Disciplinary Societies* (Washington, DC: National Academy Press, 2000), http://www.nap.edu/catalog/9831.html#toc (accessed June 2, 2007).

31. Eliot Marshall and Erica Goldman, "NIH Grantees: Where Have All the Young Ones Gone?" *Science* 298, no. 5591 (October 2002): 40.

32. National Science Board, *Undergraduate Science, Mathematics, and Engineering Education: Role for the National Science Foundation and Recommendations for Action by Other Sectors to Strengthen Collegiate Education and Pursue Excellence in the Next Generation of U.S. Leadership in Science and Technology,* NSB 86-100 (Washington, DC: National Science Foundation, 1986).

33. National Science Foundation Science Resources Statistics, *Academic Research and Development Expenditures: Fiscal Year 2005,* NSF-07-318 (Arlington, VA: National Science Foundation, 2007), table 1, p. 8.

34. National Science Board, *Science and Engineering Indicators 2006,* O.21 and fig. O.44.

35. Ronald G. Ehrenberg, Michael J. Rizzo, and George H. Jakubson, "Who Bears the Growing Cost of Science at Universities?" NBER Working Paper no. W9627 (April 2003).

36. Ibid., 5.15 and figs. 5.8, 5.9.

37. Between 1983 and 2003, the percentage of academic R&D expenditures going to the top ten universities decreased from 20 percent to 17 percent, while the percentage going to institutions not in the top one hundred universities increased from 17 percent to 20 percent. The percentage of academic R&D expenditures going to the top 11–20 universities and to the top 21–100 universities remained about the same. Ibid., 5.16–5.17.

38. See Council on Governmental Relations, "F&A Rate Computation," in *University Facilities and Administrative Costs: What They Are and Why They Are Important* (Washington, DC, October 2002), sec. 5.

39. Council on Governmental Relations, "Brief History of Indirect Costs," in *University Facilities,* sec. 2.

40. Charles A. Goldman, Traci Williams, David Adamson, and Kathy Rosenblatt, *Paying for University Research Facilities and Administration* (Santa Monica: RAND Corporation, 2000), summary; Arthur Bienenstock, "A Fair Deal for Federal Research at Universities," *Issues in Science and Technology,* Fall 2002, 34–35.

41. See National Institutes of Health, Office of Extramural Research, "NIH Research Grants to Domestic Institutions of Higher Education FY1992–2002," table available at NIH Office of Extramural Research, http://grants.nih.gov/grants/award/research/rgfanda9202.htm (accessed May 26, 2007).

42. Goldman et al., *Paying for University Facilities,* 18, 73–75.

43. See Office of Management and Budget, *Circular A-21: Cost Principles for Educational Institutions,* http://www.whitehouse.gov/omb/circulars/a021/a021.html (accessed June 2, 2007).

44. For a timeline of modifications to Circular A-21 up to 2000 see Goldman et al., *Paying for University Facilities,* appendix A.

45. Jeffrey Mervis, "Kennedy Resigns as Indirect Costs Controversy Mounts," *The Scientist* 5, no. 16 (August 19, 1991): 3; Stanford University News Service, "Stanford President Kennedy to Step Down Next Year," news release, July 29, 1991, http://news-service.stanford.edu/pr/91/910729Arc1241.html (accessed May 6, 2007).

46. See National Science Foundation, "Important Notice to Presidents of Universities and Colleges and Heads of Other National Science Foundation Grantee Organizations," Notice 128, http://www.nsf.gov/bfa/dias/policy/docs/in128.pdf (accessed May 6, 2007); Kelly Fields, "NSF Eliminates Requirement That Universities Share Cost of Research It Sponsors," *Chronicle of Higher Education,* October 15, 2004.

47. U.S. Patent and Trademark Office, Glossary, http://www.uspto.gov/main/glossary/index.html#i (accessed May 6, 2007).

48. "The University of Texas System History of Board of Regents' Rules and Regulations and Other Policies on Intellectual Property Rules, 1985–Present," no. 7, part 2, chap. 5, sec. 2.4, was deleted & renumbered part 2, chap. 12, http://www.utsystem.edu/ogc/IntellectualProperty/contract/IPhist-RR7.htm (accessed May 7, 2007).

49. See University of Michigan, "Board of Regents," Bylaws sec. 3.10, http://www.regents.umich.edu/bylaws/contents-detail.html (accessed June 2, 2007).

50. Elia Powers, "9 (Suggested) Commandments of Research Licensing," *Inside Higher Ed,* March 7, 2007, http://www.insidehighered.com/news/2007/03/07/tech (accessed January 3, 2008).

51. U.S. General Accounting Office, *Technology Transfer: Administration of the Bayh-Dole Act by Research Universities,* GAO/RCED-98-126 (Washington, DC: U.S. General Accounting Office, May 7, 1998), http://www.gao.gov/archive/1998/rc98126.pdf (accessed June 2, 2007).

52. Council on Governmental Relations, *University Technology Transfer: Questions and Answers* (Washington, DC: COGR, 1993).

53. Association of University Technology Managers, *AUTM U.S. Licensing Survey: FY2005,* (2007), 5, 13, 42–46, http://www.autm.org/events/File/US_LS_05Final(1).pdf (accessed June 2, 2007).

54. National Institutes of Health, Office of Technology Transfer, *NIH Response to the Conference Report Request for a Plan to Ensure Taxpayers Interests Are Protected* (Washington, DC: U.S. Department of Health and Human Services, July 2001), http://www.ott.nih.gov/policy/policy_protect_text.html (accessed June 3, 2007).

55. See Eyal Press and Jennifer Washburn, "The Kept University," *Atlantic Monthly,* March 2000, http://www.mindfully.org/GE/The-Kept-UniversityMar00.htm (accessed May 7, 2007).

56. For additional discussion of this matter, see Bok, *Universities in the Marketplace;* Donald G. Stein, ed., *Buying In or Selling Out?* (New Brunswick, NJ: Rutgers University Press, 2004); and David L. Kirp, *Shakespeare, Einstein, and the Bottom Line: The Marketing of Higher Education* (Cambridge: Harvard University Press, 2003).

57. See "Statement of James Love—President, Essential Inventions, Inc. NIH Meeting on Norvir/Ritonavir March-In Request, May 25, 2004," http://www.essentialinventions.org/legal/norvir/may25nihjamie.pdf (accessed May 7, 2007).

58. Garrett Hardin, "The Tragedy of the Commons," *Science* 162, no. 3859 (December 13, 1968): 1243–48.

59. M. A. Heller and R. S. Eisenberg, "Can Patents Deter Innovation? The Anticommons in Biomedical Research," *Science* 280, no. 5364 (1998): 698–701.

60. See Stanley Williams, before the Senate Committee on Commerce, Science, and Transportation, Subcommittee on Science, Technology and Space, *Hearing on Nanotechnology,* 107th Congress, 2nd sess., September 17, 2002, http://commerce.senate.gov/hearings/091702williams.pdf (accessed May 7, 2007), 5.

61. See *Technology Transfer of Federally Funded R&D,* PCAST Subcommittee on Federal Investment in Science and Technology, February 26, 2003 (Draft Summary Report), http://www.aau.edu/research/PCAST.pdf (accessed May 7, 2007).

62. National Science Foundation, *National Science Foundation: A Brief History,* chap. 4.

63. National Science Foundation, Funding, "Crosscutting and NSF-wide Active Funding Opportunities," http://www.nsf.gov/home/crssprgm/ (accessed May 7, 2007).

64. Tobin L. Smith, "Institutional Impacts of Government Science Initiatives," in *Nanotechnology: Societal Implications II—Individual Perspectives,* ed. Mihail Roco and William S. Bainbridge (New York: Springer Science and Business Media, 2006), 223–32.

65. For a further discussion of the changing nature of the interface between the federal government and research-intensive universities, see David Guston and Kenneth Keniston, eds., *The Fragile Contract: University Science and the Federal Government* (Cambridge: MIT Press, 1994).

66. Geiger, *To Advance Knowledge,* chap. 6.

67. Bush, *Science—the Endless Frontier.*

68. John T. Casteen III, "Earmarking of Science: Definitions, Interpretations, and Implications," introductory remarks, workshop hosted by the American Association for the Advancement of Science, the Association of American Universities, and the National Academy of Sciences, and the National Association of State Universities and Land Grant Colleges, October 3, 2001, http://www.aaas.org/spp/cstc/pne/events/earmarking/earmarking.pdf (accessed May 28, 2007). See also Jeffrey Brainard, "Boston U. Chancellor and Washington Insider Spar over Whether Earmarking Does Good," *Chronicle of Higher Education,* May 18, 2001, 30; Jeffrey Brainard, "Research-Universities Group Considers Stance on Earmarks," *Chronicle of Higher Education,* November 2, 2001, 30; Savage, *Funding Science in America.*

69. Casteen, "Earmarking of Science."

70. Guston and Keniston, *The Fragile Contract.*

71. See University of Michigan, Research Administration, Policies, Jerome B. Wiesner Symposia, http://www.research.umich.edu/vpr/3.29.99.wiesner.html (accessed May 9, 2007).

72. House Committee on Science, *Unlocking Our Future,* 8.

73. For additional background see Office of Science and Technology Policy, National Science and Technology Council, Documents/Reports, Archives 1993–2001, "Renewing the Federal Government–University Research Partnership for the 21st Century, Main Page and Background," http://www.ostp.gov/html/rand/index.htm (accessed May 9, 2007).

74. Some of the issues in this section here were discussed in more detail in two successive editorials by *Science* editor-in-chief Donald Kennedy in August 2004. See Donald Kennedy, "Academic Health I," *Science* 305 (August 20, 2004): 1077; Donald Kennedy, "Academic Health II," *Science* 305 (August 27, 2004): 1213.

75. David Willman, "NIH to Ban Deals with Drug Firms," *Los Angeles Times,* February 1, 2005.

76. Nils Hasselmo, Milton Goldberg, and C. Peter Magrath, transmittal letter to Neal Lane to accompany comments submitted by AAU, COGR, NASULGC, and FDP on National Science and Technology Council Report, *Renewing the Federal Government–University Research Partnership for the 21st Century* (Washington, DC: Office of Science and Technology Policy, April 1999).

77. Federal Financial Assistance Management Improvement Act of 1999, P.L. 106-107, November 20, 1999.

78. Marvin Parnes, before the House Energy and Commerce Committee, Subcommittee on Technology, Information Policy, Intergovernmental Relations and the Census, *Federal Grants Management: A Progress Report on Streamlining and Simplifying the Federal Grants Process,* 108th Congress, 1st sess., April 29, 2003.

79. Goldman et al., *Paying for University Facilities,* chap. 7, p. 3.

80. Office of the Federal Register, *Federal Register* 56, no. 192 (October 1, 1991): 50224.

81. Council on Governmental Relations, *Report of the Working Group on the Cost of Doing Business* (Washington, DC: COGR, June 2, 2003), 2.

CHAPTER 7

Federal Laboratories

The Role of Federal Research Laboratories in U.S. Science Policy

In addition to supporting research at universities, the federal government supports a network of hundreds of intramural research facilities, centers, and laboratories. By one estimate, of a total of sixteen thousand public and private laboratories that compose the U.S. R&D system, seven hundred are owned by the federal government, approximately one hundred being of a size and scope to be "significant contributors . . . to the national innovation system."[1]

These government-owned laboratories and scientific facilities play an important role in the U.S. R&D infrastructure and perform a wide range of functions. In FY2002, federal research laboratories and centers performed over $34 billion in U.S. R&D.[2] The total research performed by these federal research facilities, some government operated and others operated for the government by industrial, academic, and nonprofit organizations, amounted to approximately 30 percent of all federal R&D spending.[3] As many as one hundred thousand scientists and engineers who produce thousands of scientific and technical papers each year are employed at these laboratories.[4]

The diversity of these federal research laboratories in size, scope, and mission makes them difficult to understand and equally challenging to describe in a book such as this. In describing both industrial and federal R&D laboratories, Michael Crow and Barry Bozeman note that "knowledge of R&D laboratories as a whole is virtually impossible, except at the most superficial level. The result is that there hasn't been much in the way of attempts even to map the system, much less to make sense of it."[5] In the following chapter, we provide the reader with an introductory understanding of the federal laboratory system and outline some of the policy challenges confronting these labs.

The Role of the Federal Laboratories: What Makes Them Unique?

To the outsider, the role played by this complex network of government-owned laboratories and research facilities may be hard to distinguish from that played by university or industry laboratories: a visitor to a DOE national lab on any given day, for instance, will typically encounter university scientists and engineers, as well as industry technical staff. Indeed, the ability of outside university and industry researchers to use these government-owned facilities, while at the same time maintaining strong connections to their academic and industrial communities, is one of the greatest unrecognized strengths of the federal lab system.

These government laboratories do, however, play a very specific and unique role in the conduct of science in the United States.[6] Specifically they

1. *Support large-scale research efforts that require significant capital expenditures, unique scientific facilities, and specialized staffing.* These requirements often exceed what private entities and universities can or will provide. Thus these federal laboratories serve as a national resource and provide equipment and facilities used by both university and industry researchers. For example, almost half of all users of DOE's scientific facilities—most of which are at DOE national laboratories—are university researchers.[7]

2. *Conduct classified research of a sensitive nature.* Such research often has clear implications for national and homeland security. The federal laboratories have been developed to support national security and classified research on behalf of the government. Many universities and businesses are not equipped, and do not want, to undertake such sensitive research.

3. *Perform research that is needed to fulfill the missions, and support the regulatory functions, of specific federal agencies.* This is often high-risk, goal-oriented research that may have a significant national payoff. In some instances, such as with laboratories run by the EPA and NIST, they conduct research to help the government set national standards, assess risk, or to implement regulations aimed at protecting the public good.

4. *Manage long-term research programs.* Agency programs may require continuity and a strong connection to the sponsoring agency over several years. Agency laboratories and laboratory personnel provide continuity and institutional memory.

How Federal Laboratories Differ from Universities and University Labs

Universities tend to carry out basic research that leads to the creation of new fundamental knowledge, without concern for immediate impact or without a specific application in mind.[8] The federal laboratories, on the other hand, adapt basic knowledge to specific government needs such as nuclear weapons, national and homeland security, energy research, environmental quality, and public health. In other words, while university research is free to be undirected and truly exploratory, research at federally owned labs is, for the most part, mission based and directed.

University-based research is also generally carried out by small research groups of faculty and students, often using modest facilities that may be contained in a few rooms. Government laboratories, on the other hand, can accommodate much larger projects and facilities, and maintain teams of expert staff—resources that are often used by university faculty and students for projects that would not be feasible within the limitations of the university environment.

Universities usually do not provide the ideal setting for conducting classified government research. Universities normally pose minimal constraints on the sharing of information, and encourage student participation in new studies. The scholarly community also expects that the results of most of its research will be available for publication in open literature. The government thus often turns to its national laboratories to perform classified research.

The duration of a new study is often a factor in the decision to base a particular project at a national lab. University researchers are a fairly mobile group, moving from one institution to another. Students also come and go—at least one hopes so. Even while they are on campus, university researchers may frequently shift their attention to new questions, and must all the while manage their teaching responsibilities and administrative commitments. The university campus, while well adapted to projects that can be completed within a few years, is not usually an ideal setting for the management of long-term studies or drawn-out and labor-intensive experiments that take place over a long period of time. Clearly, the university structure is not best suited to solving a specific national problem whose solution might demand decades of commitment. Instead, the government usually assigns work on such problems to government laboratories.

How Federal Laboratories Differ from Industry

While in many ways universities and industry differ, they are alike in their differences from national laboratories. Industry scientists, like their academic counterparts, are mobile. Although industry projects are assigned, just like those at the national labs—assigned by the company, instead of the government—industrial research projects still do not function on the long time-scale that can be accommodated by national labs. In this respect they resemble universities.

Industry typically takes a well-designed object, produces quantities of that object, sells it, realizes a profit, and prepares for the next business cycle. Because of their need to make a profit, most private companies—and perhaps more importantly, their shareholders—are not willing to take on government-initiated research projects whose payoff is uncertain and is often constrained by government stipulation, or may not yield direct benefits to the company itself. Government projects are by definition focused more on broad societal benefits and national interests than on private benefits that translate into company profits. This is why, for instance, a federal laboratory may be willing to invest in new energy technologies whose uncertain prospects for profitability might be a disincentive for private industry. Finally, the clientele for federal research, especially that being conducted for national or homeland security, is often limited to one consumer—the government—with questionable potential for expansion into a broader commercial market.

Now that we have discussed the unique role these federal laboratories play in U.S. science policy, it is important for the reader to have an understanding of what constitutes a federal laboratory and the many different terms that help to categorize them.

Federal Laboratories and Research Facilities Defined

The terminology that surrounds federal research laboratories can be very confusing, especially to the newcomer to science policy. This is because those involved in science policy or in research do not always consistently define and apply the terminology they use. Moreover, when it comes to federal research laboratories and other research facilities, the terminology may vary from one federal agency to the next, with some agencies referring to these facilities as *laboratories* (e.g., DOE and DOD) and others referring to them as *centers* (e.g., NASA). Terminology may also vary within a particular agency depending on its size, the breath of the laboratory's mission, or who operates the laboratory or facility—the government or an outside contractor.

In the following section, we identify some of the key terminology associated with policy discussions related to federal laboratories and research facilities.

GOGOs versus GOCOs

In some instances, federal research facilities are fully operated by specific government agencies and their managers and directors are considered government staff. These facilities are often referred to as *government owned / government operated* facilities, or GOGOs. Examples of GOGOs include the research laboratories operated by the EPA and the NASA research centers.

Many government laboratories, including most of the Department of Energy national laboratories, however, are actually operated and managed by nongovernmental entities. Such *government owned / contractor operated* facilities are referred to as GOCOs.[9] This arrangement is common and has proven to be quite effective. GOCO contractors may include individual universities, consortia of universities, private companies, nonprofit entities, and, increasingly, some combination of the above. GOCO contracts are referred to as *management and operating (M&O) contracts.*[10]

FFRDCs

A subset of GOCOs are also designated as *Federally Funded Research and Development Centers,* or FFRDCs.[11]

FFRDCs are privately managed and operated R&D institutions that are exclusively or substantially financed by an agency of the federal government.[12] They are administered in one of two ways, either as a unit within a parent organization or as a separately incorporated organization.[13] FFRDCs fall into a special classification of federal laboratories that are not subject to Civil Service regulations. Despite their private nature, however, they are subject to personnel and budgetary controls imposed by Congress and their sponsoring agency.

The first FFRDCs were established in the 1940s and 1950 by the DOD when the Pentagon was having difficulty attracting the scientific and technological talent required to meet its broad research needs. At the time, much of the technical talent was located in companies that had a vested interest in the continued growth of specific DOD R&D projects. Rather than contract with these companies, the DOD created FFRDCs to provide sound and unbiased scientific and technical advice.

The growing need for specialized advice and technical support led to a rapid growth in the size and number of FFRDCs. By 1961, the DOD alone had established forty-one, and by 1969, the total number across all government agencies had grown to seventy-four.[14] Many of them were managed by universities or consortia of universities, while a smaller number was managed by private nonprofit corporations.[15]

The rapid growth in the size and number of FFRDCs during this period, especially those created by the DOD, raised concerns in Congress. It was believed that the Pentagon had become overly reliant upon these external institutions for advice and that the FFRDCs were undercutting the DOD's own in-house scientific and technological capabilities. At the same time, industry criticized the FFRDCs for performing work that it believed it might otherwise have received in government contracts.

As a result, Congress imposed a spending cap on the total funds the DOD could devote to FFRDCs. The spending cap along with continued congressional pressure resulted in a significant reduction in the total number of DOD FFRDCs as the Pentagon sought to increase its in-house technical capabilities. From 1964 to 1970, the DOD cut its total from more than thirty to less than fifteen.[16] In closing FFRDCs, the Pentagon maintained that the original missions of these centers had been completed and that they were no longer able to meet the changing needs of the department.[17] In the following years the DOD continued to reduce its number of FFRDCs, and by 1978 the Pentagon had eliminated all but six.[18]

Of thirty-seven total FFRDCs operating in 2005, the DOE sponsored sixteen, more than any other agency, while the DOD sponsored the second most, with nine. Other agencies that sponsor FFRDCs include the NSF, which sponsors five, the National Security Administration, the Nuclear Regulatory Commission, NIH, NASA, and the departments for Homeland Security, Transportation, and Treasury.[19] In FY2003, FFRDCs performed $11.7 billion in federally sponsored research, with over three-quarters ($9.2 billion) being conducted by DOE FFRDCs. During that same year, four FFRDCs together—DOE's Los Alamos, Sandia, and Lawrence Livermore national laboratories and NASA's Jet Propulsion Laboratory—accounted for over half of the total R&D expenditures by FFRDCs, each reporting R&D expenditures of over $1 billion.[20]

Non-DOE Federal Laboratories and Research Centers

When people refer to the national laboratories, they are often loosely referring to DOE-owned laboratories. The reader should be aware that there are times when people both inside and outside of the government refer to non-DOE research facilities as "national" laboratories. While such facilities are certainly national in scope, *national laboratories* in this book means the DOE laboratories.

Many other agencies aside from the Department of Energy, including the Department of Defense, the National Institutes of Health, the Environmental Protection Agency, NIST, NOAA, USDA, and NASA, own and oversee research laboratories, centers, and facilities for the federal government. These can range in staff from thousands of technically trained scientists and engineers to just a handful of individuals working in a small research facility. These non-DOE laboratories and research facilities represent a critical part of the U.S. scientific infrastructure and consist of a complex mix of GOGOs, GOCOs, FFRDCs, and other forms of non-FFRDC and not-for-profit research centers.

For example, the Department of Defense owns and operates major research laboratories that support the scientific and technological needs of each individual military service: the Naval Research Laboratory, the Army Research Laboratory, and the Air Force Research Laboratory. All three of these DOD laboratories are operated directly by the federal government. As mentioned earlier, the DOD also owns nine FFRDCs, including Lincoln Laboratory managed by the Massachusetts Institute of Technology for the air force, the Center for Naval Analysis managed by the CAN Corporation for the navy, and the Software Engineering Institute managed by Carnegie Mellon University for the army.

The DOD also maintains close relations with and is a primary sponsor of *University Affiliated Research Centers (UARCs)*, at Johns Hopkins University, Penn State University, the University of Washington, the University of Texas at Austin, and the University of Maryland. UARCs are research centers located within universities that partner with government agencies to solve problems of national and global significance and that receive a significant portion of their administrative and research support from the federal government. For example, the Applied Physics Laboratory (APL), a UARC at Johns Hopkins University, operates primarily under a sole-source, cost-plus-fixed-fee contract administered by the U.S. Navy Naval Sea Systems Command. This contract covers approximately 60 percent of the APL's annual budget and serves as a framework for other government sponsors to bring work to APL and fund those efforts.[21] Another UARC, the University of Maryland Center for Advanced Study of Language (CASL), serves as a laboratory for advanced research and development on language and national security. CASL was founded in 2003 with Department of Defense funding and represents a unique partnership between the university and the U.S. intelligence community aimed at providing the research necessary to improve and expand the language capabilities of the U.S. government.[22]

NASA operates several research centers aimed at helping it to fulfill its mission. Among these are the Ames, Glenn, and Langley research centers, the Marshall and Goddard space flight centers, and the Johnson and Kennedy space centers. All of these centers are owned and operated by NASA itself (GOGOs) to conduct research, develop technologies, and expand knowledge in support of space science, aeronautics, and manned and unmanned space exploration. NASA also oversees the Jet Propulsion Laboratory (JPL), an FFRDC managed and operated by the California Institute of Technology. JPL serves as NASA's leading center for robotic exploration of the solar system.[23]

The EPA, whose mission is "to protect human health and to safeguard the natural environment—air, water, and land—upon which life depends," maintains more than a dozen laboratories of its own.[24] This work is done in-house, not contracted out to industry, because industry might be the target of the regulations derived from the work. In addition, the nature and scale of EPA research are such that university sites might be appropriate only to carry out small subprojects.

Any study of the overall U.S. research effort must account for the important work of these in-house and contracted federal laboratories. Such facilities have done a great deal to advance important areas of science and technology in support of the mission agencies that sponsor them. They also share some of the complex policy issues that are experienced by the DOE national laboratories, to which we will now turn our attention.

The DOE National Laboratories

The DOE national laboratories are perhaps the best-recognized of the federal research laboratories. The DOE national laboratories and the contracting organizations that operate many of them for the government have been historically, and continue to be, critical partners with the government and the scientific community in the conduct of U.S. science.

Because of the scale and complexity of the research projects it undertakes, the DOE relies upon its laboratories and FFRDCs more than any other federal agency. In FY2005, for example, the DOE allocated $4.8 billion of its total $8 billion in R&D funding to its FFRDCs. In that same year, DOE support for its sixteen FFRDCs represented 59 percent of the total spent on all of the government's thirty-seven FFRDCs.[25] The DOE national laboratories alone employ more than thirty thousand scientists and engineers.[26]

Categories of DOE National Laboratories and Technology Centers

The DOE operates twenty-one major national laboratories and technology centers. Of these, nineteen are GOCOs (sixteen of which are also FFRDCs) and three are GOGOs (see table 7.1).[27] The DOE also operates other smaller laboratories and research facilities, some of which have very specific missions, such as the Bettis Atomic Power Laboratory and Knolls Atomic Power Laboratory, both GOCOs that support the Naval Nuclear Propulsion Program, a joint navy-DOE effort.

The exact distinction between a DOE *national laboratory* and a DOE *technology center* is quite murky. In fact, the two terms are often used interchangeably even by DOE officials. Therefore, we have chosen not to distinguish here between a DOE national laboratory and a technology center. National laboratories and technology centers, however, are categorized in additional ways that can also be quite confusing, but which can be distinguished. For

example, labs that support only one major DOE program are often referred to as *single program* laboratories, while others that support multiple programs are known as *multiprogram* laboratories. Oak Ridge National Laboratory, with its nuclear engineering, physics, and biological sciences departments, among others, is labeled a multiprogram laboratory. Fermi Laboratory, on the other hand, is a single-program facility, which is focused primarily on support for the DOE high-energy physics program.

Still other even more narrowly focused DOE laboratories, such as the New Brunswick Laboratory in Illinois, which is the federal government's nuclear materials measurements and reference materials laboratory, are referred to as *specific mission* laboratories because they focus on a very specific purpose relating to one of DOE's missions.

DOE laboratories are further categorized according to whether their primary work is defense related. For example, although a significant amount of mission-oriented basic research is conducted at Los Alamos National Laboratory, the New Mexico lab is still categorized as one of three *weapons labs* because of its historic role in the nation's atomic weapons program. California's Lawrence Livermore Laboratory and New Mexico's Sandia facility are the DOE's other two weapons labs.

The Establishment of the National Laboratories—a History

The initial DOE national labs were established to assist in building the atom bomb during World War II. These laboratories were set up as partnerships between the government and universities or companies that offered the necessary scientific expertise. Management of the Los Alamos, Argonne, and Oak Ridge labs was contracted to the University of California, the University of Chicago, and the E. I. DuPont Company, respectively. Such practices were a key part of the U.S. R&D infrastructure put into place during and after World War II.

The first national lab established was actually several labs. In 1942 and 1943, plans were put into place to build experimental reactors for fission physics research in Tennessee (Oak Ridge), Chicago (Argonne), and New Mexico (Los Alamos).[28] These facilities were created because U.S. nuclear physicists wished to replicate the findings of British researchers who had successfully demonstrated a nuclear fission reaction in 1939, and because the Allies were worried that the Germans were working on similar technology with an eye toward developing atomic weapons.

Initially, a series of feasibility studies were conducted at Columbia, the University of Chicago, Princeton, and

TABLE 7.1 Department of Energy National Laboratories and Technology Centers

Laboratory Name	Management	Lab Type	Contractor
Ames Laboratory	GOCO	Single program	Iowa State University
Argonne National Laboratory	GOCO	Multiprogram	University of Chicago
Brookhaven National Laboratory	GOCO	Multiprogram	Brookhaven Science Associates[b]
Fermi National Accelerator Laboratory	GOCO	Single program	Fermi Research Alliance, LLC[c]
Idaho National Laboratory	GOCO	Multiprogram	Battelle Energy Alliance[d]
Lawrence Berkeley National Laboratory	GOCO	Multiprogram	University of California
Lawrence Livermore National Laboratory	GOCO	Multiprogram/ weapons	Lawrence Livermore National Security, LLC[e]
Los Alamos National Laboratory	GOCO	Multiprogram/ weapons	Los Alamos National Security, LLC[f]
National Energy Technology Laboratory[a]	GOGO	Single program	DOE
National Renewable Energy Laboratory	GOCO	Single program	Midwest Research Institute/ Battelle Memorial Institute
New Brunswick Laboratory	GOGO	Specific mission	DOE
Oak Ridge Institute for Science and Education	GOCO	Single program	Oak Ridge Associated Universities
Oak Ridge National Laboratory	GOCO	Multiprogram	University of Tennessee/ Battelle Memorial Institute
Pacific Northwest National Laboratory	GOCO	Multiprogram	Battelle Memorial Institute
Princeton Plasma Physics Laboratory	GOCO	Single program	Princeton University
Radiological and Environmental Sciences Laboratory	GOGO	Specific mission	DOE
Sandia National Laboratory	GOCO	Multiprogram/ weapons	Lockheed Martin, Inc.
Savannah River Ecology Laboratory	GOCO	Specific mission	University of Georgia
Savannah River National Laboratory	GOCO	Single program	Westinghouse Savannah River Company
Stanford Linear Accelerator Center	GOCO	Single program	Stanford University
Thomas Jefferson National Accelerator Facility	GOCO	Single program	Jefferson Science Associates, LLC

Italic typeface denotes DOE laboratories and centers that are designated as FFRDCs.

[a]The National Energy Technology Lab consists of five different sites located in Morgantown, West Virginia; Pittsburgh, Pennsylvania; Tulsa, Oklahoma; Albany, Oregon; and Fairbanks, Alaska.

[b]Brookhaven Science Associates is a limited liability entity created solely for managing Brookhaven National Lab. It is a 50/50 partnership between State University of New York at Stony Brook and Battelle Memorial Institute.

[c]Fermi Research Alliance is a limited liability entity created solely for managing Fermi National Accelerator Lab. It is a partnership between the University Research Association (a consortium of eighty-nine universities) and the University of Chicago.

[d]Battelle Energy Alliance is a limited liability entity created solely for managing Idaho National Lab. It is owned by Battelle Memorial Institute.

[e]Lawrence Livermore National Security is a limited liability entity created solely for managing Lawrence Livermore National Lab. It is comprised of Bechtel National, University of California, Babcock and Wilcox, the Washington Division of URS Corporation, and Battelle and has an affiliation with Texas A&M University System.

[f]Los Alamos National Security is a limited liability entity created for managing Los Alamos National Lab. It is comprised of Bechtel National, University of California, BWX Technologies, and Washington Group International.

the University of California at Berkeley. (The latter three eventually became DOE lab sites.) Then, in 1941, a group of Princeton physicists proposed a central laboratory for nuclear fission work, which would be run by the government "in an industrial region."[29] Their proposal led to the establishment of what is today known as Argonne Lab, or the Metallurgical Laboratory (Met Lab). The work at Argonne, together with that done at Los Alamos, Oak

Ridge, and Berkeley, formed the core of the Manhattan Project and America's effort to develop the atomic bomb. Of these four labs, however, only Los Alamos was ever (and still is) considered an official "weapons" lab. While the others contributed scientifically and technically to the building of the Bomb, weapons building per se was never an official part of their mission. Thus, Argonne, Los Alamos, and Oak Ridge were all direct products of the war effort. The Berkeley lab (known as the Berkeley Rad Lab) existed before the war as part of the University of California.[30] Four other well-known national labs—Lawrence Livermore, Brookhaven, Fermilab, and Sandia—were all established after the war.

As World War II came to an end, policymakers became concerned about the future of the three war-specific labs. They had been established to build a bomb. But were they necessary now? Anxious about the potential closure of these facilities, nuclear physicists came together to plan for the "postwar organization of nuclear research."[31] They proposed that the United States "maintain and operate at least four regional laboratories" (most likely referring to Oak Ridge, Argonne, Los Alamos, and DOE's Ames Laboratory—a site involved in the Manhattan Project that was not officially established as a national lab until after the war) that would include reactor facilities and other equipment too expensive for universities to maintain on their own. These sites could be used by university and industry researchers, as well as government scientists.[32] Thus was born the U.S. system of DOE national labs.

With the approval of the Atomic Energy Act of 1946, also called the McMahon Act in honor of its chief congressional sponsor, Senator Brien McMahon (D-CT), the government formalized M&O practices by establishing the Atomic Energy Commission (AEC), the predecessor to the Energy Research and Development Administration, which would later become the U.S. Department of Energy. The AEC was created to direct the nation's ongoing efforts concerning nuclear matters, both for weapons and for civilian commercial nuclear power. To help fulfill its mission and supplement the already existing academic and industrial research infrastructure, the AEC created a system of national laboratories building on those already developed to support the war effort. Indeed, the McMahon Act specifically enabled the AEC to enter into management contracts with nongovernmental entities to help to advance research in support of its mission.[33]

The national labs were created as a coordinated system. Each lab was built so that it added to the arrangement and did not fully duplicate a capability available elsewhere. The labs were led by science policy leaders of the day such as Alvin Weinberg, the first director of Oak Ridge National Laboratory, and Ernest O. Lawrence, who directed Berkeley Laboratory. Before becoming the director of Oak Ridge, Weinberg assisted in developing the technology behind the atomic bomb while working at the University of Chicago in the 1940s. He is noted for his leadership in the area of "big science"—a phrase that he coined—science of a size and scope that could best be performed at the national laboratories.[34] Lawrence was given the Nobel Prize in 1939 for his invention of the cyclotron in 1929, an accelerator of nuclear particles that would play a critical role in the Manhattan Project and that has been critical to advancing nuclear physics ever since. He was also a central figure in bringing science and government together in the Cold War era to create the national laboratory system. For his important contributions to U.S. science, Lawrence Berkeley and Lawrence Livermore national laboratories still bear his name.[35]

In the 1950s, the labs' directors began coordinating further, meeting to organize programs and discuss common problems, forming a group that came to be known as the Lab Directors' Club. During the 1950s and 1960s, this group also provided an avenue by which the directors could present complaints and suggestions to the AEC.[36] Although the Lab Directors' Club no longer exists, its legacy can still be seen in the interdependence of the national labs system.[37]

The Administration and Management of the National Laboratories

Most of the DOE's national laboratories, including the most highly classified ones, are operated under M&O contract, by a university or university association, or by a nonprofit or for-profit partnered with a university. This might seem odd, given that national laboratories were needed initially so that the federal government could carry out research that was unsuitable for a university or industry setting. However, the federal government has always recognized the value of strong links between universities and the national laboratories, and university oversight and management of certain DOE labs over the years has effectively strengthened those ties. For their part, universities have long viewed this management as a national service.

The federal government asked the civilian contractors to manage operations of the first national labs—Argonne (University of Chicago), Los Alamos (University of California), and Oak Ridge (DuPont)—largely because it wished to employ certain leading scientists in overseeing and directing development of the nuclear bomb.[38] While Los Alamos was envisioned as a military laboratory, scientists were not willing to work under military command. The leaders of the Manhattan Project, therefore, chose at least temporarily to employ a civilian contractor, the University of California, to oversee Los Alamos. While they had imagined that the lab would eventually revert to military control, the decision by the Pentagon not to assert control over the weapons laboratory represented an important precedent that would lay the groundwork for the use of civilian contractors to manage the other DOE laboratories.[39]

Thus, universities have been partners with the federal government in administering the DOE laboratories since the labs' inception. But how does this university-government partnership work? In the recent past, in the instance of Los Alamos, the lab (a DOE entity) received funding from both the DOE and the Department of Defense (primarily because of its work on weapons), but was operated under M&O contract by the University of California, which was paid for its services. The university was responsible for hiring the laboratory director, conducting program reviews, and seeing that the mission objectives of the sponsoring agencies were met. The laboratory director reported to the university administration, which included a special board that monitored the lab's operations. A strict line was maintained, however, between university activities and laboratory research. For example, even with this relationship the University of California tended to avoid classified research, while much of the research conducted at Los Alamos is classified. This relationship has since changed slightly, with the University of California now partnering with Bechtel National, BWX Technologies, and Washington Group International under the Los Alamos National Security, LLC (LANS). This is a result of DOE introducing a competitive contracting schedule (discussed later in the chapter).

In a slightly different model, the Fermi National Accelerator Laboratory was managed from 1967 to 2006 by the Universities Research Association (URA), an incorporated group of ninety universities, with its own president, staff, and offices in Washington, DC. The DOE awarded the URA a competitive contract that included a budget for personnel and facility costs of the laboratory, along with an annual operating fee from which the association recovered its costs. Every member university pays the URA an initial entrance fee and is assessed dues thereafter. The URA uses these funds to promote work at the laboratory, award student fellowships, and support any other activity that might advance the overall laboratory mission. The URA hires a laboratory director, approves senior laboratory hires, conducts reviews of all major programs, reviews safety programs, and tracks the progress of all major construction and deployment projects. In 2007, URA joined with the University of Chicago to form the Fermi Research Alliance LLC, which now holds the contract to operate Fermilab.

Such arrangements, effective as they may be, are not without tension, especially when the lab being managed by a university is a weapons lab. In the past, UC faculty committees and students groups have called upon the university system to discontinue its management of the weapons laboratories because of their role in programs to develop and oversee weapons of mass destruction.[40] University administrators and those who support continued University of California management of the laboratories respond that universities have a responsibility for national defense, and that it is better to be involved in the associated research than to have the work done in total secrecy by the military. Moreover, administrators note that the university's own research program benefits from access—limited as it might be—to laboratory infrastructure, scientific tools, expert scientists, and the research results.

Given these possibilities for conflict, why would the government choose to have universities manage its labs? The answer, in part, is that universities have access to talent that the government needs, can provide a stream of students to help with frontier research, and are quite good at assessing scientific expertise. As mentioned previously, it was the need for top-notch scientific talent that initially led the government to seek out universities as partners in managing the labs. Less tangibly, many nonacademic scientists find it appealing to maintain a connection to the university world and the credibility that such a connection implies. The opportunity to develop such ties can also help attract top-notch scientists to a national facility.

Of course, universities are not the only entities to manage GOCOs. Many laboratories are operated by specific corporations. Oak Ridge, for example, was run by DuPont during the war, then afterward by the Union Carbide Corporation, Martin Marietta, and Lockheed Martin. These private-sector operators tend to show a keen interest in ensuring that laboratory milestones and budgets are met and that the labs themselves are run efficiently. Even so, corporate managers still recognize the benefits of strong ties to the academic community: Oak Ridge, for example, has had a historic relationship with the University of Tennessee.

In fact, we have now entered an era in which joint teams of universities and nonprofit organizations often provide laboratory oversight. The nonprofit Battelle Memorial Institute jointly manages both Oak Ridge with the University of Tennessee, and Brookhaven with the State University of New York at Stony Brook. Battelle is also the sole contractor at two other government labs, Idaho and Pacific Northwest, and partners with the Midwest Research Institute, another nonprofit, to oversee the National Renewable Energy Laboratory.

Technology Transfer and the Laboratories

What happens to the knowledge generated at national laboratories or other federal laboratories? One mechanism allowing the transfer of knowledge from these labs to the public is the Federal Laboratory Consortium for Technology Transfer (FLC). The consortium was organized in 1974, fostered by the Technology Innovation Act of 1980 (P.L. 96-480), also called the Stevenson-Wydler Act, and formally chartered by the Federal Technology Transfer Act of 1986 (P.L. 99-502). The Stevenson-Wydler Act required, among other things, that all federal laboratories, whether GOGOs or GOCOs, maintain technology transfer offices capable of identifying potentially commercializable technologies and transferring them to industry. Stevenson-Wydler thus serves as a counterpart to Bayh-Dole Act for the federal laboratories.

The FLC is responsible for moving technologies and expertise from the federal laboratory system into the marketplace.[41] How does the FLC promote technology transfer? This huge consortium provides support to the technology transfer offices of approximately 250 member federal laboratories and centers and their sponsoring federal agencies. The FLC creates and tests technology transfer processes, attempts to reduce barriers to the process, provides training, emphasizes grassroots technology transfer, and stresses national initiatives in which technology transfer can play a role. Finally, the consortium provides coordination across the national lab system for transferring government-owned intellectual property to other sectors (universities, state and local governments, and industry), with the end goal of benefiting the consumer.

The National Laboratories after the Cold War: In Search of a Mission

Like the end of World War II, the end of the Cold War gave rise to questions about the future of the national labs. Many of these facilities had designed and built the very weapons that had convinced the Soviet Union of the fruitlessness of aggression against the United States. Even many of the nonweapons DOE multiprogram labs were engaged in research that indirectly contributed to weapons development. Ironically, however, the labs may have fallen victim to their own success.

For the first few years following the end of the Cold War, many believed that the laboratories could be a major force in enhancing U.S. economic competitiveness, just as they had helped the U.S. to win the Cold War. During this time organizations, such as the Center for Strategic and International Studies (CSIS), attempted to identify the benefits the national laboratories could provide in meeting the technological and competitiveness challenges now facing the nation.[42] By turning their attention to infusing the U.S. industrial landscape with high-tech approaches to manufacturing and distribution—from the creation of "weapons" to the creation of "widgets"—laboratory proponents argued that the DOE multipurpose labs could continue to make important contributions to the nation's future.[43]

The National Competitiveness Technology Transfer Act (P.L. 101-189), passed in 1989, allowed GOCOs to enter into cooperative research and development agreements (CRADAs), which permit national laboratories to form relationships with nongovernmental entities, including private businesses, with this purpose in mind. Previously, only GOGOs had been permitted to enter into such agreements. The new law was intended to help the national labs refocus on the nation's economic security. It was envisioned as a win-win situation: businesses would benefit from the exchange of expertise and technology, and the national laboratories would benefit by justifying their continued existence in a changed world. These expectations of a new and dynamic synergy between laboratories and industry, however, were never realized. Perhaps surprisingly, one of the strongest objections was from the corporations themselves, who asked why their tax dollars should be used in a way that might benefit their competitors.

Hence the early and middle 1990s were a time of great uncertainty for the national laboratories. Calls for significant laboratory reforms during this period prompted the introduction of major reform legislation in Congress.[44] For instance, Representative George Brown, a longtime science advocate, introduced the Department of Energy Laboratory Technology Act of 1993 (H.R. 1432). This legislation was aimed at putting in place "a process of disciplined evolution for the DOE laboratories—a process through which the enormous resources of these labs are carefully directed toward meeting some of the nation's most pressing needs."[45] The bill had four main objectives: (1) to provide updated and focused missions for the laboratories; (2) to improve the organization of research, development, and technology transfer at the DOE; (3) to enhance collaboration between DOE labs and industry by streamlining technology transfer; and (4) to ensure that activities at DOE labs were regularly subjected to performance evaluations. Even though a similar bill was introduced in the Senate, it never became law.[46]

The strong interest in the labs' future also resulted in serious reviews of the laboratories' missions, operations, and management.[47] Perhaps the most notable of these reports was produced by a task force created under the auspices of the Secretary of Energy Advisory Board and chaired by Robert Galvin, the retired CEO of Motorola.[48] Formally titled the Energy Advisory Board Task Force on Alternative Futures for Department of Energy National Laboratories, the Galvin commission examined the laboratories' missions and management organization.

In its final recommendations issued in February 1995, the Galvin commission validated the important role played by the national laboratories in the areas of national security, energy, environment, and fundamental science. It expressed concerns, however, about the "top-down, command and control bureaucracies" that characterized their management system, stating forcefully that "government ownership and operation of these laboratories does not work well."[49] While the DOE multiprogram laboratories are technically GOCOs, the commission report suggested that contractors played a relatively weak management role that was dominated by government bureaucracy and congressional micromanagement, essentially making the labs act like GOGOs. The commission recommended that laboratories be "corporatized" and run as if they were private, for-profit entities. The Galvin commission suggested that by eliminating the existing M&O contracts, replacing laboratory leadership with directors who would report to corporate-like boards appointed by the president, and eliminating government red tape, DOE could streamline laboratory operations and save as much as 20 percent of the $6.8 billion the government was spending on the labs.[50]

In 1995 DOE secretary Hazel O'Leary, in response to the Galvin report, chartered a Laboratory Operations Board made up of DOE management officials and outside private-sector advisors to review and improve the operations of the DOE laboratories. In its initial review, the board found that "very substantial and pervasive changes are presently underway at the department and the DOE laboratories." While the board made several recom-

mendations, none substantially changed the process for awarding M&Os contracts, as was recommended by the Galvin commission.[51]

Responding to an NSTC interagency review of the DOE, DOD, and NASA laboratories launched by President Clinton in May 1994, the DOE went further in delineating its response to the Galvin commission's recommendations. The DOE agreed with the commission that the system of laboratory governance was broken. To help fix it, the DOE put forth a new "Management Improvement Roadmap" for the laboratories. However, the DOE took exception to the Galvin report's recommendation to "corporatize" the laboratory management system. Noted the agency in its response, the DOE is "not persuaded that the specific model envisioned [by the Galvin commission] would be either practical in the near-term, or sustainable over the long run."[52] One of the solutions pointed to as an alternative by DOE was its appointment of the new Laboratory Operations Board.[53]

The interest in examining the future of DOE labs was perhaps intensified during this period by questions about the changing missions of the DOE itself. Stated the GAO in a 1995 report: "DOE has begun to modify its Cold War organizational structures and processes to meet newer responsibilities, from environmental cleanup to industrial competitiveness. However, until a more fundamental reevaluation of DOE's missions and alternatives is undertaken—including opportunities to restructure and privatize operations—it is not clear if the Department and its missions are still needed in their present form or could be implemented more effectively elsewhere in the public or private sectors."[54]

Several of the newly elected Republican members of Congress were pushing at the time to abolish the Department of Energy, along with three other cabinet-level agencies. The abolishment of these agencies was viewed by these legislators as a way to restore fiscal responsibility and to reduce the size of federal government, part of the commitment they had made in the Republican Contract with America, a document considered to be a major factor in the 1994 Republican takeover of Congress.[55]

Despite the major push for reform of the DOE laboratories that occurred in the 1990s, significant changes were never enacted. Major legislation failed to gain approval in Congress, while the DOE survived attempts to eliminate it. In a 1998 report assessing progress in reforming the labs, the GAO concluded that "Despite many studies identifying similar deficiencies in the management of the DOE's national laboratories, fundamental change remains an elusive goal."[56]

Federal Laboratory Issues

Laboratory Management Revisited

In recent years, accusations of mismanagement and the security breaches experienced by the laboratories have again brought into question the quality of management being provided to the DOE laboratories.[57]

The DOE and its labs were severely criticized in a 2002 GAO report calling the department's contract management process "a high-risk area of fraud, waste, abuse, and mismanagement."[58] A blue-ribbon commission investigating the matter recommended in 2003 that the DOE institute a competitive contracting schedule for all fifteen of its GOCO labs.[59]

Recommendations for making the DOE laboratory contracts competitive stemmed from the belief that since the rebidding of the M&O contracts on a regular schedule had not been required, contractors had little incentive to ensure proper management of the laboratories. Indeed, some M&O contractors had not changed since the creation of the DOE national laboratory system. Some observers believe that if contractors have to compete on a regular basis, they will have an incentive to maintain higher standards in their management practices.

Some critics express concerns that competitive contracts will not address issues of cost and accountability at the labs, while others worry that the prospect of routine changes in management (because of the uncertain nature of competitive contracts) will deter top scientists from wanting to work at the national labs. In congressional hearings on this subject held in mid-2003, John McTague, former OSTP director under President Reagan, urged that care be taken in altering a system and a contracting model that, by his assessment, had functioned well overall. He urged Congress not to undertake major changes because of isolated incidents of poor management:

> The FFRDC concept has been a superb success in mission accomplishment. No other country has had anything like this success with its government-sponsored laboratories—not the former Soviet Union, not France or Germany, not Great Britain or Japan.
>
> Yet despite what is in plain sight one subset of the FFRDCs, those sponsored by the Department of Energy, have been subjected to the scrutiny of a microscope. Sometimes the microscope focuses on something really ugly, like an improper travel voucher or an inadequate safety document. Sometimes it is somewhat out of focus and may seem to show something egregious, like the apparent purchase of a Mustang automobile by a laboratory employee using government money. Better focus sometimes shows a different picture.[60]

At the heart of this issue, of course, is how the federal government should best procure the research and development efforts it needs—a matter of overriding interest for science policy.

Continued Questions about the Mission of the National Labs

A fundamental uncertainty about the mission of the DOE labs still remains. Concern about this, however, has diminished since the mid-1990s, in part because of the DOE's efforts to clarify its missions relating to national security, energy, environmental cleanup, and fundamental research.

In the post–Cold War era, the laboratories have diversified the roles they play. For example, the DOE, and not the National Institutes of Health, was the first to initiate the mapping of the human genome. The DOE took this initiative primarily because of its history in undertaking large science projects at its labs, something that the NIH was unaccustomed to and unwilling to do at first. While the NIH eventually overshadowed the DOE's role in this project in the public's eye, the DOE and its laboratories continued to play an important role. A sweeping refocus of the labs on life sciences, however, did not occur. In fact, since that time the DOE has been directed by the OMB not to explore areas of biology that are too heavily oriented to human health or medicine, areas that are perceived to better be handled by the NIH.

Increased concern about homeland security has, to a certain extent, offered a new focal point for several of the national laboratories. During the creation of the Department of Homeland Security, the interface between the labs and the new department was debated. Of course, for the DHS to fulfill its mission, it must have access to trained, experienced scientists who can help protect the nation from chemical, biological, radiological, or cyber attacks. It would be redundant and wasteful to create a whole new laboratory structure for this purpose, and for this reason, it was determined that the DOE national laboratories should, in fact, play an important role in helping the DHS to improve domestic security. But how does a new department interact with a national laboratory system that was created decades ago, and which has strong, long-standing ties to an existing agency? These struggles continue as the DHS attempts to work with and use the talent base at the DOE national laboratories.[61]

POLICY DISCUSSION BOX 7.1

Who Should Manage the National Laboratories?

Universities have been involved in the management of the DOE national laboratories and other FFRDCs since many of these facilities were developed during and immediately following World War II. Universities were an obvious manager of these facilities because university scientists were often recruited to conduct research there. Indeed, the working model for many of the laboratories was based on the idea that "scientific research is best done by teams of individuals with different fields of expertise working together."[62] Since such diverse expertise already existed on university campuses, they were the obvious choice to direct and manage the national laboratories.

As a result, after World War II the government, in most instances, turned to universities to serve as managers for the labs. Close ties to universities seemed an advantage. Meanwhile, universities saw management of the laboratories as a public service to the nation. Management of a government laboratory was also viewed as a means to increase university prestige and to allow constant interactions between top university faculty and laboratory researchers. These interactions, it was believed, would help the laboratories to perform the highest-quality science. At the same time, they would enhance research at the universities by giving faculty and students access to equipment and other resources that might not otherwise be available.

In recent years, questions have arisen over whether universities are really the best managers of the national laboratories. Given the size and growing complexity of the laboratories, are they better managed by private companies or by not-for-profit entities? Security breaches and management scandals at some of the national laboratories have fueled these questions.

Congress has also raised concerns that the management contracts have been retained by certain universities with no requirement for competition in the contract renewals. In response, the Department of Energy announced in January 2004 that it would make the contracts for several national labs competitive for the first time. For example, the University of California was not guaranteed the contract to manage Los Alamos National Lab when the previous contract expired. The goal is to award the contract to the most capable organization.

Making the laboratory contacts competitive raises many science policy questions: Are universities really the best organizations for managing national labs? How important is the linkage between universities and laboratories today? Is it as useful as it historically was? Are private companies or nonprofit entities better able to manage large government research facilities? What about consortiums or joint management partnerships between universities and private companies? Should the government manage its own labs? What are the requirements for properly managing a national lab?

Today, as concerns grow over competitiveness, energy, local and global environmental issues, and national security, one thing is for certain—our federal laboratory system and the unique research capabilities these labs provide will be an essential tool to meet these challenges.

Retaining Scientific and Technical Talent in the Federal Laboratories

Industry's higher salaries have always been available to bright scientists, but this lure has historically been offset by the prestige and other benefits of working at a federal laboratory, including access to frontier facilities and some of the world's top researchers. Over time, however, the traditional incentives for top-notch scientists to work at our federal laboratories have diminished, while the money and

benefits that private industry will pay have increased. As a result the federal laboratories are a less competitive option for bright young scientists.[63] In his article "Silence of the Labs," defense analyst Don DeYoung asserts that the federal laboratories may be losing advantages over private industry they once enjoyed, and that the laboratory system and our national security may be at risk as a result.

Providing competitive salaries at the labs is not the only problem, in DeYoung's opinion. The laboratories' ability to attract talent is also threatened by increased bureaucracy, delays in the acquisition of new facilities and equipment, decaying infrastructure, and loss of valued colleagues. Successful R&D is based on a talent pyramid, in which a few people of extraordinary ability set the organization's agenda and inspire those around them to produce at the highest possible level. An organization staffed entirely by run-of-the-mill scientists is unlikely to be as productive as one with even a few truly outstanding scientists. Thus, the loss of even some of the federal laboratories' most talented leaders could have dramatic consequences for their productivity and their effectiveness.

Indeed, DeYoung maintains that declining interest in the major federal laboratories at the Department of Defense has already resulted in the loss of critical national scientific capability. DeYoung argues that the Naval Research Laboratory, for example, lost its ability to conduct cutting-edge research in fiber optics between 1999 and 2001 in part because within a year's time twenty of the twenty-six scientists working in this area were hired away by industry. These top-flight researchers were offered salaries up to twice what they had been paid at the DOD lab.[64]

Meanwhile, many of the best young scientists are choosing careers in the private sector straight out of graduate school. Because of this, the average age of the scientists and engineers at our national laboratories is increasing, with no indication that it will level off anytime soon. Given the speed with which technology changes, any successful federal laboratory must have a continuing flow of new graduates to assure both technological and human vitality. Any trend in which young people are seeking work elsewhere, or midcareer scientists are leaving, weakens the government's capacity to conduct forward-thinking research. This should be a worrying trend to all.

Declining Discretionary Funds

Another issue of concern, especially with the DOE laboratories, is a reduction in discretionary research funds available to the laboratory directors. In the past, the directors of the DOE national laboratories had at their disposal large pools of funds that could be spent on innovative research and projects that had not specifically been appropriated by Congress or outlined in the congressional budget request. The funds, provided as a part of what is known as the *laboratory directed research and development* (LDRD) program, were aimed at stimulating exploration at the forefront of science and fostering creativity and the pursuit of exciting new areas of research. They were often used by the laboratory directors to pursue high-risk, high-payoff research and, according to many of the laboratory directors, helped retain and recruit top scientists.

In FY1999, approximately $300 million was spent by the labs as a part of the LDRD program. However, in FY2000, Congress significantly reduced the funds that could be allocated to the LDRD program, eliminating the LDRD funds for environmental management programs and reducing them by a third, from 6 to 4 percent, of the laboratories' total budget. Laboratory directors maintain that this has greatly reduced their ability to perform their missions and to conduct the basic research that underpins them. Moreover, according to a review by DOE's own Laboratory Operations Board, the reductions place the amount the laboratories can expend on research at levels well below the ratio of their total budget that industry would spend on comparable research to ensure their success in the marketplace.[65]

Unfair Competition

One complaint made by both academia and industry is that federal agencies have at times improperly diverted funding to their federal laboratories, as opposed to making competitive awards to small businesses or to universities. In these instances, businesses and universities maintain that they can perform the same work as the laboratories at a reduced cost, with increased efficiency and with better results.

In the late 1980s, this concern was the focus of two congressional hearings. The first, held in April 1987 by the Senate Committee on Small Business, Subcommittee on Innovation, Technology and Productivity, looked at whether the U.S. Army improperly requested Los Alamos National Laboratory to conduct work that could have been done by a private firm on a competitive basis. In his opening statement at the hearing, subcommittee chairman Carl Levin (D-MI) stated, "Our research indicates that the improper diversion of government work to FFRDCs is a serious problem for the small business community. Innovative and specialized small businesses, especially in the high-tech area, are vital to America's economy. The Federal Government should be encouraging, not impeding, the growth of these small companies."[66]

A second joint hearing was held in May 1987 by the Senate Committee on Governmental Affairs, Subcommittee on Oversight of Government Management, and the Senate Committee on Armed Services, Subcommittee on Strategic Forces and Nuclear Deterrence. This hearing looked at the need for the creation of a new FFRDC, the Strategic Defense Initiative Institute, which was being proposed to provide technical advice to DOD's Strategic Defense Initiative. The new institute was criticized by industry and members of Congress as anticompetitive and unnecessary given that existing research facilities and businesses could offer the same quality of support to the DOD at less cost.[67] This issue was also the focus of two subsequent GAO reports.[68]

Ultimately, the government is responsible for ensuring the work conducted for the federal government, including research, is fairly awarded and subject to laws that guard against sole-source contracting. To ensure that the public gets the best research possible at the lowest cost, it should also adhere to strict rules of competitive merit review for major research grants and awards, even when this might mean that universities and laboratories compete against each other. This being said, conflict over where work should be done—federal laboratories or industry and universities—is likely to continue.

Privatization and Outsourcing of Laboratory Functions

One question that periodically arises is whether the federal laboratories should be privatized. Or if they cannot be privatized, would it be possible to outsource some of their duties to private entities and to universities? Would either of these options be wise?

Of course, the answers to these questions depends on the federal laboratories one is examining. Given the great diversity in missions performed by these laboratories, even within the same agency, one has to look at each laboratory individually to decide if privatization or outsourcing could or should be undertaken.

Joseph Martino, a research scientist and technology forecaster, has argued that privatization of the federal R&D laboratories can and, in fact, ought to be pursued. Pointing to Britain as an example of successful privatization of government R&D laboratories, Martino in his 1996 study argues that it is both feasible and desirable to privatize federal laboratories at the DOE, USDA, NASA, NIST, and USGS.[69]

Yet most who are familiar with the multifaceted nature of the federal laboratory system, and its value to the overall scientific enterprise and to the nation—including this book's authors—do not favor privatization. While Britain successfully privatized many R&D labs, Martino fails to examine whether they have remained as productive after privatization. People like John McTague have suggested that they have not.[70]

The notion that the laboratories could be privatized misses one of the fundamental reasons for retaining a diverse inventory of federal labs: the inherent value in having a cadre of highly skilled scientists and engineers at the government's disposal to respond in times of national crisis, to fulfill agency missions, and to provide technical support to the government itself. For example, how is the government to know which private contactor can best do work for it, if it does not have its own technical laboratory staff to provide a yardstick by which to ensure that the public interest is best served?[71] Thus, while some of the functions of the laboratories might be spun out to universities and businesses, we doubt that a large-scale movement to privatize the federal laboratories can or should occur anytime soon.

Policy Challenges and Questions

In this chapter, we have provided the reader with a basic understanding of the complex system of federal laboratories in the United States. These labs range from very small research facilities composed of two to three staff to the DOE national laboratories that employ thousands of people each. The wide diversity of federal laboratories with various mission and management structures can remind one of Hungarian goulash: you are never quite sure of its makeup, but you know it tastes pretty good. Indeed, our system of federal laboratories has served the nation well since the end of World War II and played a major role in U.S. scientific and technological preeminence.

Clearly, the federal laboratories provide unique capabilities for the conduct of science that are not available elsewhere. Moreover, they are an essential means to ensure that the government has at its disposal well-trained scientific and technical talent to meet critical national security needs and to assist in carrying out the missions of the federal agencies that oversee them. Nonetheless, the federal laboratories themselves are little understood and often taken for granted.[72]

With the end of the Cold War and the apparent emergence of the United States as the world's sole superpower, doubts have arisen about whether we still need our DOE national laboratories. This question should concern anyone interested in the future of U.S. science and U.S. science policy. In a time when their missions are likely to be questioned, an ongoing challenge for both the DOE national laboratories and other federal labs will be to carve out a niche for themselves that cannot and that *should* not be fulfilled by either private companies or universities. This niche may become increasingly relevant to policy in an age when the missions of the federal laboratories are less clear and the traditional distinctions between the roles of federal laboratories, universities, and industry are blurring.[73] Policies should clarify the specific roles of the laboratories and optimize their strengths through promoting innovative collaboration between the labs, universities, and industry.[74]

We have looked at many questions surrounding the federal laboratories, yet many still remain: How many laboratories are needed, and which types? What should be done to retain the best of our current scientists, and ensure a steady stream of the best new talent? Are changes needed to the laboratory management system? Is the GOGO model better than the GOCO model? Can we reduce impediments to collaboration with industry and universities and increase the transfer of technology? What is the proper balance between security and openness at the labs?

Such questions must be addressed by future generations of scientists and policymakers if the strength of the U.S. national laboratory system is to be maintained. We cannot leave such questions unanswered, for the labs are a vital component in the intricate structures that have advanced and will continue to advance U.S. science.

NOTES

1. Michael Crow and Barry Bozeman, *Limited by Design: R&D Laboratories in the U.S. National Innovation System* (New York: Columbia University Press, 1998), 5–13. This is the most recent survey of the specific numbers on the federal laboratory system the authors could find.

2. National Science Board, *Science and Engineering Indicators 2004*, NSB-04-01 (Arlington, VA: National Science Foundation, 2004), 4.25.

3. National Science Board, *Science and Engineering Indicators 2006*, 4.26.

4. The one hundred thousand figure comes from the Federal Laboratory Consortium for Technology Transfer, *Federal Technology Transfer 2006* (Cherry Hill, NJ: FLC Management Support Office, 2006), i.

5. Crow and Bozeman, *Limited by Design*, 11.

6. The National Science Foundation describes these four rationales for federal laboratories as scale, security, mission and regulatory requirements, and knowledge management. See National Science Board, "Rationales for Federal Laboratories and FFRDCs," *Science and Engineering Indicators 2006*, 4.22.

7. Raymond L. Orbach, "Transformational Science for Energy, Environment, and America's Competitiveness," FY 2008 Office of Science Budget Presentation, U.S. Department of Energy, February 5, 2007, 12.

8. This excluded university engineering programs, where a significant amount of attention may, in fact, be devoted to developing specific technological applications.

9. In some rare instances, federal government research facilities might be contractor-owned, contractor-operated (COCO) facilities. See Office of Technology Assessment, *A History of the Department of Defense Federally Funded Research and Development Centers*, OTA-BP-ISS-157 (Washington, DC: U.S. Government Printing Office, June 1995), 13.

10. The Federal Acquisitions Regulations (FAR 17.601) defines an M&O contract as "an agreement under which the Government contracts for the operation, maintenance, or support, on its behalf, of a Government-owned or -controlled research, development, special production, or testing establishment wholly or principally devoted to one or more major programs of the contracting Federal agency."

11. Note that GOGOs *cannot* be FFRDCs.

12. The NSF defines FFRDCs as "R&D institutions that are exclusively or substantially financed by an agency of the federal government and are supported by the federal government either to meet a particular R&D objective or, in some instances, to provide major facilities at universities for research and associated training purposes. Each center is administered either by an industrial firm, a university, or another non-profit institution." National Science Foundation, *Federal Funds for Research and Development, Fiscal Years, 2001, 2002, 2003*, vol. 51, NSF 04-310 (Arlington, VA: National Science Foundation, 2004), 6.

13. General Accounting Office, *Competition: Issues on Establishing and Using Federally Funded Research and Development Centers*, GAO/ NSIAD-88-116FS (Washington, DC: U.S. General Accounting Office, May 1998), 1.

14. Office of Technology Assessment, *Department of Defense Federally Funded Research and Development Centers*, 33.

15. Bruce C. Dale and Timothy D. Moy, *The Rise of Federally Funded Research and Development Centers*, SAND2000-2212 (Albuquerque, NM: Sandia National Laboratories, September 2000), 8.

16. A listing of the total number of DOD and other FFRDCs that existed for each year from 1950 to 1995 can be found in Office of Technology Assessment, *Department of Defense Federally Funded Research and Development Centers*, appendix B.

17. Cosmo DiMaggio and Michael E. Davey, *The Strategic Defense Initiative Institute: An Assessment of DOD's Current Proposal* (Washington, DC: Congressional Research Service, Library of Congress, May 30, 1986), 4–6.

18. Dale and Moy, *Rise of Federally Funded Research and Development Centers*, 15.

19. National Science Foundation, Division of Science Resources Statistics, "Master Government List of Federally Funded Research and Development Centers," FY 2004, NSF 05-306, February 2005, http://www.nsf.gov/statistics/nsf05306/ (accessed May 21, 2007).

20. National Science Board, *Science and Engineering Indicators 2006*, 4.24.

21. Johns Hopkins University, Applied Physic Laboratory, "Working with APL, Sponsors Relationships, 2007," http://www.jhuapl.edu/aboutapl/working/working.asp (accessed May 21, 2007).

22. University of Maryland, *Center for Advanced Study of Language Annual Report, 2006* (College Park, MD: University of Maryland, 2006), http://www.casl.umd.edu/CASLannual report2006.pdf (accessed June 9, 2007).

23. NASA, "NASA Education, NASA Centers," http://education.nasa.gov/about/nasacenters/ (accessed May 21, 2007).

24. Environmental Protection Agency, "About EPA," http://www.epa.gov/epahome/aboutepa.htm (accessed June 9, 2007).

25. National Science Board, *Science and Engineering Indicators 2006*, 4.21.

26. U.S. Department of Energy, "National Laboratories and Technology Centers," http://www.doe.gov/organization/labs-techcenters.htm (accessed June 9, 2007).

27. Ibid.

28. Westwick, *The National Labs*, 27–31.

29. Jack M. Holl, *Argonne National Laboratory, 1946–1996* (Urbana: University of Illinois Press, 1997), 6.

30. Westwick, *The National Labs*, 11.

31. Ibid., 33.

32. T. R. Hogness to William Bartky, June 13, 1945, Ernest O. Livermore papers, Bancroft Library, University of California Berkeley, quoted in ibid.

33. See U.S. Department of Energy, *The National Laboratories of the U.S. Department of Energy—Creating Technology for America's Energy Future* (Washington, DC: U.S. Department of Energy, 2004), 5.

34. Associated Press, "Obituary: Alvin Weinberg, 91, Directed Oak Ridge Lab," *New York Sun,* October 20, 2006, http://www.nysun.com/article/41966 (accessed May 13, 2007).

35. See American Institute of Physics, *Lawrence and the Cyclotron,* AIP Center for the History of Physics On-line Exhibit Hall, http://www.aip.org/history/lawrence/youth.htm (accessed May 13, 2007); see also Rory Richard, "Father of America's National Labs Subject of New Web Exhibit," American Institute of Physics Inside Science New Service, May 2, 2002, http://aip.org/isns/reports/2002/042.html (accessed May 13, 2007).

36. Westwick, *The National Labs*, 110–16.

37. Ibid., 274–75.

38. The first two of these arrangements continued long after the war. Oak Ridge's management, however, has changed several times over the years, most recently in 2000, when it was taken over by a partnership between a university and a non-profit corporation.

39. Westwick, *The National Labs*, 30–31.

40. "Student Governments Pass Vote on Bills Calling for UC-Lab Severance," http://www.ucnuclearfree.org/student-gov-res.html (accessed June 9, 2007); "UCSC Student Assembly Resolution against UC Managed Nuclear Weapons Labs," http://www.indybay.org/newsitems/2006/04/19/18164461.php (accessed June 9, 2007); University of California Academic Senate, "Report of the University Committee on Research Policy on the University's Relations with the Department of Energy Laboratories," http://scipp.ucsc.edu/~haber/UC_CORP/doereport.html (accessed June 9, 2007); Academic Council, "Report of the Special Committee of the Academic Senate on the University's Relations with the Department of Energy (DOE) Laboratories," http://scipp.ucsc.edu/~haber/UC_CORP/jendrese.htm#EXECUTIVE_SUMMARY (accessed June 9, 2007).

41. The Federal Technology Transfer Act of 1986 also established CRADAs—cooperative research and development agreements—allowing GOGOs to enter into such agreements with nongovernmental entities.

42. CSIS, *National Benefits from National Labs: Meeting Tomorrow's National Technology Needs* (Washington, DC: Center for Strategic and International Studies, 1993).

43. Leah Beth Ward, "From Weapons to Widgets: It's Martin Marietta's Turn to Produce a Peace Dividend at the Sandia Labs," *New York Times,* October 24, 1993, F5. See also Phillip H. Abelson, "A Multipurpose National Laboratory," *Science* 261, no. 5128 (September 17, 1993): 1503.

44. For a complete summary of major laboratory reform bills introduced during the 104th Congress, see William C. Boesman, *Restructuring DOE and Its Laboratories: Issues in the 105th Congress,* CRS Issue Brief 97012 (Washington, DC: Congressional Research Service, Library of Congress, January 31, 1997).

45. Representative George E. Brown Jr. (D-CA), opening statement in introducing the Department of Energy Laboratory Technology Act of 1993, March 23, 1993, in Scott Veggeberg, "In Hot Pursuit of Post–Cold War Survival, Weapons Labs Seek Industrial Partnerships," *The Scientist* 7, no. 12 (June 14, 1993): 1.

46. Veggeberg, "Hot Pursuit"; David J. Hanson, "Congress Moves to Reorganize Department of Energy Labs," *Chemical and Engineering News,* June 13, 1993, 31–32.

47. Billy Goodman, "Uncertainty Marks DOE Scientists' Efforts to Adapt, as Their Labs Take on New Missions, New Objectives," *The Scientist* 8, no. 15 (July 25, 1994): 1.

48. Task Force on Alternative Futures for the Department of Energy National Laboratories, *Alternative Futures for the Department of Energy National Laboratories* (Washington, DC: Secretary of Energy Advisory Committee, U.S. Department of Energy, February 1995).

49. Ibid., 53.

50. Graeme Browning, "Energy Labs Overdue for an Overhaul?" *National Journal,* May 13, 1995, 1170–71.

51. "First Report of the External Members of the Department of Energy Laboratory Operations Board, October 26, 1995," http://www.seab.energy.gov/sub/frst_rep.html (accessed June 9, 2007).

52. Audrey T. Leath, "DOE Responds to the Galvin Report Recommendations," *FYI: The AIP Bulletin of Science Policy News,* April 27, 1995, http://www.aip.org/fyi/1995/fyi95.060.htm (accessed May 20, 2007).

53. Ibid.

54. General Accounting Office, *Department of Energy: A Framework for Restructuring DOE and Its Missions,* GAO/NSIAD-88-116FS (Washington, DC: U.S. General Accounting Office, May 1998), 2–3.

55. Other agencies targeted for elimination by the 104th Congress included the departments of Education, Commerce, and Housing and Urban Development. See Jeffery B. Gayner, "The Contract with America: Implementing New Ideas in the U.S.," Heritage Lecture no. 549, October 12, 1995, http://www.heritage.org/Research/PoliticalPhilosophy/HL549.cfm (accessed May 19, 2007).

56. General Accounting Office, *Department of Energy: Uncertain Progress in Implementing National Laboratory Reforms,* GAO/RCED-95-197 (Washington, DC: U.S. General Accounting Office, August 1995), 15.

57. David Malakoff, "Top Official Resigns as Congress Pushes for Management Changes," *Science* 285, no. 5424 (July 2, 1999): 18.

58. U.S. General Accounting Office Report to the Secretary of Energy, *Contract Reform: DOE Has Made Progress, but*

Actions Needed to Ensure Initiatives Have Improved Results, GAO-02-798, September 2002, http://www.science.doe.gov/opa/PDF/d02798.pdf (accessed December 15, 2007). See this document for additional background on this issue.

59. For additional detail, see *Competing the Management and Operations Contracts for DOE's National Laboratories,* Report of the Blue Ribbon Commission on the Use of the Competitive Procedures for the Department of Energy Labs, U.S. Department of Energy, November 24, 2003, http://www.seab.energy.gov/publications/brcDraftRpt.pdf (accessed December 15, 2007).

60. House Committee on Science and Technology, "Testimony of John P. McTague, Before the Energy Subcommittee of the Committee on Science U.S. House of Representatives, Hearing on Competition for DOE Laboratory Contracts: What Is the Impact on Science? July 10, 2003," http://gop.science.house.gov/hearings/energy03/jul10/mctague.htm (accessed June 9, 2007).

61. General Accounting Office, *Homeland Security: DHS Needs a Strategy to Use DOE's Laboratories for Research on Nuclear, Biological and Chemical Detection and Response Technologies,* GAO-04-653 (Washington, DC: U.S. General Accounting Office, May 2004).

62. Committee on National Laboratories and Universities, Policy and Global Affairs, National Materials Advisory Board, Board on Manufacturing and Engineering Design, and Division on Engineering and Physical Sciences, *National Laboratories and Universities: Building New Ways to Work Together—Report of a Workshop* (Washington, DC: National Academies Press, 2005), 1.

63. Guy Gugliotta, "Federal Scientists Face Trade-Offs: Science Fields Offer Prestige, Few Perks," *Washington Post,* May 8, 2000, A21.

64. Don J. DeYoung, "The Silence of the Labs," *Defense Horizons* 21 (January 2003): 1–8.

65. Laboratory Operations Board, External Working Group on LDRD Program, *Review of the Department of Energy's Laboratory Directed Research and Development Program* (Washington, DC: U.S. Department of Energy, January 27, 2000).

66. U.S. Senate Committee on Small Business, Subcommittee on Innovation, Technology and Productivity, *Diverting Government Work from Small High Technology Firms to FFRDCs,* S. Hrg. 100-145, 100th Congress, 1st sess., April 7, 1987.

67. U.S. Senate Joint Hearing, Committee on Governmental Affairs, Subcommittee on Oversight and Government Management, and Committee on Armed Services, Subcommittee on Strategic Forces and Nuclear Deterrence, *Need for and Operation of a Strategic Defense Initiative Institute,* S. Hrg. 100-137, 100th Congress, 1st sess., May 6, 1987.

68. General Accounting Office, *Competition: Issues on Establishing and Using Federally Funded Research and Development Centers,* GAO/NSIAD-88-22 (Washington, DC: U.S. General Accounting Office, March 1988); General Accounting Office, *Competition: Information on Federally Funded Research and Development Centers,* GAO/NSIAD-88-116FS (Washington, DC: U.S. General Accounting Office, May 1988).

69. Joseph P. Martino, *Privatizing Federal R&D Laboratories,* Policy Study No. 219 (Los Angeles: Reason Public Policy Institute, November 1996), http://www.reason.org/ps219.html (accessed May 20, 2007).

70. House Committee on Science and Technology, "Testimony of John P. McTague."

71. See DeYoung, "Silence of the Labs," 2–3.

72. Michael Telson, former high-level DOE staff member and chief financial officer, noted at a 1996 conference while discussing DOE's laboratories, "it's the tragedy of the commons applied to laboratories: because they're everybody's property, they're nobody's property." See Michael Telson in *Science: The Endless Frontier 1945–1995: Learning from the Past, Designing for the Future—Part III,* September 20–21, 1996, Conf. Transcript, 198, "Design Area Eight: Federal Laboratories and Other Assets of the Last 50 Years," http://www.cspo.org/products/conferences/bush/partthree/complete96.pdf (accessed May 20, 2007).

73. In talking about the increasing difficulty in differentiating the roles of universities, federal laboratories, and industry, Barry Bozeman has noted that "assumptions such as universities are for basic research or industry is for development and commercialization of technology run at odds with the configuration of research resources that we have in the United States." See his remarks in ibid.

74. For a discussion of best practices and future challenges with respect to national laboratory–university collaborations, see Committee on National Laboratories and Universities, *National Laboratories and Universities: Building New Ways.*

CHAPTER 8

Industry

The Role of Industrial R&D

Industries are the sustainable driving economic engines of most all advanced societies. The competitive environment in which they design, test, and manufacture products places a huge premium on efficiency and innovation. R&D is indispensable. It may not be surprising, therefore, that the majority of the nation's research and development efforts are concentrated in industry.

In the mid-1800s, huge numbers of people migrated to urban centers to work in factories. This industrial revolution was a major step away from the agrarian way of life that had predominated for thousands of years.[1] We are now witnessing the transition into a new era—the so-called knowledge revolution.[2] Though society continues to refine the processes developed during the industrial revolution, we are also developing remarkable new abilities to create, package, and convey information. Many of the new information-age industries have been fueled by cutting-edge R&D. At the same time, established industries are increasingly devoting their R&D efforts to creating new products, meeting demand, and responding to government regulations and standards.

Vibrant industries provide not only products but also a large percentage of the country's jobs. Therefore, the health of industry is a major concern of the federal government. This chapter reviews industry's role in advancing our national R&D objectives, and R&D's role in ensuring strong industries. We will also examine the influence of national science policies on manufacturing and commerce, and the special impact of high-technology industries on the economic health of the nation.

Even a cursory appreciation of American's industrial R&D efforts can give rise to several science policy questions: Should the federal government be involved in industry R&D? Does federal investment defy the tenets of a free market? Do government efforts to foster collaborative research by competing companies violate antitrust laws? How do we ensure a supply of trained personnel to staff our industries? When federal investments result in the development of successful products, should the government share in the returns? We will consider these and related questions in the pages that follow.

What Distinguishes Industry in the National Research Enterprise?

Industry makes products that perform some function, such that people will want to buy them. Unlike basic research, which is conducted without concern for its potential applications, applied research has a practical focus and is more immediately relevant to industrial activity.

Even so, the distinction between basic and applied research is not always clear. Basic research on electricity, such as that conducted by Wilhelm Roentgen in 1895, led to the discovery of x-rays and their penetrating and imaging properties, which in turn led to the use of x-rays in medical research and treatment applications. When, one might ask, did Roentgen realize that his quest to understand the behavior of electrons had a ready application in medicine? If we further consider that the quest for better medical x-ray instrumentation led to better tools for exploring the fundamental properties of x-rays themselves, we can see that basic and applied research are often indistinguishable. Indeed, many argue that it is fruitless to try to separate the two.[3]

Regardless of how one characterizes the distinction between basic and applied research, industry's role in R&D is strikingly different from that played by universities and national laboratories. While universities focus much of their attention on advancing the frontiers of knowledge, and the national labs focus on the creation of knowledge to advance mission-based applications, industry is dedicated to devising new and better products that will earn profits and help their producers compete in a free and open market. As we have seen, the knowledge used by industry in this process often has its roots in publicly funded research performed at universities or national labs.[4]

Some nations provide their industries with guidance on, and even financial support for, R&D. For instance, the Japanese government invests heavily in industrial R&D, viewing it as critical to global competitiveness. Japan's top science and research agency, the Ministry of Economy, Trade, and Industry (METI), provides government funds directly for efforts to "bootstrap" its domestic industries up to world standards. Such investments helped the Japanese electronics industry catch up with and eventually surpass U.S. industry in developing semiconductor technologies and consumer electronics.[5]

Unlike Japan, the U.S. government tends to take a hands-off approach to industrial R&D. The U.S. government does not provide funding for industrial R&D as it does for university research. This distancing is partially due to the nature of U.S. capitalism, which is characterized by a resistance to direct government interference in the market.

When policies are established that support specific areas of industrial R&D (e.g., DOE efforts to support renewable energy technologies) or promote high-risk, early-stage innovation (e.g., the DOC's Advanced Technology Program), some policymakers are quick to warn against straying from science policy into industrial policy, thus putting the government in the position of picking commercial winners and losers. Instead of addressing the nation's scientific and technological needs, these critics claim, such policies represent "corporate welfare." The one area in which legislators seem to have shown a much greater tolerance for government involvement is national or homeland security, an area in which the U.S. government itself is often the primary consumer (e.g., of high-tech military equipment).

History of Major Industrial Laboratories

Industrial laboratories played a very important historical role in U.S. R&D throughout the past half century. For example, the noted Bell Laboratories, founded in 1925, was responsible for numerous inventions and discoveries, including the transistor, the laser, and the UNIX computer operating system. It also made major advances in radio astronomy. Six Nobel Prize winners were associated with the laboratory during its life. Through a number of steps in the late 1990s, the laboratory was spun off from AT&T. With a more focused research mission, the entity is now owned by the French conglomerate Alcatel and goes by the name Alcatel-Lucent.

An additional example of a large industrial scientific laboratory with industrial roots was the David Sarnoff Research Center operated by RCA. Many breakthrough contributions may be attributed to that laboratory as well, but it ceased to exist when General Electric purchased RCA in the late 1980s. Though some of the components survived (e.g., Lockheed Martin purchased the government contract portion of the laboratory), there was a loss of the ability that existed within the original center to carry out long-range basic and applied research of the type that had helped propel it to prominence.

It is interesting to note that the elimination or transformation of industrial laboratories of this scale was often associated with the realization by corporations that the reductions or closure of their research centers would likely result in an immediate improvement in the company's bottom line (since the operating costs for advanced research would vanish). Any downsides would appear much later, given the length of time usually required for discoveries to result in final products. Moreover, with the shrinkage of the average tenure of a CEO to just a few years, it is known that major expenditures for basic research by a current CEO would, at best, likely only help the next CEO. Incentives for taking individual actions to invest in the long-term health of an industry have diminished, and industry has become less motivated to make significant investments in long-range research. Policymakers need to be aware of the changes in trends in management turnover rates and the resulting impact on the nation's overall R&D capacity.

POLICY DISCUSSION BOX 8.1

When Is Government Interference in Industrial R&D Legitimate?

National science and technology policy most often affects industrial R&D in the areas of global economic competitiveness, national security, environmental impact, public health, and disaster preparedness. In some instances, when it is viewed as being in the government's interest to do so and when normal market forces alone will not adequately ensure public safety or national and homeland security, policies may be enacted that directly attempt to influence the nature of the R&D in which industry engages and the products it produces.

One such effort, Project BioShield, was announced by President George W. Bush in his 2003 State of the Union address. Spawned by concerns that emerged after September 11, 2001, and the subsequent anthrax attacks on Capitol Hill, Project BioShield is an initiative aimed at encouraging companies to develop and make available new drugs and vaccines against pathogens and chemicals that might be used in a biological and chemical attack.[6] Over a ten-year period close to $6 billion is to be dedicated to this initiative, largely in the form of government contracts to biotech and pharmaceutical companies for the development and purchase of the next generation of biological attack countermeasures. These drug and vaccine purchases will be added to the Strategic National Stockpile (SNS).

The initiative includes other components to promote the development of treatments and vaccines, including legislation to expedite NIH research on relevant medical countermeasures and the process of FDA approval in times of crisis. The project will be overseen jointly by the secretaries of homeland security and health and human services. Both the House and Senate versions of the bill passed with great bipartisan support, demonstrating the urgency of this issue. The bill was signed into law by President Bush in July 2004.

Congressional approval of Project BioShield raises some interesting questions concerning the role that government should play in encouraging industry to invest in areas where normal market forces may not ensure public welfare and homeland security.

Of course, for the purposes of preserving our national security, the primary customer for almost all new weapons systems and military devices is the government. In this sense, the government is the only market for new products in these areas. In what other instances does it make sense for the government to create an artificial market? Was there a legitimate need in the instance of Project BioShield for the federal government to set aside funds specifically to purchase the drugs and vaccines from biotechnology and pharmaceutical companies? Is this a form of "corporate welfare"? Do corporations have a social responsibility to meet the needs of society even if there is no clear profit to be had? At what point is it appropriate for the government to provide incentives for the development and stockpiling of products, such as vaccines, to ensure public welfare and security, when regular market forces are unlikely to do so? When does it make sense for the federal government to promise to be the primary consumer of a particular product? When should market forces be allowed to operate freely? In the case of vaccines, what would happen if the government did not present itself as a consumer? Can you think of other instances where it might be necessary for the government to act in a similar manner?[7]

Types of Industry

Almost all industries conduct research and development in one form or another. While national science policy is not designed to build a better mousetrap, it is concerned with creating an environment in which private industry can do so. And this concern must extend across all sectors of industrial R&D.

High-tech industries occupy a special place in this scheme, inasmuch as they involve a great degree of early risk, albeit with a promise of comparably high returns. The Organization for Economic Cooperation and Development (OECD) identifies four industries as belonging to the high-technology category: aerospace, communications, pharmaceuticals, and computers and office machinery.[8] These areas are distinguished by a high degree of innovation, a high level of value added to their products, and positive spillovers to other industries—through products and services that enhance productivity and create new jobs.

Many other industrial sectors, however, are also affected by national science policy. Legal constraints on automobile emissions, for example, force the automotive industry to search for technological solutions to pollution. A national science policy that promotes new initiatives in nanotechnology could help the industry create technologies like fuel cells, which might enable automobile manufacturers to meet federal requirements and remain competitive.

Because of its historical responsibility for tracking U.S. trade, the Bureau of the Census has been instrumental in creating a classification system that takes account of leading-edge technologies. The list is referred to as Advanced Technology Products categories.[9]

- *Biotechnology:* the medical and industrial application of advanced genetic research, for the creation of drugs, hormones, and other therapeutic items, for both agricultural and human uses.
- *Life science technologies:* the application of nonbiological scientific advances—such as nuclear magnetic resonance imaging, echocardiography, and novel chemistry—to medicine.
- *Opto-electronics:* the development of electronics and electronic components that emit or detect light, including optical scanners, optical disk players, solar cells, photosensitive semiconductors, and laser printers.
- *Information and communications:* the development of products that process information, including fax machines; telephone switching apparatuses; radar devices; communications satellites; central processing units; and peripherals such as disk drives, control units, modems, and computer software.
- *Electronics:* the development of electronic components (other than opto-electronic), including integrated circuits and multilayer printed circuit boards, as well as surface-mounted components, such as capacitors and resistors.
- *Flexible manufacturing:* the development of products for industrial automation, including robots, numerically controlled machine tools, and automated guided vehicles.
- *Advanced materials:* the development of materials such as semiconductor materials, optical fiber cable, and videodisks that enhance the operation of other technologies.
- *Aerospace:* the development of aircraft and associated products, including military and civilian airplanes, helicopters, spacecraft (with the exception of communications satellites), turbojet engines, flight simulators, and automatic pilots.
- *Weapons:* the development of technologies with military applications, including guided missiles, bombs, torpedoes, mines, missile and rocket launchers, and some firearms.
- *Nuclear technology:* the development of nuclear energy production apparatus, including reactors and parts, isotopic separation equipment, and fuel cartridges. (Note that nuclear medical apparatus is considered a life sciences product rather than a nuclear energy technology.)

A running profile of the health of America's industrial activities in these areas can be generated by comparing exports and imports in these various categories against what is happening in other countries. Of course, any calculations of market demand have to take into account the fact that some industries, especially in defense, depend largely on government demand, as opposed to true market forces

In the sections that follow, we describe and compare in greater detail five industrial sectors that contribute significantly to, and that rely upon, U.S. R&D.

The Aerospace Industry

R&D in the aerospace industry focuses on the development of safer, faster aircraft for commercial, defense, and space use. This includes not only planes, but technologies used by pilots and air traffic controllers, as well as communications, observation, and global positioning satellites. Space research and travel have traditionally been the sole province of NASA scientists and engineers, although the spacecraft themselves are built by private firms. This may be changing, however: *SpaceShipOne,* a commercially developed vessel, flew successfully in September and October 2004.[10] Its development and launch were part of a ten-million-dollar competition sponsored by the X PRIZE Foundation, which is dedicated to driving entrepreneurial spirit and jump-starting cross-disciplinary technological innovation to meet some of the greatest societal challenges of today. General aircraft technology was spurred in the twentieth century by a similar prize, the Orteig Prize of twenty-five thousand dollars, won by Charles Lindbergh in 1927.

The United States has historically led this field. However, European and Asian nations are stepping up their efforts: with the help of European Union subsidies, Airbus, the EU's largest aircraft maker, surpassed Boeing in the production of commercial jets in 2004, though in 2006 Boeing recovered the lead. Parity between the two aerospace companies would have been unthinkable a decade ago. We note that China, too, is dedicating government funding to its aerospace industry.[11] A very significant issue in trade policy is the appropriateness of governments "tilting the playing field" by providing subsidies to their aerospace industries.

American aerospace R&D work is often funded under contract from government agencies like NASA, DOD, and the Federal Aviation Administration (FAA), each of which has its own budget for sponsored research. In recent years, however, the NASA budget for aeronautics research has remained flat, a trend that is causing concern in the industry.[12] Aviation companies also advocate for increases in the FAA's R&D budget, and increased attention to aeronautics in the DOD's Research, Development, Test, and Evaluation budget. As Clayton M. Jones, chairman, president, and CEO of Rockwell Collins, and John W. Douglass, president and CEO of the Aerospace Industries Association, noted in a 2004 issue of *Aviation Week and Space Technology:* "Although we do believe in free markets, it is evident that there is no free market for aerospace products. Every country that produces aerospace products receives a variety of government assistance in the form of R&D funding, favorable loan terms, and the like. To survive in the global marketplace, U.S. aerospace requires a level playing field."[13]

Industry leaders have also expressed concerns over the lack of an integrated vision for the future of national aerospace efforts, and the very slow transmission of government research to the private sector. The lack of a comprehensive vision has real-world implications: limited government funding has forced the industry away from high-risk but potentially high-payoff research, while innovative aerospace research initiatives—with the exception of the media-friendly but scientifically low-priority manned mission to Mars—are neglected by the Congress and the White House. This lack of vital federal support for long-term research means that manufacturers are relying on their suppliers for R&D, who in turn are outsourcing their research to subcontractors.[14] The Aerospace Industries Association is the primary trade association and works with Congress on these and other issues and others important to its members.

The Auto Industry

As in the aerospace industry, R&D in the automobile industry is focused primarily on development. Major automotive companies are spending on the order of five to seven billion dollars each, annually, on R&D, most of which is directed toward improving safety and comfort, fuel efficiency, and overall vehicle design. The Big Three automakers are currently focused on hydrogen fuel-cell and hybrid vehicles, but several other initiatives are simultaneously under way. Ford's VIRTTEX laboratory, for example, allows scientists and engineers to duplicate highway conditions and measure a driver's ability to deal with traffic conditions while using a cell phone or a navigation system, or any of the other electronic devices with which new vehicles are equipped.[15]

The automobile industry is one of the largest of America's manufacturing sectors. Unlike the aerospace industry, however, automakers are not heavily reliant on the government as a customer. Instead, the government influences industry through indirect routes, such as by forcing innovation through fuel efficiency requirements.

U.S. car manufacturers, in particular the Big Three Company (i.e., General Motors, Ford Motor, and Chrysler), have over the years sponsored a great deal of university research: to cite just one instance, in 1999 Ford Motor Company pledged $20 million over five years to MIT for research partnerships, while the General Motors Foundation gave MIT a grant of $1.5 million over five years for both research and student support.[16]

The American auto industry is now facing competitive challenges similar to those confronting the aerospace industry. This results in almost all of the automakers' R&D being linked to some specific end-product, while they rely more and more on universities for long-term research in the physical sciences, as well as some of their shorter-term engineering needs.[17] While the overall R&D budget for the industry, then, is large—globally it is 3–5 percent of revenues—it is nonetheless heavy on the "D." As Bernard Robertson noted when he was Daimler-Chrysler senior vice president of engineering technologies and regulatory affairs: "The vast majority, actually, of the R&D funds spent by the auto industry are in what I think most of us would consider development, the next platform, the next iteration of a product. That's not to say there isn't a lot of innovation in it, but I think we'd all agree it's D and not R."[18]

The industry's research efforts are largely concerned with regulation, in particular fuel emissions standards. This fact, along with consumer demand, is responsible for the increasing industry interest in fuel-cell development. Thus, while federal R&D budget policies are government's primary avenue of influence over the aerospace industry, it is agency regulations that play the greatest role in the automotive sector.[19]

The automotive industry, like many other sectors, is becoming increasingly reliant on partnerships, either with universities (e.g., the University of Michigan's Automotive Research Center), government (e.g., the Partnership for a New Generation of Vehicles, or PNGV), or among automakers themselves (e.g., the U.S. Council for Automotive Research, or US CAR). The Alliance of Automobile Manufacturers promotes the interests of the industry as a whole, and works with Congress to address issues important to its member companies.

The IT and Computing Industry

According to the U.S. Department of Labor, the IT (information technology) industry is one of the fastest-growing sectors in American business, with output projected to grow more than 65 percent between 2002 and 2012.[20] Cyberinfrastructure is becoming a linchpin for conducting research across the board—not just in IT or computer science, but in biomedical research, particle physics, political science, history, and almost every conceivable field. Higher bandwidth networks, increased CPU speeds, more robust databases, and increased capacity to dynamically share computational resources are all frontiers that promote expanded R&D in almost all areas of science and engineering, as well as pose exciting R&D challenges within the field of computer science itself. A myriad of companies have emerged to push the boundaries in these areas. Some manufacture advanced hardware; others develop advanced software, either for systems operation or specific applications. The Hewlett-Packard company, as an example, invests about $4 billion annually in its R&D activities.[21]

The IT industry is currently confronting several challenges, including making hardware and software accessible to the disabled, ensuring cybersecurity, and coming to terms with the open-source software model.[22] This sector also supports "eHealth" initiatives, which use information technologies to handle health care administration, transactions, and delivery of services, and "eVoting," or electronic voting. IT is an area where the nation is failing to produce enough trained specialists to meet its needs. The public policy advocacy group representing the IT industry is the Technology CEO Council.

The Pharmaceutical Industry

U.S. pharmaceutical companies are often collectively referred to as "big pharma." R&D in this industry is focused on drug discovery and development, including therapeutics for both humans and animals. Pharmaceutical companies are research intensive. In fact, because pharmaceutical companies are so concerned with proprietary information, they probably fund more basic research than most other sectors: to ensure that they hold exclusive intellectual property rights, it is usually in their best interests to make discoveries themselves.

The Drug Price Competition and Patent Term Restoration Act of 1984 governs much of the ownership of intellectual property in the pharmaceutical industry. More commonly known as the Hatch-Waxman Act, this law amended the Federal Food, Drug, and Cosmetics Act, permitting generic drug manufacturers to submit a drug application without redoing the clinical tests that simply duplicate tests the brand-name manufacturer performed, and allowing brand-name manufacturers to request patent extensions for their drugs.[23] The aim was to get cheaper drugs (i.e., generics) to market faster, while still providing an incentive for pharmaceutical companies to invest in new product development. Citizens' groups and generic drug manufacturers continue to claim that big pharma has abused Hatch-Waxman's extension clause by tinkering with existing medications just enough to label them as new and hence subject to patent protection; in return, big pharma claims the slow FDA approval process means that they need extended IP protection in order to recover the costs of bringing a drug to market.

On average, a given drug takes eight to ten years to travel the pipeline from discovery to market, including both preclinical and clinical trials. Since the passage of the 1992 Prescription Drug User Fee Act (PDUFA), pharmaceutical (and biotech) companies have paid an annual fee directly to the FDA to help cover the agency's administrative costs.[24] By paying the costs of additional staffing, the fee is intended to allow more efficient processing of drug approval applications, and thus get drugs to the public more quickly, yet safely. In recent years, big pharma has been beset by concerns about the transparency of clinical trial data, in particular those deriving from sponsored trials. Debate has focused on whether pharmaceutical companies should publish all such studies, including those that may reflect poorly on their products. GlaxoSmithKline, the maker of the antidepressant Paxil, came under fire in 2004 for not publishing study data supposedly suggesting that the drug may lead to suicidal tendencies in children and adolescents.[25]

The pharmaceutical industry invests more than $32 billion annually in R&D.[26] Comparison with the much lower investment numbers in other sectors points to the centrality of research to the pharmaceutical industry. While big pharma does not, like the aerospace industry, depend heavily on the government as a buyer, it does rely on government intervention, for example, regarding the question of generic manufacturers' ability to sell copies of brand-name drugs. The Pharmaceutical Research Manufacturers of America (PhRMA) promotes the industry, working with Congress to address issues of importance to its member companies.

Small Tech Companies

Small technology-based companies are often founded on the basis of R&D innovations, and in particular the "R." Because such firms are often unable to afford large manufacturing facilities, they may license larger companies to further develop and manufacture the prototype technologies they develop. Sometimes a small company will actually be acquired by a larger firm seeking the IP rights to these technologies. Esperion Therapeutics, a biopharmaceutical company that focused on the development of therapeutics targeted at high-density lipoprotein for use in treating cardiovascular disease, was recently acquired by Pfizer Inc. As Esperion founder, CEO, and president, Roger Newton, noted, "The acquisition will enable us to utilize Pfizer's skills and apply the resources necessary to develop our pipeline of compounds to benefit patients with atherosclerosis."[27] Pfizer has a strong cardiovascular disease research group, so acquisition of the technologies, including a couple of Esperion's compounds in early clinical trials, fit well in its existing program.[28]

A study sponsored by the U.S. Small Business Administration found that small-firm success is linked much more closely to scientific research than that of large firms. Several other studies have also shown that small businesses and start-up companies are more innovative than larger companies.[29] Moreover, smaller companies are much more active in the kind of risky basic research that larger, more established firms may be reluctant to invest in. The difference stems at least in part from the nature of public versus private holdings, and the role of shareholders. Large firms that may have traditionally pursued long-term research now seem to have to cut these programs back in order to maintain a competitive edge.[30] In any case, sound science policy has to be sensitive to the fact that smaller companies are, by and large, more reliant on basic research. Keeping the U.S. start-up and entrepreneurial system strong is key to maintaining the nation's scientific and thus technological and economic prowess.

The Robustness of U.S. Industry

Between 1980 and 2003, the United States consistently led the world in design and manufacture of high-technology products, generating about 42 percent of annual global output in 2003.[31] This industrial activity contributes significantly to the high living standards enjoyed in the United States.

As we know, the bulk of American R&D is now conducted by private industry. This has not always been the case. Indeed, industry R&D expenditures only overtook federal outlays in 1980 (see fig. 5.2, chap. 5). Many policy analysts debate the meaning of this crossover. Some see it as a remarkable testament to the effects of historically high levels of government investment in basic research

and knowledge production, arguing that these investments strengthened industry to such a point that it is now able to take over responsibility for a robust system of applied research. Others claim that the crossover is a sign of the government's declining commitment to front-end research, visible as a shrinking proportion of total GDP. These critics argue that we should not assume industry can replace government support of R&D, especially in basic research. Indeed, they note, of the investments made by industry in R&D, only a small fraction is directed toward basic research (see fig. 8.1).

After September 11, private investment in R&D declined slightly, and then remained at about the same level through 2004. It is too early to tell whether this is just a temporary dip in the otherwise steady increase of industrial R&D investment or a serious and lasting change. If indeed the trend is toward less industry investment, then the United States will need to identify other funding sources in order to maintain its tradition of steadily increasing R&D investment over time.

Between 1988 and 2001, the output of high-technology manufacturing industries grew at an annual rate of 6.5 percent, while output from other manufacturing industries grew just 2.4 percent a year.[32] The United States enjoyed market share gains in each of the OECD's four areas during the 1990s, except aerospace.[33] America's loss of global market share in the aerospace sector is partly due to the aggressive activities of European competitors, providing a good example of how a country can quickly lose its preeminence in a given field.

The role of R&D in a nation's economy can often be assessed by comparing the country's imports and exports of technology.[34] Since 1980, the United States has exported more high-technology products, as a percentage of GDP, than products from other manufacturing sectors. Furthermore, the export of these goods as a percentage of GDP has been increasing since 1984.[35] This illustrates the very significant role of high-technology products in America's overall trade balance. In many respects these numbers testify to the wisdom of U.S. investments in science and technology, which creates U.S. jobs that are primarily funded by revenues received from our export client countries. This indicates why there should be concern if the downturn in investments after 2000 continues.

Intellectual property (IP) activity is another common measure of R&D's role in a nation's economy. IP activity includes everything from the issuing of patents to the number of resident inventors, to royalties received for the licensing of technologies to other countries, to payments made to other countries for licensing their technologies.

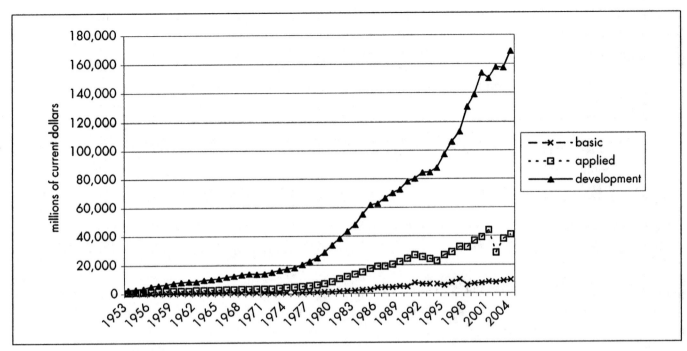

FIG. 8.1   U.S. industry expenditures, by type of research: 1953–2004

The United States has traditionally had a net intellectual property trade surplus. This gap may close, however, as other countries increase their R&D investment and output. For example, the share of U.S. licensing receipts coming from Asian nations declined from approximately 39 percent in 1999 to 26 percent in 2001.[36] In 2001, Japan and South Korea accounted for about 54 percent of the United States total receipts, while in 2003, these two countries accounted for about 45 percent.[37]

It should be obvious by now why piracy of intellectual property is a major government concern. Just because a technology is patented in the United States does not mean that this patent protection will be recognized or enforced in another country. Pharmaceutical companies in some African and Asian countries are manufacturing AIDS drugs whose compounds or technologies are the intellectual property of Western pharmaceutical companies. The World Trade Organization's Doha Declaration of November 2001 effectively permitted the compulsory licensing of technologies in the name of "WTO Members' right to protect public health and, in particular, to promote access to medicines for all"—in other words, one country's intellectual property could be requisitioned by another country facing a public health crisis. This declaration was made in response to the need to guarantee worldwide access to vaccines and therapeutics otherwise protected by patent, in particular those for HIV/AIDS.

Much of America's success as a leader in high-technology sectors is due to the innovativeness of U.S. scientists and engineers. But many other factors are also involved. For one thing, U.S. capitalist culture enables innovation. Economist Paul Romer has remarked that "The United States is considered a large, open market. These characteristics benefit U.S. high-technology producers in two important ways. First, supplying a market with many domestic consumers provides scale effects to U.S. producers in the form of potentially large rewards for the production of new ideas and innovations. Second, the openness of the U.S. market to competing foreign-made technologies pressures U.S. producers to be inventive and more innovative to maintain domestic market share."[38]

Industry-Government-University R&D Initiatives

Though industry, universities, the national laboratories, Congress, and the executive branch all share the desire to advance science and technology, divergences in the details of their respective missions sometimes impede full-scale cooperation. For example, university research often involves the free sharing of ideas, whereas industry focuses more heavily on proprietary knowledge, and the national labs are often used for classified research. Moreover, while industry is focused on the creation of near-term profits, universities and national labs are more likely to support longer-term research and development.

Numerous attempts have been made to identify areas of common interest to industry, universities, the national labs, and the federal government. Some of these are described in the following sections. It is important to note that these legislative actions, government policies, and nongovernmental initiatives share a concern for the link between R&D investment, innovation, and economic growth. Many of them are aimed at moving the fruits of basic research out of university labs and into the marketplace, either by creating direct incentives or encouraging cooperation.[39]

The environment in which we live is dynamic, and almost every major policy issue needs to be periodically revisited. Oversight is often assumed jointly by Congress and executive branch offices, including the OSTP and the OMB.

Four such initiatives receive more detailed attention in the sections that follow.

The SBIR and STTR Programs

The Small Business Innovation Research (SBIR) program was created by Congress in 1982 as part of the Small Business Innovation Development Act (P.L. 97-219).[40] The SBIR's specific aims were to support and stimulate U.S. technology innovation, to use small companies to meet federal R&D needs, to increase private-sector commercialization of innovative ideas derived from federally funded research, and to foster and encourage participation in technology development by businesses owned by historically socially disadvantaged people.

Under law, federal entities with extramural R&D budgets over $100 million are required to administer SBIR programs. Any agency or department that does so is required to set aside 2.5 percent of its annual R&D budget for the SBIR program. The goal is to encourage small companies to conduct innovative research or R&D that has potential for commercialization and public benefit. As of 2003, the SBIR program had awarded more than $12 billion in federal funds to small businesses.

The 1982 act was reauthorized in 1992 by the Small Business Research and Development Enhancement Act (P.L. 102-564) and again in 2000 by the Small Business Reauthorization Act (P.L. 106-554). The 2000 reauthorization extended the program to 2008, strengthening and

increasing its emphasis on commercial applications of SBIR project results.

Title II of the 1992 reauthorization established the Small Business Technology Transfer (STTR) program. Its goal was to promote the transfer of technology developed by nonprofit research institutes (e.g., universities) through small business entrepreneurship. Through the STTR program, awards are made to small businesses engaged in cooperative R&D ventures with a nonprofit research institution. To promote the delivery of the resulting technologies to the market, small business awardees and partnering institutions are required to sign an agreement describing their allocation of intellectual property rights.

Federal entities with extramural R&D budgets over $1 billion are required to administer STTR programs. These agencies and departments are required to set aside 0.30 percent of their annual R&D budget for the purpose.[41] The original 1992 act was reauthorized in 1997 by the Small Business Reauthorization Act (P.L. 105-135) and again in 2001 by the Small Business Technology Transfer Program Reauthorization Act (P.L. 107-50). The 2001 act extends the program until 2009.

These two programs have some differences. First, under the SBIR program, a principal investigator must be primarily employed by the small business at the time of award and for the duration of the project period. Primary employment is not stipulated under the STTR program. Second, the STTR program requires that collaborators be drawn from both a nonprofit research institution and a small business, and that these partners have a formal collaborative relationship. Both programs segment funding into three phases, but funding is distributed differently across each phase.

A 1998 review under the leadership of former Intel chairman Gordon Moore concluded that the SBIR program had strong support inside and outside Washington, but that, given its size and complexity, more study was needed to assess its success.[42] The Moore study also noted confusion over what is really meant by the "commercialization" step referred to in the authorizing legislation, and the difficulty of evaluating a program administered by different agencies, each employing its own guidelines.

Though the program enjoys significant bipartisan support, it is not without its critics. Many scientists see the SBIR as a tax on their research funds—and believe they could make better use of these funds than the SBIR researchers do. Others see the program as veering dangerously close to government interference in the market.[43]

The Advanced Technology Program

The Advanced Technology Program (ATP) was established within the National Institute of Standards and Technology via the Omnibus Trade and Competitiveness Act of 1988. It was aimed at a growing U.S. trade imbalance and the need to increase U.S. global competitiveness in light of growing Japanese dominance of certain markets. While initially authorized in 1988, the ATP program was not actually funded until 1990. Its aim is to "assist U.S. businesses to improve their competitive position and promote U.S. economic growth by accelerating the development of a variety of pre-competitive generic technologies by means of grants and cooperative agreements."[44] ATP-funded projects, then, must meet the current needs of U.S. industry, not those of the government itself. In 1994, the program began accepting proposals in five categories: biotechnology, electronics, IT, advanced materials and chemistry, and manufacturing.

For-profit firms take the lead on ATP projects, although often in partnership with academia, independent research organizations, and national labs. The program imposes strict cost-sharing rules: large firms pay 50 percent or more of total project costs, while small and medium-sized companies pay at minimum all indirect costs associated with any one ATP project. ATP funding is not for development projects; awards are granted only to high-risk research, in keeping with the program's mission to provide incentives for firms to undertake research they would otherwise not pursue. All proposals are rigorously peer-reviewed.

As of mid-2003, more than half of the ATP proposals receiving funding were from individual small businesses or joint ventures led by a small business.[45] Over half of the approximately seven hundred projects selected by the ATP since 1990 have included one or more universities as either a subcontractor or a joint-venture member. A provision in the law that precluded universities from taking the lead on ATP projects, and therefore claiming primary rights to the associated intellectual property, led some universities to refrain from active participation in the ATP program. This provision was recently modified by the America COMPETES Act (P.L. 110–69), which might lead to increased university ATP participation in the future. This legislation also modified the ATP in other ways and renamed it the Technology Innovation Program (TIP).

The ATP program was supposed to enable the federal government to act as a pseudo–venture capital firm, investing in emerging technologies that were too risky for private investors but that held great commercial and social promise. As the president of one recipient firm noted, "it's one of the few programs that really tries to transition technology to the nonmilitary sector."[46] Applicants are asked to detail how their product will be commercialized and to demonstrate the product's social benefits.

These laudable goals have not kept the program from becoming the perpetual target of some in Congress who believe that ATP involves the federal government too actively in the private sector. These legislators have repeatedly sought to zero out or greatly reduce ATP funding during the annual appropriations process. As a result, there have been some years, such as FY 2005, when new ATP funding competitions have been put on hold and no new ATP awards have been made. Funding for ATP peaked in 1995, when the program received $341 million.

Opponents view ATP and research tax credits as "corporate welfare." Stephen Moore, director of fiscal policy at the Cato Institute, defines corporate welfare as "the use of government authority to confer privileged or targeted benefits to specific firms or specific industries."[47] Moore argued in June 1997 testimony that "the explicit purpose of programs like the ATP . . . is precisely to provide targeted benefits to specific firms and industries."[48] This argument overlooks the positive spillovers resulting from government-supported industrial research that flows out to other firms, academia, and the general public.

SEMATECH

SEMATECH (SEmiconductor MAnufacturing TECHnology) is a collaborative project of the Semiconductor Industry Association (SIA) and the Semiconductor Research Corporation (SRC), two of the leading associations supporting semiconductor research. The semiconductor industry (originally called the integrated circuit, or IC, industry) originated in the United States in the late 1950s and grew rapidly through the late 1970s, at which time Japan's expanding semiconductor R&D capabilities enabled it to appropriate some of America's world market share.[49] By 1986, the Japanese had about 45 percent of the world semiconductor market, while the United States had 42 percent.[50]

In response to this shift, SIA and SRC began exploring possibilities for cooperation. SIA, a trade organization, had been established in 1977 to gather reliable information on the industry and represent the sector's interests to the government. SRC, a research consortium, was created in 1981 to assist joint university-industry research in advanced semiconductor technology. The SEMA-TECH consortium that resulted from SIA-SRC collaboration brought together fourteen U.S.-based semiconductor manufacturers and the federal government to find solutions to common manufacturing problems, leverage resources, and share risks. In September 1987, Congress authorized $100 million in matching funds. Austin, Texas, was chosen as the site for the group's R&D facility. As U.S. semiconductor firms began regaining their place in the world market during the 1990s, SEMATECH's board of directors decided to wean the organization off of federal funds by 1996.

SEMATECH has continued to serve its membership and the industry as a whole through advanced technology development in areas such as lithography, front-end processes, and interconnection, as well as through its interactions with an increasingly global supplier base.[51] In 1995, six U.S. and seven non-U.S. companies jointly formed a subsidiary—the International 300mm Initiative—which was designed to move the industry to the new standard of a 300mm silicon wafer.[52] In 1998, five of the seven foreign firms from the 300mm group joined another subsidiary of SEMATECH, known as International SEMATECH. In recognition of its increasingly international membership, SEMATECH took on the subsidiary's name in 1999, and has henceforth been known as International SEMA-TECH.[53] In 2003, International SEMATECH implemented a modified business model designed to include regional governments (e.g., Texas and New York State) in cooperative semiconductor R&D.

Impressions to the contrary, international collaboration does not run counter to America's national interest. First, an international effort does not exclude a country-wide focus by each member nation, just as a nationwide initiative does not inhibit regional or state efforts. Second, given that the United States has only a tenuous lead in the highly competitive semiconductor industry (as Japan's ascendancy proved in the 1970s), foreign nations would probably form some kind of consortium with or without the United States; by participating, the United States benefits from the knowledge shared among the group's member states. Finally, in light of the industry's push to establish international standards, membership gives U.S. manufacturers an important voice in setting the global agenda for future R&D.

Whether SEMATECH has been successful is a matter of opinion. Certainly, government intervention helped the United States regain its leadership in the market. Over time, however, the semiconductor industry and SEMA-TECH itself have become more globally oriented, and even U.S. companies have begun to move certain operations overseas. In this environment, it has become difficult to clearly identify individual companies as inherently American.

Additional R&D Policy Issues in Industry

Several additional policy issues affect R&D in industry, including the extent to which companies are allowed to collaborate on targeted research, the impact of mergers and acquisitions, and how the federal government treats their expenditures for R&D for tax purposes.

Cooperative R&D in Industry

In modern R&D, it is true as often as not that some of the most basic problems are simultaneously being studied by the labs of multiple corporations, using similar approaches. For example, both General Motors and Ford Motor Company have major efforts under way to improve fuel efficiency. While even more collaboration between auto companies on such shared R&D interests would presumably be faster and more efficient, our entire economic system and its associated laws (e.g., intellectual property and antitrust legislation) require a certain degree of separation between competitors.

Companies are often hesitant to work together on R&D because of a reluctance to share any resulting intellectual property (IP). And even if they were inclined to cooperate, antitrust laws would prohibit them from setting uniform prices for any products that might result. Two firms collaborating on product development are in a potentially favorable position to coordinate pricing to their benefit, and to the consumer's harm.

In some cases, governmental incentives or matching funds might effectively be used to encourage greater private-sector collaboration on key national science and technology priorities, without intruding on pertinent statutes or harming the public interest. The government has in some instances (e.g., the ATP program) tried to foster collaboration and cooperation among U.S. industry, national laboratories, and universities. American universities may be the neutral ground on which these collaborations can happen. If industry-government interactions on R&D are mediated through universities, the government is less likely to be accused of picking winners and losers in the private sector.

R&D and Corporate Mergers

The 1990s saw dramatic growth in the number of private-sector mergers and acquisitions. Large companies have merged with one another or acquired smaller start-ups. This trend has been spurred by intense economic competition, shareholders' expectations of increased returns on investment, and a relaxation of government regulation. In some instances, a merger or acquisition can produce savings by using a single set of resources to fill the direct research needs of two companies, or by achieving significant synergies in areas indirectly related to research. However, mergers and acquisitions also can lead to a less competitive marketplace, limiting incentives for large companies to "innovate" and ultimately disadvantaging the consumer.

For example, the 2000 acquisition of the pharmaceutical company Warner-Lambert/Parke-Davis Pharmaceutical (WLPD) by its competitor Pfizer Inc. was driven in part by Pfizer's interest in WLPD's research. The merger expanded the firm's research capacity, bringing Pfizer and WLPD scientists with complementary knowledge together under one "administrative roof." Whether this merger will ultimately be positive or negative for the development of new drugs and therapies is difficult to predict.

The acquisition of smaller firms by larger ones is often fueled by additional considerations: the larger company may see an opportunity to increase its capacity in a specific area of research, while the smaller firm may increase earnings and improve opportunities to develop and market its discoveries. As noted earlier, small businesses and start-up companies tend to be more willing to take on high-risk research, while larger companies are in a better position to exploit the results of such efforts.

Tax Investment Credits

Policies affecting entrepreneurship have an enormous impact on U.S. economic growth. In the United States, the regulation of commerce has shifted focus over time: from policing industry in the nineteenth century, to protecting consumers in the 1970s, and most recently to stimulating robust global competition.

The adoption of tax policies that encourage R&D investment has been a central component in this latter effort, as reflected in the R&D tax credit, which was first introduced as a part of the Economic Recovery Tax Act of 1981, in an effort to provide incentives for private R&D investment.[54] But why do firms need incentives to invest in R&D? Shouldn't the lure of profit be incentive enough?

The problem is risk. There is no guarantee that research will produce profitable innovations that will directly be returned to the company. Put in economic terms, the positive spillovers from R&D investment—for example, new knowledge and research tools—are such that the rate of return to society as a whole is greater than the rate of return to the company itself. This suggests that, left to their own devices, firms will not invest in R&D to the socially optimal level.[55]

The R&D tax credit—now also called the R&E (research and experimentation) tax credit—allows firms to deduct research and development expenditures from their taxes. The credit is available to firms of any size. Income tax credit is given for any "qualified research expenditure" (QRE), as well as external "basic research" payments.[56] QREs include in-house research expenditures for new product and process development, including salaries but not market research or modifications to existing products. External "basic research" includes a firm's payments under contract to colleges, universities, and other qualified not-for-profit research institutions.[57]

Why have an R&D tax credit? As we have seen, the pure benefits of research may not be sufficient in themselves to encourage industry investment. The theory behind the credit is that because they are paying less in taxes, companies claiming the credit will invest more in research activities. And analyses suggest that this is indeed true. In fact, a study by Bronwyn H. Hall, an economist from the University of California at Berkeley, the National Bureau of Economic Research, and the London Institute for Fiscal Studies, suggests that a 1 percent reduction in the cost of R&D (i.e., a tax credit) increases outlays in R&D by about 1 percent in the first two years after a policy is in effect, and possibly more thereafter.[58] Some experts have also suggested that the United States should make the credit permanent in order to compete with other countries who already offer similar programs.[59] Critics, however, question whether the government should provide subsidies—which is what the R&D credit effectively is—to for-profit entities. They claim that market forces should ultimately determine whether industries engage in R&D. Others argue that federally funded research can be more effectively conducted at universities and national labs.

Some members of Congress from both sides of the aisle have refused to endorse the industry R&D tax credit, which they see as a corporate subsidy. Although the tax credit was implemented in 1981, no legislation making it permanent has ever been enacted. Just in the period 1981 to 2004 alone, the credit had to be renewed ten times, each time after significant debate over whether to make the policy permanent.

Industry has mounted a major campaign to extend and make the credit permanent. A coalition of more than eighty-five trade and professional associations and more than one thousand companies engaged in U.S.-based research, from a broad range of industrial sectors, is lobbying Congress and the executive branch on this matter.[60] Many individual trade groups, including the National Association of Manufacturers (NAM) and the Technology CEO Council, also advocate for making the R&D tax credit permanent.[61] Even the President's Council of Advisors on Science and Technology (PCAST) endorsed the move, in a report focused on sustaining the nation's innovation systems.

Government's power to stimulate industrial R&D through financial incentives creates a significant risk of disenfranchising one constituent in favor of another; therefore, policymakers exercise great care when making such decisions. As we noted earlier, the ATP and SBIR/STTR programs are sometimes labeled corporate welfare by opponents who object to distributing funds to selected industries. But because the tax credit is available to any firm investing in R&D, regardless of its size or sector, many policymakers see it as leveling the playing field for the large firms who are ineligible for SBIR/STTR or ATP support. A second set of objections to the credit center on the loss of potential government revenues, which some claim could have been invested in direct government support for precompetitive research by private firms, universities, and national laboratories.

To the lay observer, the trade-off between federal tax credits and direct investment of federal dollars may seem inconsequential. In fact, however, the choice carries significant implications. Tax incentives are likely to stimulate R&D that is in the firm's own interests, whereas government grants of a comparable size give policymakers the leverage to promote research that benefits society and national interests overall. Direct federal funding also allows the inclusion of stipulations, suggesting, for example, that companies provide students with internship and co-op opportunities designed to train the next generation of scientists and engineers.[62]

The relative advantages and disadvantages of these options deserve further research. Their successful design and deployment will require more detailed consideration of innovation's impact on the American economy, and more careful evaluation of the balance between federally funded precompetitive research and industry R&D. Finally, future generations of policy analysts will need to untangle the connections between intellectual property rights, economic gains, and overall societal benefit.

Industry, the Environment, and Regulations

As understanding of environmental hazards has grown—in part as the result of science—the government has put in place stronger regulations to ensure public safety and to protect the environment. Industry may view government regulation as anything from a minor inconvenience to a life-threatening affliction. The issue is not only the choice of materials used in manufacturing, but the scale of production: although a product may contain only a minuscule amount of a given chemical, its environmental impact can be enormous when the item is mass-produced.

To illustrate, consider the impact of federal limits on automobile emissions, commonly referred to as CAFE, or corporate average fuel economy, standards. The government's goal is presumably to set allowable emissions levels at their lowest possible point, consistent with what is known about their effect on human health and the environment, as measured against what might be technically achievable by the industry. The challenge of meeting these standards sets in motion a huge R&D effort within the industry. The technical challenges may be daunting, and may be made even more so if management insists that regulations be met without significantly increasing the cost of the final product.

American firms are generally quite good at meeting such regulatory challenges. However, on occasion, they have chosen to fight regulations instead. In fall 1998, for example, Congress passed an omnibus appropriations law for FY1999 that included a last-minute provision quietly introduced by Senator Richard Shelby.[63] This two-line provision, buried deep within a very large and comprehensive piece of legislation, came to be known as the "Shelby Amendment."[64] It required that OMB Circular A-110 be changed to permit anyone, including private firms, to request the raw data underlying federally funded scientific studies, in accordance with the Freedom of Information Act (FOIA).

Senator Shelby included this provision at the behest of industry leaders, who opposed the EPA's tightening of ambient air-quality standards—a regulatory decision based on an almost twenty-year Harvard-led study of the morbidity and mortality effects of particulate matter in six U.S. cities. Industry stakeholders endorsed the amendment, arguing that it provided them with a fair opportunity to challenge scientific studies that prompted costly regulations, while scientists expressed concerns about increased paperwork, unfair exploitation of intellectual property by industry and academic competitors, unwarranted access to personal information about research subjects, and even deliberate use of the amendment to reinterpret data and harass researchers. The bill's final version does not require scientists to make data publicly available while their research is still ongoing, and the OMB has interpreted the law to allow protection of proprietary data.

In 2003, additional regulations were proposed by the OMB that would have prohibited the recipients of federal research funds from simultaneously taking part in studies upon which federal regulations are, or might be, based.[65] That is, if a researcher from the University of Michigan had a grant from the Department of Energy for hydrogen fuel-cell research, that researcher would not be permitted to participate in any fuel-cell study commissioned by a federal agency, such as the EPA. The proposed regulations were met with strenuous resistance from the scientific community, on the grounds that they threatened the integrity of the peer-review process.[66] As a result, OMB substantially modified its final proposal.[67]

Industry often finds it challenging to chase regulatory limits: as soon as they demonstrate that they can achieve one goal, regulatory agencies often raise the bar still further. When an existing industry reaches an impasse with federal regulations, entrepreneurs can sometimes exploit the opportunity: for example, several firms have emerged to address alternative energy needs after existing energy companies proved resistant to change. In such instances, economic policy, science policy, and public demands converge to advance science and technology.

Industry S&T Activities beyond R&D

While certainly industrial R&D efforts are aimed at increasing profits, this does not mean that private R&D will not have a broader social impact; indeed, the development of a single product can very quickly affect work in other sectors. But is industry also obligated to support basic research of the type normally conducted in universities? Should the private sector be expected to contribute to national science and technology efforts, even when those efforts might not directly serve the enterprise's commercial goals?

Any firm is first interested in ensuring its own survival. Beyond that, however, most enterprises recognize the value of collective investment in protecting their economic environment: including both a customer base that can afford their products and a pool of well-trained employees who can produce them.

In order to sustain this complex ecosystem, most large companies maintain not only research and development arms, but also community outreach divisions and public policy units. The R&D arms are often headed by senior scientists or engineers who report to top company officials, usually at the level of vice president or higher. Such units may comprise anywhere from a few scientists, engineers, and technicians to a staff of several thousand. The facilities may be housed in a single complex located on the corporate campus, or spread around the world. Their budgets may be relatively fixed, or tied to company profits by a strict formula.

Outreach units are often responsible for interacting with the communities served by the company, with an eye toward investments that would be beneficial to both community and company. Contributions to educational programs, including those that support science education, are a common element of many such efforts. Some firms are also teaming up with universities to provide undergraduates with summer internships in key scientific and engineering disciplines. Others may fund programs at science museums. Amgen has funded the design of the "Human Body Connection" display, which explores human anatomy, physiology, and evolution and how medicine and technology can affect people, at the Museum of Science in Boston. Not only are these high-tech companies dependent on a steady stream of bright young scientists and engineers, but, from a public relations perspective, such sponsorships cast the company in a positive light. If properly managed, such partnerships can enhance the overall level of U.S. science education while simultaneously meeting specific corporate needs.

Most large corporations also have a sizable government relations staff. Such units may be located in Washington, DC, at company headquarters, or around the world, and allow corporations to share their views on key policy issues, including those related to science policy, with members of Congress, agency officials, and executive branch officials. The topics of concern may range from the need for stronger science education initiatives, to R&D tax credits, to government-industry research partnerships, to support for universities' basic research. The private sector thus has the potential to significantly influence American science policy, especially since industry leaders often maintain close relationships with key government officials.

Growing Role of Corporate and Private Foundations in Supporting R&D

A few foundations with an interest in specific areas of science have the capacity to contribute to scientific research in these areas at a level comparable to that of federal agencies. Examples include the Gates Foundation (with an endowment of $33 billion available for charitable contributions in 2006) and the Wellcome Trust (with an endowment of $28 billion in 2006). Such entities are often the result of very successful industrial enterprises that created foundations as a result of corporate decisions or the private decisions of their successful leaders.

Several policy issues will arise in the decades ahead, including an assessment of whether the federal government should adjust its priorities to take into account the level of support specific areas of science are receiving from private sources. Should private foundations be asked to partner with federal agencies to increase the overall national R&D budget, but with general priorities for distribution of the funds being set using federal road maps?

How Much Should Industry Invest in Research versus Development?

Industry leaders are forever wrestling with the question of how to balance research and development investments.

While investment in development almost always brings a fairly quick return, new technologies do not appear out of thin air. Basic research, as we have noted, serves as the foundation for most innovation. But research efforts may not lead to commercially valuable discoveries. Even if a discovery does successfully translate to the market, the time from start to finish can be very long; ten years is not uncommon in the pharmaceutical industry, and even longer times are sometimes seen in other sectors. The payoffs, as we noted earlier, can be significant, but can a company afford to take the risk? Private start-ups, which are not beholden to shareholders, are sometimes better equipped for this kind of high-stakes research than larger, well-established, and publicly traded firms. In fact, this capacity for risk may be start-ups' greatest social value.

Corporate leaders are not the only ones concerned with the balance between investments in the "R" and the "D." The public stands to benefit from these efforts, too. There is no doubt that wise research investments can yield significant returns. But there is no simple formula for determining the appropriate level of industry investment in research.

POLICY DISCUSSION BOX 8.2

Paying Twice?

Industry has been at times criticized on the grounds that the public is "paying twice" for goods or services: because firms make use of academically derived knowledge, which was originally funded by the taxpayer, their selling the end product or service means, some argue, that consumers are being double charged. This is the argument repeatedly put forward by Senator Ron Wyden (D-OR), especially with regard to biomedical research, pharmaceutical and biotechnology companies, and the price of prescription drugs. Senator Wyden has presented amendments calling into question the high cost of drugs that are based on findings from research funded with public money. He has suggested that some of the revenues from such drugs should rightfully be returned to the federal treasury, not tucked into the pockets of pharma CEOs. Others, including some policymakers, respond that Senator Wyden fails to recognize that a firm, after taking some knowledge derived from a university, must expend large sums to take that knowledge down the long path toward a commercial product or service.

A specific example is when Abbott Laboratories increased the price of its anti-AIDS drug Norvir approximately 400 percent in December 2003. Some of the fundamental research that went into the development of Norvir was funded with a grant from NIH. When Abbott announced its price increase, there was great public outcry for several reasons, among them that some of the early research had been funded with public money. Some policymakers and members of the public strongly believed that Abbott Laboratories was being irresponsible, that the company was making consumers "pay twice" for this much-needed drug.

Are consumers "paying twice" when research findings from a publicly funded study are used to develop a product that a company sells for a profit? Do companies have a responsibility to return to the government a portion of the profits made from a product developed from research funded, at any point, by public money? Are such government recoupment policies really feasible? Could they actually be counterproductive? If so, why? Do you agree with Senator Wyden? How would you have handled the Abbott Laboratories/Norvir case if you were a policymaker?

While many forward-looking industries make significant contributions to basic research and will continue to do so, the bottom line for policymakers is that the health of basic research must be a key responsibility of government. Pressures from shareholders to deploy corporate resources to produce short-term gains severely limit industry investment in pure basic research.

Should the Federal Government Invest in Industrial R&D?

Given everything we have just said in the last section, the response to this question might seem an obvious "yes." But the answer is not so simple. Any efforts by the government to fund private R&D raise as many questions as they resolve: which R&D should the government fund, and in what industry? Any government activity that seems to benefit one company over others is problematic. Consider the case of two semiconductor companies that are racing to develop the fastest chip, using two different technologies. Should the government invest in research related to one company's technology, and not the other's? Doing so would fly in the face of the fundamental idea that market forces should determine which products succeed and which do not.

In some instances, government investment seems to fulfill the basic requirements of fairness, neutrality, and need. For example, federal programs, such as PNGV and Project Freedom Car, have brought national laboratories, automakers, and universities together to work on the development of next-generation automobiles. These programs are strictly precompetitive and address issues of concern to the nation as a whole. Clearly, science policy specialists, and public policy experts in general, need to study the many issues involved in funding such collaborations, and develop clearer criteria for deciding which cooperative ventures most deserve public support.

Policy Challenges and Questions

Industry is much more closely entangled with science policy than one might initially realize. Many regulations governing industry activities are set in accordance with the existing state of science and technology, and need to be routinely recalibrated as our knowledge advances. To a great extent, U.S. national and homeland security interests can only be met with the help of industry and its scientific and engineering man- and womanpower. The nation's economic health depends on the ability of American high-tech firms to compete in the global marketplace, and they in turn have to count on the availability of a well-trained cadre of scientists and engineers, as well as a research environment in which they can carry out their work. Industry in many ways is more reliant on the quality of the nation's science education enterprise than is any other sector in American life. Schools train students, but firms receive them after training and often end up being the ultimate site for the deployment of what the students have learned or not learned. Firms thus take a deep interest in the quality of American science education and the national policies that govern it.

The years ahead will bring many new national science policy questions of particular concern to U.S. industry. Conversely, firms' R&D decisions will help determine the need for new science policies. Consider the following: How can we improve American science education? What can be done to ensure that companies are not overly constrained from passing along legitimate R&D costs to consumers? At the same time, should these companies be allowed to use publicly funded basic research to develop new products that will be sold at a profit? How can the government encourage corporations to cooperate in pathbreaking R&D without violating antitrust laws? How can competitors share ownership of intellectual property generated through joint R&D? What will be the impact if more and more companies decide to move their R&D operations from the United States to foreign countries? Such questions must be addressed by future generations of scientists, policymakers, and academic and industrial leaders if the United States is to maintain its leadership in science and technology.

NOTES

1. The industrial revolution and the high population densities it encouraged led to new problems. Communicable diseases flourished in the new city centers, destroying untold lives. Crime and social strife became more common. But new problems also meant new solutions: the new effort to fight disease led to the study of public health and epidemiology. Sociologists and psychologists emerged to study social movements and their effects on individuals. Many of these fields in the new realm of social science adopted the scientific method, collecting and analyzing data and testing models of human behavior. Social science disciplines serve as a major bridge between the physical and natural sciences and society, and thus play a major role in national science policy. Indeed, national science policy largely revolves around topics such as science's relationship to the environment, the economy, education, and public health.

2. Peter Drucker is given credit for coining the phrase *knowledge work*. He used the term in his 1959 book, *Land-*

marks of Tomorrow: A Report on the New "Post-Modern" World (New York: Harper Press). The phrase *knowledge economy* is attributed to OECD's use starting in the 1990s. Others may be credited, for example, Paul Romer, an economist and leading proponent of the new growth theory of economics.

3. "No, a thousand times no; there does not exist a category of science to which one can apply the name applied science. There are science and the application of science, bound together as the fruit of the tree that bears it." Louis Pasteur, *Revue Scientifique*, circa 1871, "Sticky Wicket: What Would They Say Now?" *Journal of Cell Science* 114, no. 20 (2001): 3576.

4. A study conducted by Francis Narin, Kimberly Hamilton, and Dominic Olivastro ("Increasing Linkage") found that 73 percent of science papers cited in patent applications are based on research funded by government or nonprofit agencies. This discovery supports a belief long held by scientists and economists that scientific research performed in academic and governmental research laboratories and centers is the driving force behind the new industrial technologies that create economic growth.

5. See Gregory Tassey, Strategic Planning and Economic Analysis Group, *Comparisons of U.S. and Japanese R&D Policies*, Japan Information Access Project Special Report (National Institute of Standards and Technology, March 1998), http://www.nist.gov/director/planning/r&dpolicies.pdf (accessed March 9, 2007), 3–4.

6. For additional background see "President Details Project BioShield," White House, press release, Office of the Press Secretary, February 3, 2003, http://www.whitehouse.gov/news/releases/2003/02/20030203.html (accessed March 10, 2007).

7. Ibid.

8. OECD uses R&D intensities to identify high-technology industry. R&D intensity is usually calculated as a ratio between R&D expenditures and production (or output). Recent calculations examined these variables across twenty-two different industry sectors. Indirect R&D intensity can be determined by using technical coefficients in the calculations. In designating the five industries it has as high-technology, OECD took into account both direct and indirect R&D intensities for thirteen nations: the United States, Japan, Germany, France, the United Kingdom, Canada, Italy, Spain, Sweden, Denmark, Finland, Norway, and Ireland. For details concerning the methodology see OECD, *Knowledge-Based Industries* (Paris: Directorate for Science, Technology, and Industry/Economic Analysis Statistics, 2001). Also see Benoit Godin, "The New Economy: What the Concept Owes to the OECD," *Research Policy* 33, no. 5 (2004): 679–90; and National Science Board, *Science and Engineering Indicators 2006*, chap. 6, n. 4.

9. For additional background see U.S. Census Bureau, Foreign Trade Statistics, press releases, "U.S. International Trade in Goods and Services" (November 2000), explanation, p. 27, http://www.census.gov/foreign-trade/Press-Release/2000pr/11/explain.pdf (accessed April 25, 2007).

10. See *CNN*, Science and Space, "SpaceShipOne captures X PRIZE," October 4, 2004. In October 2006, the X PRIZE Foundation announced a competition to develop technology to sequence a hundred human genomes in ten days. See http://genomics.xprize.org.

11. Clayton M. Jones and John W. Douglass, "Viewpoint," *Aviation Week and Space Technology*, February 23, 2004, http://www.aia-aerospace.org/aianews/articles/2004/oped_02_23_04.pdf (accessed April 25, 2007).

12. See Aerospace Industries Association, Issues and Policies, Aerospace Research and Development, Press Briefing FactSheet, "Five-Year R&D Plan for American Aerospace (2004–2008)," http://www.aia-aerospace.org/issues/subject/aero_rd_factsheet.pdf (accessed April 25, 2007).

13. Jones and Douglass, "Viewpoint."

14. Committee on Science, Engineering, and Public Policy, *Capitalizing on Investments in Science and Technology* (Washington, DC: National Academy Press, 1999).

15. For additional background see the Ford Company, "Innovation, Safety, Helping You Avoid Accidents, Driver Distraction," http://www.ford.com/en/innovation/safety/driverDistractionLab.htm (accessed April 29, 2007).

16. See Lawrence S. Gould, "MIT: The Auto Industry's Other Research Lab," *Automotive Design and Production*, October 1999.

17. Committee on Science, Engineering, and Public Policy, *Capitalizing on Investments.*

18. Bernard I. Robertson speaking at a U.S. Department of Commerce, "Innovation in America Corporate R&D Roundtable," January 24, 2002, chaired by Deputy Secretary Samuel Bodman, Department of Commerce, http://www.technology.gov/reports/TechPolicy/p_CorpR&D_Innov3.htm#Spendi (accessed April 26, 2007).

19. This is not to say that agency regulations do not affect the aerospace industry. Indeed, the Federal Aviation Administration sets safety standards to direct R&D efforts.

20. Department of Labor, Employment and Training Administration, "Industry Profiles, Information Technology," http://www.doleta.gov/BRG/IndProf/IT.cfm (accessed April 26, 2007).

21. For additional background see Hewlett Packard, http://www.hp.com (accessed April 29, 2007).

22. Open source software refers to the ability to view and modify source code, allowing software to be altered and changed. In addition there is no legal restriction on redistributing the modified software.

23. The ability of drug companies to request an increase in the number of years over which one of their patents is active is referred to as a *patent-term extension*, where *patent-term* refers to the length of time over which a patent is active and can be enforced.

24. For additional background see Susan L.-J. Dickinson, "FDA User Fees to Speed Drug Review," *The Scientist* 6, no. 24 (December 7, 1992): 3; "Discussions Produce Set of Recommendations for PDUFA Renewal," *PRNewswire.com*, March 13, 2002, http://www.prnewswire.com/cgi-bin/stories.pl?ACCT=105&STORY=/www/story/03-13-2002/0001686324 (accessed April 26, 2007).

25. For additional background see Tanya Albert, "Lawsuit Claims Glaxo Hid Paxil Findings," *amednews.com*, June 28, 2004, http://www.ama-assn.org/amednews/2004/06/28/gvsa0628.htm (accessed April 26, 2007); "Spitzer Sues GlaxoSmithKline over Paxil," Associated Press, June 2, 2004, http://www.msnbc.msn.com/id/5120989/ (both accessed April 26, 2007).

26. For additional background on this and other issues see http://phrma.org (accessed April 26, 2007).

27. "Pfizer to Acquire Esperion Therapeutics to Extend Its Research Commitment in Cardiovascular Disease," *SECinfo*

.com, December 21, 2003, http://www.secinfo.com/dsvr4.2dNu
.d.htm (accessed April 29, 2007).

28. For additional background see "Pfizer to Buy Esperion Therapeutics for $1.3 Billion," *Bloomberg.com*, December 21, 2003, http://quote.bloomberg.com/apps/news?pid=10000103& sid=apU2qcYCmkO4&refer=us (accessed April 25, 2007).

29. See Zoltan J. Acs and David B. Audretsch, "Innovation in Large and Small Firms: An Empirical Analysis," *American Economic Review* 78 (1988): 678; Albert N. Link and John Rees, "Firm Size, University Based Research, and the Returns to R&D," *Small Business Economics* 2 (1990): 25.

30. R. R. Nelson, R. S. Rosenbloom, and W. J. Spencer, "Conclusion: Shaping a New Era," in *Engines of Innovation: U.S. Industrial Research at the End of an Era*, ed. R. S. Rosenbloom and W. J. Spencer (Boston: Harvard Business School Press, 1996).

31. National Science Board, *Science and Engineering Indicators 2006*, 6.12 and fig. 6.4.

32. The global economic activity of the OECD identified high-technology industries as particularly strong between 1995 and 2000. See National Science Board, *Science and Engineering Indicators 2006*, 6.4. Note that these percentages are based on inflation-adjusted averages.

33. National Science Board, *Science and Engineering Indicators 2006*, 6.13–6.15.

34. The National Science Board's *Science and Engineering Indicators* provides an overview of the statistics on science and engineering in the United States. It uses exports/imports and intellectual property as two mechanisms of reporting on the health of U.S. industrial R&D.

35. National Science Board, *Science and Engineering Indicators 2004*, 6.11; and National Science Board, *Science and Engineering Indicators 2006*, 6.15.

36. See National Science Board, *Science and Engineering Indicators 2004*, 6.13–6.15.

37. Ibid., 6.15; and National Science Board, *Science and Engineering Indicators 2006*, 6.23–6.24.

38. Paul M. Romer, "Why, Indeed in America? Theory, History, and the Origins of Modern Economic Growth," *American Economic Review* 86, no. 2 (1996): 202–6.

39. Charles F. Larson, president of the Industrial Research Institute, gave a talk entitled "Basic Research and Innovation in Industry" in spring 2000. This was subsequently published as "The Boom in Industry Research" in *Issues in Science and Technology* (Summer 2000): 27.

40. For further discussion see NIH, Office of Extramural Research, "Small Business Information Research (SBIR) and Small Business Technology Transfer (STTR) Programs," http://grants1 .nih.gov/grants/funding/sbirsttr_programs.htm; National Science Foundation, "Small Business Information Research (SBIR) and Small Business Technology Transfer (STTR) Programs," http://www.nsf.gov/eng/iip/sbir/; U.S. Small Business Administration, "Office Technology—SBIR/STTR," http://www.sba .gov/sbir/indexsbir-sttr.html (all accessed April 29, 2007); David B. Audretsch, Barry Bozeman, Kathryn L. Combs, Maryann Feldman, Albert N. Link, Donald S. Siegel, Paula Stephan, Gregory Tassey, and Charles Wessner, "The Economics of Science and Technology," *Journal of Technology Transfer* 27, no. 2 (2002): 155–203.

41. This percentage had been 0.15 percent until FY2004.

42. See Charles W. Wessner, ed., *SBIR Program Diversity and Assessment Challenges: Report of a Symposium* (Washington, DC: National Academies Press, 2004), 13–14.

43. Charles W. Wessner, ed., *The Small Business Innovation Research Program: Challenges and Opportunities* (Washington, DC: National Academies Press, 1999).

44. *Federal Register* 55, no. 142 (July 24, 1990): 30145.

45. See Advanced Technology Program (ATP), "Overview, About ATP, General Information, How ATP Works," http:// www.atp.nist.gov/atp/overview.htm; and Marc G. Stanley, Acting Director, Executive Briefing, "In Partnership with NIST and the Nation," February 2002, http://www.atp.nist.gov/atp/ presentations/atp_overview/slide1.htm (both URLs accessed April 25, 2007).

46. Douglas Brown, "Another Year, Another Threat to Advanced Technology Program," *SmallTimes.com*, February 6, 2003, http://www.smalltimes.com/document_display.cfm?document _id=5456 (accessed April 25, 2007).

47. Stephen Moore, before the Senate Committee on Governmental Affairs, *The Advanced Technology Program and Other Corporate Subsidies*, 105th Congress, 1st sess., June 3, 1997, http://www.cato.org/testimony/ct-sm060397.html (accessed April 25, 2007).

48. Ibid.

49. It was with SIA's encouragement that Congress approved the formation of the National Advisory Committee on Semiconductors. For further discussion see www.sia.org. Additional information concerning SEMATECH can be found at www.src.org; and Audretch et al., "Economics of Science and Technology."

50. Ibid.

51. Lithography is a surface or planographic printing process that is based on the aversive properties of water and grease. A portion of the semiconductor industry focuses tools and techniques for using the process on smaller and smaller surfaces as well as to do it more efficiently and effectively. *Interconnect*, in the context of the semiconductor industry, refers to increasing the speed and power of circuits by increasing the amount of devices in a given space. The industry is continually aiming to stretch the capacity, and a new approach being tried is 3-D interconnect technology.

52. Here 300mm refers to the size of the chip or wafer—300 millimeters. See Danile Seligson, "Planning for the 300mm Transition," *Intel Technology Journal*, Fourth Quarter 1998.

53. See Robert B. Reich, *The Work of Nations: Preparing Ourselves for the 21st Century* (New York: Vintage, 1992).

54. P.L. 97-34.

55. See Bronwyn H. Hall, "R&D Tax Policy during the 1980's: Success or Failure?" NBER Working Paper no. 4240 (1993).

56. There is a formulary for calculating the base levels.

57. See T. A. Watkins and L. Paff, "A Test for R&D Complementarities in Bio-Pharmaceutical and Software Industries using R&D Tax Price Changes," Lehigh University Health and Bio-Pharmaceutical Economics Working Paper Series 2004-1, presented February 2004; R&D Credit Coalition, "Research and Experiment Tax Credit Summary of S.627 and H.R. 1736 Investment in America Act of 2005," May 2005, http://www .investinamericasfuture.org/PDFs/FINALBillSummary 2005.doc; R&D Credit Coalition, "Research and Experimentation Tax Credit," April 2007, http://www.investinamericasfuture.org/PDFs/ Talking_Pts_4-26-07.pdf (both URLs accessed April 29, 2007).

58. Hall, "R&D Tax Policy"; see also Bronwyn H. Hall and John van Reenen, "How Effective Are Fiscal Incentives for R&D: A Review of the Evidence," *Research Policy* 29, nos. 4–5 (2000): 449–69; Coopers and Lybrand Tax Policy Economics Group, "Economic Benefits of the R&D Tax Credit," a study prepared for the R&D credit coalition in January 1998; James R. Hines Jr., "Taxes, Technology Transfer, and R&D by Multinational Firms," in *Taxing Multinational Corporations*, ed. Martin S. Feldstein, James R. Hines Jr., and R. Glenn Hubbard (Chicago: University of Chicago Press, 1995).

59. See "OECD Science, Technology, and Industry Scoreboard 2005: Towards a Knowledge-Based Economy," sec. A.12, http://miranda.sourceoecd.org/pdf/a12.pdf; *OECD Science, Technology, and Industry Outlook 2006* (Paris: OECD, 2006), 69, http://213.253.134.43/oecd/pdfs/browseit/9206081E.PDF (accessed April 29, 2007).

60. R&D Credit Coalition, Invest in America's Future, http://www.investinamericasfuture.org/; and "Participants," http://www.investinamericasfuture.org/members.html (both accessed June 30, 2007).

61. See National Association of Manufacturers, "Tax and Domestic Economic Policy," http://www.nam.org/s_nam/; and Technology CEO Council, "Reports, Choose to Compete" (October 2005), http://www.cspp.org/documents/choosetocompete.pdf (both accessed April 29, 2007).

62. Catherine L. Mann spoke on the necessity of providing firms appropriate incentives to train and hire U.S. workers. The rationale for this is similar to that of the R&D tax credit. Catherine L. Mann, "Globalization of IT: Economic Gains & Policy Challenges," presentation, American Association for the Advancement of Science Annual Forum on Science and Technology Policy, Plenary Session on Globalization, Offshoring, and Impacts on U.S. Science and Engineering, April 2004, http://www.aaas.org/spp/rd/mann404.pdf (accessed April 29, 2007).

63. Eugene Russo, "Debating Shelby: Tension Continues to Mount over Principles of Data Access," *The Scientist* 15, no. 7 (April 2, 2001): 14.

64. P.L. 105-277 provided that the director of OMB amend Circular A-110 to require federal awarding agencies to ensure that all data produced under an award be made available to the public through the procedures established under the Freedom of Information Act. It also provided that if the agency obtaining the data did so solely at the request of a private party, the agency may authorize a reasonable user fee equaling the incremental cost of obtaining the data.

65. Office of Management and Budget, "Proposed Bulletin on Peer Review and Information Quality," *Federal Register* 68, no. 178 (September 15, 2003): 54023–29.

66. See, for example, American Association for the Advancement of Science Council, *Resolution on the OMB Proposed Peer Review Bulletin*, March 9, 2004. See also Anthony Robbins, "Science for Special Interests," *Boston Globe*, December 7, 2003; Rick Weiss, "Peer Review Plan Draws Criticism," *Washington Post*, January 15, 2004, A19; Andrew Schneider, "White House Seeks Control on Health, Safety," *St. Louis Post-Dispatch*, January 12, 2004.

67. Office of Management and Budget, "Final Information Quality Bulletin for Peer Review," *Federal Register* 70, no. 10 (January 14, 2005): 2664–77. For a complete history of the OMB Bulletin on Peer Review, see "AAAS Policy Brief: OMB Bulletin on Peer Review" (Washington, DC: AAAS Center for Science, Technology, and Congress), updated January 24, 2005, http://www.aaas.org/spp/cstc/briefs/peerreview/index.shtml (accessed February 5, 2007).

CHAPTER 9

The States

What Is the States' Role in Science Policy?

The fifty American states are significant consumers of scientific and technical information, and the universities and businesses residing in each state greatly contribute to the creation of scientific and technical information. The role of the states in science policy and investment in R&D, however, has been overshadowed by the federal government and industry.[1] Indeed, until recently, states were not viewed as playing a significant role in science policy. They were more often seen as the beneficiaries of federal grants or the homes of federal facilities. Indeed, given competing priorities, the states themselves rarely saw any benefit in making their own investments in scientific research.

This began to change in the 1980s, and by the 1990s states were becoming intimately involved in research efforts. A 1992 report on the subject by the Carnegie Commission on Science, Technology, and Government noted, "In the late 1940s, the debate revolved around the appropriateness of a new federal role in research. This time it involves new roles for the states, too, in maintaining the national capability in science and technology and in pursuing industrial excellence, environmental quality, health care, education reform, and other domestic goals. The states are growing strong and sophisticated enough, many believe, to take a greater, more independent role in pursuing these peaceful but still fundamental national goals."[2]

In the beginning, interest the states had in science was linked to the promotion of economic growth. Except in education (e.g., efforts to legislate the teaching of creationism), states avoided setting specific science policies, especially those regulating research agendas. This, too,

has begun to change, with more state governments considering policies to constrain or promote certain types of scientific research. Recent actions taken by several states to authorize, forbid, or regulate the use of human embryonic stem cell (hESC) and cloning are just two of the better-known examples. This chapter will explore states' increasing involvement in national science policy and acknowledge the states as ever-more important players in the science policy arena, alongside universities, national labs, industry, and the general public.

Shifts in State Priorities

Prior to 1970, states did not devote much energy to creating S&T and high-technology economies. North Carolina was the only state that maintained an S&T-based state economic agency of any kind. Today, however, almost every state has an office assigned to take the lead in developing R&D activities.[3]

Momentum for state initiatives in science and technology began to pick up in the late 1970s and through the 1980s. With traditional manufacturing dying out and jobs moving overseas, the nation's governors developed a compelling interest in diversifying their economies. Many took note of those few states whose traditional excellence in research and cultivation of scientific facilities was attracting high-tech development and significant economic growth, especially in areas such as information technology and biotechnology.[4] These states were well equipped to quickly retool and create knowledge-based economies and to benefit from the massive high-tech boom of the 1990s; others followed suit, but more than a few states are still trying to catch up.

According to a 1999 National Science Foundation report, state R&D support has tended to focus on three goals: (1) enhancing the research capacity of state colleges and universities, and encouraging more university-industry partnerships; (2) supporting entrepreneurs and high-tech start-ups in an effort to encourage more "home-grown" businesses; and (3) facilitating the incorporation of new technologies into existing product lines and facilities in an effort to enhance efficiency and productivity.[5]

R&D for Economic Growth: States and Academia

For many reasons state governments did not significantly support research prior to the early 1970s. Some critics have suggested that academia's near-total dependence on federal research funding ruled out opportunities for state support.[6] Conversely, many governors and state legislatures did not see the value of investing in university-based research, and scientists did little to demonstrate the potential benefits.[7] Even so, state governments were all the while contributing, albeit indirectly, through their support of state universities, and particularly faculty teaching salaries, which cannot be paid using federal grant dollars.

Governors and state legislatures have more recently devised innovative ways to provide state-level research funding. Most states now have established entities dedicated to supporting local R&D. In many states, this responsibility has been linked closely to state economic development organizations, which have incorporated R&D funding directly into their economic development plans.[8] Many state officials are recognizing that state investments in R&D can help leverage additional federal dollars, which further influence state economic development.

The Bayh-Dole Act played an influential role in emphasizing the value of basic research to local economies. Universities' newfound intellectual property rights under Bayh-Dole stimulated technology-transfer activities that often led to new regional start-ups. These new businesses created jobs, boosting the local economy.

States were also attracted by the potential for new partnerships with the federal government, universities, and industry. In 1998, the governors of all fifty states and the five U.S. territories showed unprecedented bipartisan support for federal research investment in the form of an open letter to Congress. The governors noted:

> There is no doubt that the results of this [federally funded] research have made a marked improvement to our quality of life: improving our long-term health, economic strength, national security and overall standard of living. Such things as the polio vaccine, laser eye surgery, the Internet, cellular phones, and cancer therapies—were all made possible through basic research. As governors, we realize the benefits extend far beyond quality of life issues. The product of this research is, and will continue to be, a driving force behind a strong American economy. It creates jobs, increases productivity in the workforce, and provides the training groups for our country's next generation of high skilled workers. . . . We are united in our support for university-based research. Our future prosperity depends on it. It improves our quality of life at home and helps us compete globally.[9]

After a copy of the letter was e-mailed to the alumni of one participating university, the school's government relations representative noted, "I have worked in governmental relations . . . for 14 years, and in that time I have never before seen such a well-coordinated and bipartisan effort by our nation's governors to support the university research mission."[10]

In December 2006, the National Governors Association (NGA) announced the formation of the Innovation America task force, with the intent of joining efforts with the academic and business communities to strengthen the competitiveness of the United States in the global economy.[11] The Innovation America initiative is aimed at assisting governors in the development and implementation of short- and long-run strategies for enhancing innovative capacities of all states. Because of the traditional role of state governments in education and economic development, the initiative focuses on making improvements in science, technology, engineering, and math (STEM) education at all levels; making it easier for the postsecondary education system to support states' high-growth industries and workforce needs; and creating supportive state polices that encourage business innovation and entrepreneurship. At the February 2007 meeting of the nation's governors the message was that the states have a huge role to play in making America more innovative, along with the federal government, business, and universities. As Arizona governor Janet Napolitano, the NGA chairperson, said, "Governors are best-suited to take the lead in promoting innovation, but creating an innovative nation will require cooperation between decision makers at the state and federal levels."[12]

The interest of the governors in making their own state investments in scientific research is likely to grow as more studies link state and regional economic growth and prosperity to innovation and education. For example, in its 2005 annual report, the Federal Reserve Bank of Cleveland released a study that showed that innovation and education were the most important factors in determining growth in state per capita income. The study called into question the long-held view that manufacturing is the most important source of state and regional wealth.[13] It confirms that the United States cannot compete in the new global economy based upon cheap labor costs and points to the importance of research, higher education, and innovation as the keys to U.S. competitiveness.[14] In response, states and their governors are looking at ways to become smarter and more innovative in order to keep high-wage, high-value jobs from going abroad and increasingly creating their own state-based research initiatives.

A More Active Role in S&T: "Homegrown" Industries

After the Cold War ended, defense industries began to downsize their operations, spurring states to seek new sources of employment; at the same time, the nation's shift in priorities from defense to the economy, environment, education, and health care—areas in which the states have considerable latitude—offered states an opportunity to increase their influence on S&T policy.[15] The U.S. and global economies began their long shift away from traditional manufacturing and toward knowledge and information. This "new" economy is built on high-tech industries that require skilled workers, and has prompted many states to reevaluate their assets. For example, in Michigan, the big three automakers—Ford Motor, General Motors, and DaimlerChrysler—have long been a huge source of blue-collar jobs. Technological advances and economic changes, however, have led to downsizing and outsourcing. To counter this trend, Governors John Engler and Jennifer Granholm both launched initiatives intended to attract high-tech businesses and skilled workers to the state.

Encouraged by the success of regional S&T clusters in California's Silicon Valley, Massachusetts's Route 128, and North Carolina's Research Triangle, many states, including Michigan, have begun investing in their own S&T plans. What is common to these three best-known regional clusters is that each is built around one or more research universities.

Regional "Clusters" of Innovation

The knowledge-based economy functions very differently from the earlier industrial age. In this new era, states' economic development is driven largely by their capacity to innovate and adapt to new environments. The change also suggests that states—and the country as a whole—

will come to depend on providing and developing services rather than raw goods.

In response, many states have developed regional environments, or *clusters* of innovation, that help to nurture new industries, rather than working to attract existing industries and companies. In a 1999 discussion of states' economic strategies, the Progressive Policy Institute emphasized this need to shift away from a traditional "hunting and gathering" mode of industrial recruitment to the new "gardening" approach, in which states develop the knowledge and technology base that will enable them to promote growth from within.[16] This emphasis will better attract high-tech firms seeking to expand, and may even draw in out-of-state start-ups.

This approach stresses the value of an environment in which new ideas and companies flourish. Clusters of high-tech firms and supporting businesses are thought to provide an excellent foundation. The high-tech companies in a given cluster often tend to be in the same industry or field; for example, Silicon Valley was originally based on silicon chip and electronics innovation.[17] Such clusters, as we noted earlier, include not only start-ups and existing high-tech companies but also universities. Massachusetts's Route 128, for example, has the advantage of proximity to MIT, Harvard, Boston University, and Tufts. Innovative clusters are based on the idea that "tacit" knowledge tends to be "sticky": that is, knowledge gained from campus research tends not to diffuse rapidly or evenly across boundaries.[18] The close geographic grouping of linked companies and institutions—suppliers, universities, high-tech enterprises—is supposed to give firms within the cluster better access to this knowledge, and thus a competitive edge.[19]

Clusters tend to attract highly skilled workers. Regions with multiple firms and many professional opportunities have an easier time attracting the brightest and the best, especially when so many families now include two working partners.

In order to attract high-tech businesses and promote start-ups, most states provide tax incentives and work to foster *venture capital* (VC) and *angel investment*.[20] More and more states, working alongside state universities, have begun to establish state venture capital funds that can support start-up businesses with the potential for sustainable growth. The funds are one element in states' efforts to take a much more active role in their own growth.

An Increasing Interest in Bioscience

States have a tendency to chase "the next big thing" when it comes to science. Usually they express this interest by launching initiatives aimed at promoting a new area of research, so as to attract companies and skilled workers. Biotechnology is an area currently attracting state interest, especially stem cell research and regenerative medicine. The life sciences revolution and the resulting biotech boom in some states (e.g., Massachusetts and California) have caused many other states to do things that encourage the growth of biotech companies. By March 2002, forty-one states had created bioscience initiatives—investing in the biotech workforce, improving the business climate, or facilitating access to capital. In addition, at least thirty-five states had established biotech trade associations, while five had created state-backed "seed" funds for early-stage technology development, which were targeted exclusively at bioscience companies.[21]

A June 2004 study conducted by the Battelle Memorial State Science and Technology Institute for the Biotechnology Industry Organization (BIO) found that state interest in biotech has increased dramatically in recent years. According to the study, fourteen states had identified the biosciences as a major economic development area worth pursuing as of 2001. Only three years later, in 2004, the study found that this number had grown to forty, and that all fifty states had some type of initiative aimed at increasing the bioscience sector of their economies.[22]

Nanotechnology is also the subject of much current attention from state governments, in light of analysts' predictions that nanotechnology will be a $1 trillion market by 2015, and could provide as many as two million new jobs.[23] By the end of 2004, at least thirty states had spending initiatives targeted at encouraging nanotech development. For example, Georgia pledged $45 million for a new nanotechnology research center at the Georgia Institute of Technology, while an anonymous donor offered another $36 million for the project.[24]

While many state initiatives are devoted to providing tax incentives or attracting venture capital through state incubators, states have also been spending their own funds to attract S&T firms. Their ability to do so has been greatly enhanced by the windfall of new revenues resulting from the 1998 settlement reached between forty-six states and the tobacco companies.

The Tobacco Settlement as a Source of Funding

The tobacco settlement was a landmark agreement: the tobacco companies agreed to make annual payments to the states in perpetuity, as a means of reimbursing them for years of smoking- and tobacco-related health care costs.

The settlement, referred to as the Master Settlement Agreement, is the largest U.S. civil settlement to date. It has been estimated that the payments to the states will exceed $200 billion over the next twenty-five years. The settlement also included provisions for payments to compensate tobacco farmers for their anticipated loss of income.[25]

This contractual agreement between the states and the tobacco companies was negotiated by the states' attorneys-general. It does not limit states' use of the settlement money. President Clinton subsequently signed into law the FY1999 Emergency Supplemental Appropriations Act, which lifted any federal claims on the settlement funds, thereby allowing the states to keep the entire sum.[26]

States and their governors suddenly found themselves in the unfamiliar position of looking for creative ways to spend the windfall. At the time, state economies were relatively stable and many were growing. Thus, in 2000, when the first payments were made, it seemed reasonable to use the money for new programs and initiatives. In light of increasing interest in the biosciences, several states used their funds to launch life sciences and biomedical research initiatives.

As the tech bubble burst and the economy started to stagnate, states began diverting settlement funds previously committed to tobacco- or health-related research to instead balancing state budgets. Indeed, the U.S. Government Accountability Office predicted that the percentage of tobacco funds used to meet state budgetary shortfalls would increase from 36 percent in FY2003 to 54 percent in FY2004, making this the largest single use of settlement funds. Meanwhile, the amount spent by the states on health-related programs, including Medicaid and life sciences research, was predicted to decrease from 24 to 17 percent over the same period.[27] Such health-related services had earlier been receiving the greatest percentage of state settlement funds, with a total of 38 percent committed for such purposes between FY2000 and FY2001.[28]

A Case Study: Michigan's Life Sciences Corridor

In 1999, John Engler, then governor of Michigan, signed legislation that directed $1 billion of the state's tobacco settlement toward life sciences research. Engler announced that the state would spend $50 million per year over twenty years to support a new Michigan Life Sciences Corridor (MLSC).

The goal was to establish a corridor of biotech companies in the stretch between the Van Andel Research Institute in Grand Rapids, in the southwestern corner of the state, and the University of Michigan and Wayne State University in the southeast.[29] Michigan State University, located in East Lansing, near the center of the corridor, would serve as the midpoint. The premise was that state funding would allow Michigan to move from the second tier of about eleven biotech states to a leading position, eventually rivaling California, Massachusetts, North Carolina, and Maryland.[30] Support from the Life Sciences Fund would be awarded to industry and university applicants annually, through a competitive, peer-reviewed process administered by the Michigan Economic Development Corporation, an agency of the state government. All proposals had to be for life sciences projects that showed some potential to benefit the state's economy. Proposed projects had to allow for formal collaboration between a nonprofit research institution and a for-profit entity.

Funds were awarded for the first time in 2000. The four major research institutes along the corridor—Van Andel, the University of Michigan, Wayne State, and Michigan State—grouped together to submit what came to be called the Core Technology Alliance (CTA) proposal, which was premised on the need for a vibrant research infrastructure to support life sciences research, including core facilities in genomics, proteomics, bioinformatics, structural biology, and animal models. The CTA's success and subsequent funding for the core institutions have united researchers and administrators from these institutions in a common effort. The CTA provides services to university researchers as well as businesses, and was designed to be self-sustaining after the fifth year of MLSC support.

The MLSC fund, however, has not been without controversy. Public health advocates have complained that this is an inappropriate use of the tobacco settlement, which they contend should be devoted to smoking-cessation programs and tobacco-use prevention. Advocates won a spot on the November 2002 ballot for a proposal to reallocate the tobacco money, such that 90 percent would be used for prevention and cessation programs. The proposal was defeated for several reasons, including the change it would have required in the state constitution, and the drastic reduction that would have been necessary in the state's merit-based scholarship program.

The MLSC has also led to negotiations between the state and the awardees, including universities, about intellectual property rights, milestones, and even indirect cost rates. The state created the MLSC to spur economic development, and university officials and other awardees have sometimes had to remind the government that science is a long-term investment. This debate has had the unanticipated benefit of prompting university officials to collaborate on better metrics for technology-transfer activities.

The advent of a new state administration in January 2003, led to concern about the MLSC and the fund's future. Governor Jennifer Granholm, Engler's successor, chose to build on the MLSC by creating the Technology Tri-Corridor, focused on the life sciences, homeland security, and advanced automotive technologies. The Tri-Corridor is intended to move Michigan away from its traditional manufacturing base toward more information- and knowledge-based service industries. In spite of participants' hopes, however, funds designated for the new corridor were diverted to help counter the state's fiscal crisis.

New Technologies for Mature Industries

It is not only new industrial sectors that rely on new technologies. For example, one of the most mature of industries, the auto industry, must continually draw upon advances in technology to meet new regulatory requirements for increased fuel economy, reduced emissions, and improved safety. Most persons are not aware that companies such as Procter and Gamble spend huge sums on research and development as they create the normal everyday products we use. Indeed, that company notes that it "has more Ph.D.s working in labs around the world than the combined science and engineering faculty at Harvard, MIT, and Berkeley."[31] The art of designing fiber patterns that go into the making of toilet tissue and diapers is a clear example of how technology has an impact on almost every aspect of our lives—even on basic human processes that have been with us for millennia. Even the steel-manufacturing industry must pay close attention to technology changes. A Carnegie Mellon study has concluded that "The steel industry of the future will require new technologies to reduce capital costs and environmental concerns. But the U.S. industry spends less on research than many of its international competitors."[32] One must ask if the lack of focus on technology improvements has been one of the sources of the massive loss of market share experienced by the U.S. steel industry in recent decades.

Cascading Effects of National Science Policy

Actions taken at the national level can sometimes create opportunities at the state level, and state-level initiatives in turn affect municipality and local efforts. The following are three examples in which a national initiative affects state action, or a state initiative affects local policy.

States and Big Science

It is easy to overlook the states' role in national S&T efforts. Research grants, after all, come mainly from federal sources, most basic research is done in universities, the private sector handles the bulk of the development work, and the national labs have few direct ties to the states. But state competition for big-science facilities has a significant impact on the nation as a whole.

States have the power to acquire land needed for big projects (via condemnation, if necessary); to provide relief from ordinances that might impede a project; and to provide enormous cash incentives. For example, the State of Texas offered $1 billion from its own funds to attract the Superconducting Super Collider project, toward an anticipated total cost of $5 billion, before that project failed.

But why would the project be worth so much to Texas that it would risk $1 billion of its own money? Huge high-tech projects like the Superconducting Super Collider can have an enormous impact on the host state's economy through the creation of "clean," high-paying jobs that will attract large numbers of educated citizens and enrich the surrounding communities. In the case of the Super Collider, Texas expected to benefit far in excess of the $1 billion it planned to spend. Many states are willing to take such extraordinary steps to attract the right combination of ventures. Nor does the process always work from the top down. Some large research facilities were created by local economic forces and later grew into national assets, while others exist because states sought them out, even sometimes from other nations, with little federal involvement.

Capitalizing on Homeland Security

The federal government's efforts to encourage state participation in homeland security initiatives are providing still other opportunities for state involvement in S&T efforts. The DHS's Homeland Security Centers of Excellence Program funds university-based centers that are each focused on a single identifiable threat: biological, chemical, nuclear and radiological, explosive, or cyber. The program also funds research aimed at increasing our understanding of terrorist behavior. Each center must involve a consortium of universities, the aim being to foster collaboration among outstanding academic researchers from across the country and concentrate their talent on meeting new homeland security challenges.

The first three centers of excellence that were funded were for the University of Southern California and partners, for Texas A&M and partners, and for the University of Minnesota and partners. Of course, every state in the country was initially keen to host a center of excellence, which was expected to bring in millions of additional research dollars to the state: the USC center was set to receive $12 million over the course of three years, while the Texas A&M center was to receive $18 million and the University of Minnesota $15 million. Winning one of these awards was expected to create jobs, not just directly in the centers but also in support services, and to increase tax revenues (income, sales, vehicle, etc.), the bread and butter of state budgets. State governments have therefore largely supported their universities' efforts to compete for the homeland security research centers. It is not at all clear, however, that these state investments in homeland security research will pay off.

The Impact of R&D-Intensive Entities on Local Education

States also benefit from the presence of high-tech industries in less tangible ways, perhaps most notably is the impact such companies can have on the quality of local education. It is well known that parents play a major role in the quality of local science education. Parents from a community with high-tech industries are more likely to push for advanced science and mathematics courses than those who live in an area where science and math knowledge are less valued. Such considerations often factor into local and state government decisions on the need for tax incentives or other enticements to attract new industries or "big science" projects.

Companies, through their community outreach programs, also can contribute financially to improve local science and mathematics education. A firm can justify such charitable contributions to their stockholders, since these local education projects presumably train a skilled future workforce. Such contributions can lead to support for "Saturday Morning Science" classes, tours of industry laboratories, loans of lecturers, and summer internships, to mention only a few examples.

More generally, the presence of high-tech industries or enterprises can light the fires of cultural inspiration. Parents' and politicians' interest in cultivating this spark creates a welcoming and competitive environment that, in its own way, fosters the success of the nation's high-technology enterprise.

Increased Involvement in Establishing Science Policy

With states taking a greater interest in high-tech's contribution to their economies, legislatures have begun paying more attention to creating their own policies regulating scientific research. On the one hand, more and more states have implemented policies promoting research. On the other, some states are responding to scientific research that challenges the moral and ethical values of some of their citizens by passing laws that constrain the conduct of certain types of research. The tension between these two tendencies is perhaps most evident in the areas of cloning and stem cell research.

Reproductive and Research Cloning

For most of its history, the idea of human reproductive cloning was regarded as an invention of science fiction. However, this changed with Scottish researchers' 1997 announcement of Dolly, a lamb cloned by inserting the nucleus of one of her mother's cells into a denucleated egg cell. While this was regarded as a landmark accomplishment in science, it also unearthed deep social concerns about the meaning of individuality. Concerns about the ethics of human cloning were further fanned in December 2002, when an obscure religious sect claimed to have successfully cloned a human being. This claim ultimately proved false.[33]

Most people agree that human reproductive cloning should not be pursued. Reproductive cloning is the insertion of a somatic cell nucleus into an egg cell without a nucleus *for the purpose of producing another being*—an exact biological replica of the individual from whom the somatic cell nucleus came. The National Academy of Sciences reaffirmed this belief in 2002 when it concluded that "human reproductive cloning should not now be practiced."[34]

However, there is no such unanimity when it comes to research cloning, also known as therapeutic cloning or more accurately somatic cell nuclear transfer (SCNT). The National Academy of Sciences, backed by a large coalition of researchers, patient advocates (e.g., actor Christopher Reeve and former first lady Nancy Reagan), and university administrators, maintains that SCNT (i.e.,

the insertion of the nucleus from a somatic cell into a enucleated egg cell, producing an early-stage embryo to be *used for research or medical purposes only*) should be allowed.[35] Many individuals, however, oppose the practice on the basis that creating an embryo specifically to be used for research purposes is really destroying a person.

While several federal bills have been introduced that would prohibit some or all forms of human cloning, Congress has been slow to enact legislation, largely because of disagreements between the House and Senate. The House has voted to ban both human reproductive cloning and SCNT, while the Senate has opposed such a complete ban by a slim margin, in favor of legislation that would permit SCNT. This is not to suggest a total absence of relevant policies: in 1997, President Clinton issued a memorandum explicitly banning the use of federal funds for human reproductive cloning, that is, cloning for the purposes of producing another human being. Prior to that, in 1993, the FDA issued regulations on the use of cloning for clinical purposes. The initial policy mandated FDA oversight of research cloning, but, after the 1997 cloning of Dolly, the policy was amended to include oversight on the use of cloning technology to reproduce another human being.[36] The FDA claims this jurisdiction on the basis that "human reproductive cloning involves a 'biological product'" and thus such efforts require premarket, or agency, approval.[37]

With Congress deadlocked on the issue, the states have acted on their own. California was the first to do so, passing legislation in 1997 that specifically prohibited reproductive cloning. Since then, several other states have enacted similar laws, and at least five (Arkansas, Iowa, Michigan, North Dakota, and South Dakota) have extended their laws to prohibit SCNT.[38] Comparable bills have been introduced in several additional states, and are likely to be approved.

Embryonic Stem Cell Research

One use of SCNT is to produce embryonic stem cells, which can, in turn, be used for medical therapies and research. Researchers throughout the United States and around the world are interested in working with embryonic stem cells, in particular human embryonic stem cells (hESC), which, when grown under proper conditions, can be induced to develop into various types of specialized human cells and tissues. These could eventually be used to treat a wide variety of diseases and conditions, such as cancer, heart disease, diabetes, Parkinson's disease, multiple sclerosis, and spinal cord injuries. While stem cells can be found in some adult tissues, most scientists believe that the stem cells from embryos are more powerful and the gold standard.[39] The recent scientific advances to induce pluripotency in adult stem cells will be extremely useful, but there is still a need to push forward stem cell research on all fronts, including embryonic stem cell research. Researchers do not fully understand how this new induced pluripotent cell functions, and there are technical hurdles to overcome before they can be used clinically. In addition, human embryonic stem cells will likely be a better model system for answering some questions.[40]

Embryonic stem cells are most often obtained from one of three sources: existing embryonic stem cell lines; unused in vitro-fertilized embryos; and embryos created by SCNT. Many pro-life groups are opposed to the use of any of these sources, suggesting that hESC research is likely to result in an increase in the loss of unborn life. Some states have enacted laws that regulate or restrict research using cells or tissue obtained from these sources.

In response to pressure from both sides, President George W. Bush announced a federal human embryonic stem cell policy in his August 9, 2001, televised address to the nation. The policy allows the use of federal funds for research on hESC lines derived before, but not after, that August 9 date. This has led to a decrease in federal funding for hESC research, as well as the number and quality of available lines, severely slowing research in this area in the United States.[41]

Some states, including California and New Jersey, have fired back, passing legislation aimed at fostering hESC research. In the spring of 2004 the governor and legislature of New Jersey passed a measure establishing the Stem Cell Institute of New Jersey, to be funded by a public-private partnership.[42] In November of that same year, California passed a referendum to establish and fund a state stem cell research initiative using state funds. Several other states have made similar moves. These efforts are fueled by the expectation that states at the forefront of hESC research will attract the biotechnology industry and reap a windfall in economic returns.[43]

A Case Study: California and Human Embryonic Stem Cells

In February 2001, even before the Bush policy was put into place, the California state legislature introduced a bill to legalize hESC use in research. This bill was passed by both chambers in August 2002 and signed into law in September, making California the first state to pass legislation specifically permitting hESC research.[44]

POLICY DISCUSSION BOX 9.1

States Involvement in Stem Cell Research: Too Much or Too Little?

Since August 2001, when President George W. Bush put in place his policy restricting embryonic stem-cell research to a limited number of stem cell lines, we have witnessed an "unprecedented" movement where states are enacting their own state specific stem cell policies. In 2005 alone, states considered over 180 bills or resolutions relating to stem cell research.[45]

On the one hand, some states have taken steps to promote and provide funding for embryonic stem cell research because they do not think President Bush's policy does enough. Some of the states that have taken major steps to support stem cell research through statutes, regulations, or executive orders include California, Connecticut, Illinois, Maryland, Massachusetts, New Jersey, and Wisconsin.[46]

On the other hand, some state leaders believe that President Bush's policy is too permissive. These states have enacted policies that restrict or, in some cases, ban embryonic stem cell research and or somatic cell nuclear transfer, a technology that facilitates the creation of patient-specific embryonic stem cells. States that have taken such actions include Arkansas, Iowa, Louisiana, Michigan, Nebraska, North Dakota, South Dakota, and Virginia.[47]

Daniel Perry, president of the Coalition for Advancement of Medical Research, has said that such a "patchwork quilt" of state laws is no way to guide important medical research.[48] Do you agree or disagree? How much latitude should states have to determine their own policies on important national science policy issues such as hESC stem cell research? Do states have a right to restrict particular types of research permitted by the federal government? Should states be allowed to fund research for which federal funds are specifically restricted? Does it harm our national interests if we have divergent state policies in areas such as stem cell research, as opposed to one unified and agreed-upon national policy? What steps might be taken to mitigate such negative effects and to better unify state science policies in such areas?

Because the Bush policy is only applicable to research supported by federal dollars, a grassroots campaign to identify and secure state funding for hESC research was started in California. This initiative was presented to voters in the November 2004 state election as Proposition 71 (the California Stem Cell Research and Cures Initiative). Proposition 71 sought to establish a California Institute for Regenerative Medicine, supported by $295 million in state funds over a period of ten years, derived from the sale of $3 billion in public bonds. The proposal specifically prohibited human reproductive cloning, but allowed somatic cell nuclear transfer, or SCNT. Because SCNT is also a step in reproductive cloning process, however, opponents claimed the proposition would legalize human cloning.[49]

The proposition was backed by a broad-based coalition, including Nobel Prize–winning scientists, patients and their families, patient advocates, business groups, university researchers, community organizations, and political leaders. California governor Arnold Schwarzenegger also supported the initiative. The opposition brought together an unlikely alliance, including feminists (who feared that the demand for embryos would prompt the treatment of women as "egg farms"), fiscal conservatives, and evangelical Christians. Indeed, the proposition required an amendment to the state constitution, mandating funding for the initiative even at the expense of other programs, giving even the staunchest supporters pause.[50]

In the end, Proposition 71 passed.[51] It now stands as the largest state-supported scientific research program ever to be created in this country. With experts saying that Proposition 71 could create an average of five thousand to twenty-two thousand new jobs per year, other states have followed suit, and more may do so.[52]

Where Are We Now?

Federal Programs Encouraging State S&T Investment

The federal government is playing an ever-more important role in facilitating collaborations among the partners in science policy. But, as G. Wayne Clough, president of the Georgia Institute of Technology, noted at a 1998 National Innovation Summit, "States have played a vital leadership role in creating new R&D partnerships, and the federal government should find ways to support this trend."[53] By providing matching funding, for example, the federal government might be able to stimulate more

state programs like the Michigan Life Sciences Corridor Fund.

Tension between Federal and State Science Policies

The conflict over hESC research is, of course, just one example of the growing tension between federal and state science policies. In the absence of federal legislation, states are taking it upon themselves to regulate stem cell research. The complicating issue is that the states are not acting as a collective—some are imposing bans or restrictions, while others are promoting hESC research. Some observers are concerned that this patchwork approach will make it difficult for scientists to collaborate with fellow researchers from other states.

The Haves and the Have-nots

As recognition of S&T's value to state economies has spread, so too has states' interest in receiving their fair share of federal research dollars. Those states that have traditionally not received a sizable percentage of these funds, in conjunction with researchers from universities in these states and their members of Congress, have argued that federal S&T funding should not automatically flow to the traditional recipients. The critics argue that the dominance of certain states in competition for federal funds has given those states an unfair advantage in future competitions, by providing them with high-quality research infrastructures, top scientists, top-ranked research universities, and other assets that can contribute to a research proposal's success. Because those recipient states have a head start, they will arguably always be more competitive.

In response to these concerns, the NSF established the Experimental Program to Stimulate Competitive Research (EPSCoR) in 1979. EPSCoR is premised on the idea that "universities and their science and engineering faculty and students are valuable resources that have the potential to influence a state's development." By supporting S&T infrastructure improvements at the state and institutional levels, EPSCoR is intended to "significantly increase the ability of EPSCoR researchers and institutions to compete successfully for federal and private-sector R&D support."[54] The program's long-term goal is to develop state R&D infrastructures and allow all states to compete equally for federal research funding.

Somewhat problematically, the program offers no incentive for qualified states to "graduate" from EPSCoR support into the "regular" pool of funding applicants.

Furthermore, while EPSCoR grants are rigorously peer-reviewed, the program's funding standards are, out of necessity, lower than those for general RFPs. As a result, some question the program's effectiveness.

Policy Challenges and Questions

As states have come to view S&T investment as an economic engine, they have also begun pressuring universities and other research institutions to demonstrate a return on their investments. Indeed, states often have unrealistic expectations about the speed with which research investments will yield significant economic returns. Moreover, they also fail to realize that many of S&T investments, especially those promoting basic research, will never be successful in terms of job creation or number of start-ups.

This once again highlights the difficulty of demonstrating the return on public research investments. States, like the federal government, are now faced with the challenge of evaluating R&D programs and determining how much money should be expended on them. These assessments are likely to be especially problematic because of the state's strong emphasis on job creation and economic growth. It can be difficult even to decide which jobs should be counted: Do part-time jobs count? How high-paying does the job need to be?[55] We have already seen the problems that can result from any effort to link economic growth directly back to investment in a particular R&D program, or, for that matter the difficulty of collecting the necessary data.

As states step up their investments in the most prominent research areas, such as biotechnology, they risk ignoring other underdeveloped areas that might promise greater opportunities. How should states balance their R&D portfolios? Nanotechnology and energy-related fields might well be among the driving forces of the economic future, and states would be wise to diversify their interests rather than putting all their eggs into the biotechnology or information-technology basket. Should different states focus on different areas so as to not "step on the toes" of another state? Can every state have a Research Triangle or Silicon Valley? What are good metrics for states to use in determining whether their investment in R&D has paid off?

NOTES

1. Carnegie Commission on Science, Technology, and the Government, *Science, Technology and the States in America's Third Century* (New York: Carnegie Commission on Science, Technology, and Government, September 1992), 5.

2. Ibid., 16–17.

3. John E. Jankowski, National Science Foundation, Division of Science Resources Studies, *What Is the State Government Role in the R&D Enterprise?* NSF-99-348 (Arlington, VA, 1999), 1, http://www.nsf.gov/statistics/nsf99348/pdf/nsf99348.pdf (accessed December 15, 2007).

4. Michael McGeary and Philip M. Smith, *State Support for Health Research: An Assessment* (Washington, DC: Lasker Foundation, October 26, 2001), http://www.laskerfoundation.org/ffpages/reports/m1.htm (accessed June 21, 2007).

5. Ibid., 1.

6. Deborah Shapley and Rustum Roy, "Appendix: Invited Response from Pat Choate," in *Lost at the Frontier: U.S. Science and Technology Policy Adrift* (Philadelphia: ISI Press, 1985), 175.

7. Ibid.

8. Jankowski, *State Government Role,* 1.

9. Letter to Congress, signed by all fifty-five state and U.S. territory governors, September 10, 1998.

10. Deborah Kallick, "Governors Urge Support for Research Mission," *UCLA Today,* September 28, 1998, http://www.today.ucla.edu/1998/980928governors.html (accessed March 24, 2007).

11. For additional detail see National Governors Association, Innovation America, NGA 2006–2007, http://www.nga.org/Files/pdf/06NAPOLITANOBROCHURE.pdf; National Governors Association, "NGA Announces Innovation America Task Force," news release, December 5, 2006, http://www.nga.org/portal/site/nga/menuitem.6c9a8a9ebc6ae07eee28aca9501010a0/?vgnextoid=84d37f5e00f4f010VgnVCM1000001a01010aRCRD; National Governors Association, "Innovation America Tops Governors Meeting Agenda," news release, February 20, 2007, http://www.nga.org/portal/site/nga/menuitem.6c9a8a9ebc6ae07eee28aca9501010a0/?vgnextoid=44e95e38f0ad0110VgnVCM1000001a01010aRCRD; National Governors Association, "NGA Welcomes Governors to 2007 Winter Meeting," news release, February 24, 2007, http://www.nga.org/portal/site/nga/menuitem.6c9a8a9ebc6ae07eee28aca9501010a0/?vgnextoid=37c77131fa5e0110VgnVCM1000001a01010aRCRD; National Governors Association, "NGA Winter Meeting Convenes in Washington," news release, February 24, 2007, http://www.nga.org/portal/site/nga/menuitem.6c9a8a9ebc6ae07eee28aca9501010a0/?vgnextoid=a7d77131fa5e0110VgnVCM1000001a01010aRCRD (all sites accessed March 24, 2007).

12. National Governors Association, "Governors Focus on Federal Role in Innovation and Competitiveness," news release, February 27, 2007, http://www.nga.org/portal/site/nga/menuitem.6c9a8a9ebc6ae07eee28aca9501010a0/?vgnextoid=083e450163be0110VgnVCM1000001a01010aRCRD&vgnextchannel=759b8f2005361010VgnVCM1000001a01010aRCRD (accessed March 24, 2007).

13. Federal Reserve Bank of Cleveland, "Altered States: A Perspective on 75 Years of State Income Growth," 2005 annual report, 7–22. See also John Cranford, "Political Economy: Knowledge Is Money," *CQ Weekly,* July 10, 2006, 1834.

14. In 2006, the federal banks of Chicago and Cleveland both hosted conferences to discuss the roles universities play in

regional economic growth and in the development of industry innovation clusters. See "Fed Considers Connection between Universities, Economic Growth," *SSTI Weekly Digest,* March 5, 2007; Richard H. Matoon, "Can Higher Education Foster Economic Growth?" *Chicago Fed Letter,* August 2006.

15. Carnegie Commission on Science, Technology, and the Government, *Science, Technology and the States,* 5.

16. Robert D. Atkinson, Randolph H. Court, and Joseph M. Ward, *The State New Economy Index: Benchmarking Economic Transformation in the States* (Washington, DC: Progressive Policy Institute Technology and New Economy Project, July 1999), 36, http://www.neweconomyindex.org/states/1999/index.html (accessed June 21, 2007).

17. Michael E. Porter, "Clusters and the New Economics of Competition," *Harvard Business Review,* November–December 1998, 77–90.

18. Jaffe, "Real Effects"; D. B. Audretsch, "Agglomeration and the Location of Innovative Activity," *Oxford Review of Economic Policy* 14, no. 2 (1998): 18–29; B. D. Beal and J. Gimeno, "Geographic Agglomeration, Knowledge Spillovers, and Competitive Evolution," INSEAD Working Paper no. 2001/26/SM, August 2001.

19. Porter, "Clusters."

20. Venture capital (VC) is usually provided by a firm that invests its shareholders' money in young, rapidly growing companies with the potential to develop into significant economic contributors, even though initial investment involves high risks, and payoffs are unlikely in the short term. In exchange, the VC firm gets stock in the start-up and, as a shareholder, has some say in who will be on the management team. Angel investment is similar to VC in that it is money invested in a start-up company. But an angel investor is a single individual, not an organized firm, and tends to expect a higher rate of return on the investment. Angels are usually successful businesspersons who can provide a fledgling company with numerous connections, including connections to other funders. It is important to highlight the fact that both VC firms and angels are important to the process: both invest in high-risk ventures, providing what is sometimes referred to as the preseed or seed money to get a young company up and going.

21. Kristen Bole, "Smokin': Tobacco Windfall Spurs Biotech Investment," *Bio-IT World,* March 7, 2002, http://www.bio-itworld.com/archive/030702/smoking.html?page:int=-1 (accessed March 24, 2007).

22. Battelle Technology Partnership Practice and SSTI, *Laboratories for Innovation: State Biosciences Initiatives 2004* (Columbus, OH: Battelle Memorial Institute, 2004), http://www.iowabiotech.com/econ_dev_reports/OregonBioscience Initiatives2004.pdf (accessed June 21, 2007); Battelle Technology Partnership Practice and SSTI, *Growing the Nation's Bioscience Sector: State Bioscience Initiative 2006* (Columbus, OH: Battelle Memorial Institute, 2006), http://www.bio.org/local/battelle2006/battelle2006.pdf (accessed June 21, 2007).

23. Mike Toner, "Nanotechnology: Small Wonders," *Atlantic Journal-Constitution,* Sunday Home Edition, December 5, 2004, 1B.

24. Ibid.

25. For additional information see C. Stephen Redhead, "Tobacco Master Settlement Agreement (1998): Overview, Implementation by States, and Congressional Issues," *CRS Report for Congress RL30058* (Washington, DC: Congressional Research Service, November 5, 1999), http://www.ncseonline.org/NLE/CRSreports/Agriculture/ag-55.cfm; and Michigan Nonprofit Association and the Council of Michigan Foundations, *Michigan in Brief, 2002–03,* 7th ed. (Lansing: Public Sector Consultants, 2002), 240–42, http://www.michiganinbrief.org/edition07/About_files/MIB_2002.pdf (accessed June 21, 2007).

26. The Emergency Supplemental Appropriations Act, P.L. 106-31 (May 21, 1999).

27. Budget shortfalls jumped to 44 percent in FY2004, and the amount for health-related programs dropped to 20 percent. See Government Accountability Office, *Tobacco Settlement: States' Allocation of Fiscal Year 2004 and Expected Fiscal Year 2005 Payments,* GAO-05-312 (Washington, DC: U.S. Government Accountability Office, March 2005), 3–4.

28. Ibid.

29. A nonprofit research institution dedicated to improving human health, the Van Andel Research Institute is part of the Van Andel Institute, founded by Betty and Jay Van Andel in 1999.

30. Tim Martin, "Michigan Struggles in the Life Sciences Race," Associated Press, July 12, 2004.

31. Procter and Gamble, "Management's Financial Commitment to R&D," http://www.proctergamble.com/science/mngmnt_commit.jhtml (accessed June 21, 2007).

32. R. M. Cyert and R. J. Fruehan, *The Basic Steel Industry,* Sloan Steel Industry Competitiveness Study, Carnegie Mellon University, Pittsburgh, PA (Washington, DC: U.S. Department of Commerce, Office of Technology Policy, December 1996), http://www.technology.gov/Reports/Steel/cd91a_Contents.pdf (accessed June 21, 2007).

33. Alissa Johnson, "Attack of the Clones," *State Legislatures,* April 2003, 30.

34. Committee on Science, Engineering, and Public Policy and Board of Life Sciences, *Scientific and Medical Aspects of Human Reproductive Cloning* (Washington, DC: National Academies Press, 2002), http://www.nap.edu/books/0309076374/html/ (accessed June 6, 2007), 2.

35. Therapeutic cloning is also called *research cloning, somatic cell nuclear transfer,* and *cell nuclear replacement.*

36. U.S. Food and Drug Administration, "Application of Current Statutory Authorities to Human Somatic Cell Therapy Products and Gene Therapy Products," *Federal Register* 158, no. 197 (October 14, 1993), http://www.fda.gov/cber/genadmin/fr101493.pdf (accessed June 21, 2007); and statement by Kathryn C. Zoon, Director, Center for Biologics Evaluation and Research, Food and Drug Administration, Department of Health and Human Services before the Subcommittee on Oversight and Investigations Committee on Energy and Commerce, U.S. House, March 28, 2001, http://www.fda.gov/ola/2001/humancloning.html (accessed June 21, 2007).

37. Committee on Science, Engineering, and Public Policy and Board of Life Sciences, *Scientific and Medical Aspects,* 82.

38. National Conference of State Legislatures, "State Human Cloning Laws," April 18, 2006, http://www.ncsl.org/programs/health/genetics/rt-shcl.htm (accessed June 21, 2007).

39. Junying Yu, Maxim A. Vodyanik, Kim Smuga-Otto, Jessica Antosiewicz-Bourget, Jennifer L. Frane, Shulan Tian, Jeff

Nie, Gudrun A. Jonsdottir, Victor Ruotti, Ron Stewart, Igor I. Slukvin, and James A. Thomson. "Induced Pluripotent Stem Cell Lines Derived from Human Somatic Cells," *Science* 318, no. 5858 (December 21, 2007): 1917–20; Kazutohsi Takahashi, Koji Tanabe, Mari Ohnuki, Megumi Narita, Tomoko Ichisaka, Kiichiro Tomoda, and Shinya Yamanaka, "Induction of Pluripotent Stem Cells from Adult Human Fibroblasts by Defined Factors," *Cell* 131, no. 5 (November 30, 2007): 1–12; Christopher Scott, "The Six Degrees of Stem Cell Research," *The Stem Cell* blog, November 21, 2007; and Susan Solomon and Zach Hall, "Game Over? No Way," *The Stem Cell* blog, December 3, 2007, http://thestemcellblog.com/.

40. For an introduction to the science, ethics, and politics of human embryonic stem cells see Christopher Thomas Scott, *Stem Cell Now* (New York: Plume, 2006); Russell Korobkin, with Stephen R. Munzer, *Stem Cell Century: Law and Policy for a Breakthrough Technology* (New Haven: Yale University Press, 2007); and Eve Harold, *Stem Cell Wars: Inside Stories from the Frontlines* (New York: Palgrave Macmillan, 2006).

41. There is much anecdotal evidence that President Bush's policy is having a negative effect on hESC research in the United States. Two published studies support this notion: Aaron Levine, "Geographic Trends in Human Embryonic Stem Cell Research," *Politics and the Life Sciences* 23, no. 2 (2005): 40–45; Jason Owen-Smith and Jennifer McCormick, "An International Gap in Human ES Cell Research," *Nature Biotechnology* 24, no. 4 (2006): 391–92. Another study suggests the situation might not be so bad: Anke Guhr, Andreas Kurtz, Kelley Friedgen, and Peter Loser, "Overview of Cell Lines and Their Use in Experimental Work," *Stem Cells* 24, no. 10 (2006): 2187–91.

42. David Kocieniewski, "McGreevey Signs Bill Creating Stem Cell Research Center," *New York Times*, Health, May 13, 2004; "McGreevey Creates Nation's First State-Supported Stem Cell Institute," State of New Jersey Office of the Governor, press release, May 12, 2004, http://www2.umddnj.edu/scinjweb/News_And_Reports/press_release_mcgreevey.html 9 (accessed June 27, 2007); Alison McCook, "Stem Cells in New Jersey," *The Scientist* 5, no. 1 (August 19, 2004): 4.

43. One *Time* story on this matter, which appeared in May 2004, stated that "billions [of dollars] are at stake in the race for medical cures." The story quoted one stem cell proponent who suggested that the economic benefits of state stem cell research initiatives could result in up to $70 million in tax revenues from new jobs created even before any cures were discovered. See Margot Roosevelt, "Stem Cell-Rebels," *Time*, May 17, 2003, 49–50.

44. Andis Robeznieks, "States, Scientists Seek Alternative Funding for Stem Cell Research," *Amednews.com*, March 15, 2004, http://www.ama-assn.org/amednews/2004/03/15/prsb0315.htm (accessed June 21, 2007). New Jersey was the second state to pass such legislation.

45. Gregory Lamb, "State Laws Bypass Research Ban," *Christian Science Monitor*, February 1, 2006, http://www.csmonitor.com/2006/0201/p13s01-stss.html (accessed June 21, 2007); National Conference of State Legislatures, "State Embryonic and Fetal Research Laws," updated August 14, 2006, http://www.ncsl.org/programs/health/genetics/embfet.htm (accessed November 12, 2006).

46. Susan Stayn, "A Guide to State Laws on hESC Research and a Call for Interstate Dialogue," *Law and Policy Report*, November 1, 2006.

47. Martin Kasindorf, "States Play Catch-up on Stem Cells," *USA Today*, December 17, 2004. http://www.usatoday.com/news/nation/2004-12-16-stem-cells-usat_x.htm (accessed June 21, 2007).

48. Ibid.

49. Kathryn Jean Lopez, "Braveheart Stands Athwart a Brave New World," *National Review Online*, November 1, 2004, http://www.nationalreview.com/interrogatory/gibson200411010950.asp (accessed June 21, 2007); Rich Deem, "Arguments against Proposition 71: The California Stem Cell Research and Cures Initiative," http://www.godandscience.org/doctrine/stemcell.html (accessed June 21, 2007).

50. Kevin Drum, "Political Animal," *Washington Monthly*, September 26, 2004, http://www.washingtonmonthly.com/archives/individual/2004_09/004775.php (accessed June 21, 2007).

51. Because of a lawsuit, the initiative has had a slow start in awarding funds. In addition, the establishment of a new state agency to oversee the initiative has required time for establishing the policies governing the research to be funded by the initiative.

52. Laurence Baker and Bruce Deal, "Economic Impact Analysis: Proposition 71—California Stem Cell Research and Cures Initiative," *Analysis Group Consultants*, September 2004, http://www.yeson71.com/documents/Prop71_Economic_Summary.pdf (accessed June 21, 2007).

53. Herbert E. Hetu and F. Clifton Berry, Jr, eds., *Competing through Innovation: A Report of the National Innovation Summit* (Washington, DC: Council on Competitiveness, 1998), 21.

54. See National Science Foundation, *Experimental Program to Stimulate Competitive Research: Workshop Opportunities, Program Solicitation*, http://www.nsf.gov/pubs/2006/nsf06613/nsf06613.htm (accessed March 24, 2007), introduction.

55. Steve Bunk, "Lotteries and Tobacco Money: Basic Research Bonanza," *The Scientist* 15, no. 17 (September 3, 2001): 6.

CHAPTER 10

The Public

Why Does Broad-Based Scientific Literacy Matter?

Astronomer Carl Sagan once wrote, "Everybody starts out as a scientist. Every child has the scientist's sense of wonder and awe."[1] Yet we as a nation have not always nurtured and sustained that sense of awe. The result is a culture in which even educated people sometimes brag about their inability to understand basic scientific principles.

With science and technology determining the prospects for improving our social health and welfare, all members of society ought to appreciate science's role in their lives. Indeed, one of the goals of a democratic society should be to ensure that its citizens are fully aware of the world in which they live.

In a discussion of scientific illiteracy and democracy, George Dvorsky, president of the Toronto Transhumanist Association (a nonprofit that advocates the ethical use of technology to improve human health and capabilities), characterized the problem as follows: "Most of those who live in the West, particularly North Americans, are guilty of an anti-intellectual bias. Scientists are supposed to be nerds, right? And who wants to be a nerd? This sentiment, combined with a general suspicion of science and the predominance of aggressive technological and pseudoscientific memes, has resulted in much of the scientific illiteracy that now pervades our society."[2]

But what makes a person scientifically literate? Clearly, we do not all need to be able to understand Stephen Hawking's latest research in cosmology. At the other end of the spectrum, surely we can expect that all citizens know the earth is not the center of the universe.

Many states, scientific societies, and associations—such as the American Association for the Advancement of Science (AAAS) and the National Research Council—have developed benchmarks for assessing science literacy.[3] These guidelines often focus on ways of thinking and problem-solving consistent with the scientific process, rather than mastery of specific scientific facts. Unfortunately, studies conducted by the NSF show that roughly 70 percent of American adults do not understand the scientific process, while many believe in pseudoscience, such as astrology.[4] These and other aberrations pose a serious challenge for the nation. It will never be able to attain its full potential if its citizens are not able to understand the environment in which they live.

Science and the Public

There are many reasons to be concerned about the public's understanding of scientific issues. For example, these views can have a measurable influence on the assignment of federal research dollars, on the amount of federal support devoted to specific types of research, and the contents of the K–12 science and mathematics curriculum. The public is a critical partner with the federal government and the scientific community in the conduct and support of scientific research.

Members of the public can exert enormous influence on the formulation of new policies and regulations. Indeed, public citizen and advocacy groups routinely affect the proposal of new legislation, its wording, how lawmakers vote on it, and even how agency regulations are used to enforce it. Patient advocacy groups' effectiveness in winning increased funding for breast cancer and AIDS research forcefully illustrates the power that these groups can exert over science and science policy.

In the United States, members of Congress, as the public's representatives, must approve all federal expenditures for scientific research. Since legislators depend on their constituents for reelection, congressional views on science are often shaped by, and reflective of, popular views on the subject. While most members of Congress do not come from strong scientific backgrounds, they tend to support science because of their conviction that it is important to the national interest. Some members, such as those on the House Science and Technology Committee, become well informed about science policy through their committee assignment. Others are forced to rely on their limited knowledge of the issues. They may be inclined to base their decisions on personal attitudes, convictions, and beliefs, even when those convictions might be opposed by some significant percentage of their constituency, or may find it necessary to pay close attention to the views of voters back home.[5] These two types of congressional voting behavior have been referred to as *attitudinal* —voting based on one's own beliefs—and *representative*—members voting based on constituent views. These have also been referred to as "the two faces of democratic accountability."[6] Representative voting is common on controversial matters that provoke significant popular debate, including human cloning; embryonic stem cell research; climate change; support for social or behavioral science; science education; and homeland security.

Closer to home, parental views on science and the scientific professions can be a significant factor in guaranteeing the nation an adequate supply of future researchers. The pathway for training young people to become practicing scientists is very long and arduous, often spanning twenty years. Young people are not likely to embark upon this journey without substantial family encouragement from an early age. And the likelihood of such support often depends upon the public's general sense of the prestige and income potential attached to a career in science. Public attitudes toward science therefore play a fundamental role in ensuring the nation's overall capacity for scientific research.

Public Attitudes toward Science

Scholars and policymakers track the American public's views on science and technology quite closely. For example, the NSF's Division of Science Resources conducts a biannual survey and publishes the results in its congressionally mandated report, *Science and Engineering Indicators*.

The data offer important insights into the public's attitudes toward science. Take, for instance, public responses to the question of whether science's benefits outweigh its drawbacks, over a period of several decades. The response is consistently yes in all but a few time periods. A closer examination of these reveals that the deviations are in response to specific events, including the Chernobyl reactor explosion, the Three Mile Island reactor event, and the explosion of the *Challenger* spacecraft. This suggests that the American public sometimes associates failures in major technological systems with failures in the underlying science. It is not clear, of course, to what extent this association is a fair one: one could argue that greater scientific knowledge is needed to prevent such failures. This tendency to confuse the terms *science* and *technology* is not uncommon, and may be aggravated by the media's and scientists' tendency to use the terms interchangeably.[7]

For example, many Americans would probably view building the International Space Station as a major U.S. science project. In fact, it is much more accurately described as a technological endeavor. The station was intended to serve as an orbiting laboratory, whose low gravity and oxygen-free environment are conducive to experiments that would be impossible on earth. So, while the Space Station is a tool for the conduct of science, it is not a "science project" in itself. In fact, cost and size restraints have actually greatly constrained the amount of actual science that has been able to be conducted on the Space Station.

The public's view of science is shaped by several loosely correlated, sometimes even contradictory, opinions. Some people, for example, may decry the use of animals in scientific research while vigorously embracing medical treatments that were initially tested on animals. Others may oppose the use of genetically altered (or modified) food, while lamenting human starvation that might, in fact, be alleviated using such biotechnology.

The public is similarly conflicted about the role of science and technology in homeland security. Some citizens blame science for creating biological agents that could be used for acts of terror, like the anthrax powder that was mailed to Capitol Hill in the fall of 2001. Others look to new detection technologies and mass-inoculation strategies to ensure that subsequent acts can be effectively managed or avoided. It is possible, and even likely, that some members of the public hold these opposing views simultaneously.

In spite of the importance of public attitude toward science, many scientists never give much thought to whether others understand how their work benefits society. But in fact, almost 20 percent of the public actually questions whether science's benefits outweigh its harms.[8] When so many in the public share this concern, the issue clearly deserves serious consideration from the scientific community.

Public concerns about science have historically focused on several key issues: the development of weapons of mass destruction; the enabling of large-scale, potentially environmentally harmful manufacturing techniques; the replacement of human workers by robotic assembly methods; the use of advanced systems for collecting personal information; the increasing pace of life; and fraud, contradictory claims, and apparent conflicts between science and religion. Fears of *scary science,* including the production of genetically modified foods, recent advances in cloning, and, still more recently, concerns regarding nanotechnology, are also having an effect on the public's overall feelings about science and technology.[9]

Correlations of individual support for science with the person's level of knowledge about science have revealed that the percentage of the public believing that "we depend too much on science and not enough on faith" significantly increases as the ability to answer basic scientific questions decreases. In other words, there is a direct correlation between knowledge of science and evaluations of its importance relative to faith.[10] In one survey exploring this issue, respondents who could only answer a small number of scientific questions correctly were almost twice as likely to favor faith over science as those who could answer most of the scientific questions correctly.

Americans and Europeans differ demonstrably on several related issues. Given high levels of correct answers on the associated knowledge tests, both groups show a correspondingly high level of belief in science's social worth. The United States leads in all categories, except that Europeans have a higher expectation that science and technology will make their work more interesting in the future. However, regardless of proficiency in answering the knowledge questions, Americans and Europeans diverge significantly on the question of whether science's benefits outweigh its risks, with almost half the population of Europe rejecting this view.[11]

Public Opinion and the Environment: A Case Study

The environment is one area in which the public seems to be particularly torn over the role of science and technology. On the one hand, many people blame science and technology for making possible the large manufacturing and other industrial facilities that cause environmental problems. On the other hand, they view science and technology as tools for remediating these problems.

Scientists and policymakers have to walk the tightrope of public opinion; but the rope is far from stable. According to a poll conducted by the Gallup organization, Americans' concern about the quality of the environment has declined since 2001, to ninth on a list of twelve concerns behind health care, terrorism, the economy, drug use, illegal immigration, hunger and homelessness, and unemployment.[12] Specifically, 14 percent of respondents expressed concern about the environment in 2001; the number dropped to 9 percent by 2003. In 2005 only 1 percent of respondents named the environment as the most important concern facing the country.[13] Opinions about the relationship between environmental protection and economic growth have shifted over time: the number of respondents favoring environmental protection declined between 1998 and 2004, while the number supporting economic growth has increased (see fig. 10.1). More recently, this trend may be shifting.

A 2005 survey conducted by scientists at the MIT Laboratory for Energy and Environment suggested that the public did not thoroughly understand climate change or the threat of global warming, and did not believe that taking action to reduce their impact was a high priority. The twelve hundred respondents ranked "global warming" an average of sixth on a list of ten environmental problems, well behind other environmental issues such as water pollution and the disposal of toxic waste. When asked where the environment ranked on a more general list of twenty-two of "the most important issues facing the U.S. today," respondents ranked it thirteenth. These results suggest that advances in U.S. climate policy are not likely to be driven by public opinion, but instead will require strong leadership, and scientific awareness among elected officials. But if policymakers do not sense urgency from their constituents, they have less incentive to be concerned about curbing climate change and may put their emphasis on defending the nation's borders or creating more jobs. This poses a tough challenge for policymakers: doing the right thing from a scientific perspective may not be the wisest choice from a political perspective.[14] A recent example is the passage of the Energy Independence and Security Act of 2007 (P.L. 110–142), which requires the corporate average fuel efficiency (CAFE) for passenger cars to increase by 40 percent, or 35 miles per gallon by 2020. This the first increase in these standards in over two decades.

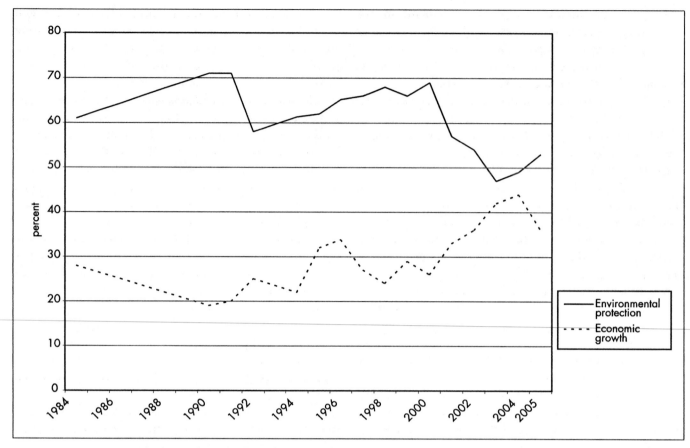

FIG. 10.1   Public priorities for environmental protection versus economic growth, 1984–2005.

Responses to the following question: With which one of these statements about the environment and the economy do you most agree—protection of the environment should be given priority, even at the risk of curbing economic growth (or) economic growth should be given priority, even if the environment suffers to some extent? (Data from D. K. Carlson, "Public Priorities: Environment vs. Economic Growth"; Gallup Poll News Service, April 12, 2005, http://www.gallup.com/poll/content?ci=15820&pg=1; and National Science Foundation, *Science and Engineering Indicators 2006*, chap. 7, fig. 14.)

A public that understands scientific concepts is better-equipped to understand the role of human factors in climate change, to appreciate science and technology's role in stewardship and remediation. Indeed, efforts to improve the environment may well occupy the scientific workforce for generations to come. To cite just one example, the automakers will be hiring many scientists and engineers in the decades ahead in order to meet emissions and mileage requirements, and create new, more efficient engines.

Pseudoscience

What is *pseudoscience*? It is not scientific fact. Any information presented as scientific fact must adhere to strict principles that, among other things, should be testable by experiment. If, for example, physicists claim that any object dropped near the surface of the earth accelerates at a rate of thirty-two feet per second per second, then anyone with the necessary equipment should be able to test this principle. In contrast, it is impossible to test or evaluate the claim that aliens have landed on earth undetected. Nevertheless, a notable number (30 percent) of citizens believe that the sighting of alien spaceships is, indeed, a fact. Similarly, in a recent survey more than 60 percent of respondents reported that they believed some people possess "psychic powers or ESP," regardless of the lack of scientific evidence to date, while 32 percent believe that some numbers are "particularly lucky for some people."[15]

Claims that are presented as fact without supporting evidence are classified as pseudoscience. Why should we trouble ourselves over pseudoscience? Shouldn't individuals be allowed to believe whatever they wish? We would argue that modern democracies work best when their leaders act based on facts, and not on superstitions or ungrounded belief. For example, if the president of the United States bases a decision to launch an attack on the location of the stars, rather than actual intelligence, the results could be disastrous. While we are not aware of any instances of pseudoscientific presidential decisions, First Lady Nancy Reagan relied on astrological indicators to determine who could meet with her husband on any given day.[16]

The state of public preparedness for life in a technological age can be measured by the extent to which they reject pseudoscience. Significant levels of public belief in pseudoscience are one indication that science education is inadequate.

This is not to say that the interpretations of scientific observation are always correct. Scientific findings frequently contradict each other, sometimes as a result of ineptitude, sloppiness, or the uncertainties of the measurement process, but more often because of differing legitimate interpretations of the scientific results. It is thus understandable why the public can sometimes be confused about the meaning of scientific findings and what can be derived from them.

Scientists therefore need to be very careful in announcing results, and should make sure to explain their true significance to the public. Researchers should also take time to explain the vetting processes that should be employed to test a result's validity. In other words, the scientific community needs to do more to help the public understand the scientific process itself—the reality that the research of today builds on what was done yesterday, and that science can often produce more questions than answers. Finally, scientists need to be aware of how unethical behavior in scientific research can damage the public trust and provoke policies that impede scientific progress.

The Growing Tension between Science and Religion

The influence of religion and moral values in modern American politics seems to be growing, as evidenced in the 2004 presidential election, when one in five voters polled said that "moral values" were the determining factor in their vote.[17] This lends credence to analysts' suggestion that elections are now being determined by moral and religious divisions, as opposed to the views on the economy

or foreign policy. If this is true, one has to wonder how it will affect science policy. Recent heated debates over stem cell research and the teaching of intelligent design combined with growing concerns within the scientific community regarding the usage of political—as opposed to scientific—litmus tests in the selection of key scientific advisors suggest that the cultural divide between science and religion may be widening in twenty-first-century America.

The growing divide is further illustrated by recent polling numbers showing that a majority of Americans reject the notion of evolution, and believe that "God created humans in present form."[18] A clash may seem inevitable. However, one must question the extent to which the division really represents popular opposition to science. Certainly, the substitution of belief for proven fact has the potential for great ramifications in a wide variety of public policies.

Is Public Understanding Necessary for Public Support of Science?

As we have noted, in a democracy like the United States, the public has a strong influence on policy decisions; yet the vote of an individual with no scientific knowledge counts exactly as much as that of a physics professor or a biologist—as it should.

Despite high levels of scientific illiteracy in the United States among the general population, and despite doubts whether the benefits of science and technology outweigh the harmful effects, the public and most policymakers still tend to support federal research funding.[19] In one survey, 83 percent responded positively to the statement, "Even if it brings no immediate benefits, scientific research that advances the frontiers of knowledge is necessary and should be supported by the Federal Government."[20] This response has remained stable over a period of many years. Further, 40 percent of respondents believed that government was not spending enough on scientific research.[21] Thus, even though the general public may not understand science, or may even at times question or disagree with its findings and applications, they do, in fact, support public funding of scientific research.

In Europe, a corresponding survey revealed an even higher level of support, with 57 percent agreeing that "my government should spend more money on scientific research and less on other things."[22] This finding is particularly interesting given Europeans' aforementioned skepticism about the value of science.

The Public's View of Leaders in Science

When members of the public hold science leaders in high esteem, they are also likely to support federal research investments. In 2004, the public expressed more confidence in the leadership of the scientific community than in the leadership of five other sectors, and second only to the military (see fig. 10.2).

Scientists have been ranked at the top of a Harris Poll that asks people to list twenty-three different professions and occupations in order of prestige, in every year since the question was first asked in 1977, with 57 percent of those surveyed in 2003 considering science as a profession with "very great" prestige. In comparison, only 30 percent of respondents assigned members of Congress this status, and just 17 percent did the same for lawyers. However, scientists' ranking has dropped an alarming 9 percentage points since 1977, from a high of 66 percent. This is comparable to the drop experienced by doctors and athletes; only lawyers have experienced a greater decline.[23] One group whose rating has risen is the teachers.

The Media

The media play a critical role in shaping public perceptions of science. In fact, most of the information the public and policymakers receive about science comes from the press. Of course, when the media fails to do its job well, it can distort information, and cause confusion about facts and policies. It has been noted, in this regard, that while almost every American newspaper has a daily astrology column, very few have even a weekly science column.[24]

There is a feeling in the scientific community that the media do not do a good job of conveying scientific information to the public. This concern is often attributed to the sense that news reporters themselves may lack a basic understanding of science, and thus are not in a good position to report on it clearly and accurately. Conversely, while journalists generally have a great deal of respect for scientists, many complain that scientists are "so intellectual and immersed in their own jargon that they can't communicate with journalists or the public."[25]

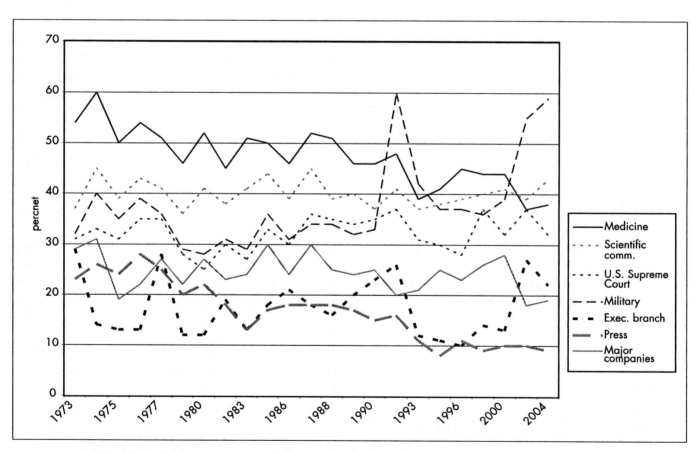

FIG. 10.2 Public expressing confidence in leadership of selected institutions, 1973–2004. (Data from J. A. Davis, T. W. Smith, and P. V. Marsden, *General Social Survey* 1972–2004 Cumulative Codebook, University of Chicago, National Opinion Research Center [2005]; and National Science Foundation, *Science and Engineering Indicators 2006*, chap. 7, fig. 19.)

In a 1998 study of media science coverage conducted by Jim Hartz, a former cohost of the *Today Show,* and Rick Chappell, the director of science and research communications and an adjunct professor of physics at Vanderbilt University, only 11 percent of the scientists surveyed said they had a great deal of confidence in the press, while 22 percent said they had hardly any. The numbers are much worse for television, with 48 percent expressing very little confidence in TV journalism, and only 2 percent expressing a great deal.[26] Hartz and Chappell stated, in joint testimony to the House Science Committee in 1998, that scientists believe reporters do not "understand many of the basics of their methods, including peer review, the incremental nature of science, and a proper interpretation of statistics, probabilities and risk."[27]

Despite this gap in confidence, and perhaps comprehension, many journalists express reservations about their own abilities to cover scientific and technological issues. Indeed, of the over seven hundred journalists who responded to Hartz and Chappell's survey, 72 percent agreed that few members of the news media understand science and technology, while 69 percent said that members of the media were more interested in sensationalism than scientific truth. While 89 percent disputed the notion that science reporting is biased, 62 percent acknowledged that "the biggest problem with science reporting is that it only tells a small part of the whole story."[28]

Some journalists have taken steps to increase the quality of their science coverage.[29] Groups such as the National Science Writers Association and the Council for the Advancement of Science Writing offer workshops, seminars, and annual awards for excellence in the field.[30] Even so, the quality of science stories is constrained by several factors, the two most important perhaps being scarcity of time and of space. The competition for scoops, the influence of editors who do not like or value science stories, and other media bias all limit science coverage still further.[31] Perhaps one of the most daunting challenges of all is the need to explain science stories to a readership that often lacks even basic scientific understanding.

As the primary conduit for information about current events, the media helps frame our understanding of government science policies. Unfortunately, the media tend to emphasize political controversy, rather than the scientific advances themselves. Journalism and communications scholar Matthew Nisbet has found that most media coverage of stem cell research has little to do with scientific activities and successes in the field. Instead, coverage has increased and decreased with the intensity of political and moral debate. During particularly contentious periods, the responsibility for coverage often shifts from science

reporters to less science-savvy political reporters. Nisbet concluded that "media attention (and therefore public attention) can be expected to be greatest . . . when the media define an issue not in terms of science, but as either a political conflict or a matter of morality and ethics."[32]

This phenomenon might explain the tremendous media coverage of *Sputnik* after its 1957 launch. During a 1997 panel discussion on the relationship between science and the news media, held in conjunction with the fortieth anniversary of *Sputnik,* Shannon Brownlee, a distinguished science journalist and former senior editor for *U.S. News and World Report,* stated that *Sputnik* provided journalists with a great story: "There's a reason that the things we produce and put on pages are called stories—not because they're fiction, because they are filled with a lot of facts. But they're packaged in the form of stories, and stories thrive on conflict. They are like horse races, and what *Sputnik* did was, it provided this wonderful sort of conflict and it was larger than a conflict between two countries. It was the conflict between good and evil."[33]

In the absence of a *Sputnik*-like event, it will be difficult for science to attract much media attention. However, given the degree to which our new information- and service-based economy relies on science, scientists should engage with the media in a dialogue about how it covers and reports the relevant science and policy issues. The scientific community also needs to better understand the sources from which the general public gets its information on these matters.

Where Does the Public Get Information about Science and Technology?

Where *does* the American public get its information about science and technology? In a recent NSF survey, about 40 percent of households indicated that television was their leading source.[34] When an American adult wishes to retrieve reference information about science and technology, on the other hand, he or she most often turns to the Internet. The decline in evening news viewership over the past two decades may create a growing concern that the public is turning to sources with even less reliable information about science and technology (see fig. 10.3).

Television

The public's reliance on television for scientific information can be viewed in both a positive and a negative light. Though many programs go to great lengths to ensure accuracy in their science reporting, television is principally an entertainment medium. Therefore, there

POLICY DISCUSSION BOX 10.1

Mass Media Coverage of Science: How Does Framing of Science Affect Public Policy?

Political analysts have often expressed concerns that media coverage of U.S. political campaigns focuses far too much on the "horse race" as opposed to "horses." The specific complaint is that the media tends to give more attention to polls and who *appears* to be winning the election as opposed to informing voters of the specific positions and views of each candidate on key issues. The result is the voters are then left uninformed and often turned off to the political process.[35]

Interestingly enough, a similar phenomenon occurs in the media's coverage of science. Just as candidates hope the media will disseminate information about their stands on critical issues, scientists often hold out hope that the mass media will serve as a tool for public education about science. Unfortunately, for both political candidates and scientists, this is often not the case.

In fact, while the media has the potential to inform people about science, it can and often does confuse and distort public understanding of science or the politics surrounding major controversies over science.[36] Journalism and communications scholar Matthew Nisbet has found that the times when science receives the greatest attention and coverage in the popular press are when issues are not defined in terms of scientific progress or major discoveries, but instead when science is at the heart of a significant political conflict or at the center of a particular moral or ethical debate.[37]

For example, media coverage tends to increase when politicians object to and call into question scientific findings indicating climate change is occurring despite the broad scientific consensus. Likewise, coverage increases when there is a major vote in Congress on whether stem cell research should or should not be permitted, while in the meantime information concerning actual advances in stem cell research again receive much less coverage. In such instances, the manner in which the media presents, or *frames,* a particular issue can greatly influence public attitudes toward, for example, nuclear energy, embryonic stem cell research, climate change research, intelligent design, nanotechnology, and agricultural biotechnology. And oftentimes this framing is much more focused on the nature of the controversy than on accurately reporting scientific facts.[38]

With these things in mind, are there steps the scientific community might take to ensure better media coverage of science and public policy controversies where science is involved? How might the media work to improve its coverage of science and related policy issues? What responsibility do the media have to be a source of information and education versus a source of entertainment? How might scientists do a better job of framing issues in the media such that the public gains a better understanding of, and appreciation for, science?

is a great temptation to emphasize the aspects of science that increase a program's entertainment value. This tendency can, and often does, lead to distortions of facts. In fact, television stations devote a significant portion of programming time to subjects such as alien abductions, ghosts, and monsters, which can lead to the impression that such phenomena are associated with scientific fact.[39] This observation has an analog with the reporting on health issues in local news, where it has been claimed that a significant amount of the content provided is wrong and possibly even dangerous.[40]

When major events, such as volcanic explosions, space shuttle launches, and hurricanes occur, they are covered on regular news programs. While there is entertainment value in the reporting of these crises, good news reporting will include some of the science underlying the situation.

Educational programs, such as PBS's *Nova* and various programs on the Discovery Channel, often cover scientific topics in ways that appeal to viewers. Television is particularly effective in conveying information about science and technology to such a wide audience because of its ability to illustrate scientific observations dynamically. Even TV programs that are not focused on science, such as weathercasts, may include bits of scientific information, and affect viewers' scientific understanding. Many stations also devote a significant amount of news time to reports on health and medicine, which by their nature include scientific information.

Televised movies transmit both positive and negative images of scientists and engineers to the public. For instance the movie *Apollo 13* shows NASA engineers struggling to bring the malfunctioning spacecraft and its as-

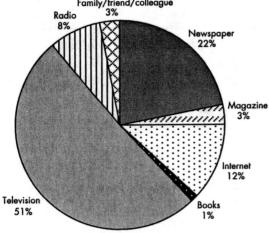

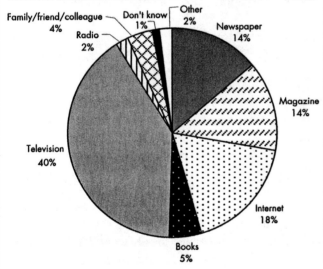

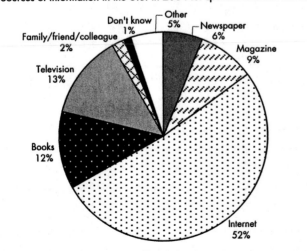

tronauts safely back to earth. Likewise, the popular *CSI* television shows have boosted enrollment in forensic science classes nationwide.[41] Some of these depictions may be more dramatic than accurate. Either way, they are relatively scarce: a study tallying the occupations of characters in prime-time TV entertainment from 1994 to 1997 found that only 2 percent were scientists, with 75 percent of these being white males. Happily, while scientists were not very often seen in TV dramas, they were more likely to appear as heroes than members of any other occupation.[42]

The Print Media

Twenty-eight percent of the households sampled in the aforementioned NSF study listed print media as a leading source of information about science and technology (see fig. 10.3).[43] Several studies have confirmed these results, with readership of newspapers and magazines falling by about 20 percent over the past decade. That said, several million science magazines are sold in the United States each month, including titles like *Popular Science, Science News, Discover, Scientific American,* and *Natural History.* These publications appeal to readers with a long-standing interest in science and technology, keeping them up to date on recent developments. Indeed, those with an interest in science rely on magazines for information to a much greater degree than does the public at large. Notably, print media consumers have a higher average level of education and family income, suggesting socioeconomic differences in the appreciation of science.

The Internet

The Internet is emerging as the source of choice for information on many subjects, although interestingly not (yet) on current events. The increasing sophistication of search engines such as Google makes it possible for an individual to filter millions of documents in a few seconds. The Internet might thus be viewed as the high-tech version of the old family encyclopedia.

FIG. 10.3 Sources of Information in the United States in 2004. Detail may not add to total because of rounding. Categories with less than 0.5 percent response not shown.
(Data from University of Michigan, *Survey of Consumer Attitudes* [2004]; and National Science Foundation, Science and Engineering Indicators 2006, chap. 7, fig. 1; and app. tables 7-1, 7-2, and 7-3.)

However, the Internet is just a medium, like TV or the newspaper. There is no guarantee that information found there is accurate or trustworthy. A 2002 survey found that 50 percent of respondents felt that most of the information on the Web was reliable and accurate.[44] As might be expected, individuals who make heavy use of the Internet develop their own sense of what sources can and cannot be trusted.

Overall, public access to computers continues to increase. Twice as many people have access to computers today as was the case just two decades ago. The implications for public policy will be enormous, serving as the main avenue for citizens' contact with government and stimulating journalism, education, and direct public participation in research.

Books and Other Sources

Though American citizens receive only a small portion of their science information from books, books have been important in educating a certain segment of the population about complex scientific principles and issues. Rachel Carson's *Silent Spring*, for example, provided much of the impetus for the original environmental movement.[45] Stephen Hawking's *A Brief History of Time* sold over nine million copies, and encouraged many readers to think deeply about subjects like black holes, perhaps for the first time in their lives.[46] Carl Sagan's *Dragons of Eden* and *Cosmos* have been read by a wide cross-section of the population, sparking broad public discussions. Such titles as *The Fabric of the Cosmos* and *The Elegant Universe* by physicist and string theorist Brian Green, and *The Physics of Star Trek* by theoretical physicist Lawrence Krauss, forcefully demonstrate how books can capture the public imagination and increase popular interest in science.[47] And the number of popular books on science has been growing.

Popular Institutions

Science and technology museums, zoos, aquaria, and natural history museums play an enormous role in encouraging public interest. One of the best ways to learn about science is through observation: a person can appreciate a killer whale, for example, by seeing it up close, more than one could from reading about it, even in a beautifully illustrated book. Similarly, museum exhibits on nuclear fission can drive home simple lessons about the underlying processes much better than any textbook description ever could.[48]

In times of budget difficulties, municipalities and governments at the state and national level often find it easy to make cuts in museum funding. From the perspective of national science policy, such actions are shortsighted.

The Scientific Community, Policymakers, and the Public

The scientific community has its own obligation to ensure that the public is well informed on science-related issues, and to engage with policymakers. The interactions, or lack thereof, among these three groups can significantly influence the shape of science policy.

The Scientific Community

Inasmuch as scientists are themselves members of the public, issues that deeply concern the public also often resonate in the scientific community. For example, both the public at large and scientists in general have expressed concern over the safe storage of radioactive waste.

However, the opinions of scientists and the populace as a whole can also diverge. While scientists might be relatively comfortable with work in a frontier area like human embryonic stem cell research, the general public may feel greater apprehension.[49] It is easy for scientists pursuing such frontier research to become totally consumed by the search for new scientific knowledge and solutions to familiar social problems (e.g., major diseases, hunger, or pain), with little thought for how their research might be perceived or viewed by the public or even their colleagues. This tendency perpetuates the image of scientists as oblivious to the concerns of the outside world. Scientists, in turn, sometimes look upon the public as unknowledgeable, or even incapable of appreciating the importance of their research. Such a divide is very dangerous. It can weaken congressional will to appropriate funds for certain areas of research, and lead to undue restrictions and regulations.

Often, of course, the difference of opinions is not a question of black and white. Human embryonic stem cell research, for example, actually has two public constituencies. Some individuals and groups, in particular patient advocates, are arguing for lifting the restrictions imposed by George W. Bush in August 2001.[50] At the same time, other segments of the public oppose stem cell research on the grounds that scientists are "playing God" and violating key tenets of the right to life.[51]

With the massive amount of research now being done using recombinant DNA (rDNA) techniques, it is hard to imagine that similar concerns accompanied this research in its early days, but the case is worth considering for how it was resolved. Many communities, including university towns like Cambridge, Massachusetts, and Bloomington, Indiana, expressed concern that local DNA researchers would create new organisms that could then escape, destroying the environment and threatening public health. Scientists in these communities made an enormous effort to inform the nearby public about the potential risks and rewards of their research, proceeding only after they had convinced the public that the actual risks were in fact minimal.

The scientific community convened to discuss the controversy at the International Congress on Recombinant DNA Molecules, held in Asilomar, California, in February 1975.[52] At this meeting, researchers agreed to a self-imposed moratorium on certain types of rDNA research, until the potential hazards could be fully evaluated and proper safeguards put in place. This self-regulation dissuaded Congress from enacting more stringent restrictions, which could have greatly impaired the advancement of rDNA studies.[53] The story offers one model for resolving future debates over human embryonic stem cell research and other controversial types of science.

Scientists have also encountered public resistance to research protocols, regardless of the field. For example even a single mistreatment of a lab animal may enrage community or national activists leading to anything from peaceful campus protests to attacks on animal research facilities. These battles may either strengthen or weaken legislative support for the affected research. If nothing else, they bring attention to important differences of opinion and perception.

Many scientists recognize the importance of informing the public of the nature of their work and regularly take part in outreach activities. Several professional societies even have committees charged with explaining the essence of new scientific endeavors, enumerating the issues relevant to proposed science policies, and advocating for science education. Organizations like the American Physical Society also honor members who have made special efforts to support science funding or interpret scientific information to the public.[54]

Policymakers

Most policymakers believe that the goal of public policy is to deploy government resources for the benefit of society, and to make laws that protect and enhance public well-being. Correspondingly, they tend to believe that science policy should allow for the advancement of research, while balancing the interests of the general public. The difficulty comes in the need to evaluate different interests. There is rarely a single, correct policy solution to any problem. Instead, legislators make trade-offs, with attention to the costs and benefits. As we have described, prevailing popular opinions often play a major role in determining policy outcomes. However, what appears to the public to be a good short-term solution may be the absolute worst course toward a given long-term goal, and the move that might make the most political sense for the legislator's career might not actually be the best policy for the nation as a whole. This gets to the question of how policymakers evaluate what is actually in the public interest, and if the public always knows what is best. Of course, where science is concerned, there are certainly instances when what is viewed as good for science and the scientific community is not always viewed as being in the public interest. When there is a real or perceived crisis, policymakers generally try to meet the challenge quickly. But there are always questions about how best to address a public need, and the views of the scientific community and policymakers may not be aligned.

Policymakers fully recognize that the public is not always able to make the best decision about what it really "needs." But they also understand that democracy requires that the people be allowed to make this determination for themselves. Policymaking thus entails a very delicate balancing act. Any new initiative must take into account both public opinions and the potential impact on the scientific community. In the long term, the advancement of publicly supported science will continue only as long as the public believes that its investment produces an improvement in quality of life.

The scientific community is too often oblivious to legislators' balancing act. While lawmakers certainly need to consider science's best interests when making public policy, they also have to accommodate innumerable other points of view. Policies designed to solve problems having nothing to do with science may inadvertently affect the conduct of research, further aggravating the situation. Indeed, such cases can be highly problematic for the scientific community.

Quite often scientific debates enter the public domain when issues of moral values, ethics, and religion become involved. Even questions that may seem straightforward, such as "What is the origin of the universe?" can mushroom into life-and-death arguments. In a democracy where religious freedoms are highly valued, policymakers face a very sensitive challenge in trying to balance science against other public priorities. They also, however, have an obligation to promote general understanding and acceptance of science and the value of knowledge.

Scientists' Lack of Engagement

Where does the combination of public's influence on science policy and public scientific illiteracy leave the individual scientist? What actions can he or she take to improve the situation? While it is easy for the scientific community to bemoan the public's and policymakers' lack of understanding, lawmakers can just as rightfully say that scientists do not understand the intricacies of politics. And to some extent the accusation is a fair one: on the whole, scientists spend too little time trying to influence public opinion and policy-making to science's advantage.

Many scientists, including senior laboratory officials and university administrators, do not even know the names of their congresspersons or senators, and would never think to invite them or their staff to visit a lab. As we have suggested, too many researchers assume that the value of their work is obvious—an entitlement mentality that has provoked some to refer to the scientific community as a group of "welfare queens in white coats."[55]

The same sense of entitlement—alloyed, perhaps, with a certain wariness of public opinion—has led to scientists' general underinvolvement in science education and community outreach. In a survey by Sigma Xi—an international honor society of scientists and engineers—74 percent of the organization's members said they did not have time to participate in outreach activities, and almost 50 percent said they did not know how. Over two-thirds did not believe that their involvement would make any difference.[56]

This lack of engagement might also account for many Americans' claim to know no scientists personally. In a 2005 national survey conducted by Research!America, a nonprofit organization advocating for medical and health research, only 18 percent of respondents said that they personally knew a scientist living in their community.[57] Researchers in the life and biomedical sciences were the best known in their communities, while those in the physical and engineering sciences were least known.[58]

Perhaps a starting point for the scientific community is to realize that they have to talk to be heard. Given the important role the media plays in shaping public opinion, which in turn shapes the views and positions of policymakers, it is particularly important that scientists work to better understand and engage the media. Likewise, it is important that scientists engage directly with policymakers, who themselves are likely to have only limited scientific backgrounds and understanding.

In their guide for scientists on how to talk with the media, Richard Hayes and Daniel Grossman note, "With a better understanding of how news stories are produced, and of the pressures on those gathering the news, scientists will be more likely to get their points of view across in a way that provides reporters with what they need. Effective communication increases the odds that today's story will be told well, and can also make [the scientist] a prized source for the harried reporter who'll want to call on [them] in the future."[59]

Of course, even those scientists who do try to engage with policymakers and the public sometimes run into difficulties. Scientists tend to be overly reliant on jargon and technical terms when explaining their work: in other words, they tend to speak as if they were talking to other scientists, making it difficult for nonspecialists to understand the relevance of the research.

A series of focus groups conducted by the Science Coalition, a Washington-based organization composed primarily of leading research universities, and whose aim is to expand and strengthen the federal government's investment in basic research, revealed that the voters and opinion elites looked much more favorably upon research when it was described as "university research" than when it was referred to as "basic" or "fundamental" research. This is most likely because the public generally trusts and can relate to universities. Almost everyone knows someone who has attended a university, and most people are aware that their local state universities are engaged in research. The public is less likely to be familiar with the idea of basic or fundamental research, because these terms are common only within the research community.[60]

More bridges need to be built across this communications gap, and the traffic needs to move across the gap in both directions. Not only should the public have a better understanding of science and technology, but the scientific community should also be more engaged and more attuned to lay concerns. Although American scientists have sustained strong overall public support, improvements are clearly possible.

Policy Challenges and Questions

The opinions of nonscientists have a powerful impact on national science policy. Understanding this, one might ask whether the scientific community is devoting sufficient effort to educating the public about its work. One might also wonder whether enough is being done to help the public distinguish between science and pseudoscience. Should scientists more aggressively reach out to the public, to the media, and to policymakers? Should they pay closer attention to the policy-making process? We believe the answer is yes.

Of course, one might reasonably expect something from the public and their leaders in return. Are federal, state, and local governments making all possible efforts to educate the general public about science? Should the media take a more active role by reporting on new advances, rather than focusing on political controversies? Why is so little attention being paid to science and technology policy in presidential elections? It is our hope that this book will help all of those concerned to address these questions.

NOTES

1. See National Academies Center for Science, Mathematics, and Engineering Education, *Every Child a Scientist: Achieving Scientific Literacy for All* (Washington, DC: National Academies Press, 1998), 1.

2. See George Dvorsky, "Scientific Ignorance Dooms Democracy," *Betterhumans.com,* December 22, 2003, http://archives.betterhumans.com/Columns/Column/tabid/79/Column/289/Default.aspx (accessed May 1, 2007); Toronto Transhumanist Association, http://toronto.transhumanism.com (accessed June 5, 2007).

3. American Association for the Advancement of Science, Project 2061, "Benchmarks for Science Literacy: A Tool for Curriculum Reform," http://www.project2061.org/publications/bsl/default.htm (accessed May 1, 2007).

4. See Paul Recer, "Study: Science Literacy Poor in US," http://www.space.com/scienceastronomy/generalscience/us_science_020501.html (accessed June 5, 2007).

5. John W. Kingdon, *Congressmen's Voting Decisions,* 2nd ed. (New York: Harper and Row, 1981).

6. Sean M. Theriault, *The Power of the People: Congressional Competition, Public Attention, and Voter Retribution* (Columbus: Ohio State University Press, 2005), 1–8, http://www.ohiostatepress.org/Books/Book%20PDFs/Theriault%20Power.pdf (accessed May 3, 2007).

7. See John Williams, "Processes of Science and Technology: A Rationale for Cooperation and Separation," in *Technology Education in the Curriculum: Relationship with Other Subjects,* PATT-12 Conference Proceedings, March 15, 2002, http://www.iteaconnect.org/Conference/PATT/PATT12/Williams.pdf (accessed May 29, 2007).

8. See National Science Board, *Science and Engineering Indicators 2006,* 7.23.

9. See ibid., 7.39; "Clone the Clowns," *The Economist,* March 1, 1997, 17–18; Ricki Lewis, "Dolly out, Human Genome in . . . under the Heading Scary Science," *The Scientist* 14, no. 22 (November 13, 2000): 30; Alexandra Cosgrove, "Bioethics and the Food We Eat," *CBS News.com,* October 19, 2000, http://www.cbsnews.com/stories/2000/10/05/tech/main238841.shtml (accessed June 5, 2007); Margaret E. Kosal, "Nanotech: Is Small Scary?" *Bulletin of the Atomic Scientists* 60, no. 5 (2004): 38–47; Alan H. Goldstein, "The (Really Scary) Soldier of the Future," *Salon,* October 20, 2005, http://www.salon.com/tech/feature/2005/10/20/soldier/index_np.html (May 3, 2007).

10. National Science Board, *Science and Engineering Indicators 2004,* 7.24–7.25 and fig. 7.11.

11. Ibid., 7.22–7.24 and fig. 7.10.

12. Ibid., 7.29; National Science Board, *Science and Engineering Indicators 2006,* 7.26.

13. National Science Board, *Science and Engineering Indicators 2004,* 7.29; National Science Board, *Science and Engineering Indicators 2006,* 7.26. However, in 2005 more than half of Americans said "too little" in response to the question, "Do you think the U.S. government is doing too much, too little, or about the right amount in terms of protecting the environment?" National Science Board, *Science and Engineering Indicators 2006,* 7.27.

14. See Nancy Stauffer, "Climate Change Poorly Understood, MIT Survey Finds," *MIT Tech Talk,* March 16, 2005, http://web.mit.edu/newsoffice/2005/climate-0316.html (accessed May 3, 2007).

15. See National Science Board, *Science and Engineering Indicators 2004,* 7.22; National Science Board, *Science and Engineering Indicators 2006,* 7.22.

16. "Astrology in the White House," *Time,* May 16, 1988, 1, http://www.time.com/time/covers/0,16641,19880516,00.html (accessed May 3, 2007).

17. Kevin Eckstrom and Michele Melendez, "Election Shaped by 'Moral Values': Abortion and Gay Marriage Turned Out to Be the Top Issues," *DeseretNews,* November 4, 2004, http://deseretnews.com/dn/view/0,1249,595102883,00.html (accessed May 3, 2007).

18. "Poll: Creationism Trumps Evolution," *CBSNews,* November 22, 2004, http://www.cbsnews.com/stories/2004/11/22/opinion/polls/main657083.shtml (accessed May 3, 2007); Jeffrey Jones, "Most Americans Engaged in Debate about Evolution, Creation," Gallup Organization, October 13, 2005, http://poll.gallup.com/content/default.aspx?ci=19207 (accessed May 3, 2007); polling report, "Science and Nature," http://www.pollingreport.com/science.htm (accessed May 3, 2007); Dan Vergano and Cathy Lynn Grossman, "The Whole World, from Who's Hands?" *USA Today,* November 10, 2005, http://www.usatoday.com/tech/science/2005-10-10-evolution-debate-centerpiece_x.htm (accessed May 3, 2007).

19. See National Science Board, *Science and Engineering Indicators 2006,* 7.39.

20. Ibid., 7.25.

21. Ibid.

22. Ibid.

23. Humphrey Taylor, "The Harris Poll® #57: Scientists, Firemen, Doctors, Teachers and Nurses Top List as 'Most Prestigious Occupations,'" Harris Poll, October 1, 2003, http://www.harrisinteractive.com/harris_poll/index.asp?PID=406 (accessed May 3, 2007).

24. Jim Hartz and Rick Chappell, *Worlds Apart: How the Distance between Science and Journalism Threatens America's Future* (Nashville, TN: First Amendment Center, 1997), xii.

25. Ibid., 31.

26. Ibid., 27–28.

27. Rick Chappell and Jim Hartz, before the House Committee on Science, *Communicating Science and Engineering in a Sound-Bite World,* 105th Congress, 2nd sess., May 14, 1998.

28. Hartz and Chappell, *Worlds Apart,* 32–33.

29. James C. Peterson, *Citizen Participation in Science Policy* (Amherst: University of Massachusetts Press, 1984), 11.

30. See National Science Writers Association, http://www.casw.org/ (accessed June 5, 2007); and Council for the Advancement of Science Writing, http://www.nasw.org/ (accessed June 5, 2007).

31. Jay A. Winsten, "Science and the Media: The Boundaries of Truth," *Health Affairs,* May 23, 1985, http://content.healthaffairs.org/cgi/reprint/4/1/5.pdf (accessed May 7, 2007).

32. Matthew Nisbet, "The Controversy over Stem Cell Research and Medical Cloning: Tracking the Rise and Fall of Science in the Public Eye," April 2, 2004, http://www.csicop.org/scienceandmedia/controversy/ (accessed April 21, 2007).

33. Hartz and Chappell, *Worlds Apart,* 143.

34. National Science Board, *Science and Engineering Indicators 2006,* 7.7.

35. Liz Harper, "Poll Crazy In Campaign Coverage," Online NewsHour Extra Feature, MacNeil/Lehrer Productions, October 2004, http://www.pbs.org/newshour/extra/features/july-dec04/polls_10-20.html (accessed June 5, 2007).

36. Matthew C. Nisbet, Deitram A. Scheufele, James Shanahan, Patricia Moy, Dominique Brossard, and Bruce V. Lewenstein, "Knowledge, Reservations, or Promise? A Media Effects Model for Public Perceptions of Science and Technology," *Communication Research* 29, no. 5 (2002): 584–608.

37. Nisbet, "Controversy over Stem Cell Research."

38. For additional discussion of framing science, see Matthew C. Nisbet and Chris Mooney, "Framing Science," *Science* 316, no. 5821 (April 6, 2007): 56; and Matthew C. Nisbet and Chris Mooney, "Thanks for the Facts. Now Sell Them," *Washington Post,* April 15, 2007, B03.

39. Ibid., 7.8–9.

40. Carol Cruzan Morton, "Prescription for More Reliable Health News Reporting," *Focus Online,* June 9, 2006, http://focus.hms.harvard.edu/2006/060906/medical_reporting.shtml (accessed May 7, 2007).

41. Heather O. Milke, "CSI: GW," *GW Magazine,* Spring 2004, http://www.gwu.edu/~magazine/archive/2004_spring/docs/feature_csi.html (accessed May 7, 2007).

42. George Gerbner and Brian Linson, "Images of Scientists in Prime Time Television: A Report for the U.S. Department of Commerce from the Cultural Indicators Research Project," in Carol Ann Meares and John F. Sargent, principal authors, *The Digital Work Force: Building Infotech Skills at the Speed of Innovation* (Washington, DC: U.S. Department of Commerce Technology Administration, Office of Technology Policy, June 1999), http://www.technology.gov/reports/itsw/Digital.pdf (accessed May 7, 2007).

43. See National Science Board, *Science and Engineering Indicators 2006,* 7.7.

44. See National Science Board, *Science and Engineering Indicators 2004,* 7.10.

45. Ibid., 7.11.

46. Stephen Hawking, *A Brief History of Time* (New York: Bantam, 1998); National Science Board, *Science and Engineering Indicators 2004,* 7.11.

47. Lawrence M. Krause, *The Physics of Star Trek* (New York: HarperCollins, 1995); National Science Board, *Science and Engineering Indicators 2004,* 7.11.

48. Lars Kongshem, "The Discoverers: Science Museums Like San Francisco's Exploratorium Have a Lot to Teach Schools about the Way Children Learn," *Executive Educator,* November 1995, http://www.kongshem.com/words/discover.html (accessed May 7, 2007).

49. See, for examples, Bill Joy, "Why the Future Doesn't Need Us," *Wired,* April 2000, http://www.wired.com/wired/archive/8.04/joy.html (accessed May 29, 2007); Dylan Evans, "Mean Machines," *The Guardian,* July 29, 2004, http://film.guardian.co.uk/features/featurepages/0,4120,1270918,00.html (accessed May 29, 2007).

50. President George W. Bush, "President Discusses Stem Cell Research," August 9, 2001, http://www.whitehouse.gov/news/releases/2001/08/20010809-2.html. See also Gerald Fischbach and Ruth Fischbach, "Stem Cells: Science, Policy, and Ethics," *Journal of Clinical Investigation* 114, no. 10 (2004): 1364–70.

51. Ray Bohlin, "What Are Stem Cells and Why Are They Important?" *Probe Ministries,* 2001, http://www.probe.org/docs/stemcells.html (accessed May 19, 2007); National Institutes of Health and National Human Genome Research Institute, "Understanding the Human Genome Project," http://www.genome.gov/Pages/Education/Kit/main.cfm?pageid=6 (accessed June 5, 2007); Rick Scheeler, "Stem Cells in Plentiful Supply," *RNC "For Life" Report,* May–June 2001, http://www.rnclife.org/reports/2001/may-june01/may-june01.shtml (accessed May 10, 2007); Paul Berg, "Asilomar and Recombinant DNA," *NobelPrize.org,* August 26, 2004, http://nobelprize.org/medicine/articles/berg/index.html (accessed May 10, 2007).

52. Paul Berg, David Baltimore, Sydney Brenner, Richard O. Roblin III, and Maxine Singer, "Summary Statement of the Asilomar Conference on Recombinant DNA Molecules," *Proceedings of the National Academy of Sciences* 72, no. 6 (1975): 1981–84. See also *Perspectives in Biology and Medicine* 44, no. 2 (Spring 2001).

53. Fischbach and Fischbach, "Stem Cells."

54. See American Physical Society, "Programs, Prizes Awards and Fellowships, Awards Medals and Lectureships," http://www.aps.org/programs/honors/awards/index.cfm (accessed June 5, 2007).

55. Sharon Begley, "Gridlock in the Labs: Does the Country Really Need All Those Scientists?" *Newsweek,* January 14, 1991, 44.

56. Christine P. Brown, Stacie M. Propst, and Mary Woolley, "Report: Helping Researchers Make the Case for Science," *Science Communication* 25, no. 3 (2004): 294–303.

57. Stacie Propst, Research!America, "Successfully Advocating for Science," Georgetown University, January 31, 2005.

58. Ibid.

59. Richard Hayes and Daniel Grossman, *A Scientists Guide to Talking with the Media: Practical Advice from the Union of Concerned Scientists* (New Brunswick, NJ: Rutgers University Press, 2006), xiii.

60. Bill McInturff and Elizabeth Frontczak, *Presentation to the Science Coalition* (Washington, DC: Public Opinion Strategies, March 13, 2001), slides 22–26.

Science Policy Issues in the Post-*Sputnik* Era

CHAPTER 11

Science for National Defense

What Is the Role of Defense Research in U.S. Science Policy?

Concerns about national defense were the primary driving force behind American investment in research after World War II and throughout the Cold War. Noted President Harry Truman in a special message delivered to Congress in September 1945 just after the Japanese surrender in World War II: "No government adequately meets its responsibilities unless it generously and intelligently supports and encourages the work of science in university, industry, and in its own laboratories."[1] In a second message delivered to Congress just over two months later, Truman urged Congress to create a new Department of National Defense. In so doing he called for the systematic allocation of scientific resources, noting, "No aspect of military preparedness is more important than scientific research."[2]

During the Cold War, investments made by the Department of Defense's Defense Advanced Research Projects Agency (DARPA) and each of the military service branches in federal research laboratories and at universities resulted in important technologies for military and civilian use. In addition to supporting the fundamental knowledge underpinning the military technologies, these investments also played a critical role in supporting some of the nation's top scientific talent, including more than sixty-five Nobel Prize–winning researchers.[3]

Defense research continues to be an important element of our national security today and will be tomorrow. New challenges now face the military, such as cyberterrorism, information warfare, biological and chemical weapons, improvised explosive devices, and the proliferation of weapons of mass destruction; all will require the development of new and more sophisticated technologies to protect the military and the nation from these threats. The knowledge required to generate these technologies is critically dependent upon sustained investments in long-term, high-risk, defense-oriented research.

One of the challenges for policymakers with regard to defense research has, and will continue to be, balancing short-term military requirements with the longer-term need to invest in the fundamental knowledge and human talent base required to ensure that the armed forces will have access to cutting-edge technologies in the future. This is a delicate balance, one that must be continually monitored. Wrote Caspar Weinberger when he was secretary of defense in the Reagan administration:

> We face the danger of losing our edge because we have not adequately replenished the reservoir of scientific concepts and knowledge to nourish future technologies . . . we must systematically replenish the scientific reservoir, using the unique and diverse strength of the United States scientific community. . . . Given the relatively long lead time between fundamental discovery and applying such knowledge to defense systems, the true measure of our success . . . may not be apparent for several decades. When the "moment of truth" arrives, we cannot afford to be found wanting.[4]

Indeed, government support for basic and applied defense research programs has often been referred to as the "seed corn" for the nation's future military capabilities.[5] In what follows we will examine the nature of defense research in more detail, looking at what defense research is and why it is important, its historical relationship to civilian R&D, the advisory structures that support it, and finally, some of the key policy issues that surround it, including how much funding it should receive compared to other military and defense needs.

Defense Research Defined

Defense (or military) research is research dedicated to ensuring the armed services' technical superiority and the nation's security. The goal of defense research is to promote "advances in fields that are likely to contribute to national defense, and in doing so, to foster a competitive technology base for the U.S. military."[6] The military engages in a broad spectrum of activities relating to research, development, testing, and evaluation (referred to in defense circles as RDT&E) of weapons and other military-related technologies.

In order to understand what is meant by defense research, it is important to understand how the DOD itself categorizes its activities in these areas. Research support within the DOD is provided through the Research, Development, Testing, and Evaluation (RDT&E) section of the agency's budget, which is broken into seven categories, labeled 6.1 through 6.7. The 6.1, 6.2, and 6.3 accounts, along with defense medical research, are commonly referred to as *defense S&T*. These are the defense funding accounts that are most relevant for students of science policy. *Basic* and *applied* defense research fall into categories 6.1 and 6.2 respectively, while *advanced technology development* falls into category 6.3. By definition, novel research proposals that might have military applications are supported by 6.1 basic research and to some extent by 6.2 applied research. To provide an idea of the magnitude of the defense R&D, the FY2007 RDT&E program was funded at $78.2 billion, with about $13.7 billion of that total dedicated to defense S&T (categories 6.1 through 6.3).[7] As one can see from these funding levels, a majority of the total RDT&E budget is *not* spent on defense S&T, but instead on operational and advanced systems development, weapons testing, and evaluation activities (6.4–6.7).

The basic and applied research categories (6.1 and 6.2) are considered DOD's *technology base*. As mentioned earlier, these DOD accounts are often referred to as the "seed corn" of the agency's technological capabilities. Early-stage research and the creation of fundamental knowledge is the primary focus of technology-based labs and university-based defense researchers. Here testing of new ideas can go on for years before leading to an innovation that can be developed for military use. *Advanced*

technology development work (6.3) is aimed at taking the initial steps required to move ideas and emerging technologies from lab to field. More advanced development and testing R&D activities supported as a part of DOD's RDT&E efforts are broken down into the following funding accounts: advance component development (6.4); system development and demonstration (6.5); management support (6.6); and operational systems development (6.7).

The Importance of Defense Research

Government-supported military research has resulted in significant military, as well as civilian, applications. These applications have helped to arm and protect the troops abroad in times of war. At the same time, they have resulted in the creation of new products and new industries. Past investments in defense research have, for example, given rise to advances, such as the Internet, radar, digital computers, integrated circuits, cryptology, wireless mobile communications, multimedia connections, composite materials, satellite navigation, and the global positioning system (GPS).[8] Little does the average person realize, however, as they surf the World Wide Web, have their groceries scanned at the supermarket, or use their computer mouse, that all of these technologies stem from research first sponsored by the military for military purposes.

One scientific advance resulting from defense-funded basic research was the laser (a word that stands for "light amplification by stimulated emission of radiation"). The basic concepts leading up to the development of the laser came from a microwave research program at Columbia University that received its funding from the military.[9] Interestingly, when it was invented, many dubbed the laser "a solution in search of a problem"; even its potential for significant military applications was not originally known.[10] Today the military uses lasers to guide bombs and to help troops aim accurately. Meanwhile, lasers are also used for multiple civilian purposes, including compact disk players, laser printers, laser pointers, and laser eye surgery. Combined with transistors, lasers are also central to the billion-dollar fiber optics industry.

While the great majority of defense research centers on the physical sciences, engineering, and computing, the military also invests significantly in other areas such as the social and behavioral sciences and the health and biological sciences. As will be discussed in more detail later in this chapter, in 1992 Congress directed the DOD to join the NIH's efforts to study breast and prostate cancer on the grounds that a huge number of military personnel are stricken with these diseases each year, at an astronomical cost.[11] The military is also interested in the health and biological sciences because it wishes to improve the stamina of its personnel, heal wounds on the battlefield, protect troops against infectious diseases, and better understand the nature of chemical and biological weapons.

A Glimpse of Current Military Research

A wide array of cutting-edge military R&D projects are annually undertaken and sponsored by the DOD through its military services, DARPA and the Office of the Secretary of Defense. DOD research projects are supported through several different channels and funding mechanisms. Sometimes contracts are awarded to researchers who have bid for a particular project, or from whom the DOD specifically seek research services. In other instances, grants are awarded through peer review. Even when funding is awarded through peer review, however, the DOD, knowing in advance what technology it wants to develop, will issue a *request for proposals* (RFP) to a particular subset of researchers with expertise in the area. If the project involves classified research, then the number of eligible scientists is limited still further. In what follows we provide a glimpse of some current and past defense research projects.

The Virtual Soldier

Rapid advances in digital electronics are transforming military strategy and deployment. For example, DARPA is currently at work on a plan to implant a microdot in every soldier's dog tag containing all relevant medical information. Were the soldier to be injured, medics would have immediate access to necessary records. The chip could be updated every time the soldier had routine medical examinations, CT scans, x-rays, or surgery, guaranteeing the currency of the recorded data. Since every person's vital organs are unique in shape, size, and location, such person-specific information could mean the difference between life and death.[12] Beyond its military usages, this technology could have significant civilian benefits: any citizen carrying such precise and up-to-date medical information would be at an enormous advantage in case of a medical emergency.

DARPA and the Internet

The history of the Internet reveals a wonderful interlacing of contributions from numerous sectors—universities, international laboratories, NSF centers, and the military.[13] The Internet's origin goes back to 1969 DOD efforts to design a fail-proof system of wartime communications in case of a nuclear attack. The Internet was also advanced because of the limited number of large and powerful research computers and the interest among scientific researchers in being able to access those computers through a remote communications network.

Full exploitation of the Internet requires ongoing R&D, much of which is taking place in the open environment of university and industry labs and for civilian purposes. However, the DOD continues funding defense-specific R&D on the Internet's particular benefits to the military in areas such as developing next-generation protocols that will enable net-centric military operations.[14]

Remote Surgery

The Internet has enabled the high-speed transmission of digital information across continents and oceans. This capability, combined with refinements in robotics and medical techniques, allows doctors to operate on patients thousands of miles away. In theory, doctors located far from battle could remotely operate on critically wounded soldiers at the front.[15]

Every aspect of this process needs improvement, however—greater robotic precision, better feedback about what the robot is "feeling," guaranteed stability of the Internet connection, better video resolution, and so on. Again, a military breakthrough in this area would have an enormous impact on the civilian population. Even today, many rural areas have only limited access to medical specialists. Life-saving operations may not be possible without costly and time-consuming trips to metropolitan hospitals. Lives—civilians' as well as soldiers'—will be saved when robotic surgery becomes a reality.

Unmanned Aerial Vehicles

Unmanned aerial vehicles, commonly referred to as UAVs, are remotely controlled or preprogrammed aircraft that fly without a pilot onboard. UAVs have long been used in reconnaissance and intelligence-gathering missions. Today, technological advances enable UAVs to be fitted with weaponry such as air-to-ground missiles and used in military attacks. UAVs have gained increasing attention as they have become heavily used in recent military conflicts in Afghanistan and Iraq.[16]

Despite the increased capabilities of today's UAVs, most lack the flexibility and range of performance that would enable them to be used autonomously to complete an entire mission in which operating conditions may be subject to change. Moreover, they are not able operate well in teams where groups of UAVs might be required to fulfill a particular military operation.

DOD continues to support research to help improve the performance of UAVs. This basic research is laying the groundwork for significant improvements in UAV capabilities. The scientific discoveries that will be generated from this ongoing research are likely to enhance their use in search and rescue, reconnaissance, tracking and targeting, and long-range operations, and help them to perform missions in environments where they are not currently able to be used, such as urban areas.

Transportation, Energy Research, and Fuel Efficiency

The logistics of military transportation are complex and constitute a research field of their own. Troop mobilizations require that equipment components be marshaled from bases around the world, arriving at predetermined stations for final assembly and battlefield insertion within a few hours or days.

The military conducts extensive research on the challenge of efficiently fueling trucks and other vehicles for training and wartime deployment. The army alone has over 240,000 trucks, and Major General N. Ross Thompson III, who has headed the Army Tank-Automotive and Armaments Command, has remarked that "in the long term, the Army will be able to save billions of dollars in logistics-related costs by consolidating the number of truck types it uses currently and by adopting hybrid-electric vehicles."[17]

Indeed, the Army spends almost $3.5 billion a year just on gasoline, so that even modest efficiencies have a significant economic impact. But there are also more practical reasons to reduce the military's reliance on gasoline. Gasoline and water constitute about 60 to 90 percent of the material, by weight, that must be transported to the battlefield.[18]

To solve these energy challenges, the DOD and its military service branches spend significant amounts of money on energy R&D projects: approximately $250 million in FY2006 alone.[19] These programs focus on research areas such as fuel cells, onboard batteries, advanced propulsion, hybrid-diesel technology, improved fuel efficiency, lightweight materials, reengineering of vehicles, and high-performance turbine systems.[20] DOD research in energy research parallels that being sponsored by civilian agencies such as the U.S. Department of Energy.

Cultural and Adversarial Decision Modeling

In the future, the success of our military in battling global terrorism and other military threats will depend upon our ability to understand and predict how our adversary might react in particular situations. It will also require a strong understanding of how local cultural, political, and economic forces might affect military strategies and which strategies might be more or less successful. For example, U.S. military failures in both Iraq and Afghanistan have been attributed in part to the failure to take such anthropological factors into account in developing our military strategy to respond to the threats posed by the irregular enemies we have faced in these regions.[21]

The importance of having better information concerning social and cultural factors and in applying that understanding to defeating our adversaries on the battlefield has been taken to heart by the Department of Defense. Recently it has begun to sponsor new research aimed at building better models to determine and respond to cultural factors that influence the attitudes and behaviors of our military adversaries.[22] This research could be used not only to help develop better military strategies for dealing with nontraditional military threats in the global war on terrorism, but will also be useful in developing new training aids and tools for U.S. military personnel and helping to make decisions in the field when dealing with civilian populations.

Who Conducts Military Research

Contrary to popular belief, much American defense research is done by scientists and engineers at universities or in industry, under grants or contracts from the research offices of the military services, the Office of the Secretary of Defense, or DARPA. This academic and private-sector workforce is significantly augmented by government scientists at the nation's government-owned, government-operated DOD laboratories, the most notable being the Naval Research Laboratory, the Army Research Laboratory, and the Air Force Research Laboratory.

While the air force, army, and navy all have R&D-specific facilities, significant weapons development work is also done at three of the major DOE laboratories: Los Alamos (New Mexico), Sandia (New Mexico), and Lawrence Livermore (California). These DOE weapons labs have long played a role in nuclear weapons R&D, with Los Alamos being the site where the first atomic bombs were created.

These military and DOE labs, however, are not the only places where military R&D takes place. Several FFRDCs and other laboratories also conduct significant DOD research, including Lincoln Laboratory, an air force FFRDC managed by MIT; the DOD's Command, Control, Communications, and Intelligence FFRDC, managed by the nonprofit MITRE Corporation; the Applied Research Laboratory (ARL), a DOD-designated navy university-affiliated research center (UARC), at Pennsylvania State University; and the Applied Physics Laboratory, a not-for-profit center (also a UARC) and division of the Johns Hopkins University. Some of the nonweapons DOE national labs also engage in military R&D.

Universities also conduct a considerable amount of 6.1 basic and 6.2 applied defense research, as we have mentioned, but do very little development work. After the NIH and the NSF, the DOD is the largest funder of university research, with almost 55 percent of the agency's basic research funding and a substantial portion of its applied research monies spent at universities.

In many academic fields, the DOD is the leading federal supporter of research. For example, in FY2003 it was the largest government sponsor of engineering research at academic institutions, providing 37 percent of the total (NSF provided 35 percent of the academic engineering funding). During this same year, DOD supported 70 percent of all federal funding for mechanical engineering, 66 percent for electrical engineers, 42 percent for materials engineering, 29 percent for computer science, and 28 percent for ocean sciences.[23] It is also worth noting that some of DOD's funding supports the training of students in the form of scholarships and graduate fellowships aimed at attracting and retaining students in science and engineer fields critical to national security.

Industry is also a major player in the defense R&D world, with the government acting as the primary contractor for entire industrial sectors, such as the aerospace industry. Most of the military R&D conducted by industry is focused on development work, as opposed to basic research.

Finally, many small businesses exist primarily to enhance the military's R&D capabilities. Consequently,

most citizens would need only to drive a few miles from their homes before encountering some entity deeply involved in defense R&D.

Defense R&D and Universities

The relationship between universities and military research has always been complex. Students and faculty more often than not see universities as places of unfettered learning, oriented toward improving the world around us. In contrast, those concerned with national defense are focused on national security and foreign policy objectives, even when their achievement might require killing and destroying. But the distinction is not as simple as it appears. Universities have a long history of support for military research objectives, as in the case of the Manhattan Project. Difficulties may arise when universities are called to help achieve those objectives that enjoy little campus support.[24]

During the Vietnam War, antiwar students and faculty routinely confronted campus researchers involved in military research. These confrontations led to permanent changes in the way defense research is pursued on university campuses today, for in an effort to send a message to campuses concerning increasing war protests and to reduce unrest over campus-based defense research, Congress passed an amendment that specifically restricted the nature of research that could be conducted on university campuses with DOD funding.[25]

The Mansfield Amendment, passed in November 1969 as part of the Defense Authorization Act of 1970 (P.L. 91-121), required that the DOD only support basic research with a "direct and apparent relationship to a specific military function or operation."[26] In one single step this amendment removed a DOD capability that had allowed the agency to remain flexible in its relations with the academic community, while at the same time procuring the research it needed.

The amendment pushed the DOD away from its support of basic research, toward work with a more direct emphasis on application and development for specific military goals. While program officers and researchers were sometimes able to rewrite project proposals to suit the new requirement, much basic research—conducted largely at universities—that previously would have been supported by the DOD could no longer be funded in this way. The result was another step in the shift from ongoing basic to short-term applied research: as the National Academies noted in a report on computing, "the Mansfield Amendment, and the mood that gave rise to it, had the longer-term impact of shortening the time horizons for government research support in general and defense research in particular."[27]

As a case in point, the Office of Naval Research was for many years one of the largest supporters of American physics research. This support was predicated on the sense that as-yet unspecified benefits would accrue from the work of bright scientists and engineers, regardless of whether it was directed to some specific military mission. After the enactment of the Mansfield Amendment, however, ONR support simply ceased. As former Foreign Service officer Robert G. Morris wrote:

This provision of questionable wisdom betrayed its author's lack of understanding of science. It required that all DOD-sponsored research, whether basic or applied, have a clear relationship to a military objective. This requirement clearly flew in the face of science as a curiosity-oriented, investigator-initiated pursuit of knowledge. Only imaginative and inspired descriptions of Defense-sponsored research projects after that time enabled the department to continue supporting any basic research at all of the type that had contributed to its support of the discovery of the laser. It is hard to remember now that for twenty years or so after the discovery of the laser in 1957 it was considered brilliant but of little or no practical use.[28]

The Mansfield Amendment caused the migration of many defense-related projects from university campuses to off-campus research centers. While many of these centers maintained ties to the originating universities, often via the continued involvement of faculty and students, the amendment fundamentally altered the post–World War II relationship between the DOD and the academic community.

A significant challenge for future national science policymakers and military planners will be how precisely to define the relationship between university research and national security interests. This issue becomes even more relevant—and complicated—when recent homeland security concerns are added to the picture.

Defense R&D and Industry

The U.S. military is tied to a group of industries often referred to collectively as the *military-industrial complex.* This "complex" includes companies like Grumman Aviation, Lockheed-Martin, Boeing, General Dynamics, and so on, all of which derive a significant percentage of their income by manufacturing equipment or designing systems for the armed forces. These companies play a vital role in the nation's military efforts, generating and testing ideas for future weapons systems. A great deal of the nation's R&D know-how rests in such companies, and any comprehensive analysis of the nation's defense R&D capacity has to take their resources into account.

These companies are still subject to the pressures of marketplace survival, however. A huge order from the air force for one hundred fighter jets might be cause for elation in an aircraft manufacturing company. Yet the same company could find it difficult to survive in a period of relative peace or tightened fiscal policy. Such boom-and-bust scenarios have a demonstrable impact on the nation's science and engineering workforce: it is precisely this kind of ebb and flow that has led to stories of PhD scientists driving taxicabs in the Pacific Northwest.

Of course, stories about five-hundred-dollar toilet seats being charged by industry to the DOD still abound. Some such contracting abuses are most likely the result of simple greed. In other instances, however, they represent one part of the system being charged at an excessive rate to make up for losses sustained elsewhere. One such "elsewhere" derives from companies' inability to more directly recover their R&D investments.

Defense R&D and Federal Agencies Other Than DOD

It would be misleading to suggest that the DOD is responsible for all military and national security R&D. The emerging threats of nuclear weapons launched from space, for example, are being addressed by tapping the experience and facilities at NASA. Indeed, many launches made by NASA are associated with a military mission. NASA, an agency created to specifically respond to *Sputnik,* also has strong historical roots in ensuring national security.

As we discussed earlier, the DOD also maintains a close relationship with some national laboratories. Los Alamos, Lawrence Livermore, and Sandia, for example, while operated by the Department of Energy, are largely devoted to defense-related projects.

How did weapons labs come to be situated in a civilian agency? Recall that the researchers involved in the creation of the atomic bomb came from academia, not the military. The lab system arose from Manhattan Project facilities that, after the war, were overseen by the new Atomic Energy Commission (AEC), the precursor to the DOE.

The national labs' role in military research highlights what many policymakers believe to be the inherent benefit of civilian participation in defense research: these lawmakers claim, as did Vannevar Bush, that high-quality scientific research will enable the United States to retain its military supremacy, and that this work should not be entrusted solely to a military institution.[29]

Advisory Structures for Input on Military R&D

The DOD obtains advisory information about its operations, including research and development, from multiple sources, some within the department, and others outside it.

The Defense Science Board and Other Military Service Research Advisory Boards

While government has long been sensitive to the need for independent advisory boards, this principle was formally acknowledged with 1972 passage of the Federal Advisory Committee Act.[30] The act recognized that outside experts are an efficient means for obtaining insights into agencies' plans. Indeed, by meeting for one day at a time, a few times a year, outside advisors can have a major influence on the direction an agency takes in pursuit of its mission.

The Defense Science Board (DSB) is just one of more than fifty such committees maintained by the DOD alone. Under its charter, the DSB advises "the Secretary of Defense, the Deputy Secretary of Defense, the Under Secretary of Defense for Acquisition, Technology and Logistics, and the Chairman of the Joint Chiefs of Staff on scientific, technical, manufacturing, acquisition process, and other matters of special interest to the Department of Defense."[31] The DSB charter explicitly precludes the board from providing advice on individual DOD procurements. The membership is made up of approximately forty civilian experts, each of whom typically serves a four-year term. Several members serve ex officio, as a result of their roles on other federal advisory boards.

The titles of the DSB's recent studies offer some insight into their priorities: a 2003 report was *DOD Roles and Missions in Homeland Security,* and one published in 2004 was *Preventing and Defending against Clandestine Nuclear Attack.*[32] The board also investigates such issues as the adequacy of the country's smallpox vaccination

plans and the status of the DARPA's technology strategy. Topics studied by the board are normally assigned (or "tasked") by the DOD. The recommendations stemming from these studies are strictly advisory, and their advice may be ignored, or even publicly refuted. For example, the Bush administration rejected the board's suggestion that the level of troop strength in Iraq was insufficient.

Though the DSB is the most visible DOD external advisory body on matters of science and technology, the army, navy, and air force all have their own scientific advisory boards, and there is also an Intelligence Science Board. These external advisory bodies are composed of leaders from both industry and academia.

The JASON Defense Advisory Group

JASON, a reference to the Greek mythology character Jason, is an independent committee operating loosely under the auspices of the Institute for Defense Analyses and was formally established in 1960 to keep the DOD in direct contact with top-notch civilian scientists.[33] JASON membership is by invitation only and is restricted to elite scientists, many of whom are physicists, although in the last decade or so this has expanded to include experts in other fields, including biologists.[34] The membership is self-selecting; that is, the DOD has no say in which scientists are invited to participate in the group. At any given time, most if not all JASON members are academics, usually only tenured faculty and senior researchers. Very few people know of the enormous influence that JASON has historically had over major DOD decisions and projects.

The JASONs, which receives funding from the DOD, meets for one to two months each summer to produce an annual report. The members conduct original research that is often devoted to evaluating an existing DOD or DOE project, or recommending a new one. Given the nature of its studies and analyses, the panel requires access to top secret government information.

JASON's independence has been threatened over the last several years, however. In 2002, the DOD canceled its contract with JASON, and only an appeal from members of Congress, especially Representative Rush Holt (D-NJ), reinstated their funding. In a March 2002 letter to the chair of the House Armed Services Committee, Holt noted:

> I have met with JASON in the past and know personally that JASON is truly a valuable source of advice for the Department of Defense. The decision to disband the

group deprives our country of something that America badly needs—a creative, independent approach to the challenging security problems facing our nation.[35]

This strong legislative support was an interesting reversal from the Vietnam era, when JASON became unpopular for criticizing the government's proposed bombing campaign, suggesting instead an anti-infiltration barrier across the demilitarized zone and the Laotian panhandle that relied primarily on recently developed land mines and sensors. At the same time, however, JASON also came under public criticism for having any role in the Vietnam conflict.[36]

Military versus Civilian R&D

In order to understand the role of defense R&D in national science policy, it is important to distinguish between civilian and military (or defense) types of federally supported research.

Military research is likely to be focused on the DOD's national security mission, such as weapons development and deployment, battlefield communication, minimization of combat equipment, and the mechanics of human cognition and performance under extreme conditions. Civilian research, on the other hand, is focused on improving citizens' quality of life and enhancing economic development. Certainly, research done in one sector may be applied in the other. For example, the high-quality GPS systems now available for sale in most electronics stores were originally developed for military use.

Every country makes choices that determine the percentage of its national R&D budget spent on defense and security issues versus civilian goals. Two countries with the same R&D expenditures per GNP may divide their appropriations very differently: one country might invest a large proportion of its electronics research funding in projects for battlefield use and potential military applications, while the other may invest primarily in developing consumer or medical products that enhance economic competitiveness. Analyses of the two countries' economies would of course show significant electronics investment in both cases, but the very different outcomes can only be understood if we distinguish between military and civilian R&D and look at the respective national science policies in detail.

Each nation has to assess its own priorities when making this determination. Often the degree to which a country emphasizes R&D for the purposes of national defense depends upon the degree to which the country emphasizes overall defense spending.

Interestingly enough, however, one should not assume the inverse: when a country invests more money in national defense, research spending may not increase. In fact, at least in the United States, investments in longer-term research often *decline* during wartime, when more money is spent on development and deployment of existing technologies to meet near-term military needs.

The Historical Linkage between Civilian Science and National Defense

Vannevar Bush's push to establish a national science program was premised on the need for significant civilian involvement in both civilian and defense-related research after World War II. Indeed, Bush declared that "military preparedness requires a permanent, independent, civilian controlled organization, having close liaison with the Army and Navy, but with funds direct from Congress and the clear power to initiate military research which will supplement and strengthen that carried out directly under the control of the Army and Navy."[37]

An examination of early NSF budgets is revealing. In 1945, the proposed budget for the National Research Foundation, the NSF's originally proposed name, included a line item for national defense research: $10 million in 1945 dollars ($108.5 million in 2005 dollars).[38] Although many may think that the administrative mixing of civilian and military research is a new phenomenon, this mixing is in fact a direct outgrowth of the NSF's original mission—which was, of course, created based upon Vannevar Bush's vision.

Spillover from Military to Civilian R&D

One of the most common buzzwords in science policy circles of the last decade is *dual use,* which refers to products or processes that can be used for both civilian and military purposes. Dual use is simultaneously one of our nation's greatest hopes—in that the products of our defense R&D will spill over into new civilian applications—and our greatest fears, inasmuch as hostile nations may develop potentially dangerous capabilities or systems masked behind more benign applications. Is Iran developing nuclear capabilities in order to meet its energy needs, or to develop atomic bombs? Such questions can be difficult, if not impossible, to answer.

Even though we have described the differences between military and civilian R&D, it is important to recognize the considerable overlap between the two. These similarities run deeper than the sharing of individual applications like GPS or the Internet. Civilian and military research both draw upon the nation's pool of scientists and engineers. Civilian research could lead to technologies that have military applications, just as military research could result in new civilian technologies. As others have suggested, it would therefore be a mistake to attempt to clearly distinguish between civilian and military purposes in R&D, as the Mansfield Amendment attempted to do. As discussed in the next section on defense conversion, another reason not to draw clear lines between civilian and military R&D is that major shifts in world events can greatly influence the perceived need to achieve military goals versus civilian goals and to ensure our international economic competitiveness.

Defense Conversion

With the fall of the Berlin wall and the unexpected end of Cold War in the early 1990s, very quickly the threat that had been posed by the Russians was replaced with a new perceived economic challenge to America's security from Japan. With this change came public pressure to downsize the military and to transform the defense industrial complex from supplying weapons and warheads to addressing the country's economic and social needs.

The process of transforming individual defense firms and plants to either dual use or commercial operations, and domestic as opposed to military R&D, was often referred to during this period as *defense conversion*. Among other things, defense conversion consisted of community adjustments to defense cutbacks and downsizing; conversion of defense plant operations to production of non-military goods; retraining of laid-off defense workers; transfer of defense technologies to nondefense industries; and shifting previous defense R&D investments to focus on dual-use products and process technologies.[39] One of the key challenges that came with defense conversion was determining how the nation could put to "good use on the civilian side research talents and institutions that were formerly devoted to defense."[40]

This was a time of great angst for the DOE national laboratories, whose primary reason for existence had been removed. It also posed the challenge of ensuring that expenditures made by the government for defense R&D were reapplied in innovative ways to advance economic growth and international industrial competitiveness and to retain the scientific and technological talent that had been developed to support our military.

One important lesson from this period is that we must ensure that both our defense and civilian R&D efforts are integrated and well supported. Both efforts should continually work jointly to reinforce and support both the military and economic security.

DOD Research Budget Trends and the Future

What do recent DOD budget trends look like? Defense S&T funding began to grow in the Carter administration, and then picked up speed with the Reagan administration's military buildup. The bulk of Reagan-era S&T increases were devoted to advanced technology development (6.3), primarily for the Strategic Defense Initiative, also known as "Star Wars."

As the Cold War came to an end, Congress, concerned that a drop in short-term weapons development might have an adverse impact on the DOD's long-term developmental capacity, slightly increased S&T funding, including funding for basic research. After a few years, however, faced with internal budget trade-offs and growing deficits, Congress found itself unable to sustain this level of investment (in real terms). Thus, total RDT&E funding declined during the 1990s. While the decline was initially tied to efforts to control federal deficits, and not directly linked to the end of the Cold War, it continued as the United States began down the path of defense conversion.

Contrary to the budget declines of the 1990s, the beginning of the twenty-first century, marked by the election of George W. Bush as president, brought with it new all-time highs for the DOD RDT&E budget. For instance, in the FY2006 budget, RDT&E funding was over $70 billion, an increase of over $30 billion from its FY2001 budget.[41] Despite these overall increases in defense R&D, very little of these additional funds were devoted to defense S&T; the new funds went primarily to the development of large weapons systems. Within defense S&T, funding for basic research (6.1) has remained essentially flat, in constant dollars, since the late 1980s (see fig. 11.1).

In many ways, defense R&D spending policies tend to reflect the priorities of the Republican and Democratic administrations that control the White House and Congress at the time: a Reagan-era ramp-up, a Clinton-era decline, and increased spending during the presidency of George W. Bush.

However, as we noted earlier, while the DOD is receiving larger and larger percentages of the total R&D budget, fundamental and applied research is taking a back seat to development. Indeed, when inflation is taken into account, basic and applied research funding has actually declined over the past decade. The relatively stagnant funding level for the defense basic (6.1) and applied (6.2) programs is a science policy issue of growing concern, one we will discuss in further detail subsequently.

Defense Research Policy Issues

Several policy issues relating to defense research are worthy of further examination. In this section we briefly examine some of them.

Government Spending on Defense Research

One of the most important questions relating to defense research is how much should be spent on defense S&T and more narrowly on the basic and applied research that comprise the defense tech base.[42] While no precise answer is possible, based on a complex analysis of anticipated national needs, the DSB suggests that approximately 3 percent of the total defense budget invested in 6.1 through 6.3 research is the level required to ensure an adequate flow of knowledge and technologies. This conclusion was affirmed by the Pentagon in its 2001 Quadrennial Defense Review, which outlines the major policies to be pursued by the DOD.[43]

The health and stability of the DOD's S&T program have been the focus of much congressional attention in recent years, thanks to a growing recognition that American military power may be directly dependent on S&T investments.

Yet even if the government decides that 3 percent of the DOD total should be spent on S&T, how much of

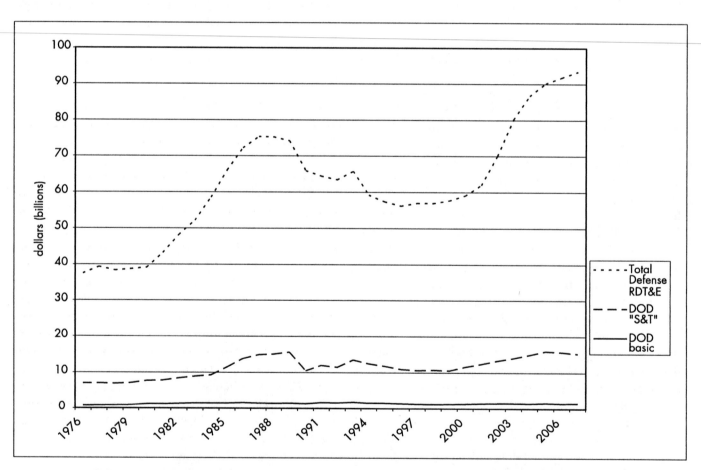

FIG. 11.1 Trends in defense R&D, 1976–2007. Includes conduct of R&D and R&D facilities. Constant dollar conversions based on OMB's GDP deflators from the FY2008 budget. (Data from "Historical Data on Federal R&D, FY 1976–2008," revised March 2007, http://www.aaas.org/spp/rd/tbdef08p.pdf; "Historical Table–Trends in Basic Research, FY 1976–2008 [budget authority in billions of constant FY2007 dollars]," revised April 2007, http://www.aaas.org/spp/rd/tbbas08p.pdf; AAAS Reports I through XXXII, based on OMB and agency R&D budget data.)

POLICY DISCUSSION BOX 11.1

Defense Research Funding: How Much Is Enough?

Policymakers often ask, "What is the right amount of the federal budget to spend on research?" Different arguments have been made to justify different spending levels. Some argue that research spending should be tied to a particular share of GDP. Others argue that research funding at current levels is sufficient so long as funding grows at the rate of inflation and accounts for rising research costs. Still others maintain that constant funding levels for research are sufficient. The same discussions about how to determine funding levels for research also occur within federal agencies.

In trying to address this issue for the Department of Defense, the DSB notes, "There is no magic formula to determine the optimum level of S&T funding." The DSB, however, issued reports in both 1998 and 2002 recom-

mending that 3 percent of the total DOD budget be devoted to defense S&T (basic research, applied research, and advance technology development). The DSB recommendation was based upon the average percentage spent on R&D by commercial industries in terms of their total sales spent on R&D.[44] This figure was also included in the 2001 *Quadrennial Defense Review Report*.[45]

Do you think it is reasonable to use the percentage of total sales devoted by industry to R&D spending as the basis to determine science and technology funding for agencies such as the Department of Defense? What other means might the government use to determine how much is optimal to spend on research? On development? How does one ensure that short-term and more immediate military needs such as personnel costs, procurement of new and better weapons systems, and expense related to troop deployments do not squeeze out funding for longer-term defense S&T research?

that amount should go to support the 6.1 basic research that will drive future military advances in areas such as communications, detection, tracking, and surveillance?

Specific concerns have been raised about recent declines in funding for defense basic research over the last two decades. In the early 1980s, basic research accounted for nearly 20 percent of total defense S&T funding. By 2005, that figure had declined to less than 12 percent, as funding for applied research (6.2) and advanced technology development (6.3) increased.[46] As noted earlier, funding for basic research (6.1) has stayed relatively unchanged in constant dollars. These concerns were highlighted further in a 2005 assessment of basic research by the National Academies committee on Department of Defense Basic Research, which found that from 1993 to 2004, funding for 6.1 basic defense research had declined by 10 percent using DOD inflation indexes and 18 percent using the consumer price index.[47] Concerns about support for basic research were also raised by the industry-led Task Force on the Future of American Innovation in 2006 in a report it released comparing U.S. innovation to that of other countries. Referring to concerns raised by the Defense Science Board about U.S. investments in areas of basic science critical for national defense, such as high-performance computing and semiconductors, the report stated, "If the nation does not reinvigorate its investment in the creation of new fundamental knowledge for national se-

curity, the United States will not have the most advanced weapons systems and military technologies."[48]

It is always difficult, as noted in our earlier discussion of R&D economics, to justify public funding of basic research, in part because it is so difficult to draw a clear, single line from such research to marketplace innovation, and in part because of the lag time from initial creation of knowledge to its use. But additional challenges exist in justifying defense basic research. The military's foremost interest is ensuring the safety and well-being of its personnel in combat and training. While the military may appreciate the vague but significant connections between basic research and end product, it still tends to focus on R&D that meets immediate needs. The research community thus finds it challenging to justify support for basic defense research. When they succeed, it is often by reminding policymakers of basic research's role in generating existing military technologies.

The Expansion of Medical Research at DOD

As we have noted, the military has long maintained an interest in medical research, particularly studies on subjects like infectious disease, public health, radiobiology, and wound healing. The military also has a keen interest in using basic life science research to understand soldiers' sustenance, performance, and survival, including pathogenic mechanisms of action and methods for protecting soldiers against the effects of chemical and biological agents. This biological research is often combined with engineering work: for example, DARPA's Defense Sciences Office is currently sponsoring a collaborative project in which biologists and engineers explore techniques for growing and stabilizing living cells, in order to allow the recovery of cells and tissues under the stressful conditions characteristic of military operations, including sleep and caloric deprivation. Another area of particular interest to DOD is the creation of improved blood clotting factors and synthetic blood substitutes.

Since FY1992, however, the DOD has also become involved in research areas not traditionally associated with national security. In 1992, Joint Appropriations Conference Committee Report No. 102-328 established the DOD Breast Cancer Research Program (BCRP), providing $25 million for breast cancer screening and diagnosis research that fiscal year.[49] Around the same time, the National Breast Cancer Coalition (NBCC), an advocacy group, lobbied Congress for over $200 million for breast cancer research. In addition, Senators Tom Harkin (D-NY) and Alfonse D'Amato (R-NY) also both introduced legislation to have funds transferred from the DOD to the NIH's National Cancer Institute (NCI), based on the notion that, with the Cold War over, DOD funds could more effectively be put to peacetime use. Both the Harkin and the D'Amato bills failed.[50]

With help from the NBCC, however, Harkin and D'Amato joined to introduce an amendment to the FY1993 DOD Appropriations Act transferring funds within the agency, to the peer-reviewed BCRP. Their plan was to find a way to transfer the funds to the NCI after the legislation passed. This plan changed, however, when the director of the NCI stated, "The NCI is a battleship and a battleship doesn't turn on a dime."[51] The money stayed in DOD, and the Army Medical Research and Material Command head, showing enthusiasm for the research, said: "We've been given a mission and our job is to achieve whatever we've been given. . . . Ladies, we're going to win this war."[52]

Shortly thereafter, the DOD created the Congressionally Directed Medical Research Programs (CDMRP) to administer the Breast Cancer Research Project. The CDMRP now oversees, in addition to the BCRP, research programs in women's health, osteoporosis, neurofibromatosis, prostate cancer, ovarian cancer, chronic myelogenous leukemia, and several other specified areas. Between FY1992 and FY2004, Congress appropriated nearly $1.65 billion to CDMRP for breast cancer research

alone, leaving the CDMRP second only to the NCI in its support for breast cancer research.[53]

Anyone can be diagnosed with cancer, or any other disease, the reasoning goes, so why shouldn't the DOD conduct research on a disease that can affect its soldiers? On the other hand, some have asked, why isn't such research being entrusted to the NIH, in keeping with that agency's mission? Some of the critics have also expressed concerned that support for the NIH and CDMRP has, at times, grown faster than support for the physical sciences. As we have stated repeatedly, a balance of funding across all scientific disciplines is a goal policymakers should strive for.

Classified versus Unclassified Military Research

By its nature, military R&D is often classified. There is clear strategic value in denying the enemy details that could be used in producing comparable weapons systems or developing countermeasures to render the systems ineffective. Classification is the principal means by which the military retains control over information.[54]

Researchers whose projects are classified cannot publish outcomes in the open literature or discuss their work with individuals who do not have the same level of clearance or higher. The events of September 11, 2001, provoked a discussion of the creation of a new category for information that is considered sensitive but that does not reach the level of classified, which might, for example, prevent a researcher who had found a way in the lab to synthesize the ebola virus, *de novo,* from publishing his or her results.[55] Government officials argue that the open publication of such information puts national security at risk. Opponents note that the definition of what is sensitive is fuzzy at best, and that expanded controls put censorship powers in the hands of a few bureaucrats. Concerns have also been raised since September 11 regarding the government's tendency to overclassify information and research.[56]

Many universities have simple, straightforward policies prohibiting the conduct of classified research on campus. Those that do accept contracts for classified research usually require that this work take place in specified off-campus facilities, such as MIT's Lincoln Laboratory. Even so, classified research poses challenges to university administrators. Indeed, conducting military research on campus can lead to a clash of cultures since secrecy, highly valued by the military, is anachronistic in universities. How, for example, can a doctoral student involved in a classified project publicly defend his or her dissertation? Must every member of the dissertation committee apply for security clearance? The review associated with obtaining security clearance can cost the government thousands of dollars for each single case.

Yet it is clear that national interests can be served when universities and the government work collaboratively toward national security goals. In most instances, this collaboration can be accomplished without classifying or restricting the dissemination of research results, even when that research is conducted with national defense objectives in mind. To ensure that the results of defense-related research is not unnecessarily classified or restricted through other means, an ongoing dialogue aimed at reconciling the contradictory demands for military secrecy and scientific transparency is essential.

DOD Support for Discovery-Oriented and Exploratory Basic Research

A concern that emerged recently within the scientific and engineering community regarding defense research is that the DOD is increasing emphasis on short-term research with clear payoffs while moving away from its historical support for long-term, open-ended research and exploration.[57]

This concern has been perhaps most strongly expressed with regard to DARPA research. DARPA was originally established in 1958 in direct response to *Sputnik* as part of the U.S. effort to address the national security challenge posed by the Soviet Union. DARPA was specifically designed to "be an anathema to the conventional military and R&D structure and, in fact, to be a deliberate counterpoint to traditional thinking and approaches."[58] DARPA was to serve as the "technological engine" for the DOD and to supply technological options for the entire department and its military branches.[59] Indeed, DARPA has long been praised for its effectiveness at supporting high-risk, high-payoff basic research and the outside-the-box thinking that resulted in the Internet and major advance in other areas including hypertext systems, graphic user interfaces, stealth technologies, and advanced robotics.

In recent years, however, DARPA has been criticized for moving away from support of high risk, high payoff basic research and instead emphasizing more narrowly focused, application-oriented research promising immediate results. There has also been concern that the traditional autonomy and flexibility granted to DARPA program officers to fund such research has been diminished. Moreover, academic scientists complain that DARPA projects, which now require progress reports every couple of months, are not conducive to training students. Perhaps the most vocal

criticism of DARPA has come from academic computer scientists and electrical engineers, who have seen significant funding decreases from DARPA for their research.[60]

An ongoing policy challenge for the DOD and Congress will be to ensure a proper balance between funding for research aimed at meeting short-term military needs and that which is more discovery oriented and exploratory in nature. It is our view, however, that DOD must continue to play a significant role in support of such unfettered basic research and not focus too heavily upon support for research with immediate and predictable payoffs; to do so would be a mistake for both science and national security. As Gerard A. Alphonse, president of Institute of Electronics and Electrical Engineers-USA stated in a letter he wrote to the *New York Times* featured on the IEEE Web page, "Any near-term payoffs will be long forgotten 20 years from now when our foes pull a technology surprise, as the Soviets did with Sputnik in 1957, and we find that our nation's technology cupboard is bare."[61]

Policy Challenges and Questions

Debate continues in Congress, the Pentagon, and the scientific community over the amount of funding needed to ensure the health of the DOD's science and technology efforts. Indeed, no formula can determine the ideal percentage of the military budget that should be devoted to long-term research, since basic research rarely has obvious applications, military or otherwise, in its early stages.

The importance of such research investments to our national defense and to our economic security, however, should not be taken for granted. Former Speaker of the House Newt Gingrich (R-GA) and Representative Bart Gordon (D-TN) teamed up in January 2007 to sound an alarm concerning the potential impacts of failing to adequately support defense research:

> Investments made in fundamental scientific research after World War II and during the Cold War have been essential to making our fighting men and women today the best equipped in the world. These previous investments and the new knowledge they generated also made enormous contributions to our economic vitality. But our commitment to that defense-oriented fundamental research—the kind of research that pays off not in a year or two but in the long run, sometimes decades in the future—has eroded. If we do not renew this commitment, it will harm our global economic competitiveness as well as the effectiveness and safety of our troops.[62]

POLICY DISCUSSION BOX 11.2

The DARPA Model: Can It Be Replicated to Address Non-Defense-Related National Goals?

At the same time DARPA's effectiveness is being called into question, the DARPA model is being pointed to as one that might be replicated by other agencies as a means to address other critical needs. For example, DARPA inspired the creation of the Homeland Security Advanced Research Projects Agency within the Department of Homeland Security. (It, however, ended up very different from DARPA, with DHS officials making it clear that the two ought not to be compared.)[63] In its *Rising above the Gathering Storm* report, the National Academies proposed creating a DARPA-like entity, a so-called ARPA-Energy, to help create and develop new energy technologies required to reduce U.S. dependence on foreign oil and to enhance energy self-sufficiency.

DARPA's historical success and the potential replication of the DARPA model to address other national goals present two sets of questions worthy of further examination by students of science policy. First, was DARPA really as successful as has been suggested? If so, what was it about the DARPA model that made it successful? Can some of those characteristics be applied to the research conducted by other agencies? Second, keeping in mind that the primary customer for the technologies flowing from DARPA was the Department of Defense itself, is it reasonable to think that a DARPA model would work to address other national needs in areas such as energy, where the government is not the sole consumer and user? How would the market forces that drive energy use affect the effectiveness of a DARPA-like entity for energy? Is creating an ARPA-E a good idea?

While the uncertainty inherent in investment in fundamental research is a fact of life in university research, it is subject to greater skepticism from organizations like the military, which is concerned with meeting immediate operational needs. The loose but certain connection between basic research and technological advances needs to be better understood by the military and by those making funding decisions on its behalf. At the same time, the scientific community must understand the military's attention to knowledge that can help it achieve concrete objectives.

As we have discussed, defense research provokes many important science policy questions and challenges for both policymakers and the research community. Some of these questions we have tried to address in this chapter, while others remain to be tackled. To briefly summarize: How can new technologies be more quickly developed and adopted? How can a strong defense research program be retained in the face of other competing and seemingly more immediate military needs? How can we make better use of the private sector to meet the military's technological needs? How, given their competing priorities and concerns, can we sustain a healthy relationship between universities and the national defense? Is the military the proper place to fund medical research? What is the role of DOD in providing mechanisms for training scientists and engineers for military research? What is the obligation of universities and colleges to train scientists and engineers for military research? Should more investment be made in DARPA (or a DARPA-like agency) to foster the development of innovative technologies such as the World Wide Web? Do we have the proper mechanisms in place to generate the fundamental knowledge required to face the military challenges of the future?

NOTES

1. President Harry S. Truman, *Special Message to the Congress Presenting a 21-Point Program for the Reconversion Period,* September 6, 1945.

2. President Harry S. Truman, *Special Message to the Congress Recommending the Establishment of a Department of National Defense,* December 19, 1945.

3. National Science and Technology Council, *Discovery, Education and Innovation: An Overview of the Federal Investment in Science & Technology* (Washington, DC: National Science and Technology Council, Summer 2000), http://www.ostp.gov/NSTC/html/dei/dod.htm (accessed June 30, 2007).

4. Caspar W. Weinberger, "The R&D Key," *Defense* 83 (February 1983): 2.

5. John D. Moteff, *Defense Research: A Primer on the Department of Defense's Research, Development, Test and Evaluation (RDT&E) Program,* 97-316 SPR (Washington, DC: Congressional Research Service, Library of Congress, July 14, 1999), 2.

6. National Academies Committee on Department of Defense Basic Research, *Assessment of Basic Research,* ix.

7. Koizumi, "R&D in the FY 2008 Department of Defense Budget," table II-2, "R&D in the Department of Defense," 138.

8. See Office of the Under Secretary of Defense for Acquisition, Technology and Logistics, Director of Research, http://www.acq.osd.mil/ddre/research/index.html (accessed May 10, 2007). See also William Berry and Cheryl Loeb, *Defense and Technology Paper: Breakthrough Air Force Capabilities Spawned by Basic Research* (Washington, DC: Center for Technology and National Security Policy, National Defense University, April 2007), 3–5.

9. See Office of the Under Secretary of Defense for Acquisition, Technology and Logistics, Director of Research, "Success Stories," http://www.acq.osd.mil/ddre/research/success.html (accessed May 10, 2007); Charles H. Towns, *How the Laser Happened: Adventures of a Scientist* (New York: Oxford University Press, 1999).

10. Mario Bertolotti, *The History of the Laser* (Philadelphia: IOP Publishing, 2005), chap. 14.

11. See U.S. Department of Defense, *Congressionally Directed Medical Research Program FY 2006 Annual Report* (Fort Detrick, MD: U.S. Army Medical Research and Materiel Command, September 30, 2006).

12. Noah Shachtman, "Virtual Soldiers? Dream on, DARPA," *Wired*, August 14, 2003, http://www.wired.com/news/medtech/0,1286,60016,00.html (accessed May 16, 2007).

13. See Hilary Poole, Tami Schuyler, and Theresa M. Senft, *History of the Internet: A Chronology, 1843 to the Present* (Santa Barbara: ABC-Clio, 1999).

14. U.S. Department of Defense, Office of the Assistant Secretary of Defense for Public Affairs, "News Release: Next-Generation Internet Protocol to Enable Net-Centric Operations," no. 413-03, June 13, 2003, http://www.defenselink.mil/releases/release.aspx?releaseid=5457 (accessed June 30, 2007).

15. "Researcher Brings Space Age to Surgery Equipment, Procedures," *Science Daily*, August 23, 2006, http://www.sciencedaily.com/releases/2006/08/060822101545.htm (accessed June 30, 2007).

16. Elizabeth Bone, *Unmanned Aerial Vehicles: Background and Issues for Congress* (Washington, DC: Congressional Research Service, Library of Congress, April 25, 2003).

17. Sandra I. Erwin, "Army's Next Battle: Fuel, Transportation Costs," *National Defense Magazine*, April 2002, http://www.nationaldefensemagazine.org/issues/2002/Apr/Armys_Next.htm (accessed May 16, 2007).

18. Ibid.

19. Scott C. Buchanan, "Energy and Force Transformation," *Joint Force Quarterly* 42 (2006): 51–54.

20. See ibid., 52–53, for a list of DOD programs by military service branch.

21. Tony Corn, "Clausewitz in Wonderland," *Policy Review: Web Special*, September 2006, http://www.hoover.org/publications/policyreview/4268401.html (accessed June 30, 2007); Colin S. Gray, *Irregular Enemies and the Essence of Strategy: Can the American Way of War Adapt?* (Carlisle, PA: Strategic Studies Institute, U.S. Army War College, 2006).

22. Yudhijit Bhattacharjee, "Pentagon Asks Academics for Help in Understanding Its Enemies," *Science* 316, no. 5824 (April 27, 2007): 534–35.

23. National Science Board, *Science and Engineering Indicators 2006*, appendix table 5.9.

24. For example, on April 12, 2006, a motion was brought before the University of Oregon Faculty Senate by a group of faculty members who opposed the university's efforts to seek increases in DOD research funding on the grounds that it promoted the growth of militarism and was therefore unethical. In the end, the motion was tabled. See the minutes from the meeting of the University Senate, University of Oregon, April 12, 2006, http://www.uoregon.edu/~uosenate/dirsen056/12Apr06minutes.html (accessed June 30, 2007).

25. Smith, *American Science Policy*, 81–82.

26. Richard Jones, "OSTP Director Marburger Discusses Astronomy, Physics, and the Budget," *FYI: The American Institute of Physics Bulletin of Science Policy News*, January 24, 2002; Defense Authorization Act, P.L. 91-121, November 19, 1969.

27. *Funding a Revolution: Government Support for Computing Research* (National Academy Press, January 1999), chap. 4, "The Organization of Federal Support: A Historical Review," http://www.nap.edu/readingroom/books/far/ch4.html (accessed May 16, 2007).

28. Robert G. Morris, *Science and Technology in United States Foreign Affairs* (Ashland, OR, 1999), chap. 16, http://www.geocities.com/rgmscitech/s_t_16.html (accessed May 16, 2007).

29. Zachary, *Endless Frontier*, 257–58; Bush, *Science—the Endless Frontier*.

30. Federal Advisory Committee Act, P.L. 92-463, October 6, 1972.

31. Charter for the Defense Science Board, Section II, "Objectives and Scope," http://www.acq.osd.mil/dsb/charter.htm (accessed May 16, 2007).

32. Defense Science Board, *DOD Roles and Missions in Homeland Security*, vol. 2, part B (Washington, DC: Office of the Under Secretary of Defense for Acquisition, Technology, and Logistics, September 2004); Defense Science Board, *Preventing and Defending against Clandestine Nuclear Attack* (Washington, DC: Office of the Under Secretary of Defense for Acquisition, Technology, and Logistics, June 2004). For a list of other DSB reports see the Federation of American Scientist Web page Defense Science Board, http://www.fas.org/irp/agency/dod/dsb/ (accessed July 1, 2007).

33. Ann Finkbeiner, *The Jasons* (New York: Viking, 2006), xi–xxx.

34. The original JASON consisted solely of physicists, and the *crème de la crème* to be sure. However, over the years, geologists, mathematicians, computer scientists, and others in the physical sciences were asked to become JASONs. In the mid-1990s biologists were first invited to join, and as noted by a physicist, JASON biology is "about the extreme limit that a physicist can interact with easily." Ibid., 202–3.

35. Richard M. Jones, "Rep. Holt Expresses Concern over DOD Decision to Disband JASON," *FYI: The American Institute of Physics Bulletin of Science Policy News*, March 27, 2002.

36. For descriptions of this and other JASON projects and reports see Finkbeiner, *The Jasons*.

37. Zachary, *Endless Frontier*.

38. See National Science Board, *Science and Engineering Indicators 2000,* vol. 1, NSB-00-01 (Arlington, VA: National Science Foundation, 2000), text table 1.5, "Proposed National Research Foundation Budget."

39. Jacques S. Gansler, *Defense Conversion: Transforming the Arsenal of Democracy* (Cambridge: MIT Press, 1995), 40–41.

40. U.S. Congress, Office of Technology Assessment, *Defense Conversion: Redirecting R&D,* OTA-ITE-552 (Washington, DC: U.S. Government Printing Office, May 1993).

41. See Kei Koizumi, *Congressional Action on Research and Development in the FY2006 Budget—Final Appropriations* (AAAS Intersociety Working Group, January 6, 2006), 28, table 4 and "Final FY2001 Appropriations: DOD Basic Research Rises 13%; Congress Allocates $9.4 Billion for S&T," *AAAS R&D Funding Update,* January 4, 2001, table A, "Department of Defense by Program," http://www.aaas.org/spp/rd/dod01f.pdf (accessed May 16, 2007).

42. Early discussions concerning this matter can be traced back to the late 1960s, when the Defense Department undertook a major study, known as Project HINDSIGHT, which attempted to assess the impact of post–World War II investments in science and technology to the advancement of major weapons systems. See U.S. Department of Defense, *Project HINDSIGHT: Final Report* (Washington, DC: U.S. Department of Defense, Office of the Director of Research and Engineering, October 1969).

43. Richard Jones, "DOD Report Calls for 3% Investment in S&T," *FYI: The American Institute of Physics Bulletin of Science Policy News,* October 18, 2001.

44. Defense Science Board, *Report of the Task Force on Defense Science and Technology Base for the 21st Century* (Washington, DC: Office of the Under Secretary of Defense for Acquisition and Technology, U.S. Department of Defense, June 30, 1998); Defense Science Board, *2001 Summer Study on Defense Science and Technology* (Washington, DC: Office of the Under Secretary of Defense for Acquisition and Technology, U.S. Department of Defense, May 2002).

45. U.S. Department of Defense, *Quadrennial Defense Review Report* (Washington, DC: U.S. Department of Defense, September 30, 2001).

46. Task Force on the Future of American Innovation, *Measuring the Moment,* 5.

47. National Academies Committee on Department of Defense Basic Research, *Assessment of Basic Research,* 19–20.

48. Task Force on the Future of American Innovation, *Measuring the Moment,* 5–6.

49. Department of Defense, Congressionally Directed Medical Research Programs, "Fact Sheet: Breast Cancer Research Program," December 15, 2003, http://cdmrp.army.mil/pubs/factsheets/bcrpfactsheet.htm (accessed May 16, 2007).

50. Christine Haran, "Three's Company: The Army, Women with Cancer, and the Medical Community Have Joined Forces," *MAMM* 3 (January 2001): 38–43, 57.

51. Ibid.

52. Ibid.

53. Department of Defense, Congressionally Directed Medical Research Programs, "Fact Sheet."

54. The levels of classification used by the DOD are *(a)* confidential—the lowest level, applied to information whose unauthorized disclosure could be expected to cause damage to the national security of the United States; *(b)* secret—a midlevel classification applied to information whose unauthorized disclosure could be expected to cause serious damage to the national security of the United States; *(c)* top secret—the highest level, applied to information whose unauthorized disclosure could be expected to cause exceptionally grave damage to the national security of the United States. See the University of California System-wide Home, Office of the President, Laboratory Management, Security Clearance Information, 2004 Security Refresher Brief, sec. 3, "Overview of Security Classification System," http://labs.ucop.edu/internet/security/brief04/#Anchor-Overview-11481 (accessed May 10, 2007).

55. In fact, researchers at SUNY Stony Brook did publish a report in August 2002 in which they describe the the de novo synthesis of the polio virus. See Jeronimo Cello, Aniko V. Paul, and Eckard Wimmer, "Chemical Synthesis of Poliovirus cDNA: Generation of Infectious Virus in the Absence of Natural Template," *Science* 297, no. 5583 (August 2002): 1016–18. The publication created some controversy within both the scientific and security communities. See Eckard Wimmer, "The Test-Tube Synthesis of a Chemical Called Poliovirus: The Simple Synthesis of a Virus Has Far-Reaching Societal Implications," *EMBO Reports* 7, S1 (2006): S3–S9; Philip Campbell, "Dual-Use Biomedical Research and the Roles of Journals," PowerPoint presentation, International Forum on Biosecurity, Como, Italy, March 20, 2005, http://www7.nationalacademies.org/biso/Biosecurity_Campbell.ppt (accessed May 10, 2007), slide number 3.

56. American Civil Liberties Union, *Science under Siege: The Bush Administration's Assault on Academic Freedom and Scientific Inquiry* (New York: American Civil Liberties Union, June 2005), 3–6.

57. National Academies Committee on Department of Defense Basic Research, *Assessment of Basic Research,* 16–18.

58. Defense Advanced Research Projects Agency, "DARPA over the Years," http://www.darpa.mil/body/overtheyears.html (accessed July 1, 2007).

59. Tony Tether, before the House Armed Services Committee, Subcommittee on Terrorism, Unconventional Threats, and Capabilities, 108th Congress, 1st sess., March 27, 2003. http://www.securitymanagement.com/library/DARPA_tether0603.pdf (accessed December 16, 2007).

60. John Markoff, "Pentagon Redirects Its Research Dollars: University Scientists Concerned by Cuts in Computer Projects," *New York Times,* April 2, 2005.

61. "DARPA Sharply Cutting Research Dollars at Universities: Scientists and Engineers Alarmed," IEEE-USA Web page, Letter to the Editor, in response to John Markoff, "Pentagon Redirects Its Research Dollars; University Scientists Concerned by Cuts in Computer Projects," April 2, 2005, http://www.ieeeusa.org/policy/features/darpa.asp (accessed July 1, 2007).

62. Gingrich and Gordon, "Invest More Now."

63. William New, "Homeland Security Research Agency Has Lofty Vision," *National Journal's Technology Daily,* January 6, 2004.

CHAPTER 12

Big Science

What Is "Big" Science?

A significant scientific contribution can result from work conducted by one investigator pursuing ideas on his or her own—the monk Gregor Mendel exploring heredity with his pea seedlings—or by huge numbers of scientists working together on large and complex research projects, as in the Hubble Telescope. Between the two extremes are small groups of researchers working as equals, like Marie and Pierre Curie studying radioactivity and nuclear energy.

This chapter is concerned with *big science,* a term coined in a 1961 article by Alvin Weinberg, then the director of the Oak Ridge National Laboratory. Weinberg used the phrase to mark the emergence of a new class of government-supported projects involving numerous researchers working in hierarchically structured groups and utilizing shared resources.[1] His article contrasted this kind of research with the more traditional individual and small-group methods, and cautioned that funding for "big science" should not lead us to neglect the pursuits of "little science"—a warning that scientists continue to discuss even today.[2]

Even though new projects seem to involve ever-larger budgets and teams of researchers, grander challenges and more expansive facilities, we still lack a more specific definition of what, precisely, counts as "big science." Perhaps, as Justice Potter Stewart suggested about pornography, one knows a big-science project when one sees it.[3] Our goal in this chapter is to provide the reader with a better understanding of what big science is.

As we will see, big-science projects pose several policy challenges. While one might assume "little science" and "big science" can coexist in harmony—indeed, while this is the shared goal—the reality is that supporters of single-investigator research and backers of big science often come into conflict. The former argues that big science inhibits individual creativity and promotes a factory-like paradigm, while the latter responds that big science is more efficient, and more reliably produces great strides in research.

Examples of Big-Science Projects

We begin with an overview of some of the most notable big-science projects. As will become obvious, these projects, while collectively distinct from single-investigator projects, also have characteristics that distinguish each of them from each other, further complicating our efforts to clearly define "big science."

The Superconducting Super Collider

The Superconducting Super Collider (SSC) was a Department of Energy project intended to equip high-energy physics researchers to explore the fundamental building blocks of matter. The project, which collapsed after expenditures of more than $2 billion, offers a clear example of the challenges surrounding big science. Indeed, a retrospective study of this project reveals serious weaknesses in the country's ability to plan and execute large multi-year research projects. What led to the SSC's demise? In order to grasp the answer, we first need to know more about the project itself.

A Brief History of the SSC

High-energy physics is, among other things, dedicated to identifying the smallest components of matter, and studying how they interact with each other. There are approximately forty-five hundred practicing high-energy physicists in the United States and probably twice that many more worldwide, with significant concentrations in Japan, Western Europe, China, and the Soviet Union.[4]

Because of its focus on the basic building blocks of matter, high-energy physics is one of science's fundamental subfields, meaning that it provides the theoretical foundations and basic principles upon which many other scientific areas stand. What are the fundamental units of matter? How many of them are there? How do they interact? What gives them their basic properties? The search for answers to such elemental questions has fueled the quest for higher-energy accelerators. Building these tools, however, is by no means a simple undertaking. Great distances are required in order to accelerate particles to sufficiently high energies to permit modern-day research. The construction of a high-energy accelerator thus requires huge expanses of real estate. The SSC, for example, would have had a circumference of fifty miles. And space, of course, was not the only consideration: the instrumentation and technical support needed to build such a huge facility was expected to cost approximately $4 billion.

A project of this magnitude poses not only technical but also policy challenges. Any publicly funded project of this size requires difficult political decisions on such issues as site, the allocation of valuable fabrication contracts, and the drain of funds that might otherwise be available for other, smaller-scale projects. While these questions are common to all big-science enterprises, no other project has touched on as many national science policy issues as the SSC.

How It Began

In the years after World War II, the American high-energy physics community pursued and was uniquely successful in developing facilities that anticipated its future needs. For example, the concept for what would eventually become the Fermi Accelerator Laboratory (Fermilab) was first put forward by a scientific panel in 1963.[5] The project was authorized by the U.S. Atomic Energy Commission (AEC) and President Johnson in 1967, and funded by Congress less than a year later.[6] On March 1, 1972, the first beam passed through the lab's main ring, making Fermilab the world's highest-energy operational particle accelerator.[7] Early and under budget, this project established a good record for the field going into pre-SSC deliberations

It was a history of precisely these kinds of successes that fueled the high-energy physics community's decision in the early 1970s to build a new facility, nicknamed ISABELLE, ten times as large as the Fermilab accelerator. In 1974, the plan was endorsed by the U.S. High Energy Physics Advisory Panel (HEPAP), an NSF and AEC formal advisory body composed of leading scientists and engineers from across the United States, which recommended that the machine be built at Brookhaven National Laboratory on Long Island.[8]

The Brookhaven project was terminated before completion for several reasons, some technical and some managerial.[9] At the time, the United States was in a race to discover the W boson, an important particle that had been predicted by theory, and ISABELLE was to be an important aspect of winning that race. When the European Organization for Nuclear Research beat the Americans to their goal by discovering the W boson first, ISABELLE quickly became "was"-ABELLE.[10] The American high energy physics community, with encouragement from OSTP, decided to abandon the partially completed ISABELLE in favor of an even larger project—the Superconducting Super Collider (SSC), which was to leapfrog all existing accelerators.[11]

This groundswell of support for a high-energy experimental physics device prompted the Department of Energy to fund the necessary preliminary R&D. Three reference designs developed by a large team of engineers and accelerator physicists were used to lay the groundwork for prototype magnet development during the summer of 1984.

President Reagan approved the SSC after a review by Secretary of Energy John Harrington. States were invited to submit proposals during the site selection process. Forty-three proposals were received, of which thirty-five were judged to meet the minimum standards for the next level of review. A committee from the National Academy of Sciences, chosen in order to ensure the project's maximum possible independence from the political process, recommended that eight proposals—from Arizona, Colorado, Illinois, Michigan, New York, North Carolina, Tennessee, and Texas—be forwarded to the Department of Energy for final review.[12] The DOE selected a site in Ellis County, Texas, near the town of Waxahachie, in November 1988. Construction costs were estimated at $4.4 billion in FY1988 dollars.[13] Immediately after the selection of the site, politicians who supported the SSC when it had appeared it might be located in their state now raised concerns over the site-selection process. Such early skepticism planted seeds of ill will and distrust that would ultimately contribute to the SSC's demise.

After a series of project reviews, the SSC's final footprint was submitted to the DOE in December 1989. Texas began acquiring the sixteen thousand acres of land needed soon thereafter. By this time, estimated overall project costs had already risen to more than $5.9 billion.[14] Major construction started in 1991. Brookhaven, Fermilab, and Lawrence Berkeley, along with industrial partners General Dynamics and Westinghouse, made significant progress in building the prototype quadrupole and dipole magnets that would be critical for the accelerator. By the fall of 1993, over fifteen miles of the machine's circumference had been bored at an average level of three hundred feet below ground. Planning of the SSC's experimental program had also progressed. Two major experimental collaborations were formed in 1991 and received approval to proceed to the next stage by preparing technical design reports. Altogether, approximately two thousand scientists and engineers, representing more than two hundred institutions worldwide, were involved.

The Demise of the SSC and What We Learned

By 1992, the anticipated cost of the SSC had risen from about $4 billion to almost $8.3 billion, and there was little reason to believe that the number would not rise even further. Critics took the DOE and the scientific community to task for mismanagement of the project and for failing to justify the increases to taxpayers, citing similar problems with other federally supported big-science projects, such as the synthetic fuels program, the space shuttle program, and the Hubble Space Telescope.

In 1992, in a briefing paper for the Cato Institute, Kent Jeffries noted:

Congress soon will be deciding the fate of the superconducting supercollider—the $11 billion Department of Energy atom smasher. After five years of skyrocketing cost estimates and increasing skepticism about the scientific management of the SSC, there is now growing support on Capitol Hill for pulling the plug on what would be one of the most expensive science projects ever undertaken by the federal government. The SSC appears to be an ill-conceived project with weak economic justification and a tremendous amount of special interest support. With federal deficit spending rising to new heights, satisfying the curiosity of a small segment of the scientific community should not be considered a high national priority.[15]

Jeffries' article typified an influential strain of argument, which criticized project supporters for failing to prove that the SSC's scientific value would outweigh the merits of the scientific projects it would displace. In his 1992 briefing, Jeffries further noted that "the SSC promises to do little more than provide permanent employment for hundreds of high-energy particle physicists and transfer wealth to Texas."[16]

Congress killed the SSC in October 1993 when the House voted for the third time in a sixteen-month period not to fund the project.[17] What went wrong? Why was the SSC abandoned, when so many other federal programs were—and still are—completed regardless of dissenting voices? Hazel O'Leary, the secretary of energy at the time, and George E. Brown Jr., a House member long involved in science policy, issued a thoughtful postmortem in an opinion piece that appeared in *Los Angeles Times* in November 1993:

The SSC suffered for having failed from the outset to incorporate international funding and participation. The Reagan and Bush administrations made critical early decisions about the technical design and site location as if the SSC were purely a national project. Only later did they proclaim it to be an international collaboration—with a goal of nearly $2 billion in foreign funding. Is it any wonder that substantial foreign funding never materialized? This shortfall eroded congressional support, which made foreign involvement even less likely, accelerating the project's downward spiral.

The obvious lesson to be learned is that foreign participation must be incorporated into large-scale science and technology projects from the very beginning, when prospective partners still have a say in why, where, when and how such projects will be pursued. Not so obvious is how we as a nation will make and keep such international agreements in the future. Although the United States has determined that it cannot fund projects of this scale alone, neither have we demonstrated that we can undertake such endeavors with others. The abrupt termination of the supercollider adds to a long list of large international projects that the United States has suddenly and unilaterally killed or drastically altered, including the Ulysses solar satellite program, the solvent-refined coal project, and the space station. This embarrassing legacy raises serious questions about the reliability of the United States in international research projects.

Although Congress intensely criticized the supercollider project for failing to receive substantial foreign funding, it was never clear that Congress was prepared to share with other nations the jobs and technological benefits that would have flowed from a true partnership. Is it realistic for the United States to want all the "good" jobs and all the critical technological components of a project like the SSC, while also insisting that other nations put billions of dollars on the table?

This raises a related concern: Political support for large projects appears to be directly proportional to the parochial benefits received, yet spreading the wealth of large scientific projects invites appropriate criticism of pork-barreling. When 25 states were competing for the SSC site, the level of political support was enormous. Elected officials nationwide—from senators to city supervisors—heralded the project as vital for the United States and also for their individual states. Once Texas was selected as the project site, however, this overwhelming interest vanished in a flash.[18]

One can argue that the SSC effort failed on many fronts. Scientists did not sufficiently explain the project to the American public: Congress and the public needed a better understanding of the scientific advances it promised, of the reasons why it should be supported in difficult budgetary times, and of the chances that it would have a positive impact on their lives. Members of Congress also appeared to be confused about the nature of international science research, failing to understand that America could not demand large international contributions while retaining absolute control over the project and the information it produced.

Like is often the case in policy-making, however, politics and personalities also played a significant role in the SSC's demise. Indeed, a major turning point for the SSC came in June 1992 when the U.S. House of Representatives first voted to terminate its funding by a vote of 232-181. Only a year earlier, a similar amendment to terminate funding for the SSC had been defeated in the House by an eighty-six-vote margin.[19] The difference this time was that the SSC vote came on the heels of a contentious House debate on a Balanced Budget Amendment (BBA) to the Constitution that had occurred only a week earlier. The push for a BBA was being led at the time by conservative Republican senator Phil Gramm (R-TX) with strong backing from others in the Texas congressional delegation, including Rep. Joe Barton (R-TX) who represented the district in which the SSC was to be located. House Democrats, who had defeated the BBA the week prior in a vote they would have preferred not to have taken, saw the SSC vote as a way to retaliate, to show the American public true fiscal responsibility without need for a BBA. Noted Robert Park of the American Physical Society in his weekly update to the scientific community after the vote:

The BBA is dead, but it reached out from the grave to take the SSC with it. Democrats were furious with the Texas delegation for leading the fight for the BBA and were delighted to have an opportunity to retaliate. Moreover, in the debate on the BBA many of them had bared their chests, growling that they could make the tough decisions to cut the deficit without a constitutional amendment. The vote to zero the SSC was the first opportunity to show constituents just how tough they could be.[20]

The SSC teaches us that any big-science project's major partners must be fully involved in planning and siting from the outset. But we need to go further, informing the public of both the difficulties inherent in budgeting for exploratory research, and the potential benefits of such ambitious endeavors. Concurrently, the scientific community needs to improve its ability to estimate costs, and to more persuasively communicate the positive social effects of its work. In the end, however, the SSC also illustrates that even with these efforts, big-science projects may fall victim to unique and unforeseen political circumstances.

The International Space Station

Like the Department of Energy, NASA has long been involved with so-called megaprojects. In fact, such projects have been a major focus of NASA since its inception in 1958.

One of NASA's best-known recent big-science projects, the International Space Station (ISS), is a complex of orbiting modules capable of sustaining astronauts in space for long periods of time and carrying out research tasks. One of the first suggestions that a space station might be possible came from rocket scientist Werner von Braun, who in 1952 imagined a station 250 feet in diameter, orbiting the earth at an altitude of one thousand miles.[21] In 1971, the Soviet Union launched the first actual space station, *Salyut 1;* in 1973, the United States launched its larger *Skylab* station. In 1984, President Ronald Reagan approved plans for a space station program, and in 1988, a group of countries (including Canada, Japan, nine European nations, and the United States) signed an agreement to work together to build an international space station, then called *Space Station Freedom.*[22] With the Cold War over, and enticed by the promise of partnership with the Russian Space Agency, these member-states signed a new agreement welcoming Russia to the fold in January 1998.[23] The first two modules of the ISS were launched into orbit in June of that year. Several more have since been added, and the first crew stepped on board in 2000.

A press release marking the anniversary of the 1988 agreement called the multilateral collaboration "the largest, most complex international cooperative science and engineering program ever attempted."[24] This is a partitioned enterprise, however, in which each nation constructs a particular module of the larger complex, rather than working together on a single module.

Many assurances have been made about the practical benefits expected to emerge from research conducted aboard the space station. Experiments on combustion, for example, were pledged to promote better understanding of global warming and environmental change.[25] The station was to offer physicists an unparalleled opportunity to study gravitational waves, atomic clocks, laser cooling, and microgravity—some of these processes having potential commercial applications. Other scientists were interested in using its capabilities to test protein crystal growth in space, where low gravity levels held the promise of fewer imperfections and thus, perhaps, purer pharmaceutical drugs and other products. Future policymakers and scientists will certainly have an interest in seeing that these projects come to fruition. For now, however, the ISS's value as a long-term research facility remains unproven. Like many big-science projects, this one has had its share of funding difficulties. As one might expect, cost has been a major issue. When President Reagan announced in his 1984 State of the Union address that NASA would build a space station within the next decade, the estimated cost of the project was $8 billion. However, by 2001, estimates for completion of just the United States' core of the station had risen to almost $23 billion.[26] The ISS's history of cost overruns has caught the attention of both the president and Congress; indeed, in 2004, President George W. Bush ordered NASA to take steps to contain ISS cost increases.

This pressure to restrain the project's growing budget has led to operational problems: while the station was originally designed to suit the needs of the scientific community, budget-driven reductions in the station's scope and functionality have led many scientists to see the project as less about research than about the post–Cold War politics. More than a few of them now view it as little more than an American effort to develop joint projects with the former Soviet Union, while at the same time providing work for American defense contractors during the 1990s when there were cutbacks in defense spending. Indeed, some U.S. policymakers have suggested that, in a time of global uncertainty, such political considerations are not negligible. Others within the scientific community, however, still truly do believe in the station's potential as a global research facility.

What Happens Now?

In the wake of the February 2003 *Columbia* disaster, questions began to surface in the media about whether the risks of the ISS and its supporting shuttle missions outweighed any scientific gains that might come from the project.[27] At the time of the disaster, others were also questioning whether the ISS should be the sole focus of NASA's manned space program. It is true that the ISS has done little to advance that program's goal of space *exploration*.

Backed by a chorus of such concerns, in 2004 President Bush announced that NASA's space exploration program would be redirected toward getting humans back to the moon, and, hopefully, on to Mars. He also instructed that work on the ISS be completed by 2010, thus fulfilling the U.S. obligation to its fifteen international partners. The station's mission, once imagined as broad and inclusive, would now be limited to research "on the long-term effects of space travel on human biology," and "the skills and techniques necessary to sustain further space exploration."[28] In 2010, space shuttle flights, which have made both Hubble and the ISS possible, will be discontinued, to be replaced by new spacecraft capable of both servicing the ISS and putting humans (especially Americans) back into the business of space exploration. The president stated: "The Crew Exploration Vehicle will be capable of ferrying astronauts and scientists to the Space Station after the shuttle is retired. But the main purpose of this spacecraft will be to carry astronauts beyond our orbit to other worlds. This will be the first spacecraft of its kind since the Apollo Command Module."

The announcement of this initiative stirred controversy in the scientific community. What does the change in priorities mean for NASA's unmanned scientific projects, such as the Hubble telescope? What of the billions already invested in the ISS? Regardless of one's position on these and related questions, President Bush's declaration certainly offers one clear example of how political decisions can fundamentally influence big-science projects.

Congress has had its say as well on the future of the ISS. In 2005 the NASA Authorization Act was passed by the Congress. Among other things this piece of legislation states that the American portion of the ISS will become a "national lab" that can be used by public and private U.S. entities to pursue research not being pursued by NASA.[29] Michael Griffin, NASA administrator, has also made statements about a role for the ISS in future NASA work, including in President Bush's "Vision of Space Exploration," stating that the ISS is a "stepping stone on the way" to the moon, Mars, and beyond.[30]

Policy Issues Raised by the ISS

Collaborative projects like the ISS are enormously complicated. Who are the partners? How much does each contribute? What happens when one or more of the partners is unable or unwilling to meet its commitments? What role will scientists play in determining the project's mission? The ISS also raises the question of whether international science projects are really intended to foster cooperative science and geopolitical relationships or are simply an extravagant scheme for sharing costs.

Should resources be redirected from existing programs to support new big-science projects? There is already a long-standing struggle within NASA over manned versus unmanned space exploration—should we divert funds from existing projects like the Mars rover, to put men on the red planet? What is the scientific value? The political or social value? And how should we allocate for engineering and technology programs—such as designing craft for space exploration—versus research-based efforts to understand the nature of the galaxies? Such tensions will only intensify as NASA embarks on its current vision to send astronauts back to the moon and beyond.

The Human Genome Project

The Human Genome Project (HGP) is the first and probably best-known big-science project in the life sciences. The project, which involved numerous researchers working with DNA sequencing machines, differs from our two previous cases in its focus on gathering and processing data, rather than on construction of a large instrument or facility. The exact cost of the HGP is difficult to ascertain, but is thought to be in the neighborhood of $3 billion. This figure actually encompasses the total projected funding over a thirteen-year period (1990–2003), including studies of human diseases; the sequencing of genomes from experimental organisms such as bacteria, yeast, worms, flies, and mice; the development of new technologies for biological and medical research; the development of computational methods for genomic analysis; and social science research on the ethical, legal, and social issues around genetic and genomic research. The actual sequencing of the human genome represented only a small fraction of the overall thirteen-year budget.[31] In addition, the project led to the creation of an entirely new institute within the NIH—the National Human Genome Research Institute (NHGRI) established in 1993.

What Was Sequenced and Why

We often hear that "genes are the functional unit of all life." Indeed, genes, in combination with environment, determine many of our characteristics and traits.[32] Genetic information is, importantly, passed from one generation to the next, and our current understanding is that a single gene is capable of coding hundreds or even thousands of variations of a given protein. A *genome* refers to all DNA, including the entire collection of genes, contained within an organism. The goal of the Human Genome Project was to sequence not just the genes but all DNA composing the human genome. The HGP was completed in rough draft form in June 2000, the final draft form released in April 2003.[33] It contains 3.2 billion "letters" of the DNA code, and it has been estimated that humans have twenty thousand to twenty-five thousand genes.

Why is the complete sequence of the human genome valuable? Almost invariably, any given disease is caused by malfunction of a protein or set of proteins. The protein malfunction can be caused by some environmental factor altering the protein or the gene encoding the protein, or the defective (or mutated) gene can be inherited.

Thus, by knowing the full sequence of the genome, and knowing where each gene is located and the arrangements of genes on chromosomes, we can better understand how to develop biotherapies and treatments for disease. It will be commonplace in the next several decades for physicians to have readily available to them patients' full genomic sequence data to use in tailoring health care to the individual. Knowing the genetic sequence also moves us forward in the continuing quest to fully comprehend the intricate workings of human physiology and biochemistry, including how the human brain functions. Finally, genomic information provides information of value to evolutionary scientists: genome sequence comparisons across species take us a long way toward learning about our ancestors and addressing one of humanity's ultimate questions—how life began and when.

Many associate the HGP with the triumph of Francis Crick and James Watson (with insight from Rosalind Franklin and Maurice Wilkins) in correctly elucidating the structure of DNA. While this was indeed a landmark finding in biology—even the beginning of "molecular" biology—numerous discoveries before and after the discovery of the double helix must also be credited with the project's success, from the determination that DNA did indeed hold the genetic material of all organisms, to the discovery of restriction enzymes (a collection of DNA scissors), to the development of early methods for sequencing. In addition, in terms of available technology—automated DNA sequencers, PCR (polymerase chain reaction), the World Wide Web, and so on—the time was right for the biological community to take on this ambitious project.

The Project's Origins

The HGP is the product of a wide-ranging collaboration involving the Department of Energy, whose discussions of the project date back to 1984; the National Institutes of Health, which began participating in 1988; numerous universities throughout the United States; and international partners in the United Kingdom, France, Germany, Japan, and China.[34]

The reason for the NIH's interest and involvement in this big-science effort is likely quite obvious. But why would the idea for the project have originated in the DOE? After the atomic bomb was developed and used in 1945, Congress charged the DOE's predecessors, the Atomic Energy Commission (AEC) and the Energy Research and Development Administration, with studying the consequences of genetic mutation, especially those mutations caused by radiation and the chemical by-products of energy production.[35] In late 1984, several key agency scientists organized a small meeting—the Alta Summit—that brought together researchers working on DNA analytical methods.[36] Their goal was to address a technical question: could new methods permit direct detection of mutations?[37]

While the summit did not lead directly to a plan for sequencing the human genome, its importance was described by Robert Cook-Deegan, an Office of Technology Assessment analyst at the time: "Many historical threads in the fabric that later became the Human Genome Project wind through that meeting, although it was not a meeting on mapping or sequencing the human genome. Through happenstance and historical accident, Alta links human genome projects to research on the effects of the atomic bombs dropped on Hiroshima and Nagasaki 40 years earlier. If genome projects prove important to biology, then historians will note the Alta meeting."[38]

In 1987, the DOE established a genome-sequencing pilot project. Agency researchers meanwhile began discussions with the scientific community at large about initiating a program to sequence the human genome. Congress finally launched the HGP in FY1988, through an appropriation of funds to the NIH and the DOE. These two agencies subsequently formalized an agreement to "coordinate research and technical activities related to the human genome."[39] Initial project planning was completed in 1990 with the publication of a DOE/NIH joint research plan, which outlined specific goals for the first five years of what was expected to be a fifteen-year project.[40]

The private sector also played a significant role. J. Craig Venter, the founder and former chief science officer of Celera Genomics, a for-profit company, entered the scene in 1999 proclaiming that his firm would sequence the genome within a year, at a fraction of the cost projected by public-sector researchers. The race this announcement launched between private and public sector efforts probably contributed to the completion of a final sequence two years ahead of schedule, in 2003. Celera's "whole genome shotgun" sequencing technique also sparked important debates about the best approach to the problem.[41] While Venter and Francis Collins, director of the NHGRI and the public-sector HGP, exchanged harsh words, it must be acknowledged that the intense—even personal—competition did lead to the early completion of a draft sequence. Scientists are only human, and share the common desire to be the "first" or the "best" in their field. The public-private race to complete the human genome sequence is but one example of the role that personal ambition and competitiveness play in science.

How the HGP Differs from Other Big-Science Projects

While the physics community has been engaged in big-science projects since World War II and the Manhattan Project, the HGP, as noted earlier, was the first massive-scale project in the life sciences. Until then, funding priorities for the life sciences had been focused on small groups and individual researchers. Now, in the aftermath of the HGP, the NIH and the NSF's biology directorate are adding more biological big-science projects to their agendas, thus altering the balance of funding, both within the life sciences and across all disciplines.

In light of the notable failures of other big-science projects, the HGP's success may lead one to wonder what enabled such a positive outcome. To begin with, the HGP concerned human life and the promise of cures and treatments. This in and of itself provided the project with a political advantage over other projects like the SSC, whose social benefits were not immediately apparent. The HGP was also not tied to a single site, and thus was not subject to the congressional battles over location that have killed other projects. Indeed, the HGP did not require large fabrication contracts—which often provide further opportunities for congressional infighting. Rather, the HGP was based in part on large numbers of people using large numbers of automated sequences to input data into a central database. For the HGP, success was largely a matter of the "time being right"—of the technology that was available and the position of the life sciences with regard to discovery and knowledge. Finally, the project opened up opportunities for cooperation among disciplines, since scientists in several fields—evolutionary biology, bioinformatics, molecular biology, microbiology and yeast biology, physiology, and neuroscience—stood to benefit from the sequencing of the human genome. Meanwhile, the data gleaned from the HGP offered enormous benefits to private-sector pharmaceutical and biotech firms.

Undergirding all this was the surprising cooperation between the DOE and the NIH. We have seen why the DOE became involved, but why did it remain so invested in a biological sciences project?[42] The answer seems to lie in the extraordinary importance of research infrastructure to large scientific projects. Owing largely to its historical national security mission, the DOE manages several of the nation's largest research laboratories; it has access to impressive computing power and maintains a large staff of skilled researchers, in a wide range of specialties. In short, the DOE was an indispensable partner in the HGP, and the cooperative venture between the DOE and NIH has become an exemplary case of interagency cooperation.

Did this collaboration actually contribute significantly to the HGP's success? Certainly, DOE's past history with big science helped to advance the project in its early stages, although the NIH is most often credited with the project's completion. While a definitive answer may be impossible, the case certainly illustrates the political, public, and scientific complexity of big-science projects.

What's Next?

The completion of the HGP has primed the scientific community, particularly the life science community, for the next phase of discovery—mapping the expression of proteins in human cells, also known as mapping the proteome. The complexity of human beings (or, for that matter, any organism) arises not from the order of nucleic acids, but from proteins, whose interactions in cells and tissues proteomics is devoted to measuring.[43] But will such big-science projects continue to appeal to researchers in the life sciences? As Joshua LaBaer noted in April 2001:

> Academia hasn't fully embraced the idea of large biological science yet. Biology will always have a contingent of cottage industry—small labs that focus on very specific questions. And at least for now, there's a very strong element in biology that wants that. In fact, I would say the majority of biologists prefer that.[44]

The data from the HGP also confront society with a host of ethical questions. Indeed, the organizers of the HGP had the foresight to include a component on the ethical and social dimensions of their work, called the Ethical, Legal, and Social Implications (ELSI) Research program. How should the new genetic information be interpreted and used? Who should have access to it? And should we allow gene sequence and genetic therapy techniques to be patented? While receiving a small portion of the overall annual NHGRI budget, this program continues to fund social scientists and humanities scholars to conduct research exploring these and other questions. One of the missions is to identify and address questions like these even as the basic research is still being done, hopefully facilitating public discussions on policy as we continue to learn from the sequence data.

Fusion Energy and the International Thermonuclear Experimental Reactor

One of the greatest hopes among supporters of nuclear research is that a cheap and reliable method of producing energy will be discovered using the inner workings of the atomic nucleus.[45] It is well known that enormous energy can be produced through nuclear fission, in which a nucleus is split apart. The same scientific framework suggests that large energies can also be released when certain nuclei are pressed together. This latter process, known as fusion, is particularly appealing because it promises to create energy with almost no harmful by-products, unlike fission, which inevitably generates radioactive by-products.

Fusion is already a fundamental component of our daily lives, in the form of the sun's energy. The aim of fusion research is to re-create that reaction on earth, and harness it as a virtually unlimited source of power. To effectively accomplish this goal, researchers need to heat plasmas—hot burning gases—and compress them into very tight spaces: so tight, in fact, that nuclei fuse. One of the most effective means so far developed to achieve this is by using very large magnetic fields.

The Policy Context of Magnetic Fusion and the International Thermonuclear Experimental Reactor

Significant efforts have been made to pursue fusion research over the last thirty years. While these efforts have significantly advanced our technical knowledge of fusion, they have not produced a commercially viable energy source. The technological requirements of such research are enormous, and the devices invented thus far have not even been able to produce as much energy as they consume. This has led to calls from the fusion research community for increased federal support, to fund larger and more complex experimental devices, which would move us closer to the ultimate goal of a viable fusion power reactor.

Many experts now believe that viable generation of commercial fusion power is three to five decades away. However, fusion scientists made similar predictions twenty years ago, claiming that all that was needed was adequate federal support. One member of Congress who chaired a key energy funding committee in the early 1990s told fusion scientists: "Twenty years ago, I would ask you when you thought that you would be able to light a light bulb with fusion energy and you would tell me in twenty years. And now, twenty years later, I ask the same question, and you provide me with the same answer."[46]

Since the early 1990s, plans have been in place to build and operate an internationally based experimental fusion reactor: the International Thermonuclear Experimental Reactor (ITER). This project is intended to "demonstrate the scientific and technological feasibility of fusion energy for peaceful purposes . . . [to demonstrate] moderate power multiplication, . . . the central fusion energy technologies in a system integrating the appropriate physics and technology and test key elements required to use fusion as a practical energy source."[47] In other words, ITER's aim is to integrate physics and technology in order to demonstrate fusion's energy-generating potential, and to test key elements in its application.

The group that participated in the initial ITER design phase included Canada, the United States, the European Community, Japan, and Russia (China, India, and South Korea have since joined, while Canada has pulled out).[48]

Unlike the SSC, all the international parties involved in the ITER project recognize that this reactor will only be built if they make a joint commitment—in terms of both funding and expertise—from the project's inception. In the project's design phase, each of the partner countries' teams was headed by a scientist from one of the other participating nations, in an effort to make this project truly international from the beginning—something that was not done for either the SSC or the ISS. The project, some suggest, is a major test of whether large-scale science can be conducted on a truly international scale, and one significant question to be worked through was who would be the host country.

ITER, like many big-science projects, has encountered bumps along the road to completion. One such bump came in 1998, when the U.S. Congress, alarmed by the project's burgeoning costs—more than $8 billion by then—directed the DOE to conduct an orderly closeout of its ITER activities. The international partners agreed to delay the construction phase for a few years while they sought ways to lower costs, but members of Congress continued to worry about costs, delays, and the overall merits of U.S. involvement. FY1999 funding was cut by about 75 percent, with the remaining monies to be used specifically for the orderly termination of ITER-related activities and closeout costs. The DOE was also instructed not to sign any agreement extending U.S. involvement in ITER without written approval from Congress.[49] The United States agreed that henceforth it would only participate in ITER as a junior international partner, and would not serve as a host site for the reactor. On February 3, 2003, however, Secretary of Energy Spencer Abraham announced that the country would reenter international negotiations on the construction phase of ITER. In doing so, he made it clear that the United States would not host the facility, which would have required further financial contributions. A DOE press release accompanying Secretary Abraham's speech indicated that "the nature and details of the U.S. participation and contributions would be determined during the negotiations" and that "the U.S. share of the construction cost is expected to be about 10 percent of the total," the minimum required for full ITER participation.[50]

Policy Challenges That Remain

The ITER project raises different issues in science policy than those posed by the SSC. ITER's potential value to society, like the HGP's, is relatively obvious. In addition, as we noted earlier, ITER planners have made an extensive effort to build a base of international collaborators, even before the project got fully under way.

At the same time, however, the collaboration's success has brought about a host of new policy challenges. How should the international host country be chosen? What happens to the ITER if one of its backer nations nullifies its commitment? And how will any scientific and economic benefits that derive from the project be shared among the participating nations?

Industry leaders, engineers, and basic scientists also differ over the project's focus. Members of industry and the engineering community argue that fusion science is well understood and that the next logical step would be to build a large test reactor. Some in the scientific community, however, maintain that significant amounts of money and time could ultimately be saved by first developing a better basic understanding of the underlying science, allowing for a project of significantly smaller scale, scope, and cost, but with comparable output.

The project has also suffered from public confusion (at least in the English-speaking world) about the distinction between "fission" and "fusion" power. We know from public opinion data that fission (what we think of as conventional nuclear power) is widely viewed with apprehension; the words *fission* and *fusion* sound similar enough in English that the public mistakenly attributes the potential dangers of fission to fusion as well. Opposition to fusion thus runs high, often stoked by the same groups that have led successful campaigns against nuclear fission power.

Ironically, fusion's other major public relations problem is tied to its potential benefits: as long as fusion seems to offer a solution to our future energy needs, its success will continue to be measured in those terms—how much energy fusion devices actually produce—leading people to overlook the significant scientific advances that have resulted in plasma physics from such research devices. The fusion research community must convince the general public and key policymakers that the project's future payoff—which might be generations away—is worth major investment now.

The Future of the ITER

At the outset there was great hope that the ITER project could avoid many of the pitfalls that ultimately doomed the SSC. International agreements were to be worked out ahead of time, and the partners would make key decisions based on scientific considerations, not politics. However, there were signs that global politics seemed push to the fore.[51] A competition for hosting the project was held

that created some tensions. Once the list had been narrowed to two proposals, the ITER partners had agreed that a site would be chosen at a key 2003 meeting, but agreement was not forthcoming. The first location was a small village in northern Japan (Rokkasho), which boasts good port access and proximity to both sea and fresh water. The second, chosen by the EU after an extensive internal review, was the city of Cararache, France, not far from Marseille and an existing French research reactor site.

The EU's own, internal selection process had led to two strong proposals, one from Cararache and another from in Vandellos, Spain. Spain had made a late move to double its contribution to the project, hoping to improve its prospects. In an effort at achieving consensus, the EU's Competitiveness Council chose the French site for the ITER's physical home, but designated Spain as the host country of the European Fusion Agency, the administrative entity overseeing the EU's contribution to the ITER. It was also decided that one of the two European ITER directors would be Spanish. Not everyone was pleased with this compromise; indeed, one Spanish official portrayed the agreement as being imposed by a "French-German axis."[52] Thus, internal dissent was dividing the Europeans even before the final site choices made it to the table.

Other nations that might have been expected to submit strong proposals were not represented at all in the final competition. For example, Canada, an early supporter of the project, dropped out of ITER negotiations just four days after the ill-fated site-selection meeting, on December 23, 2003, allegedly as a result of lack of federal financial support (Canadian involvement had been orchestrated through a nongovernmental nonprofit consortium known as ITER Canada), although opposition among environmentalists was also a factor.[53] It is clearly difficult for a private entity like ITER Canada to function without full government support.

What were the stakes in site selection for the ITER, and why would the partner nations care about location, provided that basic scientific and technical needs are met? Designation as the site country would bring tremendous prestige, not to mention political and economic benefits: the ITER is expected to cost approximately $12 billion and to create over two thousand jobs during the construction phase alone.

Reports suggested that Russia and China planned to join the EU in favoring the French site at the December 2003 meeting, while the United States, South Korea, and Japan favored the Rokkasho site. Some suggested that the United States opposed the French proposal because of tensions between Washington and Paris over the Iraq war.[54] Some also speculated that the United States wanted to see Japan chosen to strengthen the chance that the International Linear Collider, a new high-energy physics project under study, would be sited in the United States.[55]

In 2005, a final decision was made by the ITER partners to locate the facility at Cararache, France. The cost of the project is expected to be approximately five billion euros (approximately $6.4 billion) over a ten-year period.[56] It is expected to cost at least another five billion euros for its operation over a twenty-year period.

Any expectation that high-level siting and funding decisions can be made in a political vacuum is clearly unrealistic. As has been pointed out throughout this chapter, political as opposed to scientific considerations are likely to dominate such decisions. Therefore the scientific community should take into account the political aspects from the very beginning of such projects, in order to avoid unpleasant—and sometimes fatal—surprises down the road. Because the ITER project may well set the paradigm for future multinational big-science projects, it will be carefully observed in the years ahead.

The Drawbacks of Big-Science Projects

As has been suggested, there are some drawbacks to the participation in and conduct of big-science projects. For example, big-science projects (1) may not provide sufficient incentives, rewards, and recognition for young scientists; (2) often lead to publications with fifty or more authors; (3) may face major quality-assurance challenges; (4) are difficult to manage and often run by scientists with little administrative experience; (5) require enormous expense, while carrying a high risk of failure; (6) may squeeze out other, smaller projects in a particular field or discipline; and (7) may take so long that graduate students are not able to stay with them from inception through to completion.

Young Scientists and Publication

Large science projects often pose problems for young scientists, whose contributions can too easily be subsumed in the overall project. It is too easy to credit group leaders for new ideas, rather than those individuals who may actually have generated them. This may limit the incentive for young scientists to propose new ideas in large projects, or even to participate at all. With the scientific community often placing a premium on the order in which authors are listed on articles and publications (e.g., in some fields the first and last authors are assumed to have made the greatest contribution), junior researchers may not show much interest in joining a project where they could wind up being sixteenth of fifty or more authors on an article, even if they put in as much effort as the first author.[57]

Quality Assurance

As a group moves toward its scientific goals, the contributions of any individual often rely on the efforts of others within the group. If one element of the project is flawed, subsequent phases will likely be flawed as well. Given the intense scheduling pressures characteristic of big-science projects, it is often impossible for each individual's work to be scrupulously checked, creating the possibility that a pyramid of contributions may be built on a shaky foundation. Managers of big-science projects must be very careful to incorporate as many cross-checks as possible in order to ensure the quality of the final product. While opponents of big science often cite the difficulty of quality assurance as a major drawback, supporters cite the success of big-science endeavors such as the HGP to prove that such projects can be hugely successful.

Project Management

It is not easy to manage a group of more than a thousand independent scientists over a period of several years, in a coordinated quest for a single scientific goal. Scientists are inquisitive and independent by nature. Moreover, only a few have any training in management or business. A high premium must therefore be placed on identifying project leaders with a combination of political, personal, management, and scientific skills.

High Costs and High Risks

Big projects often carry a huge risk of failure. The sums of money involved are large, and project plans often penetrate deep into the unknown. This ambition translates into uncertainty when it comes to budgets and schedules. Too often this uncertainty is misunderstood as inept leadership, with accusations often reaching a crescendo at precisely the most critical phase of the project, as in the case of the SSC. Big-science researchers therefore need to educate the public and policymakers about the nature of their projects—about when to be patient, and when to become concerned.

Smaller Science Projects

While the risks and costs associated with large science projects often lead to questions about their value, some of the fiercest criticisms of these projects come from within the scientific community itself. Individual scientists may not agree with the focus of a particular project, or may favor smaller and less costly experiments. For example, the fusion research community has had to sacrifice many smaller experiments in order to pursue the ITER, leading some in its ranks to criticize the American involvement. These individuals have gone to members of Congress and testified before key congressional committees to express their concerns.

Graduate Students

Most graduate students work on their research for five to seven years, but big-science projects usually span a much longer period of time. Most students therefore will not have the invaluable opportunity to work with big-science projects from conception through to the end—an experience that is an invaluable part of their training: while a graduate dissertation typically describes a project from beginning to end, a thesis tied to a big-science project may not be able to make such delineations or to identify the student's precise contribution.

How Different Fields Seek Big-Science Projects

Scientific communities that require large facilities to pursue their work have developed internal mechanisms for assuring that their future needs of this sort are articulated in a timely manner. In high-energy and nuclear physics, for example, joint panels of the NSF and the DOE continually review the health of the discipline, carrying out detailed studies to determine which facilities will be needed in the future. These panels, known as the High Energy Physics Advisory Panel (HEPAP) and the Nuclear Sciences Advisory Committee (NSAC), are composed of distinguished scientists from across the nation, meeting several times a year. The respective agencies solicit HEPAP's and NSAC's input on relevant sections of their proposed budgets, seeking their advice on how to allocate funds in order to capitalize on existing opportunities and ensure maximum returns. Until recently, this process has worked quite well. After the failures of the SSC and ISABELLE,

however, some have begun to question HEPAP's ability to identify viable projects. Similar challenges will face the NSAC and the nuclear physics community as they discuss their next-generation nuclear physics accelerator.

Astronomy is one field with a long history of big-science projects. Every ten years, the Astronomy and Astrophysics Survey Committee publishes a report under the auspices of the National Academies that prioritizes new research initiatives for the coming decade. Members of the astronomy community for the most part accept this list as representative of their interests. Most of the other scientific disciplines lack a big-science tradition, and hence the formal structures that are necessary to guide planning and prioritization.

The Challenge of Prioritization

The establishment of research priorities involves a great deal of horse-trading, with federal agencies trying to work out funding for individual research areas, agency programs, and multidisciplinary initiatives. Naturally, the level at which one area is funded affects the amount available to another, particularly in a time of tight federal and agency budgets.

While scientists in fields such as astronomy have enjoyed some success in establishing criteria for big-science projects, setting priorities across disciplines is much more difficult. It is like comparing apples and oranges, since the big-science needs for advancing one discipline can be very different from those required to advance another. In fact, it is difficult for scientists to effectively make funding decisions for big-science projects across disciplines. It is challenging to establish reasonable criteria based on scientific value, since each scientific discipline has different goals. Certain research projects—increasingly of the big-science type—may offer the best opportunity to achieve those goals within a discipline, but what is necessary to push environmental biology to the next frontier is likely to be quite different from what will push astrophysics to the same limit. Such decisions are inherently political, and must be made by agency officials with input from the various scientific communities involved. However, as with so many political decisions, an individual's estimation of a project's quality—whether that individual is a scientist or a policymaker—will often depend upon where he or she sits at the table.

The aversion to this political complexity has driven some agencies to avoid establishing a clear and transparent prioritization process in the first place. The National Science Foundation, for example, has assigned an increas-

POLICY DISCUSSION BOX 12.1

Big Science versus Little Science: What Is the Right Balance?

The movement toward big science has been driven largely by the changing nature of science itself and the demands and needs of various scientific communities. After World War II and the science that led to the atomic bomb, physicists found the need for larger research tools (such as the atom smashers referred to as *particle accelerators*) to better understand the nature of atomic particles. While at first the need for big science was confined largely to the physics and astronomy communities, in recent years other disciplines such as those in the life and biological sciences have found an increasing need for larger and larger science projects.

Given the drawbacks and challenges involved in big science, however, some might question the value of investing in such efforts. After all, federal dollars invested in big-science projects such as the SSC or HGP are not able to be invested in smaller, individual-investigator-driven research projects. Big-science projects are tremendously costly, difficult to manage, and often beset by problems in quality assurance. Much of the work in the early stages of big-science projects is often criticized for not being really science at all, but instead focused on developing the technology or determining the correct scientific machinery required to conduct an experiment or a series of experiments. Indeed, an early criticism of the HGP was that the task was largely development—not experimental or fundamental science—and that the work would "be mind-numbingly dull."[58]

Moreover, from start to finish, big-science projects can take years and involve literally hundreds to thousands of researchers. The sheer numbers of researchers involved in big-science projects can raise major challenges in evaluating, attributing, and recognizing the work of a single doctoral student. With so many people involved, for instance, who gets credit on a scientific paper that results from the project? Certainly, the issues here are more challenging than in smaller research projects led by a single principal investigator (PI), conducted by a small group over several months, as opposed to several years.

While to some big science may seem purely developmental or esoteric, these projects are increasingly required to advance the frontiers in scientific disciplines. Thus, it would appear that big science is here to stay. That being said, big science has posed and will continue to pose challenging questions for both scientists and policymakers alike: How much should the government invest in big science compared to traditional single investigators and research conducted by small research teams? What is the "right" balance between big science and little science? How does the government ensure that big-science projects are worth the cost and determine which scientific disciplines require support of big science? What is the role of the scientific community in making such decisions? What policies might alleviate some of the drawbacks posed by large science efforts? How do you assess what younger scientists contribute to projects that involve large numbers of researchers and are managed by a leading scientist?

ing number of large-scale science projects to its Major Research Equipment and Facilities Construction (MREFC) account, an agency-wide capital asset account intended to fund science and engineering infrastructure projects with costs ranging from several tens to hundreds of millions of dollars.[59] While the account is supposed to fund infrastructure, rather than big-science projects per se, the agency's use of MREFC has allowed it to support big science without making any recommendation concerning the priority of these projects or the fields they represent.

The NSF's underlying difficulty is that its budget has not grown at a pace that permits funding of all of the projects that the National Science Board has approved based upon their scientific merit. Instead, decisions on which projects to fund in a given year have been left to the NSF director, who has no clearly defined guidelines for making such determinations. Some research communities have sometimes been left to fight for support on Capitol Hill. In fact, in some instances, such as the HIAPER—the High Performance Instrumental Airborne Platform for Environmental Research—advocates have successfully used Congress to circumvent the NSF leadership and gain funding in the final NSF appropriations bill.[60]

Congressional appropriators and members of the House Science Committee became so frustrated with the NSF's inability to state its funding priorities that they directed the National Academies to conduct a review of the NSF's approval process, and to "develop a set of criteria that can be used to rank and prioritize large research facility projects sponsored by the National Science Foundation (NSF)—particularly those funded through the Major Research Equipment and Facilities Construction account . . . [to] review the current prioritization process and report to us [Congress] on how it can be improved . . . [to] provide us with specific criteria that will lead to a prioritized ranking of competing large research facility proposals that address both scientific merit and management criteria."[61] Congressional appropriations staffers who helped draft the language requesting the review suggested that it was aimed at preventing the MREFC from becoming politicized—which seemed inevitable if a clearer and more transparent process was not established to explain MREFC funding priorities to Congress and the scientific community.

In an attempt to avoid replicating these problems, the Department of Energy released an unprecedented twenty-year facilities plan in November 2003. The plan lists the twenty-eight top DOE scientific priorities for large-scale facilities and projects over the period in question, assuming the availability of funding.[62]

The DOE plan is based on input from the department's scientific advisory boards, regarding their respective fields' need for large science projects. The document does not establish schedules for specific projects, because these will be determined in part by external factors, including federal budgets. Even the authors of this ambitious plan, however, found it impossible (or inadvisable) to establish a specific order of priorities, or to pit the interests of one scientific field against another. In many instances, they categorize projects in groups of four or five, assigning them equal scientific priority.

While the DOE and the NSF are likely to face the greatest challenges in establishing big-science priorities over the next few years, other agencies are not immune. NASA, as we have seen, has long been involved with big science, as has the NIH. Indeed, the biological community is increasingly using larger and larger scientific tools, such as the light sources managed by the DOE, in their research.

Clearly, Congress would like the scientific community and federal agencies to establish a rational and transparent mechanism for determining which big-science projects will be funded, and in what order. The tough decisions that are required, however, are inherently political, and will only become more complicated as the demand for big-science projects expands, and spreads to disciplines like biology, ecology, computing, the social sciences, and

geology. The scientific community, then, must ultimately grapple with the question of the role it wishes to play in the design of a rational prioritization process for big-science projects.

The Challenge of the International Arena

Most scientific questions are not constrained by national borders. By their nature, big-science projects are designed to achieve economies of scale, share the burden of costs, and bring together a broad collection of talent—all of which advocates for American engagement in a greater number of cooperative international ventures. But doing so will force policymakers to decide what fraction of our national science and technology resources should be invested in international activities versus domestic programs.

The decision is complicated by drawbacks inherent in international science endeavors. How, for example, do American students carry out research in a facility located in Europe or Asia, while also discharging their duties on their home campus? How should faculty and researchers continue their research at such a site while also devoting attention to their teaching, administrative, and research duties? Obviously, the internationalization of scientific research will have an impact on the nation's colleges and universities, and on our national laboratories.

The management of international projects poses its own unique difficulties. How does one achieve consensus on the design and execution of sophisticated scientific experiments that involve scientists from multiple countries? Big-science projects have frequently been stymied by disagreements among the collaborators on matters of relatively small importance.

On the other hand, once they have been established, international projects are hard to change—a characteristic that can work to their advantage. A project that is purely national in scope can be easily defunded by an agency, Congress, or the president. International projects, in contrast, are normally empowered through a signed international agreement involving many countries. The decision of any one country to withdraw is thus a violation of an international agreement: a step that most major nations would prefer not to take. Therefore, while harder to instigate, international projects once established have a layer of protection not enjoyed by strictly domestic enterprises. By the same token, of course, wariness of entering into a binding commitment can make countries less favorably disposed to new international big-science projects.

The Future of Big Science

Over the past several decades, the number of big-science projects has remained relatively small, with physicists and astronomers being the primary beneficiaries. The pace is picking up, however, and the demand for such projects has expanded to new disciplines and scientific communities. Tremendous scientific opportunities lie just over the horizon. Some of the projects designed to take us there have already been approved and are awaiting funding; others have been prioritized by a major federal agency and will likely be funded in the next few years. We are standing on the brink of a new era of big science, which makes increased attention to the relevant policies particularly important.

As science becomes interdisciplinary and more complex, the need for big-science projects may increase accordingly, placing unique strains upon science policy and science funding. Policymakers, agency officials, and the scientific community will find it particularly challenging to strike the correct balance in spending on individual investigator-driven research grants, research training, instrumentation, equipment, and large-scale scientific facilities. Interagency cooperation, as between the NIH and the DOE on the Human Genome Project, will also become ever more important—and will obviously complicate the policy picture.

What does the future of big science hold? What kinds of large-scale projects are on the horizon? At the DOE, ongoing and future projects include the siting and construction phases of the ITER, the development of Ultra-Scale Scientific Computing Capability, building the Linac Coherent Light Source, and large-scale high-energy and nuclear physics projects such as the Linear Collider and the Rare Isotope Accelerator. Other more applied and industrial-led efforts are likely to focus on hydrogen and other potential energy sources.

Now that the task of sequencing the human genome has been completed, researchers in the life sciences must be devoted to interpreting the results. Both the DOE and the NIH, through the NHGRI, are likely to continue their critical role in advancing this research and supporting continued analysis of the vast data that the project continues to generate.

Like the large-scale projects in physical science that grew out of World War II and the atomic age, the new life sciences projects—part of our new "biological" age—are likely to produce a wave of big biology projects and increasing interest in the biological application of tools traditionally used by physicists and engineers, including high-speed and high-performance computer networks and large-scale light sources. The words of Winston Churchill, although spoken more than sixty years ago and in a different context, still resound: "Now this is not the end. It is not even the beginning of the end. But it is, perhaps, the end of the beginning."[63]

And what about research in other fields? NASA's efforts will, by nature, continue to focus on big science. Now that the ISS is for the most part completed, questions arise about how best to use the facility. The debate concerning how best to launch humans into space and keep them there safely will continue. NASA will continue to spend many billions of dollars to come up with the most effective means to convey scientists from the earth to the ISS and beyond.

The *Columbia* disaster of February 1, 2003, which killed all seven crew members, has only intensified NASA's efforts to revamp its manned space program. The need for faster space vehicles is driving joint efforts by NASA and the DOE to develop new power sources, perhaps derived from nuclear fusion or plasma propulsion. Meanwhile, the need to understand the biological and physiological effects of living in space will drive much ISS-based research, and has led to plans for a new NASA Space Life Sciences Laboratory, a 100,000-square-foot facility located at the John F. Kennedy Space Center.

NASA continues its work on a hypersonic vehicle that can fly at seven times the speed of sound and is planning several unmanned projects, including more powerful earth- and space-based telescopes that build on the successes of Hubble. And any number of projects are in the works to further explore the solar system.

Finally, the NSF is engaged in numerous ambitious projects, including some discussed earlier, which may transform the agency from one that supports individually initiated science projects into one deeply involved in the development of large scientific tools and facilities. Such a shift is not uncontroversial, as evidenced in aforementioned congressional frustrations over the agency's inability to explain its priorities. The debate over who sets future priorities, and how, will go on, and will require more effective dialogue between the scientific and policy communities.

Policy Challenges and Questions

America's record of conceiving and delivering big-science facilities is far from perfect. Billions of dollars have been spent on projects that were never completed.[64] Over $2 billion was spent on the SSC before it was abandoned, while more than $200 million was spent on ISABELLE.

One could legitimately argue that such instances exemplify the failures of national science policy. Indeed, current U.S. science policy is ill suited to big science. Because the federal government has no long-term science funding structure, long-term projects begun under one administration can easily be cancelled or modified by a subsequent administration or by a Congress controlled by a different party. While the appropriations process should never be so rigid as to preclude adjustments, big projects like ISABELLE and the SSC require more stability than the current system allows. This is one of the great policy challenges of the new era.

Yet another is how to better prepare the State Department for its increasing role in international scientific projects. The failures of several such ventures have been attributed to State Department officials who did not fully understand the nature of the projects they were brokering. A solution to this problem would most likely require expansion of the State Department's scientific staff and additional powers and authority for the department's science advisor. It would also require better communication between the State Department and other major science agencies.

At the local level, many big-science projects rely on financial support from the host state and relevant industries in the region. Local entities generally anticipate that their investment will be repaid in high-tech jobs, and increased demand for services and equipment from local businesses. While local support eases the federal government's burden, it also gives local authorities leverage, introducing local nuances into the already complex mélange of federal politics.

What impact do such projects have on universities? Big-science projects, after all, are often big because the United States can fund only a few of the entities in question. Thus researchers, many of whom are university faculty and graduate students, may have to travel great distances to use the facility. As we noted earlier, these travel requirements may pose difficulties to the individual researcher; but the problem also affects the university system as a whole. Will faculty be away from campus even more than they already are? How will students take their classes and yet spend large amounts of time at the research facility? Should the normal PhD criteria be modified to reduce the expectation of original research?

Finally, federal agency staff also feel the weight of big science. These individuals take their responsibility for allocating and monitoring research funds seriously. On small grants, agencies can often rely on universities to monitor expenses on a daily basis, backed by regular audits. But on big-science projects with annual budgets reaching into the hundreds of millions, agency officials often need to be housed on site. These individuals can support or impede a project, and their appropriate role is another matter demanding review.

Clearly, the United States must improve its ability to select and follow through on big-science projects. Much can be learned from past failures, and policymakers should pursue these lessons. As more and more ventures move into the big-science category, our ability to design and implement such projects will be key to getting the most science out of the funds allocated, and to maintaining public confidence in the scientific enterprise.

The intensity of the debate over big science will only grow, as competition for limited R&D funds increases and large-scale projects become more prevalent. Policymakers will be forced to grapple with a series of questions: What is the proper balance between "big" and "little" science? Who should determine it? How can we make the appropriations process flexible enough to guide projects effectively, while also stable enough to fund long-term projects? What sacrifices are we willing to make in our domestic R&D programs to participate in large international science projects? What is the best management model for big-science projects? How should the public be kept informed about the status of major projects? How should these projects be integrated with other national needs (e.g., science education)? How should the federal government, working with various scientific communities, set priorities for which big-science projects should be pursued?

These questions will only grow as science becomes more multidisciplinary and big science is required to advance the frontiers of almost all disciplines. Big science is here to stay and will play a central role as more complex projects are undertaken. It is important that policymakers in the executive branch agencies and offices and in Congress give focused attention to addressing these questions.

NOTES

1. Alvin M. Weinberg, "Impact of Large-Scale Science on the United States," *Science* 134, no. 3473 (July 21, 1961): 161–64.

2. For further discussion of the promise and the problems posed by big science, see Alvin M. Weinberg, *Reflections on Big Science* (Cambridge: MIT Press, 1967).

3. In his concurrence to the majority opinion in the obscenity case of *Jacobellis v. Ohio* in 1964, Justice Stewart wrote that "hard-core pornography" was difficult to define, but "I know it when I see it." Judith Silver, "Movie Day at the Supreme Court or 'I know it when I see it': A History of the Definition of Obscenity," *Coollawyer.com,* 2001, http://www.coollawyer.com/webfront/pdf/Obscenity%20Article.pdf (accessed May 27, 2007).

4. This figure is an estimate based on data from a Lawrence Berkeley National Lab particle physicists' census and March 2004 electronic communication with Michael Witherell, director of Fermilab, "2007 Directory and Census of U.S. Particle Physicists, Census Summary, Historical Summary," http://hepfolk.lbl.gov/census/summary/latest/CensusTable2007.htm (accessed June 8, 2007).

5. The panel was convened to look at future needs in accelerator research and was headed by Norman Ramsey. According to the Fermilab Web site, among the panels' recommendations was that Berkeley National Laboratory construct a proton accelerator of about 200 GeV. "This machine evolved as Fermilab's original Main Ring." Fermilab, "About Fermilab," http://www.fnal.gov/pub/about/index.html (accessed June 22, 2007); Fermilab, "History, Timeline," http://www.fnal.gov/pub/about/whatis/timeline.html (accessed June 2, 2007).

6. The U.S. Atomic Energy Commission was the early precursor to the modern-day U.S. Department of Energy, which was established as a cabinet-level agency under President Jimmy Carter in 1977. More details concerning DOE's history can be found at the Department of Energy, "About the DOE, History, Origins," http://www.doe.gov/about/origins.htm (accessed June 2, 2007).

7. More information on the history of Fermi Laboratory can be found at Fermilab, "About Fermilab"; Fermilab, "What Is Fermilab, History," http://www.fnal.gov/pub/about/whatis/history.html (accessed May 28, 2007).

8. For a description of ISABELLE see John G. Cramer, "RHIC: Big Bangs in the Lab," *Analog: Science Fiction and Fact Magazine* (June 1991), http://www.npl.washington.edu/AV/altvw46.html (accessed May 27, 2008).

9. See Brookhaven National Laboratory, "The Long Road from ISABELLE to RHIC," http://www.bnl.gov/bnlweb/history/RHIC_history.asp (accessed June 17, 2007).

10. Cramer, "RHIC."

11. See "The Superconducting Super Collider Project: A Summary," http://www.hep.net/ssc/new/history/appendixa.html (accessed June 22, 2007).

12. While New York was originally recommended by the National Academies panel for consideration as a potential site for the SSC, strong local opposition forced New York officials to withdraw its best-qualified site from further consideration. As a result, the Department of Energy's official list of finalists released January 19, 1988, contained only seven states and omitted New York. See Louis Weisberg, "Finalists Asked More to Join SSC Effort," *The Scientist* 2, no. 3 (February 8, 1988): 1.

13. When President Ronald Reagan made the announcement of the building of the SSC, the price tag was $4.4. billion. The higher estimate of $5.3 billion came after the fact, and includes inflation. Victor S. Rezendes, before the House Committee on Science, Space, and Technology, *Federal Research: Superconducting Super Collider Cost and Schedule,* 103rd Congress, 1st sess., May 26, 1993.

14. John G. Cramer, "The Decline and Fall of the SSC," *Analog: Science Fiction and Fact,* May 1997, http://www.npl.washington.edu/AV/altvw84.html (accessed May 28, 2007).

15. Kent Jeffries, "Super Boondoggle: Time to Pull the Plug on the Superconducting Super Collider," Cato Institute, Briefing Paper no. 16, May 26, 1992.

16. Ibid.

17. See for example: Michael D. Lemonick, "The $2 Billion Hole," *Time,* November 1, 1993; "Congress Pulls the Plug on the Super Collider," *Los Angeles Times,* October 22, 1993.

18. Hazel O'Leary and George E. Brown, "Perspective on the Super Collider: Resuming the Pursuit of Knowledge," *Los Angeles Times,* November 21, 1993. See also Lawrence Berkeley National Laboratory, "The Future of the Superconducting Super Collider," December 10, 1993, http://www.lbl.gov/Science-Articles/Archive/ssc-and-future.html (accessed June 8, 2007).

19. Richard M. Jones, "Senate Votes $550 Million for the SSC," *FYI: The AIP Bulletin of Science Policy News,* August 4, 1992.

20. Bob Park, "Did the Balanced Budget Amendment Kill the Supercollider," *What's New,* June 19, 1992, http://bobpark.physics.umd.edu/WN92/wn061992.html (accessed June 17, 2007).

21. Werner von Braun, "Crossing the Last Frontier," *Colliers,* March 22, 1952.

22. John M. Logsdon, "Foreign Policy in Orbit: The International Space Station," *Foreign Service Journal* 78, no. 4 (2001).

23. NASA and U.S. State Department, "Space Station Agreements to Be Signed in Washington," Press Release # 98-17, January 29, 1998, http://quest.arc.nasa.gov/space/news/1998/01-29-98a.txt (accessed May 28, 2007).

24. Ibid.

25. Aerospacescholars, "Astronaut Training," http://aerospacescholars.jsc.nasa.gov/HAS/cirr/ss/3/3.cfm (accessed May 28, 2007).

26. Marcia S. Smith, before the House Committee on Science, *NASA's Space Station Program: Evolution and Current Status,* 107th Congress, 1st sess., April 4, 2001, http://www.spaceref.com/news/viewsr.html?pid=2562 (accessed May 28, 2007).

27. Matthew B. Koss, "How Science Brought Down the Shuttle," *New York Times,* June 29, 2003.

28. George W. Bush, "President Bush Announces New Vision for Space Exploration Program," White House, news release, January 14, 2004, http://www.whitehouse.gov/news/releases/2004/01/20040114-3.html (accessed May 28, 2007).

29. NASA, *Report to Congress Regarding a Plan for the Future of the International Space Station National Laboratory,*

May 2007, 4, http://images.spaceref.com/news/2007/2007.05 .iss.lab.report.pdf (accessed June 22, 2007).

30. Michael Griffin, "Why Explore Space?" January 18, 2007, http://www.nasa.gov/mission_pages/exploration/main/griffin _why_explore.html (accessed June 4, 2007).

31. Human Genome Project, "The Department of Energy and the Human Genome Project Fact Sheet," http://www .ornl.gov/sci/techresources/Human_Genome/project/whydoe .shtml#budget (accessed June 22, 2007).

32. Briefly, all cells contain strands of biological material called DNA, which consists of combinations of four different molecules that are represented by the letters A, C, G, and T. All cells within a particular organism have the same DNA sequence, their genes consisting of specific combinations of these letters. Genes carry the information necessary for the human body to make proteins, the molecules that perform biochemical activities and serve as structure and support in cells.

33. The rough draft was not published until February 2001. See J. Craig Venter et al., "The Sequence of the Human Genome," *Science* 291, no. 5507 (February 16, 2001): 1304–51; and International Human Genome Sequencing Consortium, "Initial Sequencing and Analysis of the Human Genome," *Nature* 409, no. 6822 (February 15, 2001): 860–921.

34. For a detailed account of the Human Genome Project see Robert Cook-Deegan, *Gene Wars: Science, Politics, and the Human Genome* (New York: W. W. Norton, 1994); and Leslie Roberts, "Controversial from the Start," *Science* 291, no. 5507 (February 16, 2001): 1182–88.

35. A mutation is a permanent structural alteration in DNA. In most cases, DNA changes either have no effect or cause harm, but occasionally a mutation can improve an organism's chance of surviving and passing the beneficial change on to its descendants.

36. Alta is a ski area in the Wasatch Mountains in Utah.

37. Background information taken from Robert Cook-Deegan, "The Alta Summit, December 1984," *Genomics* 5, no. 3 (1989): 661–63; Human Genome Project, "Department of Energy and the Human Genome Project Fact Sheet."

38. Cook-Deegan, "The Alta Summit," 661–63. Cook-Deegan was at the time a fellow in the congressional Office of Technology Assessment.

39. National Human Genome Research Institute, "About NHGRI, About the Institute: A History and Timeline, October 1 1988," http://www.genome.gov/10001763 (accessed May 28, 2007).

40. National Institute of Health, *Understanding Our Genetic Inheritance—the U.S. Human Genome Project, The First Five Years: Fiscal Years 1991–1995*, NIH publication no. 90-1590, April 1990, http://www.ornl.gov/sci/techresources/ Human_Genome/project/5yrplan/summary.shtml (accessed May 28, 2007).

41. Shotgun sequencing is a method that involves randomly sequencing tiny cloned pieces of the target DNA. The fundamental idea is obtaining random sequence reads with redundancy from a genome and assembling them into a contiguous read on the basis of sequence overlap. Proponents of this methodology state that it allows for more effectively sequencing of long stretches of DNA and that it can be done more cheaply than other methods. A drawback is that error rate is higher when sequencing larger more complex genomes (e.g., human). Also, for one account of the HGP see J. Craig Venter, *A Life Decoded: My Genome, My Life* (New York: Penguin, 2007).

42. See the Human Genome Project Information, http:// www.ornl.gov/sci/techresources/Human_Genome/home.shtml (accessed May 28, 2007). The U.S. Department of Energy has established the DOE Joint Genome Institute, which brings together the expertise of Lawrence Berkeley National Lab, Lawrence Livermore National Lab, and Los Alamos National Lab. The focus of the institute is to advance genomic technologies, including high throughput and computational tools, for discovering and understanding the basic principles of living organisms and relationships; see DOE Joint Genome Institute, http://www .jgi.doe.gov/index.html (accessed May 28, 2007).

43. Mark R. Wilkins et al., "Progress with Proteome Projects: Why All Proteins Expressed by a Genome Should Be Identified and How to Do It," *Biotechnology and Genetic Engineering Reviews* 13 (1996): 19–50.

44. Douglas Steinberg, "Is a Human Proteome Project Next?" *The Scientist* 15, no. 7 (April 2, 2001): 1.

45. The atomic nucleus refers to the center of an atom and is where protons and neutrons are located. In the life sciences, nucleus also refers to the center—the center of a cell, where DNA is housed.

46. As chairman of the House Energy and Water Committee, Congressman John Myers (R-IN) would often challenge the Fusion Energy Community to demonstrate results, stating that he had to justify funding a program that had yet to produce more energy than was required to run a fusion reactor. Myers would ask the community: "how do I justify spending money on this program to the Senior Citizen that comes into my office back home and is worried about receiving their social security check" (as relayed by Tobin L. Smith).

47. ITER International Agreement, signed July 21, 1992, an agreement among the European Atomic Energy Community, Japan, the Russian Federation, and the United States on cooperation in the engineering design activities for the International Thermonuclear Experimental Reactor. See ITER, "Search: ITER EDA Agreement," http://www.iter.org/index.htm (accessed June 4, 2007).

48. See ITER, "What Is ITER, and The ITER Project, Introduction," http://www.iter.org/ (accessed June 4, 2007).

49. Audrey T. Leath, "US May Consider Rejoining ITER," *FYI: The American Institute of Physics Bulletin of Science Policy News*, January 25, 2002, http://www.aip.org/fyi/2002/009. html (accessed May 28, 2007).

50. U.S. Department of Energy, "Energy Secretary Abraham Announces U.S. to Join Negotiations on Major International Fusion Project," PR-03-026, January 30, 2003.

51. "Sponsors Split on Site of Fusion Machine," *CNN. com*, December 20, 2003, http://www.cnn.com/2003/TECH/ science/12/20/fusion.plant.reut/ (accessed May 28, 2007).

52. Daniel Clery, "E.U. Puts France in Play for Fusion Sweepstakes," *Science* 302, no. 5651 (December 5, 2003): 1640.

53. "Canada Withdraws from Nuclear Fusion Project," *Spacewar.com*, January 15, 2004, http://www.spacewar .com/2004/040115193914.qm9lci1a.html (accessed May 28,

2007); Sierra Club of Canada, "Cabinet Delays Decision on ITER Fusion Reactor . . . Mcguinty Supports Billion-Dollar Boondoggle," news release, October 22, 2003, http://www.sierraclub.ca/national/media/iter-delay-03-10-22.html (accessed May 28, 2007).

54. "Sponsors Split on Site."

55. Peter Rodgers, "Time for Big Decisions," *Physics World*, March 2004, http://physicsweb.org/article/world/17/3/1 (accessed May 28, 2007).

56. ITER, "The ITER Project, a Short History," http://www.iter.org/a/index_nav_1.htm (accessed May 28, 2007).

57. In some disciplines, the order of authors is alphabetical. It is becoming more common to find a note that certain authors contributed equally to the work.

58. Roberts, "Controversial from the Start."

59. The MRE account is also known as the Major Research Equipment and Facilities Construction (MREFC) account.

60. HIAPER is an aircraft that allows scientists to conduct research studies in high altitudes of the earth's atmosphere.

61. Stated in a letter to Bruce Alberts, president of the National Academy of Sciences, dated June 12, 2002, from the U.S. Senate and signed by Barbara Mikulski, Christopher Bond, Ernest Hollings, John McCain, Edward Kennedy, and Judd Gregg, requesting a study to develop criteria NSF can use to rank and prioritize large research facility projects.

62. Department of Energy, Office of Science, *Facilities for the Future of Science: A Twenty-Year Outlook*, DOE/SC-0078, rev. ed. (Washington, DC: U.S. Department of Energy, December 2003).

63. Winston Churchill, after the British defeat of the German Afrika Korps in Egypt, November 10, 1942, http://www.ornl.gov/sci/techresources/Human_Genome/project/info.shtml#posthgp (accessed May 28, 2007).

64. Edward E. Scharff, "Some Setbacks for Synfuels," *Time*, September 14, 1981; Michael Riordan, "The Demise of the Superconducting Super Collider," *Physics in Perspective* 2, no. 4 (2000): 411–14.

CHAPTER 13

Scientific Infrastructure

What Is Scientific Infrastructure and Why Is It Important?

Infrastructure is an essential, but underappreciated, component of successful scientific research. The word *infrastructure* is sometimes used interchangeably with the word *facilities,* in the sense of brick-and-mortar buildings. However, *infrastructure* can also refer to wiring and communications systems, plumbing, even copiers; or it can mean the administrative framework that allows staff to issue salary checks and to complete and submit paperwork required to meet federal regulations.

Because modern science is ever more complex, the tools, machines, computer networks, instrumentation, and administrative structures needed to support it have also become increasingly complex.[1] Projects may need massive computing capability, high-speed Internet connections, temperature-controlled labs, containment space for biological and radioactive agents, animal facilities, machine shops, electronics shops, libraries, videoconferencing facilities, shared instruments, and more. As more complex problems are being attacked, larger and larger scientific instruments are also required, ranging from laser arrays and spectrometers all the way to particle accelerators spanning many hundreds of square miles. In addition, given the scope of modern research, special entities are required to deal with such matters as radioactive waste and animals used in research, if a research institution is to allow its scientists to conduct their work efficiently while conforming to federal and state regulations.

But although such resources are necessary for the conduct of research, it is obvious that individual researchers cannot be expected to provide them on their own. Resources such as libraries or laboratories are likely to be shared. While it may not be feasible for a single research project to hire a veterinarian and veterinary technicians to care for the small number of lab animals used in a single project, the university may hire several veterinarians and a team of technicians to support a large campus-based animal research effort. In fact, research campuses with large biological and biomedical research programs often maintain an entire unit or department to care for research animals and train new researchers in their proper care and use. Similarly, it is not feasible to expect every scientist who needs a state-of-the-art microscope to buy one for his or her lab. It makes much more sense for scientists to share access to such instrumentation and to jointly support a staff dedicated to working with that particular instrument. This shared-technology approach is a more efficient use of funding and instrumentation, since instruments in a shared facility may be used more frequently than they would be if their use was limited to researchers on a single project.

What counts as scientific infrastructure? It is easy to lump almost anything that aids research into this category. For example, good roads between a scientist's home and laboratory are certainly a shared resource that will enhance the number of hours that the researcher will be able to work. Such a definition is too broad for our purposes, however. Instead, we will adopt the definition proposed in a recent National Science Board (NSB) study on the state of the nation's scientific infrastructure:

"Research infrastructure" is a term that is commonly used to describe the tools, services, and installations that are needed for the science and engineering (S&E) research community to function and for researchers to do their work. For the purposes of this study, it includes: (1) hardware (tools, equipment, instrumentation, platforms and facilities), (2) software (enabling computer systems, libraries, databases, data analysis and data interpretation systems, and communication networks), (3) the technical support (human or automated) and services needed to operate the infrastructure and keep it working effectively, and (4) the special environments and installations (such as buildings and research space) necessary to effectively create, deploy, access, and use the research tools.[2]

Why is infrastructure important to science policy? In many ways, a nation's ability to carry out modern research depends upon the robustness of its scientific infrastructure. A strong infrastructure permits scientists to advance their fields, rather than spending time struggling with basic services and administrative tasks. A noted research biologist, for example, would be much better deployed trying to understand the dynamics of the proteome than setting up a network connection or hooking up a printer. The latter should be part of the available infrastructure.

Research infrastructure poses a host of policy issues. Who should provide infrastructure funding and coordination? How should federal agencies cooperate to meet national infrastructure needs? How should we determine the responsibilities with regard to providing for scientific infrastructure of universities, industry, the national laboratories, states, the federal government, and finally, the international community? What percentage of any single agency's funding should go to supporting infrastructure? Should this percentage differ depending upon the specific mission of the agency? Finally, would it make sense to allocate responsibility for major infrastructural components to specific agencies, or should the responsibility be spread across multiple agencies?

These are some of the questions that policymakers and those involved in research need to address. They are important because scientific infrastructure directly affects what research can be done and the efficiency with which it can be carried out.

Examples of Research Infrastructure

Answers to these questions, of course, vary: research infrastructure can take many different forms and can be organized and maintained by many different entities, in different places, for different reasons. But these facilities are all united by their purpose of supporting the research needs of a large number of scientists. Perhaps the range of

forms infrastructure can take might best be conveyed by a look at several examples.

Animal Colonies, Cell Lines, and Plant or Bacteria Strains

In addition to places where scientists can house nonhuman primates or colonies of mice or rats, facilities also exist where researchers can obtain particular strains of animals, plants, or bacteria, or particular cell lines. Jackson Laboratory in Bar Harbor, Maine, is a nonprofit organization that provides the global scientific community with critical genetic resources, including commonly used mice strains. American Type Culture Collection (ATCC), a nonprofit bioresource center, serves as a repository for cell lines, yeast and fungi cultures, and plant strains, in addition to providing related technical services, biological products, and even educational programs. The Bloomington Drosophila Stock Center maintains the scientific community's supply of fruit flies, another common research organism, including stocks from different genetic backgrounds. Newly established lines can also be deposited for maintenance and distribution through the Bloomington facility. The overarching goal of these entities, and others like them, is to authenticate and preserve particular genetic strains of useful organisms so that researchers from across the globe can start their work from a common basis.

Dedicated Research Buildings

Some research-related work can be conducted in conventionally equipped, multiple-use buildings. Experiments involving radioactive sources, chemical pathogens, select biological agents, or animals, however, may require special facilities that occupy a significant portion of a building. While research organizations often use indirect cost charges to cover the construction and maintenance of these facilities, the federal government also sometimes pays up front for these buildings. NASA, for example, has constructed space science research facilities on campuses across the United States. Such direct investment of federal funds in facilities and research infrastructure that are not on federal property helps to ensure that researchers have the capacity and the physical space in which to conduct their research. These federal investments in research buildings on nonfederal property are more than simply investments in immovable infrastructure; they are instrumental in enabling the United States to continue as a global leader in science.

Libraries and Databases

Scientific progress often happens incrementally, making it essential that we maintain a rigorous system for recording and accessing past results. Libraries have traditionally served this function, but their role is already shifting as the Internet revolutionizes the way we access and use information. Now, instead of delving into the stacks for back issues of journals and copying pages from bound books, scientists can log on to their campus library's intranet and take advantage of online journal subscriptions and the large collection of databases. With the Web taking over some of the libraries' warehousing function, librarians are learning to emphasize their expertise in locating information, whether on the shelves or on-line.

University Research Reactors

Most people do not think of research universities as housing nuclear reactors, but some do. Large research universities with degree-granting programs in nuclear engineering often house *university research reactors,* or *URRs.* These facilities and their adjoining laboratories have long provided specialized training and research facilities where faculty, postdocs, and graduate students can conduct research and get the necessary hands-on training to operate commercial or research reactors.

Despite the apparent value of this training, "URRs are on the path to extinction," as Kenneth Rogers, former commissioner of the U.S. Nuclear Regulatory Commission, noted in a 2002 article in the pages of *Science*.[3] During the 1960s, there were approximately sixty URRs throughout the United States; by 2003, that number had dwindled to twenty-seven.[4] The decline was largely due to a drop in the number of students enrolled in nuclear engineering programs through the 1980s and particularly the 1990s. Interest in nuclear technology has since rebounded, however, as a result of the growing demand for alternative energy sources.

Cyberinfrastructure

Scientists use networks to move large amounts of data from collection to analysis, to shift data among collaborating scientists, and even to control data-collection instruments remotely. The Internet and World Wide Web promote communication among scientists in modes as mundane as e-mail and as technologically advanced as network-based videoconferences. Advanced computing networks are nearly ubiquitous in any modern research environment.

It clearly makes no sense to have each researcher maintain his or her own network. Since essentially every faculty member, staff member, and student on a campus, for example, has a stake in a robust network, the responsibility for maintaining IT capabilities is assigned to the highest level of university leadership. Indeed, at most universities, industries, and national labs, computer networking is managed by a specific office or unit. Increasingly such matters are overseen by a chief information officer (CIO) who reports directly to the president or CEO. The prominence of the CIO has risen greatly as information technology has become more important to the work of universities and businesses alike, especially for the research they conduct.

Networking is not the only IT feature important to research infrastructure, however. While networking provides the physical basis for the rapid exchange of information, it is useless unless it takes into account how users communicate with each other, with their instruments, and with the network as a whole. This broader goal is often referred to as cyberinfrastructure (CI).

An efficient cyberinfrastructure has enormous implications for R&D, but realizing this dream has been a major challenge. In 2003, the National Science Foundation (NSF), recognizing the importance of cyberinfrastructure, commissioned a blue-ribbon panel whose report called for a higher priority for CI support.[5] The report also stressed that CI is made up of many components, including high-performance computation, knowledge-management, measurement, visualization, and collaborative services (see fig. 13.1). Specifically, it noted that

a new age has dawned in scientific and engineering research, pushed by continuing progress in computing, information, and communication technology; and pulled by the expanding complexity, scope, and scale of today's research challenges. The capacity of this technology has crossed thresholds that now make possible a comprehensive "cyberinfrastructure" on which to build new types of scientific and engineering knowledge environments and organizations and to pursue research in new ways and with increased efficacy. The cost of not doing this is high, both in opportunities lost and through increasing fragmentation and balkanization of the research communities.[6]

We need to make progress across the board if we want to advance the pace of scientific discovery. Progress in one area or another is not enough. Extremely fast computers might be capable of running advanced calculations, for example, but without the means to coherently present

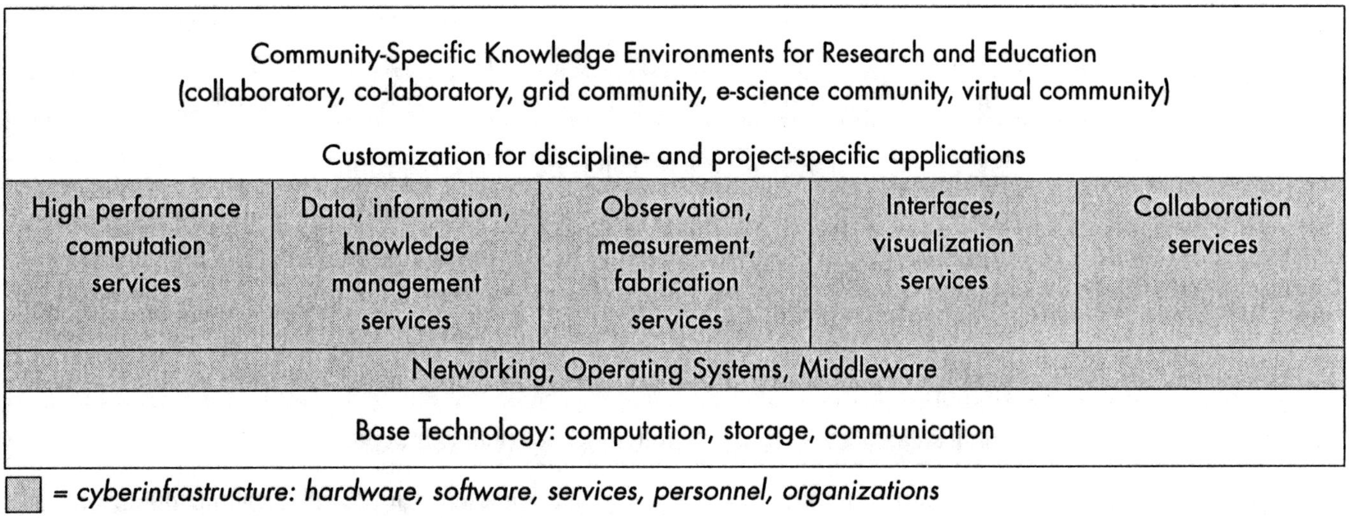

Community-Specific Knowledge Environments for Research and Education (collaboratory, co-laboratory, grid community, e-science community, virtual community)				
Customization for discipline- and project-specific applications				
High performance computation services	Data, information, knowledge management services	Observation, measurement, fabrication services	Interfaces, visualization services	Collaboration services
Networking, Operating Systems, Middleware				
Base Technology: computation, storage, communication				

▨ = *cyberinfrastructure: hardware, software, services, personnel, organizations*

FIG. 13.1 Community-specific knowledge environments for research and education. (Blue Ribbon Advisory Panel on Cyberinfrastructure "Revolutionizing Science and Engineering through Cyberinfrastructure," National Science Foundation [publication number cise051203, January 2003].)

the results to remotely located scientists who may recognize their implications (through database, word-processing, and communications and presentation software), we could never realize the full range of breakthroughs made possible by those initial calculations.

The Need for High-Speed Computation

Some CI needs are more pressing than others, and high-speed computation is at the top of the list. In spite of a more than four-hundred-fold increase in computing speeds over the past two decades (1985–2005), even faster computers are urgently needed. For example, while climate models are theoretically capable of using regional data from one point in time to predict potentially threatening conditions thirty-six hours into the future, current computers may require up to one hundred hours to complete the necessary calculations. A tornado alert is of no use once the twister has already struck. In this instance, higher speeds would meet an important societal need.

Another pressing concern is the connection between current computing capacity and the need for ever-greater capabilities. Scientists are by nature driven to push frontiers ("I've now seen what a nucleus looks like. I wonder what's inside it"). As they reach these frontiers, they create a demand for next-generation tools that will allow them to go even further. Consequently, there will always be a "technology pull" that continually advances both science and society as a whole. The definition of "adequate"

CI will always be changing. It is a perpetual process, in which funding will always be needed, and ultimate success can never be declared.

There are three rather obvious ways to improve scientific computational capabilities. First, local computing power from computer processors can be increased. Second, the ability to retrieve information from remote servers can be improved. And, third, the ability to parse problems and assign component tasks to other computers not running at full capacity can be exploited.

This third option, commonly known as "grid computing," could revolutionize global CI, by making every one of an institution's computers part of its common research infrastructure.[7] The transformation is fueled by increases in the relative speed of computer CPUs, and in the speed and pervasiveness of networks.[8] If, for example, an individual has one dollar to spend on increased computing capability, he or she can invest it in a faster computer, or can leverage that dollar by spending it on a network and middleware that will allow that individual to seamlessly use someone else's computer on the network. Though the CPU speed of computers has been growing at a prodigious rate over recent decades, the capability of fiber optic networks has been growing even faster.[9] We are rapidly reaching a point at which that hypothetical dollar might be more wisely invested in facilitating computer sharing than in upgrading a single CPU. Of course, in reality, both fronts will need to be pursued.

So, what's the big deal? The term *grid computing* assumes that the benefits of this technology are obvious, and that the technology is there to be activated. When most people think of grids, after all, they probably think of the electrical grid, which allows users to plug into an existing socket and extract power. Few people wonder where the electricity is coming from when they plug in their coffee pot in the morning. Is it from a nuclear power plant next door, or a hydroelectric plant two thousand miles away?

The tendency is to think of grid computing in the same way: as a preexisting network, there for anyone who wants to plug in. But in reality enormous technical challenges must be overcome. How, for example, in real time, can a remote computer authenticate a user and decide what portion of its resources to allocate to a request? If payment is required, who will handle the billing? What if an individual wants to use his or her own computer, without setting aside any of its computing power for remote users? How can someone keep others from hogging his or her cycles? If a researcher submits a big calculation job to the grid and the job fails, can he or she tell where it failed and whether the failure was the researcher's fault or the error of a CPU thousands of miles away? If these issues can be resolved, grid computing will offer exciting new capabilities for enhanced research infrastructure.

Collaborative Services and Tools

While plenty of individual investigators still make significant contributions to their fields, investigators are engaging in more and more collaborative efforts, ranging in size from a few scientists all the way up to several thousand. This trend is due in part to the growing complexity of the problems being studied: it would simply be beyond the reach of any one individual to take on some of the more ambitious projects currently under way. New instruments can also be enormously complex and expensive, a burden that is best distributed among individuals and their supporting institutions. The need to collaborate on big projects has sparked interest in new technologies that can help scientific interaction across a single campus or the world.

Today's researchers often need to exchange multimedia information and data. There is a growing belief that location and distance should not impede the efficient exchange of information. This belief drives the strong demand for a worldwide system of seamless communication. But who should take responsibility for coordinating that system? Who should set standards for transmission quality, reliability, and access? The development of a robust system clearly depends on an unambiguous understanding of how these responsibilities should be apportioned, from the desktop all the way to the World Wide Web.

Success is more than a matter of good hardware. Humans are accustomed to face-to-face contact, in which they benefit from physical cues that have developed over thousands of years. The nodding of the head, the glazing of the eyes, the smirk, the yawn, the fidget, the raising of the eyebrows—all are important signals in the exchange of ideas. The concept that we should be able to cooperate on complex tasks through e-mail, telephones, and the occasional videoconference is foreign, and deserves close scrutiny. How should humans adjust themselves to a new environment in which most interactions are purely verbal or textual? Various institutions are already studying this question closely, and trying to understand how humans can best interact using modern collaborative tools.[10] For example, some researchers are conducting psychological studies on the organization and management of virtual meetings.[11]

Aside from the big evolutionary questions, we also need to improve the technology itself. Videoconferencing, for example, is still not widely regarded as a reliable communications tool. Too many videoconferences are disrupted by network failures, and robust systems are still quite expensive, only feasible for large or well-funded institutions. As demand grows, however, equipment is likely to improve and costs decline: given the enormous effort being put into improving network infrastructure, the system will inevitably become more robust. It is beyond a doubt that researchers from all over will eventually have ready access to their colleagues around the world—at any time—and that it will become easier to exchange complex information. This evolution will in turn place new burdens on universities and industry to ensure that their researchers have access to state-of-the-art collaborative tools.

Such tools are not without policy implications of their own. Should priorities be established for the use of bandwidth on the World Wide Web, or should all users enjoy equal access (the so-called network neutrality issue)? Is it more important for scientists at the University of Michigan and Helsinki University to have a standing videoconference each week, with reserved bandwidth, than to allow Web surfers access to that capacity? Given that the capability already exists for enforcing priorities on the Web, should researchers be allowed any special privileges? After all, the Web was originally invented to support the collaboration of researchers.

As scientific research becomes an increasingly global activity composed of international collaborations that require regular communication, policymakers need to focus on collaborative tools that can provide robust and cost-effective linkages.

Grants Administration and Compliance

The personnel and procedures of the grants process are also part of the research infrastructure. Consider that when a scientist wishes to submit a grant proposal to the federal government (or any entity), he or she must follow a formal process. It is not enough to write a basic outline of preliminary data and propose methods for pursuing a particular line of research. Budgets must be completed and forms must be submitted documenting whether the project will use animal or human subjects, recombinant DNA, or radioactive or other hazardous materials; and, if so, it must be demonstrated that the pertinent regulations are being followed. Funders also need to know who will be participating in the project, at what level, and where. Applicants are also expected to guarantee that no duplicative funding has been secured from other sources for the same purpose, and to describe any intellectual property that may be generated by the proposed research.

It would be naive to assume that every individual researcher has the knowledge and skills to generate a proposal that addresses these complex expectations. Almost all major research institutions, whether educational or nonprofit or commercial, have specialized departments whose mission is to assist with the preparation and submission of grant proposals on behalf of their researchers. The people staffing these offices are also great resources for investigators trying to identify new funding opportunities.

Many nonscientists are not aware that research funds are awarded to institutions, not to individual investigators. Because of this, grants administration offices are also responsible for ensuring that the terms of the funding agreement are respected. Staff members in these offices keep funded projects within budget and assist with submitting routine progress reports and renewals.

Federal Agencies' Role in Support of the Nation's Scientific Infrastructure

Nearly all of the federal agencies that fund research contribute to scientific infrastructure, and some have specific programs designed solely to meet this need.

The NSF, for example, provides funds for the academic research fleet, advanced networking, digital libraries, observatories, major research equipment, national astronomy centers, the National Center for Atmospheric Research, and ocean drilling programs. These programs, which are beyond the means of any single university, are designed to support research by multiple simultaneous users. The NSF also has a fund called the Major Research Equipment and Facilities Construction (MREFC) account, which funds big-science projects that contribute to the nation's scientific infrastructure.

More specifically, the MREFC program provides funding in large amounts, which is designated for the acquisition, construction, or commission (rather than the planning, design, development, or operation and maintenance) of scientific equipment "intended to extend the boundaries of technology and open new avenues for discovery for the science and engineering community."[12] While actual funding for this account was $174 million for FY2005, this sum may be far too small, since many outstanding proposals cannot be funded. Indeed, as discussed in Chapter 12, several projects approved by the NSB to receive MREFC funding have so far gone unsupported simply because of lack of funds.

The NSF also has a separate Major Research Instrumentation (MRI) account that provides funds for scientific instrumentation in the $100,000 to $2 million range. As with the MREFC account, MRI-funded projects are supposed to enhance the quality and scope of research and promote the integration of research and learning. The MRI program also promotes the acquisition and development of shared inter- or intrainstitutional instrumentation, as well as instrumentation to be used in concert with private-sector partners.

The National Institute of Standards and Technology (NIST), which is part of the Department of Commerce, maintains several special facilities, many of which are available for use by outside researchers. Similar to the facilities at the DOE national labs and DOD research laboratories, the NIST facilities are of a kind not typically found on most university campuses or at private companies. For instance, the NIST's Building Integrated Photovoltaic Testbed can be used to evaluate integrated photovoltaic (i.e., solar) panels by comparing their energy production, operating temperature, and heat flux. The NIST's Magnetic Engineering Research Facility is designed for advancing technologies in the area of ultra-high-density data storage.

The U.S. Department of Agriculture (USDA) also maintains a significant research infrastructure in support of U.S. agriculture with more than one hundred in-house laboratories overseen by the USDA's Agricultural Research Service (ARS) and more than fifty state agricultural experiment stations throughout the nation run by the state land grant universities with significant federal funding and USDA support through the Cooperative State Research, Education, and Extension Service (CSREES). These laboratories and experiment stations are designed and equipped to conduct original basic and applied research, investigations, and experiments bearing directly on and contributing to the establishment and maintenance of a permanent and effective agricultural industry on a state and regional basis.

The National Institutes of Health (NIH) has a Division of Research Infrastructure (DRI) that sponsors a wide array of programs aimed at developing, expanding, and stimulating America's biomedical research infrastructure. The DRI falls under the jurisdiction of the larger National Center for Research Resources (NCRR), which is responsible for encouraging the development of critical research technologies and providing cost-effective, multidisciplinary resources for biomedical research. Under the auspices of the DRI, the NIH supports the remodeling or renovation of existing facilities, the construction of new facilities, the upgrading of animal facilities, and the expansion of clinical research capacity at minority colleges and universities with medical schools and doctorate-granting programs in the biomedical sciences. The NCRR also supports two programs specifically focused at providing support for research instrumentation: the Shared Instrumentation Grant Program that supports instrumentation in the $100,000–$500,000 range and the High End Instrumentation program, which supports instrumentation in the $750,000–$2 million range.

One of the nation's research infrastructure needs is for specialized animal research facilities. The NCRR, for example, funds eight National Primate Research Centers (NPRCs), a network of highly specialized sites focusing on nonhuman primate research. These centers are staffed by experienced scientists and support personnel, who provide an appropriate environment for nonhuman primate research while also promoting the development of models for studying human health and disease. The NPRCs, which offer advanced training in nonhuman primate biology and the use of nonhuman primates in research, also maintain well-established and well-documented colonies of nonhuman primates for use by outside investigators. The facilities are accessible to any funded researcher, with priority given to NIH investigators. While these centers have served as a valuable resource for progress in the biological sciences, the NIH has recently announced that it will end its support of breeding chimps (humans' closest relative) for research purposes, which many who use them in research consider a real loss.[13]

As biomedical research becomes more dependent on advanced instrumentation and computation, it will require more research infrastructure, as well as the means to support it. Researchers' desire to translate their basic findings into applicable technologies will require the increased use of lab animals and cell culture techniques, which require special facilities and staff. It is likely, then, the NIH's role in and need to provide additional support for the U.S. research infrastructure will increase.

No discussion of the role of federal agencies in support of scientific infrastructure would be complete without pointing out the significant role the DOE, DOD, and NASA play in support of our nation's scientific infrastructure. Of course, DOE supports the national laboratories, while the DOD and NASA both operate major federal research laboratories and research centers, some government owned and government operated (GOGOs) and other government owned and government contracted (GOCOs). The importance of these federal laboratories has been discussed in previous chapters (e.g., Chap. 7), so we will not spend a great deal of time discussing them here. It should be noted that in addition to their support for major federal research laboratories, the DOE, DOD, and NASA also support major scientific infrastructure at universities in the form of research equipment, facilities, and major research instrumentation.

Infrastructure in Universities, National Laboratories, and Industry

A comprehensive look at American research infrastructure requires an understanding of how universities, national laboratories, and industrial laboratories meet their infrastructural needs. While much of what is discussed here refers to university infrastructure, many of the same issues face industry and the national labs, with one exception: universities and the national labs pay for much of their infrastructure—either directly or indirectly—with federal dollars, unlike industry, which, for the most part, pays for its own research infrastructure. Still, industry scientists do also rely on the national labs, which are sometimes more cost effective and can provide valuable training and assistance.

Scientists in industry sectors need laboratories outfitted with state-of-the-art instrumentation if their companies are to be competitive. The large and expensive instruments involved—driving simulators in the automobile industry, for example, or air tunnels in the aerospace industry—enable the study of various physical phenomena. But scientists' needs are only one factor in determining the extent of industry infrastructure. Monetary swings also play an important role: in good times, expenditures may be made to upgrade research equipment; in down times, major upgrades may be postponed.

National labs often serve the special role of providing research infrastructure for projects that are too large, too complex, or too long-term to be managed by universities or industry. In fact, over half the usage at DOE user facilities is attributable to university and private-sector researchers in disciplines ranging from high-energy physics to mouse genetics to computer science.[14]

Maintaining Infrastructure at Universities and Government Laboratories

A laboratory in a hundred-year-old physics building may have been the site of great discoveries over its life span. But if the facility is not up to modern standards, plugging in a laser may blow every electrical circuit. The building may have insufficient storage for gas bottles or radioactive materials. In other words, even the best-outfitted labs must be continually updated.

Unfortunately, it is often difficult to find the necessary funds, especially at universities and national labs. Most administrators, private donors, and policymakers like to see visible results from their investments. A new thirty-million-dollar building stands as a lasting monument to the donor or key administrator who supported it. On the other hand, a thirty-million-dollar project to improve electrical wiring and air handling may not attract much public notice, although it may transform an almost unusable research building into a fantastic research facility. This poses a dilemma for infrastructure planners—how to ensure that existing facilities are updated, while also acquiring new facilities and funding new research.

Many universities have guidelines on apportioning funds for routine maintenance and renovation, with multidecade plans to modernize facilities on a rotating basis. However, chronic problems and funding shortages persist even under the best of these plans. When the economy puts a squeeze on funds for salaries and daily operating expenses, administrators often find it tempting to cut money from maintenance and renovation budgets. It may not seem particularly problematic to postpone the renovation of a lab building for a year; but after several years of postponements, a lab can become almost unusable. When it does, no top-notch researchers will want to work in it.

Policy Issues Relating to Scientific Infrastructure

Creeping obsolescence, whatever the reason for it, can have a detrimental impact on the nation's scientific enterprise. Since 1995, reports and studies conducted by government research agencies, including NSF, NIH, NASA, and DOE, have called attention to the decaying U.S. scientific infrastructure and the critical need for its renewal.[15] Policymakers must take the concerns these reports highlight seriously when setting science policy. Adequate mechanisms and funding to support research infrastructure at universities as well as at the DOE national laboratories and at other federal research facilities, such as those overseen by DOD and NASA, must not be ignored.

In what follows we highlight a few key policy issues that are often discussed in relationship to government support—or in some instances lack of support—for scientific infrastructure.

Better Coordination and Funding for Scientific Infrastructure across Federal Agencies

The amount of federal support for scientific infrastructure, especially at academic institutions, has long been a contentious issue. With the size and scale of the necessary infrastructure increasing, however, calls for additional government support and coordination for major research facilities and instrumentation have been on the rise. One such recommendation has been made by the Rising Above the Gathering Storm (RAGS) report, which calls on the federal government to establish a National Coordination Office for Advanced Research Instrumentation and Facilities to manage a fund of $500 million per year over five years to support construction and maintenance of research facilities.[16]

Advanced research instrumentation and facilities (ARIF) are instrumentation and facilities housing closely related or interacting instruments, including networks of sensors, databases, and cyberinfrastructure.[17] ARIF can be distinguished from other types of instrumentation in several ways, including their large size (they are commonly acquired by large-scale research centers and programs, as opposed to individual investigators); their significant cost (generally in the millions to tens of millions of dollars); the high level of institutional commitment and decision making involved (acquisition of ARIF requires a substantial institutional commitment, and decisions regarding such acquisitions are made at the highest levels at both academic institutions and within federal agencies);

the institutional management authority required (at universities ARIF are often managed by the institution administration); and their need for technical operation and maintenance staffing (technical staff are often assigned specifically to help maintain and operate ARIF).[18]

The RAGS report recommendation was aimed at addressing the unmet need for ARIF at universities and government laboratories. It was based in large part upon the work of another National Academies' committee that examined ARIF and policies to "enhance the design, building, funding, sharing, operation, and maintenance of advanced research instruments."[19] In its review, the National Academies' ARIF committee found a gap in support for major instrumentation and facilities projects from federal research agencies; no agency has an agency-wide program to support ARIF projects, and those that have programs to support research instrumentation rarely fund projects exceeding $2 million.[20] To address this gap, the committee recommended that each federal research agency establish a peer-reviewed, agency-wide program to support ARIF projects. The committee also encouraged agencies to establish career development and support programs to train technical support staff essential to the operation of ARIF.[21]

Picking up on another of the ARIF committee's recommendations for enhanced government coordination and cooperation of research agency activities with respect to ARIF, the RAGS committee also recommended that the National Coordination Office for Advanced Research Instrumentation and Facilities be created using existing coordination offices and efforts in networking and information technology research and development, nanotechnology, and climate change as models.[22] In July 2007, Congress passed and President Bush approved the America COMPETES Act (P.L. 110-69). This legislation will make the proposed National Coordination Office for Research Infrastructure a likely reality.

POLICY DISCUSSION BOX 13.1

Research Infrastructure: Who Should Pay?

An ongoing challenge for the scientific community is how to ensure adequate funding for the research facilities and infrastructure required to support the U.S. scientific enterprise. As the tools to conduct science have grown in size and as the instrumentation has become more complex, this challenge has grown.

In the years following World War II and immediately after *Sputnik,* the U.S. government invested heavily in the development and funding of scientific infrastructure at universities, national laboratories, and other federal research facilities. However, by the early 1970s, many federal programs that had previously existed to support construction and renovation of research facilities ended, and federal obligations for research facilities and large equipment in colleges and universities dropped significantly.[23] During this period, "the neglect of laboratory instrumentation and the erosion of the physical infrastructure for research threatened the long-term vitality of even leading universities."[24]

During the 1980s, the Association of American Universities (AAU) approved resolutions urging the federal government to restore its previous commitment to fund not only basic research, but also the physical plant and scientific infrastructure needed to perform it. These resolutions were prompted in part by a growing concern within the AAU that the lack of programs to support scientific facilities at universities was leading universities to pursue congressional "earmarks" to fund such facilities. AAU urged that proposals for research facilities be evaluated in an open, competitive review of scientific merit and urged its members to refrain from seeking congressional earmarks.[25] Despite attempts by the AAU to initiate new programs for federal support of research facilities in universities at the NSF and the NIH, these programs were never well funded, leaving universities to rely upon other sources, such as indirect costs and state funds, to support research facilities.

Today, a need continues for renewed scientific facilities and research. However, just as in the past, issues remain concerning who pays for this research infrastructure. What should the respective roles of the federal government and universities be with regard to support of scientific facilities and infrastructure? If the responsibility for funding research facilities on university campuses lies with the universities themselves, where do they acquire the necessary funds, given a limited number of revenue sources (e.g., indirect costs, student tuition, state funding)? Should such research facilities be funded by grants awarded based on scientific merit, or should universities use their own funds for such facilities? How does the federal government balance the need to maintain scientific facilities at government laboratories and at universities with the need to fund basic research itself?

Indirect Costs

Indirect costs, also referred to as facilities and administrative (F&A) costs, are often associated with infrastructure, since everything from electricity bills, to the cost of a new autoclave, to the salary of the vet tech who cares for the research mice, to new buildings and research facilities is most often paid via indirect costs. A certain percentage of funding, whether from a federal grant or from revenues, must be devoted to covering these expenses.

When a new research building is built, it is important to ask whether the capital costs will be recovered—at universities, via indirect costs charged to research grants, and at national and industry labs as indirect costs figured into future budgets. A university, for instance, assumes all of the risk when administrators decide to build a new research building, since the planning and construction expenses must be borne by the institution itself, in the hope that the money will be recovered over several decades through revenues obtained by the collection of indirect costs.

The OMB Circular A-21 caps the amount of indirect costs that can be recovered from federal research grants for administrative costs. The circular was originally written in 1958. Since that time is has undergone numerous revisions, perhaps the most notable being those made in 1991 and 1993, which resulted from inconsistencies in cost allocation practices and concerns about the misuse of funds. The federal government limited the indirect costs that could be charged to a grant for administration to 26 cents for every dollar spent on direct costs.[26]

In the end, as demonstrated by comprehensive reviews by both the RAND Corporation and the Council of Governmental Relations, an association of research-intensive universities, universities are often unable to recover their total investment in research infrastructure and in many instances subsidize a significant amount of these costs themselves.[27] The question of how new scientific buildings and research infrastructure at both universities and national laboratories should be funded must therefore be of concern to anyone involved in national science policy.

Earmarking of Research Funds to Support
Scientific Infrastructure

Why include a brief discussion of academic earmarking, or "pork," in a chapter on research infrastructure? Many universities and colleges argue that their quest to locate funding outside established processes is a direct result of the inadequate infrastructural funding provided by state and federal governments. Institutions that are not histori-cally strong in research and which lack research infrastructure often complain that they have no means to build up their scientific infrastructure outside of seeking funds directed to them through congressional earmarks.[28]

Moreover, many institutions maintain that large, established universities have an unfair advantage in the peer-review process because they are more likely to have faculty that regularly review proposals for the major federal agencies; this potentially creates biases among reviewers, who might have a tendency to give high marks to proposals from other large institutions with substantial research infrastructure and support systems. If scientists at less well known or well equipped institutions do not get grants, the institutions have no way to elevate their status and to build their research infrastructure. If they don't elevate their status, they don't get grants. Earmark-ing can thus play a powerful role in the evolution and distribution of research infrastructure—often the justifi-cation that smaller institutions have used to defend their earmarking practices.

The debate over peer review versus academic pork, re-member, goes back to the proposals forwarded by Van-nevar Bush (favoring "elitist" distribution) and Harley Kilgore (favoring equitable distribution based on geogra-phy). The question still remains whether efforts should be made to ensure that funding for research infrastructure is more uniformly distributed across all regions of the coun-try, or whether resources would be more profitably con-centrated in a few chosen centers.

Policy Challenges and Questions

The NSB study mentioned earlier in the chapter noted that the United States has underinvested in academic re-search infrastructure, and suggested that future growth in the NSF budget should be targeted for infrastructural improvements.[29] The NSF study also identified infrastruc-ture projects of a few million to ten million dollars in cost as being most in need of attention. In other words, while the NSF has initiatives dedicated to the smallest and larg-est projects, there is a serious lack of funding for midsize efforts.

While the report's recommendations were specific to the NSF, they nonetheless lend general support to the idea that (1) scientific infrastructure is vitally important to national research; and (2) policymakers need to incor-porate research infrastructural needs in their funding decisions.

Another report, from the NIH Working Group on Construction of Research Facilities (sometimes known as the Brody Report), also focused on the state of research infrastructure, particularly in biomedical fields.[30] This report found that the nation's biomedical research infrastructure was outdated and inadequate to meet the demands of this growing field. University officials in particular believed that the existing infrastructure was stretched so thin that it could eventually limit the ability to take full advantage of NIH biomedical funding. The report recommended congressional appropriation of the full $250 million that had been authorized for NIH facilities improvements in FY2002; reducing or eliminating matching requirements for facilities awards wherever feasible, in order to avoid discouraging smaller institutions from applying for funds; setting aside a portion of the Research Facilities Improvement Program budget for minority and emerging institutions; making concerted efforts to address the modernization and expansion needs of the Regional Primate Research Centers; and establishing a federal loan guarantee program to support construction and renovation of biomedical research facilities.[31]

Other analyses published since the mid-1990s concur: attention must be paid to the nation's research infrastructure, and quickly. A 1995 study by the Clinton-era NSTC indicated that university research infrastructure was in need of significant renewal, and conservatively estimated a funding deficit in the neighborhood of $8.7 billion.[32] A 1998 NSF survey estimated the cost of deferred capital projects to build, repair, or renovate university research infrastructure at about $11.4 billion.[33] And a 2003 DOE analysis found the Office of Science's labs and associated infrastructure aging and in disrepair, with at least 60 percent of the space unrenovated for at least thirty years. The DOE also identified over $2 billion in needed capital investment projects that were planned for between FY2002 and FY2011.[34]

With so many complicated challenges ahead—technical, financial, and political—the need for a coherent and forward-looking policy on research infrastructure is clear. Indeed, this is one of the most significant areas of concern for science policy: by definition, infrastructure is the foundation that supports the scientific endeavor.

Although infrastructural issues have been debated for decades, several huge questions remain unresolved. Who should pay for research buildings on university campuses? How should indirect costs be structured, so that institutions can use their own funds to construct new research facilities, with the expectation that the costs will be recovered in a reasonable period of time? What should the

policy be on the distribution of federal funds for research infrastructure? Should we distribute them geographically rather than through peer review? Which entity should be the primary investor in new infrastructural technologies? How can support for scientific infrastructure be better and more effectively coordinated between different federal agencies? Which agencies, for example, should take the lead in the development of the nation's cyberinfrastructure? Are we paying enough attention to the interface among human activities, collaborative tools, and disciplinary research?

Infrastructure is an important, and often overlooked, component of the nation's research fabric. Policymakers will need to pay increasing attention to this area in their efforts to encourage scientific progress.

NOTES

1. The growing importance of the development of research tools for the conduct of research is demonstrated by the fact that eight Nobel Prizes over the past twenty years have gone to researchers who developed novel technologies leading to new research instruments. See National Science Board, *Science and Engineering Infrastructure for the 21st Century: The Role of the National Science Foundation*, NSB-02-190 (Washington, DC: National Science Foundation, February 2003), executive summary.

2. Ibid.

3. Kenneth C. Rogers, "The Past and Future of University Research Reactors," *Science* 295, no. 5563 (March 22, 2002): 2217.

4. Ibid.; James F. Stubbin, before the House Committee on Science, Energy Subcommittee, *Hearing on University Resources for the Future of Nuclear Science and Engineering Programs*, 108th Congress, 1st sess., June 10, 2003.

5. For additional background see National Science Foundation Blue Ribbon Advisory Panel on Cyberinfrastructure, *Revolutionizing Science and Engineering through Cyberinfrastructure*, cise051203 (Arlington, VA: National Science Foundation, January 2003).

6. Ibid.

7. For background on grid computing see J. Joseph, M. Ernest, and C. Fellenstein, "Evolution of Grid Computing Architecture and Grid Adoption Models," *IBM Systems Journal* 43, no. 4 (2004): 624; Daniel Clay and David Voss, "All for One and One for All," *Science* 308, no. 5723 (May 6, 2005): 809; John Bohannon, "Grassroots Supercomputing," *Science* 308, no. 5723 (May 6, 2005): 810; John Bohannon, "Grid Sport: Competitive Crunching," *Science* 308, no. 5723 (May 6, 2005): 812; Mark Buchanan, "Data-Bots Chart the Internet," *Science* 308, no. 5723 (May 6, 2005): 813; Ian Foster, "Service-Oriented Science," *Science* 308, no. 5723 (May 6, 2005): 814–17; Tony Hey and Anne E. Trefethen, "Cyberinfrastructure for e-Science," *Science* 308, no. 5723 (May 6, 2005): 817–21; Kenneth H. Buetow, "Cyberinfrastructure: Empowering a 'Third Way' in Biomedical Research," *Science* 308, no. 5723 (May 6, 2005): 821–24.

8. Ian Foster, "The Grid: A New Infrastructure for 21st Century Science," *Physics Today,* February 2002, 42.

9. Ibid.

10. University of Michigan School of Information, Research, Technology-Mediated Collaboration, http://si.umich.edu/research/area.htm?AreaID=3 (accessed May 23, 2007).

11. Ibid.

12. National Science Foundation, *Summary of the FY 2003 Budget Request to Congress* (Arlington, VA: National Science Foundation, 2002), 15–16, http://nsf.gov/about/budget/fy2003/budget.pdf (accessed June 26, 2007).

13. Jon Cohen, "NIH to End Chimp Breeding for Research," *Science* 316, no. 5829 (June 1, 2007): 1265.

14. Department of Energy, Office of Science, *Facilities for the Future,* 8.

15. A full listing of these reports and other unmet scientific infrastructure needs can be found in National Science Board, *Science and Engineering Infrastructure,* 18–20.

16. Committee on Science, Engineering, and Public Policy, *Rising above the Gathering Storm,* 145.

17. Ibid.

18. Ibid., 146.

19. National Academies Committee on Science, Engineering, and Public Policy, *Advanced Research Instrumentation and Facilities* (Washington, DC: National Academies Press, 2006), 1.

20. Ibid., 94–95.

21. Ibid., 1–2.

22. Committee on Science, Engineering, and Public Policy, *Rising above the Gathering Storm,* 148–49.

23. Smith, *American Science Policy,* 82–84.

24. Ibid., 84.

25. The AAU approved resolutions and issued statements concerning research facilities on October 25, 1983; April 16, 1987; April 18, 1989; and June 9, 1989.

26. A brief history of Circular A-21 can be found in Goldman et al., *Paying for University Facilities,* appendix A.

27. Ibid.; Council on Governmental Relations, *Cost of Doing Business.*

28. Savage, *Funding Science in America,* 31–32.

29. National Science Board, *Science and Engineering Infrastructure.*

30. NIH Working Group on Construction of Research Facilities, *A Report to the Advisory Committee of the Director, National Institutes of Health* (Rockville, MD: National Institutes of Health, July 6, 2001).

31. Ibid. See text of the report for the full list of recommendations available at http://www.nih.gov/about/director/061901.htm#recommendations (accessed May 23, 2007).

32. National Science and Technology Council, *Final Report on Academic Research Infrastructure: A Federal Plan for Renewal* (Washington, DC: National Science and Technology Council, March 17, 1995).

33. National Science Foundation, Division of Science Resource Studies, *Science and Engineering Research Facilities at Colleges and Universities: 1998,* NSF-01-301 (Arlington, VA: National Science Foundation, October, 2000).

34. U.S. Department of Energy, Office of Science, *Infrastructure Frontier: A Quick Look Survey of the Office Science Laboratory Infrastructure* (Washington, DC: U.S. Department of Energy, April 2001).

CHAPTER 14

Scientific Ethics and Integrity

Why Care about Ethics in Scientific Research?

The term *ethics,* for our purposes, is taken to mean "the principles of conduct governing an individual or a group."[1] Why are research ethics so important that the topic merits an entire chapter in a book on national science policy? As former National Academies leaders Bruce Alberts and Kenneth Shine point out, "The scientific research enterprise is built on a foundation of trust: trust that the results reported by others are valid and trust that the source of novel ideas will be appropriately acknowledged in the scientific literature."[2] If this trust is broken, then the entire scientific enterprise is put at risk. Ensuring the integrity of science and that the science is performed ethically, morally, and in socially responsible ways should be a concern of the entire scientific community.

There are incentives in science, as in any field, to be dishonest or less than truthful. Not all people are naturally motivated to report the facts as they find them, to give credit to the work of others, or to avoid situations where they might benefit by shading their interpretation. But any introduction of suspect information into scientific study can undermine outcomes. False information can mislead scientists, doctors, engineers, politicians, and the public at large, with potentially disastrous consequences. Scientific fraud could, for instance, delay the search for a cure to a specific disease, or hinder the development of a cutting-edge medical technology. Beyond this, abuses waste public money and greatly erode public confidence in scientists, scientific research, and scientific results.

States the National Academies in its introductory guide to scientists on the ethical conduct of research, "The sci-entific research enterprise, like other human activities, is built on a foundation of trust. . . . This trust will endure only if the scientific community devotes itself to exemplifying and transmitting the values associated with ethical scientific conduct."[3] The scientific community therefore diligently imposes its own ethical standards on its members. When these standards are violated, or even sometimes when it only appears that they *may* have been violated, the federal government often becomes involved. This is even more likely to happen when the violations involve federally funded research.

There are also other times when society may impose certain ethical values on the scientific community. Such policies may be imposed by policymakers through laws or regulations that restrict the activities or nature of research in which scientists can be engaged. Sometimes these regulations might pertain to how research is conducted, such as research involving animals or human subjects. Other times, whole areas of research might be regulated by policymakers for moral or ethical reasons, such as in the use of fetal tissue or human embryonic stem cells in research.

Scientific ethics reach far beyond individual conscience and morality. Remember that scientific research builds on prior knowledge (the "pyramid of knowledge"), so any ethical breach can push a field backward, wasting irreplaceable resources. While scientists do continually check and cross-check each other's work, even the hard work of identifying and isolating bad data can require precious time and resources.

Since science is by definition the search for "truth" or knowledge, why would any scientist introduce false concepts into the field? Some introduce faulty data because of a quest for personal fame and a desire to be associated with a major discovery. Recognition can, in principle, benefit a particular researcher both financially and professionally for the rest of his or her life, and the temptation to take the low road is more than some can resist.

The associated questions relating to ethics in scientific research are numerous, and differ in their complexity: What are the appropriate mechanisms for giving credit to work done by others? What distinction can be made between maliciously misleading results and the products of sloppy research? How should credit for new discoveries be shared between senior and junior researchers? How do we manage potential conflicts of interest that could adversely affect research results? What is the role of human or animal subjects in research and how do we protect them against research abuses? What limits should be placed on scientists conducting research in areas such as cloning or human embryonic stem cells (hESC)? And finally, if a scientific discovery could be used for harmful purposes, are scientists responsible for bringing their findings to the attention of the authorities, even if it means discontinuing their research?

Questions of scientific ethics may seem to come down to a simple choice between right or wrong, but reasonable people often disagree on the right answer to any particular question. While debates on ethics have raged in recent decades, the intensity and importance of this discourse are increasing as scientists approach the boundary of understanding human nature and expand their work in controversial areas like hESC research. The following is a deeper analysis of some of the ethical issues associated with scientific research. We begin with a discussion of research integrity and scientific misconduct.

Research Integrity: Defining Scientific Misconduct

Research misconduct has been defined by the Office of Science and Technology Policy (OSTP) as the "fabrication, falsification, or plagiarism in proposing, performing, or reviewing research, or in reporting research results." Specifically, OSTP defines fabrication as "making up data or results and recording or reporting them"; falsification as "manipulating research materials, equipment, or processes, or changing or omitting data or results such that the research is not accurately represented in the research record"; and plagiarism as "the appropriation of another person's ideas, processes, results, or words without giving appropriate credit."[4] More broadly, *scientific misconduct* can be viewed as any research activity that threatens the integrity of research by violating broadly agreed upon ethical standards or codes of conduct held by the scientific community and expected in the performance of scientific research.[5] *Responsible conduct of research (or RCR)* is the term used for research behavior that is compatible with "the professional responsibilities of researchers, as defined by their professional organizations, the institutions for which they work and, when relevant, the government and public."[6] These topics will be discussed in more detail subsequently, as will be some scientific areas where the lines between what is morally right and wrong are less clear.

Scientific misconduct is not usually thought to cover honest errors or differences of opinion. However, "sloppy science" that produces poor or misleading results, and results that are knowingly fabricated or falsified, can do similar harm to the generation of new knowledge. In a discussion of the preservation of the integrity of research, experimental ineptitude is a topic worthy of further discussion.

Experimental Ineptitude

Fallacious results are sometimes produced by unscrupulous scientists attempting to advance their own agendas. Other misdirection, however, even that of whole scientific fields, may be due to sloppy practices and techniques employed by a single scientist. Inept scientists, in other words, often stumble upon spectacular, but wrong, results. Other scientists may redirect their own work on the basis of what initially seem like accurate results.

While it is no surprise that some professionals are more precise than others, or better at following established practices and procedures, premature shouts of "Eureka" can be particularly disruptive in a discipline as tightly interwoven as science. A bad artist probably will not sell very many paintings, and as a result may be led to reconsider art as a career. An underperforming scientist, however, may be involved in a long-term collaborative project, in which his or her foibles will be obscured for some time by the work of colleagues. Even the best scientists may experience lapses in judgment or make careless errors that go unnoticed.

In order to discourage sloppiness, the scientific community must respond strongly whenever it discovers improper procedures. It must also constantly remind its members and students that careful practice is the heart of the scientific enterprise. Scientific education should continually reinforce the notion that flawed or unreplicable results can be damaging. The proper training and mentoring of future scientists is also essential to ensuring sound practice. Working scientists must be certain that future generations are fully acquainted with proper methods and equipped with accurate tools.

Individuals often learn ethical behavior by example, and individuals trained by unscrupulous or careless researchers will be ill equipped to practice ethical science. Ineptitude and outright misrepresentation can perhaps best be kept in check through proper scientific mentorship and training. Such undesirable behavior is also kept in check by a strong and scrupulous peer-review system. While peer review often corrects or prevents the dissemination of flawed, inaccurate, or sloppy research, no system is fail-safe. It is therefore imperative that all scientists not only be meticulous in their own practice of the scientific method, but also teach correct procedures to their junior colleagues and students.

Plagiarism and Intellectual Piracy

To plagiarize is "to steal and pass off the ideas or words of another as one's own" or "to use another's production without crediting the source."[7] Plagiarism is a serious infringement on the norms of social discourse. The theft of someone's ideas undermines individuals' willingness to articulate their findings and beliefs in public and is akin to the theft of tangible property. Plagiarism is morally wrong, and its corrosive effects have been formally recognized in copyright law.

The practice of "borrowing" words or phrases has been exacerbated by the Web, which makes it easy to paste sentences, paragraphs, even whole pages of someone else's work directly into one's own. In science, plagiarism is most commonly manifested in publication of findings. Science, like all intellectual pursuits, operates incrementally, with each author or researcher building on prior knowledge. Therefore, there is a built-in expectation that new work will make use of and reference past ideas. This is problematic only when a researcher creates the impression that he or she is the creator, and by extension the owner, of ideas that actually originated elsewhere. Stealing the ideas of others may allow an individual to accrue prestige but corrupts the profession as a whole. Plagiarism can impede the evaluation of students and employees, undermine the credibility of journalists and scholars, and lead to public skepticism about the value of research. These are compelling reasons to promote ethical behavior.

The rules of attribution can be violated to different degrees. Intentional plagiarism is, of course, the worst form. But accidents also happen, and authors sometimes express ideas without remembering where they came from.

The end result is essentially the same in both cases, but the difference in intent is a mitigating factor. Nevertheless, even the possibility of accidental appropriation has moved professional societies and academic institutions to train students and researchers about the importance of careful sourcing and attribution.

Interestingly, the idea of the scientific journal itself stemmed from a desire to make sure that an author could have ideas vetted and presented to peers.[8] *Philosophical Transactions*, which appeared in March 1665, is widely considered the first English scientific journal.[9] Its backer, Henry Oldenburg, the secretary of the Royal Society of London, believed that scientific findings should be published not only to share findings over long distances, or to disseminate results as widely as possible, but also to protect the intellectual property rights of the discoverer. That is, Oldenburg saw the journal as a means to encourage scientists to share their discoveries without fear that their ideas would be stolen.

A discussion of seventeenth-century intellectual property concerns may seem anachronistic, but fears of intellectual piracy and plagiarism did exist even that long ago. Isaac Newton is said to have recorded and distributed his notes to trusted colleagues in the form of cryptic anagrams, avoiding publishing and the risk of plagiarism until he was ready to present the entire story of his discoveries.[10] Other scientists similarly elected to keep their findings secret until they could publish a definitive paper. The appearance of an official scholarly journal, backed by a respected organization, provided scientists with the "protection of rights" they sought. Because findings published in *Philosophical Transactions* carried the endorsement of the Royal Society of London, a scientist whose paper was published in the journal could rest assured that the society would defend him in any dispute over the originality or validity of his ideas.[11]

The peer-review process that has evolved over the subsequent four centuries requires that new papers be reviewed and critiqued by the author's professional peers. Feedback is provided anonymously, and comments and concerns generally are presented *before* a work is published. Not only has this process become a hallmark of the scientific method, it also conveys prestige, to the extent that most scientists would rather be published in peer-reviewed journals.

Such review, no matter how scrupulous, however, does not always suffice. For example, in 1989, Carolyn Phinney, a research associate at the University of Michigan's Institute for Gerontology, was working for Professor Marion Perlmutter. Phinney accused Perlmutter of lying in order to gain access to Phinney's data, and then using them as her own. Phinney sued, and the Washtenaw County Circuit Court's decision in her favor was later upheld in the Michigan Court of Appeals. The process lasted nearly a decade. In the end, Phinney was awarded damages of $1.67 million.[12] This may have been a case involving the rights of mentors and mentees to collaborative work, but it nevertheless points to scientists' protective feelings toward their own data and ideas.

Another high-profile case often cited in discussions of scientific misconduct was related not to plagiarism, but to potential piracy and misrepresentation in the use of scientific research samples. In this instance, Dr. Robert Gallo, the head of the National Cancer Institute Cancer Laboratory, claimed to be the first to prove that human immunodeficiency virus, or HIV, was the cause of acquired immunodeficiency syndrome, or AIDS. Gallo's announcement, made public in 1984 in articles that appeared in *Science,* was quickly contested by the Pasteur Institute in Paris, led by French scientist Luc Montagnier, on the grounds that his lab had been the first to isolate the HIV from the lymph nodes of an AIDS patient. Moreover, a further controversy arose when Montagnier's Paris lab claimed that Gallo had actually used a strain of the French virus sample to test for and diagnose AIDS, as opposed to a strain of the virus identified and developed in Gallo's U.S. laboratory.

In the end, it was determined that both the Gallo and Montagnier research teams deserved to share credit for discovering AIDS. While the French team had, in fact, been the first to isolate HIV from an AIDS patient, Gallo's team was first to prove the virus was the cause of AIDS. At the same time, however, Gallo was ultimately found to have committed scientific misconduct for misleading his colleagues into thinking that his NIH research team had independently grown its own virus with no knowledge of the French virus and for providing no credit to the French researchers in the publication of the findings in *Science.* In fact, even Gallo himself eventually admitted that his lab had received a sample of the virus being studied by the French and used that sample extensively to further their AIDS research.

Despite the examples already given, blatant and intentional acts of plagiarism of scientific research or piracy in scientific research results are relatively rare.[13] Of 162 determinations of scientific conduct made by the U.S. Office of Research Integrity from 1992 to 2005, only 19 actions were taken against individuals for plagiarism. Indeed, a still rare but more prevalent form of scientific misconduct is when researchers fabricate or falsify data.[14]

Still, it is critical that funding agencies, professional societies, academic institutions, and especially senior research supervisors remain vigilant against such scientific abuses and make clear the importance of proper attribution for prior work. This is clearly a policy concern, since any relaxation of norms would ultimately have a negative effect on science as a whole. Accordingly, the responsibility for addressing plagiarism and scientific piracy falls on the entire scientific community, as does discouraging the active falsification or fabrication of data, a topic we will now discuss in further detail.

Data Fabrication and Falsification

The fabrication and falsification of data can have a tremendously harmful effect on scientific programs. The good news is that instances in which scientists *knowingly* falsify or fabricate data "with the intent to deceive" are few and far between. However, notable instances—some where claims of falsification were made but not proven and others where fabrication occurred—have brought increased attention to this issue, which is worthy of further discussion.

One of the most notable cases focused on an allegation of data falsification involving Nobel laureate David Baltimore and his colleague, a junior scientist named Thereza Imanishi-Kari. Margot O'Toole, a postdoctoral fellow in Imanishi-Kari's MIT laboratory, accused her of including fraudulent data in a 1986 article in the journal *Cell.* The data in question were produced by Imanishi-Kari, but because Baltimore was the senior coauthor on the paper, and because of his name recognition, the incident has come to be called the "Baltimore Case." At the time, Representative John Dingell, chair of the powerful House Energy and Commerce Committee, held highly publicized congressional hearings and turned the case over to the U.S. attorney general's office in Maryland for possible criminal charges.

After extensive investigations by both university and the federal authorities—including the Secret Service—both Baltimore and Imanishi-Kari were exonerated with no scientific misconduct proven. Even so, the accusations and the high-profile investigation and congressional hearings led Baltimore to resign from the presidency of the Rockefeller Institute and damaged Imanishi-Kari's career. Moreover, this case motivated both the National Academies and the government through OSTP to take a closer look at the standards and polices aimed at regulating and preventing scientific misconduct.[15]

One of the lessons of this experience is that the clash between science and politics can sometimes result from a clash of personalities. The ordeal, however, actually produced a positive outcome: the drafting of a uniform policy on research conduct with input from the scientific community. Scientists learned the hard way, you might say, that unless they take self-governing measures, the federal government will act for them, and not always in a way that is helpful or that reflects well upon the scientific community.

Of course, in other cases the accused turns out to be guilty. In 2002, Jan Hendrik Schön, a researcher at Bell Labs, was found to have made up or altered data on at least sixteen occasions between 1998 and 2001. Schön claimed to have gotten results in experiments on superconductivity and molecular crystals and electronics that, had they been real, would have broken new ground in condensed-matter and solid-state physics. His work was called into question when one of his peers decided the data from his experiments seemed too perfect. Bell Labs went public about these incidents, and the findings of the investigative committee appointed to reach the bottom of the matter were made publicly available. Needless to say, Schön was fired.[16]

Schön's profuse but fraudulent publications had included multiple coauthors. Why did they not validate the work contained in a paper on which their names were to appear? Listed coauthors sometimes have little or no contact with a paper's originators; instead, they are sometimes used as *gift-authors* included because of their prestige and the hope for influence over editors.[17]

Concerns about gift authorship have been elevated because of the case of Woo-Suk Hwang, a South Korean expert in veterinary medicine.[18] In March 2004, Hwang and his research team published a paper in *Science* demonstrating that they had used somatic cell nuclear transfer (SCNT) to create a hESC line.[19] Hwang and his largely South Korean team followed up this landmark paper with another in June 2005, claiming to have used the same technique they had developed in the earlier paper to efficiently (i.e., using very few eggs) generate eleven patient-specific hESC lines.[20]

After the sole American collaborator on the 2005 paper severed all ties with Hwang and asked that his name be withdrawn from the article, an investigation was launched, and among other things, the data were shown to have been fabricated in both the 2004 and 2005 papers.[21] In fact, a review panel from Seoul National University, where Hwang had been on faculty, found that SCNT had not been used as had been claimed and that the images provided as evidence that the lines in both papers were hESC lines were faked, as were images used in the 2005

to show that the hESC lines were patient specific.[22] The case involving Hwang represents one of the biggest scientific scandals in recent history and illustrates why it is important that the scientific community and journal editors establish protocols requiring all authors to sign off on a final manuscript before it goes for publication.[23]

Institutional and Government Efforts to Ensure Integrity in Research

After serious violations were reported in the biological sciences in the late 1970s and early 1980s, Congress decided to establish misconduct policies for the Department of Health and Human Services and the NIH.[24] Section 493 of the 1985 Health Research Extension Act included provisions enabling the secretary of health and human services to require that all institutions receiving DHHS awards create a process for reviewing reports of scientific fraud, and make the secretary aware of any investigations into such allegations. The NIH was also ordered to establish a mechanism for responding to allegations of scientific misconduct reported by awardee institutions.[25]

Prior to 1986, there was no centralized location within the U.S. Public Health Service (PHS) to report instances of research misconduct, and such reports were to be filed with individual research funding agencies and institutes. In 1986, the NIH took the first step to centralize responsibilities for misconduct by assigning the duty for receiving and responding to such reports to the NIH Institutional Liaison Office.

In 1989, the PHS established the Office of Scientific Integrity (OSI), within the Office of the Director, NIH, and the Offices of Scientific Integrity Review (OSIR), within the Office of the Assistant Secretary of Health (OASH). These offices were assigned to handle misconduct reports and reviews, and OSIR was to begin to remove the jurisdiction for research integrity reviews from the individual funding agencies. In May 1992, OSI and OSIR were consolidated into one office, the Office of Research Integrity (ORI) within OASH. With the signing by President Clinton of the NIH Revitalization Act in June 1993, the ORI was officially established as an independent entity within the DHHS. This office has jurisdiction over all PHS research integrity activities on behalf of the secretary of health and human services, with the exception of the regulatory research integrity activities of the Food and Drug Administration.

Current Federal Policy

Current federal law applies a blanket policy on research integrity to all research funded by U.S. government agencies.[26] This policy is based largely on the findings and recommendations contained in a 1992 report from a panel of the Committee on Science, Engineering, and Public Policy (COSEPUP), which was appointed by the National Academies of Sciences and Engineering and the Institute of Medicine. The report included twelve recommendations on how to define scientific misconduct. It also proposed that research institutions and federal agencies adopt uniform policies on misconduct.[27]

In April 1996, the OSTP—responding in part to the COSEPUP recommendations—asked the National Science and Technology Council (NSTC) to begin formulating a government-wide policy on research integrity and misconduct. The NSTC panel included representatives from the major research agencies—NIH, NSF, DOE, NASA, USDA, and DOD—as well as from the OSTP. Representatives from nongovernmental organizations such as the American Association for the Advancement of Science, the Sigma Xi Research Society, and the National Academies also contributed to the effort. After a period of public comment and a National Academies–sponsored town hall meeting in 1999, the final draft was published in late 2000.[28] The policy laid out in that report applies to all federal agencies and to federally funded research conducted at nongovernment institutions, although those institutions are free to implement additional policies of their own choosing.

After the first draft of the policy was published in the *Federal Register* in October 1999, some researchers raised concerns about the rights of the accused. How would a fair and timely investigative procedure be ensured? What were the mechanisms for handling false accusations? Some asserted that time limits should be set on investigations, and statutes of limitations on charges. The final regulations were published in the December 2000 *Federal Register* and went into effect in December 2001.

While under the new OSTP policy all federal agencies supporting extramural or intramural research were to create agency-specific misconduct guidelines within a year, as of December 2004 only four of the fifteen agencies to which this directive applied had done so. As of April 2007, all but five agencies had issued their guidelines, while the departments of Agriculture, Commerce, Education, Interior, and Justice reported that drafts of their rules were undergoing internal review.[29]

The Role of Universities

The final NSTC policy also placed greater responsibility for preventing and detecting scientific misconduct on local research institutions. In addition, it gave these institutions much of the responsibility for investigating and adjudicating allegations of research misconduct, the premise being that the institutions themselves were closer to the source of the misconduct and thus could more readily identify and address it. The government did not want the costly task of dealing with every reported case of scientific misconduct, nor was the scientific community eager to grant the government such power, preferring instead to police itself.

Universities are often in the vanguard of new research. At the same time, they are also responsible for training future researchers. Because of this dual role, universities have a particular responsibility to ensure that research is performed ethically and with integrity. Sound practices and norms are undoubtedly best taught when young people are learning the basics of science and the scientific method, and when they still have the opportunity to learn about appropriate practices from senior researchers and mentors. In fact, today's top research institutions take research integrity very seriously. Many maintain their own policies, building on the OSTP guidelines. It is important to note that these university policies apply beyond the limits of scientific research: all faculty are accountable to policies on research integrity. These policies must cover an enormous variety of issues, and their language is the result of much work not only by the staff of the institution's relevant administrative offices, but by faculty committees, and sometimes the university regents.

Conflict of Interest

A 2001 Association of American Universities report defines conflict of interest as follows:

> Individual financial conflict of interest in science . . . refers to situations in which financial considerations may compromise, or have the appearance of compromising, an investigator's professional judgment in conducting or reporting research. The bias such conflicts may conceivably impart not only affects collection, analysis, and interpretation of data, but also the hiring of staff, procurement of materials, sharing of results, choice of protocol, involvement of human participants, and the use of statistical methods. . . .

POLICY DISCUSSION BOX 14.1

Self-Regulation versus Government Regulation: Finding the Right Balance

When policymakers force ethical standards on the scientific community, they increase the regulatory burden, leading to additional research costs and constraints. This has led many in the scientific community to advocate for self-policing first over government regulation.

Perhaps one of the best illustrations of the ability of the scientific community to police itself occurred in the 1970s, when significant public concerns began to arise about recombinant DNA research. When this research was first contemplated by researchers, members of the scientific community self-imposed a moratorium on the use of the technology in research until rules and regulations could be established. Moreover, members of the scientific community initiated the request that the National Academies look into the issues of safety as-

sociated with this new technology. They also organized what is now the famous Asilomar Conference of 1975 to discuss the ethical, legal, and social implications of this new technology. It is widely agreed that the steps taken by the scientific community to regulate itself dissuaded the government from applying its own regulations to recombinant DNA research.[30]

Yet there will always be a few bad apples, and their actions will likely cause policymakers to consider imposing outside regulations in certain instances. There will also be certain areas of science where the danger and consequences are viewed as so great that Congress and federal policymakers will be inclined to apply strict regulations.

To what degree is it reasonable to place the burden for ensuring ethical conduct of research on the scientific community itself? How much of the burden should be shouldered by the federal government? What are the potential positive and negative impacts of both options?

> Institutional financial conflict of interest . . . may occur when the institution, any of its senior management or trustees, or a department, school, or other sub-unit, or an affiliated foundation or organization, has an external relationship or financial interest in a faculty research project. . . . The existence (or appearance) of such conflicts can lead to actual bias, or suspicion about possible bias, in the review or conduct of research at the university.[31]

Researchers are human, and their judgment, like anyone's, can be clouded by opportunities for financial gain.[32]

Conflict of interest refers to a situation in which an individual's competing obligations or financial interests make it difficult for that individual to fulfill his or her duties. Similarly, there are instances where the missions of research institutions themselves may be clouded by financial interests in a research project. Potential conflicts of interest at both the individual and the institutional level are becoming increasingly common as the web of relationships and dependencies between universities, government labs, and the commercial sector grows.

Most universities have policies aimed at managing individual conflicts of interest, and are also developing policies to identify and manage institutional conflicts. Both the PHS (which includes NIH) and the NSF require the institutions to which they award funds to maintain policies to ensure that the financial interests of the institution's

employees do not compromise the objectivity with which research sponsored by the agency is designed, conducted, or reported.[33]

Researchers funded by the NIH, for example, are obliged to follow these 1995 guidelines from the DHHS's Public Health Service division:

> Prudent stewardship of public funds that support research programs requires that appropriate steps be taken to ensure high quality results. Therefore, recipient organizations must establish safeguards to prevent employees, consultants, or members of governing bodies from using their positions for purposes that are, or give the appearance of being, motivated by a desire for private financial gain for themselves or others such as those with whom they have family, business, or other ties.
>
> In addition, the institution has the responsibility for maintaining objectivity in research by ensuring that the design, conduct, or reporting of research will not be biased by any conflicting financial interest of investigators responsible for the research.[34]

Despite such policies, it has been suggested that more must be in place to guard against financial conflicts of interest that may taint the objectivity of researchers.[35] This concern has been driven, in part, by conflicts of interest that have been identified with intramural researchers, particularly at NIH. In late 2003, the *Los Angeles Times* cited several instances in which NIH intramural investigators had financial consulting relationships with a company whose therapeutic drug they had promoted in either clinical trials or journal articles.[36] This prompted the House Energy and Commerce Committee, which oversees DHHS and the NIH, to hold an investigative hearing. In addition, Elias Zerhouni, NIH director, formed a blue-ribbon panel to review NIH practices for approving its researchers' consulting relationships.[37]

The eighteen recommendations put forth in the panel's June 2004 report included limiting researchers' annual income from consulting fees from industry or academia to 50 percent of the researcher's salary; prohibiting compensation through stock options or other forms of equities; restricting the amount of time spent on compensated consulting to four hundred hours per year (writing excepted); and requiring an annual, agency-wide report documenting the number and type of outside activities that have been approved.[38] Zerhouni, on the other hand, proposed a complete ban on paid consulting work by senior NIH officials and, for other employees, a limit on consulting fees to 25 percent of an employee's annual salary.[39] The final rules, issued in August 2005, went even further, banning all NIH employees from consulting for "substantially affected organizations"—that is, pharmaceutical and biotechnology companies and manufacturers of medical devices; extramural NIH-supported research institutions; and health care providers and insurers.[40]

Universities have also come under greater scrutiny not only for how they manage individual conflicts but also for cooperative ventures with industry. For example, in 1998 University of California Berkeley struck an arrangement with the Swiss pharmaceutical and agrochemical company Novartis, entailing a $25 million Novartis grant to Berkeley's Plant and Microbial Biology Department. As part of the agreement, Novartis was given exclusive first rights to negotiate licenses on a share of the discoveries generated by the department. The agreement not only covered research funded by the Novartis grant, but also results funded by state and federal sources. The agreement enjoyed significant campus support, including support from university administrators, because it freed up funds that could be used to address infrastructure needs.[41] At the same time, the agreement was heavily criticized by others, including some Berkeley faculty, and stirred up murky waters: were the Berkeley biologists who were receiving Novartis funding working independently, outside the realm of the company's interests? Were Novartis's commercial interests influencing the scientists' work? Did the department and the university have the right to grant Novartis privileged access to discoveries flowing from publicly funded research?

The Berkeley-Novartis deal received additional attention when in 2003, Dr. Ignacio Chapela, an assistant professor at Berkeley and an outspoken critic of the Novartis deal, was denied tenure.[42] The decision was made in spite of strong endorsements from Chapela's colleagues, including an overwhelming departmental vote in his favor. It came to light that one of the members of the tenure review committee had ties to Novartis and had served on the committee that oversaw Berkeley's agreement with that company. In spite of requests from other faculty, this individual had not excused himself from Chapela's tenure review.[43] Chapela successfully appealed the decision, and was granted tenure.[44]

The Berkeley-Novartis deal and the circumstances surrounding Chapela's tenure represent an instance where issues of both institutional and individual conflict of interest needed to be considered. Was the tenure committee's decision influenced by a member's ties to Novartis? The agreement with Novartis failed to anticipate this question and many others. The company's interests, and this committee member's, were in conflict with Chapela's, and many would suggest that this faculty member should not have been making decisions affecting Chapela's career. Such agreements with major companies must be carefully scrutinized by institutions before they enter into them. Such agreements must be crafted so as to ensure that research outcomes and faculty are not put in positions that jeopardize integrity and academic freedom.[45] At the same time, they must take into account how the public and policymakers may react to such agreements, recognizing that public perception can affect the response of policymakers in Washington to such agreements.

Scientists and researchers often have multiple responsibilities, which can result in conflicts of interest. Consider an engineering professor who is supervising a graduate student working on a federally funded grant to develop a particular crystalline substance and study its structure. That same professor may also be the president of a spin-off company developed to use such materials in the commercial sector. The professor is on the one hand responsible for urging the student to explore every aspect of the substance in question and publish the results, while on the other hand responsible for developing proprietary information or trade secrets that can maximize profits for the spin-off. How should this professor decide where to draw the line? Or should the professor be allowed to be in this situation in the first place?

The opportunities for such conflicts of interest were much less likely in the past. Professors were expected to devote almost all of their time to teaching and research, while researchers in the private sector had only modest contacts with academia. The disadvantage of this strict separation was (as noted in our discussion of Bayh-Dole in Chap. 6) that much of the knowledge generated at universities was left to "sit on the shelf." Bayh-Dole did much to bridge that divide. But it also encouraged faculty to become entrepreneurial, to the extent that a strong technology transfer program is now sometimes the decisive factor in a university's efforts to recruit a star researcher. Some universities have even explicitly added economic development to their mission, alongside education, research, and scholarship.[46]

How do we navigate the many ethical pitfalls of this new interaction? It would be possible, of course, to contrive other mechanisms for moving new discoveries from university labs into the commercial sector. The intellectual property rights to technologies developed with federally funded dollars might, for example, be assigned to the federal government. Companies would then license these technologies from the government, just as they did prior to Bayh-Dole. Of course, experience has already shown that this was not an effective system for encouraging technology transfer.[47] Another option might be to establish nonprofit entities, separate from the university, that would own and license intellectual property. The Wisconsin Alumni Research Foundation (WARF) is one example, a private nonprofit organization whose mission is to support technology transfer at the University of Wisconsin. Since faculty are involved in the start-ups that come from foundation-owned technologies, however, the separation between academic and commercial worlds is to some extent artificial.

While Bayh-Dole has certainly ensured that university research is used and that industry is able to turn knowledge into products that enhance the public good, it has also created the need for more and more complex policies that manage conflicts of interest and regulate them when they arise so that they do not adversely influence scientific results. Most modern institutions, accepting conflicts of interest as inevitable, focus on oversight and management of conflict of interest. Within a university setting, for example, faculty committees typically review all cases involving the potential for conflict of interest to ensure that it does not adversely affect research.

POLICY DISCUSSION BOX 14.2

Managing Public Perception: How Much Conflict of Interest Is Too Much?

One of the challenges universities and other research organizations face in crafting effective conflict-of-interest policies is that public perception of conflict of interest is often a more serious threat than running afoul of the law. This challenge is made even more difficult because the public and many policymakers believe that any degree of conflict of interest is bad and should not exist. This differs from the view held by many university officials, who see conflict of interest as an inescapable element of ensuring that university-derived knowledge with both societal and economic value is developed by the private sector. It is not, therefore, reasonable for the public or for policymakers to expect that conflict of interest be totally eliminated. In fact, without significant incentives for academic researchers to cooperate with industry, the societal and economic benefits of federally funded research cannot be realized. Therefore, the approach taken by research administrators is to identify when and where conflicts exist and to manage them so that they do not taint the quality of research or violate reasonable ethical standards.

This raises a series of questions at the center of current discussions surrounding conflict of interest. At a time when more and more policymakers are pushing universities in their states, particularly state universities, to serve as economic drivers for their state economies, how do university officials protect the integrity of research? How do administrators balance the need to transfer knowledge to, and work more closely with, industry against the need to ensure that researchers are not captured by industrial interests? What levels of conflict should be allowed at the individual level? At the institutional level? How do university administrators manage public perception when conflicts exist? Can you actually have research where no conflicts of interest exist? What do you think?

Use of Animals and Humans in Research

Before the fruits of basic research can be used for real-life therapies and made widely available to the general public, there must be certainty that new medicines, drugs, and devices pass tests of safety and efficacy. While cell cultures and simulated computer models allow some testing, it is often necessary to continue testing on animals and in some cases on human subjects. Detailed guidelines must be established and enforced in order to preserve the dignity and safety of these research subjects. If human subjects are involved, for example, proper consent must be obtained, and the individuals must fully understand the nature of the risks of participation. All research projects involving human subjects must therefore pass a detailed approval process, designed to ensure that the safeguards are implemented. Research organizations, including universities, maintain committees often known as institutional review boards, or IRBs, which oversee all such activities. Institutional animal care and use committees exercise similar oversight regarding animals in research. Research universities also support animal-use offices that train researchers in the proper handling and care of animals and ensure that they follow federal guidelines.

Animals in Research

Many advances in medicine have come about through the use of animals in research. Indeed, the term *vaccine* means "from cows." Louis Pasteur used animals to study the devastating effects of the anthrax bacterium. The smallpox vaccine was based on a serum derived from cows. Much of the early research of Michael DeBakey (a pioneering cardiovascular surgeon) on coronary bypass surgery was conducted on animals. There can be no doubt that research involving animals has improved our understanding of diseases and helped researchers perfect medical procedures that benefit almost all people.

There is, however, considerable public opposition to the use of animals in medical and biological research. Some of these opponents are concerned about the large number of animals (indeed, millions) that are used in research in the course of any given year, many of which will die in the course of the study, often as a direct result of the experimental tests to which they are subjected. In other, rare, cases animals may be treated unethically by researchers who willfully or unwittingly disregard protocols.[48]

How are research animals used? Manufacturers of everything from cosmetics to pharmaceuticals test their products on animals. A pharmaceutical company, for example, may administer varying doses of a new drug to tens of thousands of guinea pigs, observing their behavior for signs of negative effect on mood or appetite. After a short time the animals are euthanized and dissected, in order to identify any toxic, carcinogenic, or other harmful effects. In such cases the benefits to humans are predicated on what happens to the physiology and biochemistry of the animals used in the research. There are two sides to any debate, and many different opinions about what is right and wrong. Should thousands of animals be sacrificed to prove that a new drug is safe? If not, what are the alternatives?

These ethical questions are increasingly converging with financial factors (reducing the number of animals in testing would inevitably reduce costs), pressuring scientists to find new ways to evaluate the potential impact of their discoveries. Of course, we need to know the effect of any drug on a whole organism, rather than on a single cell or tissue, so in the short term scientists will continue to test compounds on animals before trying them out on humans.

This is one of the areas of science where public opinion plays a significant role in shaping public policy. The organization People for the Ethical Treatment of Animals, or PETA, which promotes itself as the largest animal rights organization in the world, is dedicated to an animal rights (as opposed to animal welfare) philosophy.[49] PETA opposes the use of animals for food, clothing, entertainment, experimentation, or any other purpose. The group believes that animals' interests should always trump human benefits—even if those benefits appear to justify the loss of animal lives. PETA's message and its impact on public opinion have given it some influence on science policy. Other groups such as the Humane Society of the United States strive to decrease and eliminate harm to animals in research through "promoting research methods that have the potential to replace or reduce animal use or refine animal use so that the animals experience less suffering or physical harm."[50]

While PETA is often viewed as more extreme than the Humane Society of the United States, other organizations are even more militant in their tactics for eliminating the use of animals in research. For example, the Animal Liberation Front has been referred to as a terrorist threat by the FBI.[51] In June 2006, the group took credit for trying to put a Molotov cocktail on the doorstep of a University of California at Los Angeles researcher who does experimentation on animals.[52] Another extremist group, calling itself the Justice Department, sent eighty-seven razor-blade-laced threats via U.S. mail to primate researchers in the fall of 1999.[53] Indeed the tactics of some of these groups have become so extreme that in 2006, the Congress passed and the president signed legislation, the Animal Enterprise Terrorism Act (P.L. 100-374), to strengthen laws to protect scientists and their families, along with the companies and research organizations involved in animal research, against harassment, intimidation, threats, and trespassing or vandalism by those opposed to animal research.[54]

While they disagree with animal rights advocates concerning the appropriateness of using animals in research, most researchers are intimately concerned with the humane treatment of research animals. Of course, trials are reliable only if conducted on healthy animals. With this in mind, the scientific community has created two groups, the American Association for Laboratory Animal Science (founded in 1950) and the Association for the Assessment and Accreditation of Laboratory Animal Care International (1965), which are responsible for ensuring the quality of laboratory animal care. Some professional societies, such as the Society for Neuroscience and the American Psychological Association, maintain their own standards for the care and treatment of laboratory animals.[55]

The first law protecting laboratory animal welfare, the Laboratory Animal Welfare Act (or the Animal Welfare Act [AWA]), was established in 1966, and authorized the secretary of agriculture to regulate the transport, sale, and handling of dogs, cats, nonhuman primates, guinea pigs, hamsters, and rabbits intended for research or other purposes.[56] USDA was given this responsibility because it was viewed as having less strong ties with the NIH research community that used such animals. There have since been numerous amendments to this act, which extend its protections beyond lab animals. While the scientific community was, for the most part, originally against any statutory regulations on use of animals in research, over the years it has come to accept these regulations, recognizing that unfettered freedom of inquiry must be balanced with ethical treatment of animals and that the public needs assurances of this.[57]

The National Association for Biomedical Research (NABR) and the Foundation for Biomedical Research (FBR) provide the scientific community with a unified voice on this issue. These two groups were established to ensure the humane treatment of research animals, educate the public on animals' important role in medical advances, and advocate sound public policy on the use of animals in biomedical research and higher education. While they are not oversight organizations per se, they keep universities informed about new regulations and laws governing animal use, and take considerable responsibility for countering campaigns launched by the more extreme animal rights groups.[58]

Human Subjects

Should scientists be permitted to conduct research that might harm a single human being, but benefit the population as a whole? Regardless of whether the subject undergoes the testing knowingly or not, there are strong arguments against using public funds for this purpose. But the issue becomes fuzzier when viewed in terms of statistics. Suppose a given disease kills a thousand individuals out of every one million in a given year. Suppose further that a new drug has been shown to reduce that number from one thousand to one hundred, but that other research has shown that one hundred others are likely to die from a side effect of the drug having nothing to do with the original disease. Are the lives of the nine hundred people who would be saved by the drug worth the deaths of the one hundred who would fall victim to its side effects? Does the government have either the right or the responsibility to make that decision?

Regardless of the ethical dilemmas, testing on animals will always have to be followed by trials on humans. While animal testing allows the identification of most side effects, certain things can only be known by studying the drug's interactions with the human body. These clinical trials are performed on volunteers who are willing to test a new therapy or drug that has often not yet been tested on humans. Therefore, while thought to have promising potential the actual impacts—positive and negative—the therapy or drug will have on humans remain to be seen during these trials. Humans are also sometimes needed for basic research; a project on autism might use DNA samples from study participants to identify the gene or genes associated with the disorder. Policies on both clinical and nonclinical research protect subjects by requiring written consent, full disclosure of risk, assurance that any such risks will be in reasonable proportion to the anticipated benefits, equitable selection of participants, and additional safeguards for more vulnerable populations (e.g., children, the economically disadvantaged, or the mentally ill).[59] A major role of IRBs is to ensure the safety and protect the rights of human subjects participating in research studies.

Some studies using humans, especially in the behavioral and social sciences, do not require signed consent. Behavioral social science, after all, is focused on human actions and interactions in the environment. For example, if a sociologist wants to better understand what people do when they realize they have locked themselves out of their cars, that sociologist might stand in a parking lot and watch for people who have lost their keys. People who knew that they were being observed might not respond naturally. Such studies, although they may not require subjects' prior consent, must still be approved by IRBs, who ensure that there will be no risk to participants.[60]

More recently there have been growing concerns from researchers in the behavioral and social sciences because IRBs have a tendency to apply the standards used for biomedical, clinical research (where there is potential significant risk to participants) to studies in the behavioral and social sciences (where there is often minimal if any risk to participants). This can be problematic, for example when a signed informed consent is required of participants in phone interviews or online surveys. Behavioral and social scientists are not recommending, for the most part, that IRB approval not be required for their studies. Rather they are suggesting that whether IRB approval is needed be based on methodologies used—not on the mere fact that humans are involved as participants in the study. Essentially, they are calling on IRBs to evaluate realistically the actual potential harms to human subjects in the protocols they review.[61]

Indeed, most IRB approval is a necessary precondition for funding or continuation of any project that includes the use of human subjects, whether federal funding is involved or not. All federally funded research using humans must also comply with federal policy. Industry, too, must follow federal regulations, because while it may not rely on federal dollars, federal approval will eventually be needed before new products can be sold.

Clearly, human subjects' testing is necessary if we are to progress in our understanding of how our minds and bodies function, or achieve medical breakthroughs. What regulations and ethical guidelines have been put in place to ensure that mishaps are minimized and abuses prevented? Here we talk about the current regulations and practices concerning the ethical treatment of human subjects in research, and we discuss some of the abuses that have led to the current policies that protect human subjects.

The foundation for biomedical ethics dates back to atrocities perpetrated by Nazi physicians during World War II. From October 1946 through August 1947, part of the war crimes tribunal in Nuremberg, Germany, encompassed a trial in which twenty-three German physicians and scientists were accused of inflicting vile and lethal procedures on vulnerable people, including inmates of concentration camps. Fifteen were found guilty, and, of those, seven were given the death penalty.[62]

Out of this trial came what is known as the Nuremberg Code. This doctrine lays out ten standards to which physicians and scientists must conform when conducting experiments on human subjects. Among them are the notion of voluntary informed consent—that all human subjects participating in an experiment must be fully informed of the risks and benefits of their participation as well as the experimental procedures.[63] Participation should not be the result of coercion or any other undue influence, and unnecessary pain and suffering must be avoided. The Nuremberg Code details the obligation of scientists to protect research subjects from harm, and as such has served as a foundation for biomedical ethics.

The so-called Tuskegee experiment is perhaps the most infamous case of abuse of human subjects in American history.[64] In 1932, the Public Health Service sponsored a study, officially called the "Tuskegee Study of Untreated Syphilis in the Negro Male," to study syphilis, ostensibly with the goal of justifying publicly funded treatment programs for African Americans. All the men selected to participate in the study were sharecroppers from the Deep South, poor and uneducated.

Tragically, the researchers decided not to inform the participants (399 African American males with syphilis and 201 without the disease) of their true intent, but instead told them they were being treated for "bad blood," a blanket term for ailments including anemia, fatigue, and syphilis. Moreover, the afflicted participants were not given any treatment. The study, although originally intended to last only six months, went on for forty years. The syphilitic men, because they were left untreated and were not properly educated about their condition, put their wives at risk of infection (in some instances infecting them) and even produced offspring with congenital syphilis. The participants also endured regular spinal taps without being informed of the considerable risks or the reasons they were deemed necessary. Instead, the researchers lied, telling participants the spinal taps were therapeutic.

The horrors of the study came to light only when Peter Buxtun, a venereal disease interviewer and investigator for the Public Health Service in San Francisco, heard colleagues mention the Tuskegee study. Not believing what he was hearing, he requested documents and background materials (under the pretense of working on a job assignment). In 1972, Buxtun shared his findings with a reporter. The story broke in the *New York Times* on July 26, 1972.[65]

The Tuskegee study will forever influence U.S. science policy. First, and most obviously, the study inflicted tremendous harm on 399 men, their wives, and offspring.

It was an unspeakable crime, and the nation must do everything in its power to prevent anything similar from happening again. The experiment also involved the government in abuse of a vulnerable group: poor and uneducated participants were enticed by offers of free meals, free medical exams, and free burial service, without disclosure of the risks. Worse yet, the men with syphilis were allowed to go untreated long after 1947, when a cure (penicillin) was discovered. The study has become a symbol of racism and a stunning example of the unethical treatment of human beings.

At about the same time the Tuskegee experiment was being exposed, the details of another case, the Willowbrook study (1956–72), were also coming to light. The Willowbrook researchers, noting a high rate of incidence of hepatitis among residents of the Willowbrook State School for the Retarded, had injected some of the residents (including many children and adolescents) with a mild form of the hepatitis virus. The goal was to identify preventative measures for the infectious disease, including testing of a vaccine. Researchers took advantage of the wait list to become a resident of the school; families willing to enroll their disabled family member in the study were guaranteed a place in the school.[66] The Willowbrook study was not conducted in secret, as the Tuskegee experiment had been, but the researchers did not obtain informed consent when conducting this high-risk research on a very vulnerable population. The Willowbrook study was able to continue largely because there was no federal policy in place at the time that required board review of research involving human subjects.[67]

This changed in 1974, when the National Research Act was signed into law (P.L. 93-348), largely as a result of the publicity surrounding the Tuskegee study. The law created a National Commission for the Protection of Human Subjects of Biomedical and Behavioral Research, which was charged with identifying basic ethical principles for the use of human subjects in the conduct of biomedical and behavioral research, and developing guidelines to ensure that research was conducted in accordance with these principles. The commission issued a report (known as the Belmont Report) in 1979 and recommended that the Department of Health, Education, and Welfare (DHHS's predecessor) adopt the commission's recommendations in their entirety.[68] The Belmont Report, along with the Nuremberg Code and the World Medical Association Declaration of Helsinki, first adopted in 1964 to expand the Nuremberg Code by developing principles for human subject use in clinical and nonclinical biomedical research, are foundational documents of bioethics.[69]

The current Federal Policy for the Protection of Human Subjects was put into effect by the federal government in 1981 when the Department of Health, Education and Welfare (now the DHHS) established regulations pertaining only to human subjects' research sponsored by the department. These regulations (set forth in 45 CFR 46, Subpart A) were to be implemented through the Office of Human Subject Research (OHSR), which at that time resided within the NIH.[70] The regulations, which created IRBs and set the standards for human subjects research, have since been modified to address conflict of interest, IRB aptitude, informed consent, and the increase in human subject research.

In 1991, sixteen other agencies joined DHHS in adopting the policy, thus giving rise to the so-called Common Rule. The Common Rule thus represents the rules that most all federal research agencies have agreed upon to guide their federally supported research involving human subjects. The Common Rule focuses on three basic requirements: (1) informed consent of research subjects; (2) review of research proposals by IRBs; and (3) assurance from institutions concerning compliance with the regulations.[71] The adoption of uniform policies to guide all federal agencies' treatment of human subjects grew out of recommendations made in 1981 by the President's Commission for the Study of Ethical Problems in Medicine and Biomedical and Behavioral Research, which recommended that all federal agencies adopt the 1974 DHHS regulations. Currently, while legislation has been introduced in Congress, standardized rules for human subjects' protection across all federal agencies has not been required by congressional legislation.

In 2000, the OHSR was renamed the Office of Human Research Protection (OHRP) and moved from the NIH to the Office of the Secretary of Health and Human Services. This move was in part designed to address growing concerns about conflicts of interest between NIH researchers and the NIH entity governing their research using human subjects.[72] It was also stimulated by increasing questions about the quality of protections in response to the highly publicized death of eighteen-year-old Jesse Geslinger.

In 1999, Geslinger, an eighteen-year-old suffering from a rare genetic metabolic disorder called ornithine transcarbamylase deficiency, died after participating in a gene therapy clinical trial at the Institute of Gene Therapy at the University of Pennsylvania. His death was the first directly resulting from gene therapy, also sometime referred to as gene transfer, believed to be a potential mechanism for curing genetic diseases such as cystic fibrosis and sickle cell anemia and various types of cancer.[73]

In a follow-up investigation, it was discovered that Geslinger and his family had not been informed of the death of two monkeys given similar treatment in preclinical trials. In addition, investigators found the conduct of the clinical trial problematic; the University of Pennsylvania had not immediately reported two earlier adverse events and had enrolled Geslinger under medical conditions that should have excluded him.[74] Also undisclosed during the informed consent process were the financial ties between the lead researcher in the clinical trial, James Wilson, and a biotechnology company that funded about one-fifth of the institute's annual budget and had exclusive rights to Wilson's findings for commercial products.[75] Other companies had financial interests in the trial in which Geslinger participated as well, raising the issue of conflict of interest—between business decisions and medical decisions.[76]

Following the investigation, all gene transfer human trials were suspended at the University of Pennsylvania, Wilson was barred from conducting any research using human subjects, and the NIH were found at fault for not keeping better oversight on the University of Pennsylvania, and on other gene transfer trials.[77] As a result of this case and follow-up audits and reports conducted by the GAO, DHHS inspector general, and the National Bioethics Advisory Commission (NBAC), Representatives Diana L. DeGette (D-CO) and Jim Greenwood (R-PA) introduced legislation in 2002 aimed at revamping federal protections to all research involving human subjects, including that which is not funded by federal agencies. While at the time of this writing this legislation has yet to be approved by Congress, it is likely that the protection of human research subjects will continue to receive attention from Congress in the future.[78]

Human Cloning

While bacterial and tissue-culture cells are routinely "cloned" in the lab, somatic cell nuclear transfer, or SCNT, and "cloning" have become the subject of heated political debate. SCNT is simply the transfer of a nucleus from a cell into an egg whose nucleus has been removed. In fact, the first successful attempt at SCNT in mammals was the 1997 cloning of Dolly the sheep, in which DNA from an adult sheep's cell was placed into an egg from another sheep to produce a genetically identical, although time-delayed twin.[79] While SCNT was the same *technique* required to create Dolly and is often confused with the

process of *reproductive cloning*, SCNT represents only the production of an embryo at its earlier stage, stopping short of implanting the embryo into a uterus to develop naturally, the additional step required for reproductive cloning. SCNT is of interest to scientists because of its great potential for use in nonreproductive, *therapeutic* or *research cloning*, a technique through which embryonic stem cells can be derived.[80]

Almost everyone, scientists and members of the public alike, agrees that reproductive cloning should be banned. The use of SCNT to create children poses a vast number of concerns about the safety of mother and child, individuality, family integrity, and treatment of children as objects rather than unique human beings. The fear of scientists "playing god," as Nazi scientists did in the 1930s and 1940s, is also a powerful factor in opposition to reproductive cloning.

After Dolly was born, President Clinton issued a presidential directive forbidding the use of federal dollars for human cloning research, and asking the NBAC to thoroughly review the ethical and legal implications of the technology within ninety days.[81] The commission recommended a continuation of the moratorium; a request that all private firms, clinics, and professional societies voluntarily comply with the ban; enactment of legislation prohibiting anyone from creating a human child through SCNT; and the requirement that any such legislation not interfere with scientific research into therapeutic cloning, animal SCNT, or other legitimate areas of inquiry.[82] The National Academies issued a concurring report in 2002.[83]

Congress has yet to pass any form of cloning legislation, even legislation against reproductive cloning, however, because of strong differences of opinion over the relative importance of medical progress and the rights of the embryo: the very same issues involved in the debate over human embryonic stem cell (hESC) research.[84] Congress is stymied in part because the cloning debate has been subsumed within the pro-life/pro-choice debate. While there is general agreement about the need for regulations, it has become virtually impossible in the current political climate to determine what those regulations should say. There is a great deal at stake—not just progress toward treatments and cures for debilitating diseases and injuries, but also America's future status as a world leader in science (especially since several European and Asian nations allow research cloning). As Shirley Tilghman, president of Princeton University, and David Baltimore, former president of Cal Tech, noted in a *Wall Street Journal* editorial: "A broad ban [on research cloning] would stifle innovation in a key area of biomedical research. Scientists would have to leave the U.S. or switch to other fields."[85]

The President's Council on Bioethics issued a report in March 2004 recommending that Congress limit pro-creation to the union of an egg and sperm.[86] This was essentially a recommendation that Congress ban repro-ductive cloning. The council, however, remained notably silent on the issue of therapeutic cloning. It is uncertain whether the council's silence represented a tacit distinc-tion on their part between the two types of cloning.[87]

The issue of therapeutic cloning to advance research is likely to remain a contentious issue that will pit strongly held moral and religious values directly against scientific progress and potential medical breakthroughs. It is an area where despite the need to establish laws and regula-tions against reproductive cloning, the strong interest by some religious organizations to extend such laws to ther-apeutic or research cloning may well prevent the creation of effective, and arguably necessary, laws and regulations to limit reproductive cloning.

Human Embryonic Stem Cells

Human embryonic stem cell research is a relative new-comer to the world of scientific ethics.

In late 1998, a group of researchers from the Univer-sity of Wisconsin reported the first isolation and stabiliza-tion of five hESC lines. Embryonic stem cells (ESCs) are essentially blank, or *pluripotent,* cells from a human or other organism. With the right stimulation, they can be encouraged to turn into tissue-specific cells, such as brain, heart, liver, or kidney cells. This means that ESCs hold great potential for treating cell-based diseases like diabe-tes mellitus. In addition, they might be used to repair or even replace damaged tissue in, for example, the spinal cord or the heart.[88]

Their enormous medical and scientific potential not-withstanding, the use of ESCs from humans raises serious ethical concerns. What is the moral status of the embryo? Is an embryo equivalent to a person? Is it ethical to de-stroy an embryo for scientific purposes? Is it ethical to create an embryo using SCNT solely for the purpose of deriving hESCs? Does a good outcome (e.g., therapies for many) justify what some construe as an evil or at least ethically questionable means (e.g., destruction of em-bryos)? Should scientists be allowed to use the hESC lines that already exist, since what's done is done? Is it ethical to insert hESCs into domestic animal embryos to create

a chimera for the purposes of harvesting transplantable human organs?[89] Would ready availability of replace-ment organs and tissues cause people to take more risks and be less responsible about their health? Because of the potential of deriving patient-specific hESC using SCNT, ethical concerns also arise over egg donation, for exam-ple, risk to donors, informed consent, and payment and coercion.[90]

In light of such questions, and with the federal ban on human embryo research resulting from the Dickey-Wicker Amendment in 1995, President Clinton once again asked the NBAC to review the issues and release a report.[91] Harold Varmus, then head of the NIH, also consulted with the NBAC on pertinent ethical issues, and with the DHHS counsel on legal issues. Could hESCs derived with private funds be used by federally funded researchers? DHHS counsel stated that "DHHS funds can be used to support research utilizing human . . . stem cells that are derived from human embryos: the statutory prohibition on human embryo research does not apply to research utilizing human . . . stem cells because human . . . stem cells are not embryos."[92]

The NBAC final report included the recommendation that federal law be changed to allow the government to finance hESC research, including the derivation of new lines using human embryos. Ultimately, the Clinton ad-ministration and the NIH opted for a more conservative approach, asking the NIH to draft guidelines for federal funding of hESC research, but not for the actual deriva-tion of new cell lines. Under the Clinton policy scientists could use federal funds to do research using *any* hESC lines, derived at any date in time, but could not use fed-eral funds to *derive* new lines.

As so often happens, with a new administration came a new policy. In August 2001, President George W. Bush ordered that federally funded hESC research be limited to existing lines, and that no federal funding could be used either to create or even to study lines created after the August 2001 date. Moreover, any grant submitted to the NIH must include appropriate paperwork for licensing one or more of these existing lines. When the Bush pol-icy first went into effect, there appeared to be more than sixty viable and eligible lines. However, many of these have not been fully characterized, and scientists lack the knowledge needed to reproduce them in sufficient quan-tities under quality-controlled conditions, meaning that the number of authorized lines is actually much smaller. Because each line has a distinct genetic makeup, research-ers need different hESC lines to study a wide spectrum of diseases. The scientific community has therefore been for the most part unified in expressing concerns that limiting the number of hESC lines is having a negative impact on their ability to do fully advanced hESC research. There have been a few in the scientific community, and many in the pro-life community, however, who maintain that adult stem cells hold the same promise as embryonic stems cells for therapeutic purposes. Therefore, they maintain that hESC research that requires destruction of an early stage human embryo is not needed.[93] While the recent scientific advances to induce pluripotency in adult stem cells will provide powerful tools to researchers, there is still a need to push forward stem cell research on all fronts, including embryonic stem cell research (see chap. 9).[94]

There is no doubt but that the stem-cell debate will remain a contentious issue among policymakers, one where a consensus surrounding the ethical rules that should guide such research cannot be reached. Thus far, attempts in Congress to ease restrictions on federal funding of hESC research have failed, with President George W. Bush using the first veto of his presidency to overturn legislation that would have lifted current restrictions limiting the use of federal funds to those stem cell lines created before August 9, 2001. President Bush once again vetoed similar legislation approved very early in the 110th Congress.

As the research potential of the existing hESC lines is exhausted, however, universities that are typically reliant on federal dollars are looking to other sources to support hESC research. This includes private research funds from venture capital, foundations (e.g., the National Parkinson Foundation), corporations, and state research funds that are not restricted by the president's ruling. In the short term, these alternative funding sources may make continued hESC research possible. But without federal dollars, some are concerned that the United States may fall behind in this area of science, which could be costly when considering the therapeutic—and thus economic—potential of this technology.[95] Some are concerned that with hESC research being funded by nonfederal sources, scientists will not be required to follow federal policies regulating science. However, any drug or other therapeutic that results from this research will still need FDA approval, creating some incentive for privately funded and state-funded research to abide by federal policy.

While no federal law has been passed governing hESC research, members of the scientific community have made recommendations. In April 2005, the National Academies issued *Guidelines for Human Embryonic Stem Cell Research;* among the recommendations made in this report were that institutional oversight committees specific

for embryonic stem cell research, which the report called "ESCRO Committees" for embryonic stem cell research oversight, should be created and that donors of eggs, sperm, or any other cells for use in the derivation of hESC should be reimbursed only for direct expenses incurred as a result of the donation but should not receive any other payment or gifts in kind.[96] These recommendations are just that—recommendations. However, in the absence of overarching federal laws, several states and institutions have used the National Academies guidelines in crafting their own stem cell policies.

Policy Challenges and Questions

Research integrity and ethics are complex and controversial areas of science policy. Scientists must conduct themselves in an upright manner in order to maintain the public trust. As in any other profession, all community members in good standing must abide by agreed-upon practices, an ethical and moral "code of conduct." To ensure that the code is adhered to takes a commitment from leading scientists. It requires that education programs continue to be supported and expanded and the scientific community be quick to respond to alleged violations of ethical standards.

Alberts and Shine suggest that to maintain the public trust, scientific leaders must continually ask themselves the following questions: Are we setting a good example? Do we go out of our way to give credit to others on whose findings and ideas we build? Are we explicit about the contributions to our own work by students? Do we reward the scientific quality rather than the quantity of publications? Do we reward faculty who contribute to the scientific community through outstanding public service, teaching, and mentoring? When allegations of misconduct arise, are they scrupulously examined regardless of the rank or status of the scientist in question or the financial implications for the scientist or the institution? Are we contributing to the mechanisms that spread appropriate values? And, finally, do we support educational efforts that promote the high ethical standards of science?[97]

When the scientific community fails to effectively regulate itself, policymakers step in and tend to impose conservative solutions that can impede future research. Such approaches may ultimately have a negative impact on the public and on the progress of science and illustrate why the scientific community must remain ever vigilant in its attempts at self-regulation.[98]

As sponsors of intramural and extramural research, federal agencies must ensure that research is continually supported by both the public and Congress. Because science is conducted by scientists, who are after all only human, standards and guidelines must be in place—some voluntary and some mandatory—that the agencies reinforce through policies, regulations, and incentives.

Since the 1990s, federal funding in certain areas of research has been directly tied to training in research ethics. For example, all graduate students and postdoctoral fellows funded by NIH dollars must take the "Responsible Conduct of Research" course. Research funding from many other federal research agencies, however, does not require such formal training in scientific integrity and ethics. This raises a question, whether other agencies should make the award of research funding contingent upon training in these areas. Having asked that question, one can argue that such requirements are excessive, given they should be a natural part of the faculty-student mentorship process that teaches appropriate scientific techniques and behaviors.

As has been noted throughout this chapter, there are areas such as the use of animals in research, therapeutic cloning, and hESC research where the public is split in its views of what is morally acceptable. In the instance of animal research, these views are often determined by individuals' views of the nature of rights enjoyed by animals and whether those rights are equal to those of humans, as balanced against the potential value to both humans and animals of such research. In instances such as research cloning and hESC research, these attitudes are often driven by religious and moral views concerning what constitutes life. This leaves these matters much more likely to become the subject of heated political debates where no consensus can be achieved.

Regulation is undoubtedly necessary, and scientists generally respect the boundaries of their work. However, balance must be maintained between a completely unregulated environment and rules that prohibit reasonable research. An ongoing challenge and policy question to be addressed by scientists and policymakers alike is this: when is self-regulation by the scientific community of its conduct impossible or undesirable?

Clearly, scientists must be engaged in the policy-making process, communicating with policymakers and advocating for a sound and well-informed regulatory environment that does not impede research. They must also work hard to promote a regime of self-regulation so as to prevent policymakers from feeling the need to get involved.[99]

NOTES

1. *Webster's 9th Collegiate Dictionary* (Springfield, MA: Merriam-Webster, 1986), s.v. "ethics."

2. Bruce Alberts and Kenneth Shine, "Scientists and the Integrity of Research," *Science* 266, no. 5191 (December 9, 1994): 1660.

3. Committee on Science, Engineering, and Public Policy, *On Being a Scientist: Responsible Conduct in Research,* 2nd ed. (Washington, DC: National Academy Press, 1995), v.

4. Office of Science and Technology Policy, "Federal Research Misconduct Policy," *Federal Register* 65, no. 235 (December 6, 2000): 76260-64.

5. *Research misbehavior* and *questionable research practices* are two more terms sometimes used to refer to behaviors of scientists that do not fall under the narrow definition of research misconduct (fabrication, falsification, and plagiarism). For more discussion see Nicholas H. Steneck, "Fostering Integrity in Research: Definitions, Current Knowledge, and Future Directions," *Science and Engineering Ethics* 12, no. 1 (2006): 53–74; Raymond De Vries, Melissa S. Anderson, and Brian C. Martinson, "Normal Misbehavior: Scientists Talk about the Ethics of Research," *Journal of Empirical Research on Human Research Ethics* 1, no. 1 (2006): 43–50; Brian C. Martinson, Melissa S. Anderson, and Raymond De Vries "Scientists Behaving Badly," *Nature* 435, no. 7043 (2005): 737–38 ; and Lila Guterman, "Pittsburgh Panel Says Scientist Engaged in 'Research Misbehavior,' but Not Misconduct, in Stem-Cell Fraud," *Chronicle of Higher Education,* February 13, 2006, http://chronicle.com/daily/2006/02/2006021302n.htm (June 18, 2007).

6. Steneck, "Fostering Integrity in Research," 55.

7. *Webster's 9th Collegiate Dictionary,* s.v. "plagiarize."

8. Jay W. Rojewski and Desirae M. Domenico, "The Arts and Politics of Peer Review," *Journal of Career and Technical Education* 20, no. 2 (2004): 41–54; Carol Berkenkotter, "The Power and the Perils of Peer Review," *Rhetoric Review* 13, no. 2 (1995): 245–48.

9. Just two months prior, Frenchman Denis de Sallo had started a weekly scholarly journal that focused on reviews and current news in science and other scholarly activities, including law and theology. Because of this, the *Philosophical Transactions* is considered the first scientific journal.

10. An anagram is a rearrangement of the letters in a word or phrase to form another word or phrase. For example, *Old England* is an anagram of *golden land* and *gapes* is an anagram of *pages*.

11. For additional background see "The Scientific Article: From Galileo's New Science to the Human Genome," http://www.fathom.com/course/21701730/session2.html (accessed June 6, 2007); Caroline Whitbeck, "Responsible Authorship," http://onlineethics.org/reseth/mod/auth.html (accessed June 6, 2007); Robert K. Merton, "The Matthew Effect in Science II: Cumulative Advantage and the Symbolism of Intellectual Property," originally published in *Isis* 79(4), 621 and modified for Committee on Science, Engineering, and Public Policy, *On Being a Scientist.*

12. "Home for Scientific Whistleblowers," *Science* 277, no. 5332 (September 12, 1997): 1611. See also Thomas Bartlett and Scott Smallwood, "Mentor vs. Protégé: The Professor Published

the Student's Words as His Own. What's Wrong with That?" *The Education,* December 17, 2004; "Whistleblower Awarded Damages in Retaliation Case," *ORI Newsletter,* vol. 2, no. 1, Office of Research Integrity, U.S. Public Health Service, December 1993.

13. Jim Giles, "Preprint Analysis Quantifies Scientific Plagiarism," *Nature* 444 (November 30, 2006): 524–25.

14. During this same period, ORI found 162 instances of falsification or fabrication of research records or both. See Alan R. Price, "Cases of Plagiarism Handled by the United States Office of Research Integrity 1992–2005," *Plagiary: Cross-Disciplinary Studies in Plagiarism, Fabrication, and Falsification* 1, no. 1 (2006): 1–11.

15. For further discussion concerning the Baltimore case, see Daniel J. Kevles, *The Baltimore Case: A Trial of Politics, Science, and Character* (New York: W. W. Norton, 1998); and Judy Sarasohn, *Science on Trial: The Whistle-blower, the Accused, and the Nobel Laureate* (New York: St. Martin's Press, 1993).

16. See David Goodstein, "In the Matter of J. Hendrik Schön," *Physics World,* November 2002, http://www.its.caltech.edu/~dg/The_physicists.pdf (accessed June 6, 2007); "Bell Labs Scientist Dismissed for Fabricating Data," *Electronic News,* September 26, 2002, http://www.ferret.com.au/articles/d9/0c0111d9.asp (accessed June 18, 2007).

17. Columbia University, Responsible Conduct of Research, "Research Misconduct, Foundation Text, John Darsee and Robert Slutsky," http://www.ccnmtl.columbia.edu/projects/rcr/rcr_misconduct/foundation/index.html#1_B_3 (accessed June 6, 2007).

18. For additional background, see David Cyranoski, "Hwang Takes the Stand at Fraud Trial," *Nature,* November 1, 2006, http://www.nature.com/news/2006/061030/full/444012a.html (accessed April 14, 2007); Nicholas Wade and Choe Sang-Hun, "Research Faked Evidence of Human Cloning, Koreans Report," *New York Times,* January 10, 2006, Science section; Anthony Faiola and Rick Weiss, "South Korean Panel Debunks Scientist's Stem Cell Claims," *Washington Post,* January 10, 2006, A09; *Nature,* January 11, 2006, http://www.nature.com/news/specials/hwang/index.html (accessed April 14, 2007).

19. Recall that somatic cell nuclear transfer is the process by which a nucleus from a somatic (body) cell is transferred into an enucleated egg. This process is sometimes referred to as *research cloning* or *therapeutic cloning.*

20. *Patient specific* refers to the fact that the nuclei used in SCNT is to come from an individual who will benefit from the hESC derived from the SCNT clone. This means that any tissues generated from these hESC will be immuno-compatible with the patient because the hESC will have the same DNA as the patient's cells.

21. See Mildred K. Cho, Glenn McGee, and David Magnus, "Lessons of the Stem Cell Scandal," *Science* 311, no. 5761 (February 3, 2006): 614–15, for additional discussion of the research misconduct associated with the Hwang scandal.

22. Incidentally, Hwang's claim, published in the August 4, 2005, edition of *Nature,* to have used SCNT to generate a dog, was found to be valid. See Erika Check, "More Cloned Puppies Born," *Nature,* December 19, 2006, http://www.nature.com/news/2006/061218/full/061218-4.html (accessed April 14, 2007). Interestingly, while Hwang's group did not use SCNT

to create cell lines as they had claimed, they did create lines using parthenogenesis. Had they recognized their actual accomplishment, they would have been the first group to do so. See Nicholas Wade, "Within Discredited Stem Cell Research, a True Scientific First," *New York Times,* August 3, 2007.

23. Several journals in the biomedical sciences do require all authors to sign a statement that they did indeed participate in the study and have read the manuscript. Some go as far as to request that each author's role be described. For additional information see International Committee of Medical Journal Editors, "Uniform Requirements for Manuscripts Submitted to Biomedical Journals: Writing and Editing for Biomedical Publication," http://www.icmje.org/ (accessed June 6, 2007); *Journal of the American Medical Association,* "Information for Authors/Reviews, Instructions for Authors," http://jama.ama-assn.org/misc/ifora.dtl# (accessed June 6, 2007); *PLoS Medicine,* "*PLoS Medicine* Guidelines for Authors," http://journals.plos.org/plosmedicine/guidelines.php#supporting (accessed May 8, 2007). For an additional discussion of the dilution of authorship in scientific papers see Sean B. Seymore, "How Does My Work Become Our Work? Dilution of Authorship in Scientific Papers, and the Need for the Academy to Obey Copyright Law," *Richmond Journal of Law and Technology* 12, no. 3 (2006): 1–28.

24. See Goodstein, "Matter of Schön," 1.

25. *Federal Register* 54, no. 151 (August 8, 1989): 32446–51.

26. See Remarks by the Honorable Neal Lane at a Town Hall Meeting on the Proposed Federal Research Misconduct Policy, November 17, 1999, http://clinton3.nara.gov/WH/EOP/OSTP/html/00222_2.html; and Larry Hand, "Research Misconduct Defined: Comments Sought," *The Scientist* 13, no. 22 (November 8, 1999): 11; *Federal Register* 65, no. 235 (December 6, 2000).

27. See National Academies, "Scientists, Institutions Must Confront Misconduct in Science Head-on," press release, April 22, 1992, http://www4.nationalacademies.org/news.nsf/isbn/POD500?OpenDocument (accessed June 6, 2007).

28. For additional background see Office of Research Integrity, Policies/Regulations, Federal Policies, Misconduct regulations, http://ori.dhhs.gov/policies/federal_policies.shtml (accessed June 18, 2007); National Academy of Sciences, Statement on Responsible Conduct of Research from Bruce Alberts, President, National Academy of Sciences, William A. Wulf, President, National Academy of Engineering, and Kenneth I. Shine, President, Institute of Medicine, December 10, 1999, http://www4.nationalacademies.org/news.nsf/isbn/s12101999?Open Document (accessed June 6, 2007); Richard W. Hanson, President, American Society for Biochemistry and Molecular Biology, "ASBMC Comments on the Federal Government's October 1999 Proposed Federal Policy on Research Misconduct," December 1999, http://www.asbmb.org/ASBMB/site.nsf/0/87E3995AA000A50C85256C7C00535A7E?OpenDocument (accessed June 6, 2007); and Ellen Paul, "Reviewing the Proposed New Federal Research Misconduct Policy," American Institute for Biological Sciences Washington Watch, March 2000, http://www.aibs.org/washington-watch/washington_watch_2000_03.html (accessed June 6, 2007).

29. Office of Research Integrity, "Federal Policies," http://ori.dhhs.gov/policies/federal_policies.shtml (accessed June 6, 2007).

30. Berg, "Asilomar and Recombinant DNA." See also Marica Barinaga, "Asilomar Revisited: Lessons for Today," *Science* 287, no. 5458 (March 3, 2000): 1584–85.

31. Association of American Universities, Task Force on Research Accountability, "Report on Individual and Institutional Financial Conflict of Interest," October 2001, i, http://www.aau.edu/research/COI.01.pdf (accessed June 6, 2007).

32. Conflict of interest also exists in some research settings, in which scientists involved in conducting a study have a dual role. In many clinical research studies, the scientists are clinicians *and* researchers. They have an interest and duty to protect their patients, who are often the subjects enrolled in the research study. Wearing their "researcher" hat, however, these scientists also have an interest in seeing the study completed. Situations in which a research subject's health outcomes would be better if the subject did not enroll in the study being conducted are something these scientists face regularly.

33. *Federal Register* 60, no. 132 (July 11, 1995): 35810–19; National Science Foundation, "Public Health Service Regulations, 45 CFR Part 50, Subpart F and 45 CFR Part 94," sec. 510 of NSF Publication 05-131, in *Grant Policy Manual* (Arlington, VA: National Science Foundation, July 2005), http://www.nsf.gov/pubs/manuals/gpm05_131/gpm05_131.pdf (accessed June 6, 2007).

34. See National Institutes of Health Program Policy, http://grants1.nih.gov/grants/policy/coi/coi_grantees.htm (accessed June 20, 2007).

35. See, for example, Jennifer Walshburn, *University Inc.: The Corporate Corruption of Higher Education* (New York: Basic Books, 2005).

36. David Willman, "Stealth Merger: Drug Companies and Government Medical Research," *LA Times,* December 7, 2003, A1, A32–A33.

37. National Institutes of Health, Office of the Director, "NIH Statement about Outside Consulting Arrangements," press release, December 10, 2003, http://www.nih.gov/news/pr/dec2003/od-10.htm (accessed June 6, 2007).

38. National Institutes of Health, *Report of the National Institutes of Health Blue Ribbon Panel on Conflict of Interest Policies, a Working Group of the Advisory Committee to the Director of NIH,* June 22, 2004, http://www.nih.gov/about/ethics_COI_panelreport.pdf (accessed June 6, 2007).

39. Ted Agres, "NIH Needs 'Drastic Changes,'" *The Scientist* 5, no. 1 (June 23, 2004), http://www.the-scientist.com/article/display/22245/ (accessed June 6, 2007); "Conflict at NIH (cont.)," *Nature* 430, no. 6995 (July 1, 2004): 1.

40. See Elias A. Zerhouni, Director of NIH, memorandum to all NIH staff, August 25, 2005, "Ethics Rules Announcement," http://www.nih.gov/about/ethics/coipolicymemo_08252005.htm (accessed June 6, 2007); Department of Health and Human Services, 5 CRF Parts 5501 and 5502, Rules and Regulations, *Federal Register* 70, no. 168 (August 31, 2005): 51559; and National Institutes of Health, "About NIH, Conflict of Interest Information and Resources, Summary of NIH-Specific Amendments to Conflict of Interest Ethics Regulations," http://www.nih.gov/about/ethics/summary_amendments_08252005.htm (accessed June 6, 2007).

41. Press and Washburn, "The Kept University," 9; Goldie Blumenstyk, "A Vilified Corporate Partnership Produces Little Change (Except Better Facilities)," *The Education* 47, no. 41 (2001): 24.

42. Rex Dalton, "Berkeley Accused of Biotech Bias as Ecologist Is Denied Tenure," *Nature* 426, no. 6967 (2003): 591. An added twist to this story is that not only had Chapela been critical of the UC Berkeley–Novartis deal, but he and his graduate student David Quist had published a paper in *Nature* in November 2001 asserting that traces of DNA from genetically modified corn had spread to native Mexican corn. Chapela and Quist also suggested that genes from the bioengineered corn moved around and reassorted once in the native species. This finding contradicted the claims of plant biotech companies that inserted genes are stable. After the paper was published, several scientists disputed the methodology used and the findings. In August 2002, the Environment Ministry of the Mexican Government independently verified the findings of Chapela and Quist. See David Quist and Ignacio H. Chapela, "Transgenic DNA Introgressed into Traditional Maize Landraces in Oaxaca, Mexico," *Nature* 414 (November 29, 2001): 541–43; Keisha-Gaye Anderson, "Seeds of Conflict," *NOW with Bill Moyers, Public Broadcast System,* October 4, 2002, http://www.pbs.org/now/science/genenature.html (accessed June 6, 2007).

43. Tom Abate, "Critic of Biotech Corn Fears UC Won't Give Him Tenure: Junior Professor Fought School's Ties with Industry," *San Francisco Chronicle,* March 23, 2003, http://sfgate.com/cgi-bin/article.cgi?file=/c/a/2003/03/23/MN248867.DTL (accessed June 6, 2007); Charles Burress, "UC Won't Give Tenure to Critic of Biotech," *San Francisco Chronicle,* December 12, 2003, http://sfgate.com/cgi-bin/article.cgi?file=/chronicle/archive/2003/12/12/BAGVH3LEO529.DTL (accessed June 6, 2007).

44. Erik Stokstad, "Embattled Berkeley Ecologist Wins Tenure," *Science* 308, no. 5726 (May 27, 2005): 1239.

45. Jennifer Washburn, "Big Oil Buys Berkeley," *Los Angeles Times,* Opinion, March 24, 2007.

46. See John W. Bardo and Billy Ray Hall, "Communicating the University's Role in Economic Development," University of North Carolina, 2005, http://www.wcu.edu/chancellor/Communicating%20the%20University.htm (accessed December 16, 2007); Sarah Davidson, "National Conference at Cornell Explores Land-grant Mission in Economic Development," *Cornell Chronicle Online,* June 23, 2006, http://www.news.cornell.edu/stories/June06/NABC.cover.SD.html (accessed June 6, 2007); University of Michigan Research, "Guide to Understanding Conflict of Interest in Sponsored Research and Technology Transfer Agreements: Overarching Principles Guiding Conflict of Interest Evaluation," http://www.research.umich.edu/policies/um/coi/principles.html#1 (accessed June 6, 2007).

47. Howard W. Bremer, "University Technology Transfer: Evolution and Revolution," in *50th Anniversary—Journal of Papers,* Council on Governmental Relations (1998), http://www.cogr.edu/docs/Anniversary.pdf (accessed June 6, 2007).

48. It is important to note that researchers have no desire to mistreat their research animals—quite the contrary. Because data are collected from these animals, researchers want to treat and care for their animals quite well. In fact, laboratory animals probably have better medical care than most humans in that veterinary staff are on call at all times, there are strict institutional

guidelines in the handling of these animals and the housing environment, and handlers must go through specialized training.

49. See the PETA Web site for additional information, www.peta.org/about/index.asp (accessed June 6, 2007).

50. Humane Society of the United States, "Animals in Biomedical Research, Testing, and Education," *Statements of Policy,* October 22, 2005, 2.

51. Statement of James F. Jarboe, Domestic Terrorism Section Chief, Counterterrorism Division, FBI, "The Threat of Eco-Terrorism," before the House Resources Committee Subcommittee on Forests and Forest Health, February 12, 2002, http://www.fbi.gov/congress/congress02/jarboe021202.htm (accessed June 6, 2007).

52. David Epstein, "Throwing in the Towel," *Inside Higher Ed,* August 22, 2006, http://www.insidehighered.com/news/2006/08/22/animal (accessed June 6, 2007).

53. For additional information see Jocelyn Kaiser, "Booby-Trapped Letters Sent to 87 Researchers," *Science* 286, no. 5442 (November 5, 1999): 1059; Frankie L. Trull, "Biomedical Attacks," *Science* 286, no. 5442 (November 19, 1999): 1477; Ingrid Newkirk, letter to the editor, *Atlanta Journal and Constitution,* November 1, 1999, A10; Willman, "Stealth Merger."

54. Ted Agres, "Congress Passes Animal Terrorism Bill," *The Scientist* (November 16, 2006), http://www.the-scientist.com/news/home/36515/ (accessed June 6, 2007).

55. For additional information on these organizations see American Association for Laboratory Animal Science, http://www.aalas.org/; Association and Assessment and Accreditation of Laboratory Animal Care International, http://www.aaalac.org/; Society for Neuroscience, http://www.sfn.org; and the American Psychological Association, "Guidelines for Ethical Conduct in the Care and Use of Animals," http://www.apa.org/science/anguide.html (all sites accessed June 6, 2007).

56. Animal Welfare Act of August 24, 1966, P.L. 89-544, 89th Congress, Code of Federal Regulations, Title 9, chap. 1, subsection A, http://www.nal.usda.gov/awic/legislat/pl89544.htm (accessed June 6, 2007).

57. See Michael Kreger, D'Anna Jensen, and Tim Allen, eds., *Animal Welfare Act: Historical Perspectives and Future Directions Symposium Proceedings,* WARDS (Working for Animals in Research, Drugs, and Surgery), 1998, http://www.nal.usda.gov/awic/pubs/96symp/awasymp.htm#overview (accessed June 6, 2007).

58. National Association for Biomedical Research, http://www.nabr.org/; and Foundation for Biomedical Research, http://www.fbresearch.org/ (both accessed June 23, 2007).

59. U.S. Department of Health and Human Services, *Code of Federal Regulations,* Title 45 (Public Welfare), Part 46 (Protection of Human Subjects), http://www.hhs.gov/ohrp/humansubjects/guidance/45cfr46.htm (accessed June 6, 2007).

60. It is important to note that many studies in the behavioral and social sciences that use human subjects do require informed consent, even interviews of scientists asking about attitudes and perceptions of research ethics.

61. For additional discussion see Committee A on Academic Freedom and Tenure, *Institutional Review Boards and Social Science Research* (Washington, DC: American Association of University Professors, 2000); and Committee on Academic Freedom and Tenure, *Research on Human Subjects: Academic Freedom and the Institutional Review Board* (Washington, DC: American Association of University Professors, 2006).

62. Jennifer Leaning, "War Crimes and Medical Science," *British Medical Journal* 313, no. 7070 (December 7, 1996): 1413.

63. The U.S. Holocaust Memorial Museum, Commemoration of the 50th Anniversary of the Doctors Trial (the Medical Case of the Subsequent Nuremberg Proceedings), http://www.ushmm.org/research/doctors/Nuremberg_Code.htm (accessed April 14, 2007); Circumcision Information and Resource Pages, Reference Library, Bioethics and Human Rights Index, Nuremberg Code, http://www.cirp.org/library/ethics/nuremberg/ (accessed April 14, 2007).

64. For additional information, see "The Tuskegee Syphilis Experiment," *infoplease.com,* http://www.infoplease.com/ipa/A0762136.html (accessed June 6, 2007); U.S. Centers for Disease Control and Prevention, *U.S. Public Health Service Syphilis Study at Tuskegee,* http://www.cdc.gov/nchstp/od/tuskegee/time.htm (accessed June 6, 2007); James H. Jones, *Bad Blood: The Tuskegee Syphilis Experiment,* rev. ed (New York: Free Press, 1993); William J. Clinton, "Presidential Apology," transcript, May 16, 1997, http://www.cdc.gov/NCHSTP/OD/tuskegee/clintonp.htm (accessed June 23, 2007); and LinkAge 2000, Treasury, Witness to History, Mary Starke Harper, interview, July 9, 1997, http://library.advanced.org/10120/treasury/harper.html (accessed June 23, 2007).

65. Jean Heller, "Syphilis Victims in U.S. Study Went Untreated for 40 Years," *New York Times* July 26, 1972, 1, 8.

66. Jeffrey Kahn, "Informed Consent in the Context of Communities," *Journal of Nutrition* 135 (2005): 918–20.

67. See the U.S. Department of Energy, *Advisory Committee on Human Radiation Experiments (ACHRE) Final Report* (Washington, DC: U.S. Government Printing Office, October 1995), chap. 3, http://www.hss.energy.gov/HealthSafety/ohre/roadmap/achre/report.html (accessed June 6, 2007).

68. U.S. Department of Health, Education, and Welfare, *The Belmont Report: Ethical Principles and Guidelines for the Protection of Human Subjects of Research,* National Commission for the Protection of Human Subjects of Biomedical and Behavioral Research, April 18, 1979, http://ohsr.od.nih.gov/guidelines/belmont.html (accessed June 6, 2007).

69. David Magnus, Professor of Pediatrics and Stanford Center for Biomedical Ethics Director, and Henry Greely, Professor of Law and Stanford Center for Biomedical Ethics Steering Committee Chair, Stanford University, November 17, 2005, personal communication to Jennifer B. McCormick. See also National Institutes of Health, "Nuremberg Code," http://ohsr.od.nih.gov/guidelines/nuremberg.html (accessed June 6, 2007) and "Declaration of Helsinki," http://ohsr.od.nih.gov/guidelines/helsinki.html (accessed June 6, 2007).

70. This policy is delineated in the U.S. Department of Health and Human Services, *Code of Federal Regulations,* Title 45, Part 46, Subpart A.

71. Erin D. Williams, *Federal Protection for Human Research Subjects: An Analysis of the Common Rule and Its Interactions with FDA Regulations and the HIPAA Privacy Rule,* RL32909 (Washington, DC: Congressional Research Service, Library of Congress, June 2, 2005), http://ftp.fas.org/sgp/crs/misc/RL32909.pdf (accessed June 6, 2007).

72. See Society for Women's Health Research, "Our Advocacy Issues, Human Subject Protection," http://www.womens healthresearch.org/site/PageServer?pagename=policy_issues (accessed June 6, 2007).

73. See U.S. Department of Energy and National Institutes of Health, "Human Genome Project Information, Gene Therapy," http://www.ornl.gov/sci/techresources/Human_Genome/medicine/genetherapy.shtml (accessed June 6, 2007).

74. Larry Thompson, "Human Gene Therapy, Harsh Lessons, High Hopes," *FDA Consumer,* September–October 2000, http://www.fda.gov/fdac/features/2000/500_gene.html (accessed June 6, 2007). Apparently Jesse Geslinger went out for a hamburger the night before he was to be enrolled in the trial. This elevated his ammonia levels (Margaret Eaton, Stanford University, April 27, 2007, Science and Technology Studies seminar).

75. Sheryl Gay Stolberg, "Institute Restricted after Gene Therapy Death," *New York Times,* May 25, 2000.

76. Sophia M. Kolehmainen, "The Dangerous Promise of Gene Therapy," abridged version of an article from *GeneWatch,* February 2000, http://www.actionbioscience.org/biotech/koleh mainen.html (accessed June 6, 2007).

77. Sheryl Gay Stolberg, "Agency Failed to Monitor Patients in Gene Research," *New York Times,* February 2, 2000; Stolberg "Institute Restricted after Gene Therapy Death."

78. National Institutes of Health, "Human Subjects Legislation: H.R. 4697 and S. 3060: Background," http://olpa.od.nih.gov/legislation/107/pendinglegislation/humansubjects.asp (accessed June 6, 2007); "Human Subjects Legislation: H.R. 3594: Background," http://olpa.od.nih.gov/legislation/108/pending legislation/humansub.asp (accessed June 6, 2007); Andrew Kessler, American Psychological Society, Director of Government Relations, *Protection Money: Human Subjects Research Legislation,* http://www.psychologicalscience.org/observer/1001/subjects.html (accessed June 6, 2007).

79. I. Wilmut, A. E. Schnieke, J. McWhir, A. J. Kind, and K. H. S. Campbell, "Viable Offspring Derived from Fetal and Adult Mammalian Cells," *Nature* 385, no. 6619 (February 27, 1997): 810; Tim Beardsley, "A Clone in Sheep's Clothing," *Scientific American.com* (March 3, 1997).

80. Research or therapeutic cloning involves SCNT; however, the derived cell is not allowed to develop into a living organism. Rather the cell, which for all intents and purposes is an embryo, is used for therapeutic purposes, for example, derivation of human embryonic stem cells. Using SCNT and research cloning, scientists will be able to create hESC that are immunologically compatible with individual patients; thus using these hESC therapeutically will not result in immuno-rejection.

81. A copy of the directive is viewable at http://grants .nih.gov/grants/policy/cloning_directive.htm (accessed June 6, 2007).

82. National Bioethics Advisory Commission, *Cloning Human Beings* (Rockville, MD: National Bioethics Advisory Commission, June 1997), http://65.205.1.226/cloning/cloning _report.html (accessed June 6, 2007).

83. Committee on Science, Engineering, and Public Policy and Board on Life Sciences, *Scientific and Medical Aspects.*

84. As noted in chapter 9, this has not stopped some states from taking action.

85. Shirley Tilghman and David Baltimore, "Therapeutic Cloning Is Good for America," *Wall Street Journal,* February 26, 2003.

86. The President's Council on Bioethics, *Reproduction and Responsibility: The Regulation of New Biotechnologies* (Washington, DC, March 2004), http://www.bioethics.gov/reports/reproductionandresponsibility/ (accessed June 6, 2007); see chapter 10. When President George W. Bush took office, he renamed NBAL the President's Council on Bioethics and appointed new people to serve on the advisory committee.

87. Gregory M. Lamb, "In Cloning Debate, a Compromise," *Christian Science Monitor,* Sci/Tech section, April 8, 2004.

88. For a general overview of the science, politics, and ethics of human embryonic stem cells see Scott, *Stem Cell Now;* Russell Korobkin with Stephen R. Munzer, *Stem Cell Century: Law and Policy for a Breakthrough Technology* (New Haven: Yale University Press, 2007); and Eve Herold, *Stem Cell Wars: Inside Stories from the Frontlines* (New York: Palgrave Macmillan, 2006).

89. A *chimera* as used here is an organism made with cells from two different species. For a discussion on ethical issues with chimeras see Henry T. Greely, "Defining Chimeras . . . and Chimeric Concerns," *American Journal of Bioethics* 3, no. 3 (2003): 17–20.

90. David C. Magnus and Mildred K. Cho, "Issues in Oocyte Donation for Stem Cell Research," *Science* 308, no. 5729 (June 17, 2005): 1747–48.

91. In 1995, an amendment banning the creation or destruction of embryos for research purposes was included on the appropriations bill for DHHS. This amendment, known as the Dickey-Wicker Amendment for its authors Congressmen Jay Dickey (R-AK) and Roger Wicker (R-MS), has been included on the DHHS appropriations bill ever since. The ban specifically prevents the use of federal funds for any research on a human embryo, unless that research will directly benefit the embryo. Kyla Dunn, "The Politics of Stem Cells," *NOVA Science Now,* April 13, 2005, http://www.pbs.org/wgbh/nova/sciencenow/dispatches/050413.html (accessed May 5, 2007).

92. Senate Appropriations Subcommittee on Labor, Health, and Human Services, Education and Related Agencies, *Statement on Stem Cell Research by Harold Varmus,* Director, National Institutes of Health, U.S. Department of Health and Human Services, January 26, 1999, http://www.hhs.gov/asl/testify/t990126a.html (accessed June 6, 2007).

93. The value of adult stem cells versus embryonic stem cells has been a matter of significant scientific and political controversy. See, for example, Sabin Russell, "'Adult' Stem Cells Could Skirt Embryo's Ethical Dilemma," *San Francisco Chronicle,* June 25, 2005, A1; Michael Fumento, "Stem-Cell Political Science: Natures Agenda," *National Review Online,* March 28, 2002, http://www.nationalreview.com/comment/comment-fumento032802.asp (accessed May 22, 2007); Sylvia Pagan Westphal, "Adult Stem Cell Promise May Be Deceptive," *NewScientist.com,* March 13, 2002, http://www.newscientist.com/article.ns?id=dn2041 (accessed May 21, 2007).

94. Junying Yu, Maxim A. Vodyanik, Kim Smuga-Otto, Jessica Antosiewicz-Bourget, Jennifer L. Frane, Shulan Tian, Jeff Nie, Gudrun A. Jonsdottir, Victor Ruotti, Ron Stewart,

Igor I. Slukvin, and James A. Thomson, "Induced Pluripotent Stem Cell Lines"; Kazutohsi Takahashi, Koji Tanabe, Mari Ohnuki, Megumi Narita, Tomoko Ichisaka, Kiichiro Tomoda, and Shinya Yamanaka, "Induction of Pluripotent Stem Cells"; Christopher Scott, "The Six Degrees of Stem Cell Research"; and Susan Solomon and Zach Hall, "Game Over? No Way."

95. Christopher Thomas Scott and Jennifer B. McCormick, "Stem Cell Science Losing Ground," *Boston Globe,* April 18, 2006; Owen-Smith and McCormick, "International Gap."

96. National Academies Board on Life Sciences and Board on Health Sciences Policy, *Guidelines for Human Embryonic Stem Cell Research* (Washington, DC: National Academies Press, 2005), 5–13.

97. Alberts and Shine, "Scientists and Integrity," 1660–61.

98. An interesting approach being piloted at several universities, including Stanford University and Case Western Reserve University, is a research ethics consultation service or a bench-side ethics consultation service. As science progresses there will be more and more societal and ethical considerations related to research; a resource for scientists to turn to for help in addressing societal and ethical issues will be beneficial. These consultation services are designed to focus on broader societal issues but also provide advice regarding, for example, informed consent documents, recruiting human subjects, and designing experiments implanting human neural tissue into mice embryos. A major goal of these consultation services is to help scientists incorporate societal and ethical considerations into the research process—*not* to "police" science. See Helen Pilcher, "Bioethics: Dial 'E' for Ethics," *Nature* 440, no. 7088 (2006): 1104–5.

99. See Sally Smith Hughes, "Making Dollars Out of DNA: The First Major Patent in Biotechnology and the Commercialization of Molecular Biology, 1974–1980," *Isis* 92, no. 3 (September 2001): 541–75.

CHAPTER 15

Science, Technology, Engineering, and Mathematics Education

What Is STEM Education?

Science, Technology, Engineering, and Mathematics (STEM) education encompasses the curriculum, instructional materials, and delivery approaches used to train students in these fields. It spans the distance from kindergarten classrooms through graduate and continuing education—a vast system commonly referred to as the "STEM pipeline." Recent discussions about STEM education have largely been focused on instructional quality, prospective compensation for STEM workers, and access to continuing education.

STEM education also reaches beyond the classroom to include journalism, museum collections, and community outreach. Beyond teaching the details of science and the scientific method, science education requires fostering a sense of curiosity, imagination, and wonder, among both children and adults. To be successful, it must involve not just teachers and students, but parents, school administrators, school board members, and, increasingly, practicing scientists.

For their part, American universities have a long history of integrating education, research, and scholarship, especially at the graduate level. But debates have roiled over recent decades about whether universities pay enough attention to their undergraduate teaching, and whether future generations of scientists are being given the training necessary to succeed.

If one accepts the premise that all members of a democratic society should be imbued with a basic understanding of their environment, then science education must be one of the cornerstones of democracy, and should be an essential element in our educational curricula. The growing importance of science and technology to the nation's economic strength, social welfare, public health, defense, and homeland security makes a technologically skilled workforce a critical national priority.

In this chapter we explore the state of STEM education in the United States and its relationship to key policy issues.

U.S. Science Education: A Brief History

The Russian launch of *Sputnik* in 1957 captured the attention of Americans like no other event of its time. It convinced our nation of the need to greatly enhance support for science education at all levels to ensure that we would never again be caught so far behind in such a critical area to our national well being. One of the best-known by-products of this newfound conviction was the National Defense Education Act (NDEA) of 1958, a program designed to make the United States more competitive with the Soviet Union in science and technology. The act concerned itself with science, mathematics, and foreign language instruction at all levels, from elementary through graduate school and vocational training.[1] It also provided limited funding for work in area studies, geography, English as a second language, counseling and guidance, and school libraries and librarianship, and for the support of educational media centers.

Congressional approval of the NDEA marked a major shift in the federal government's educational policy. Federal authorities had traditionally avoided involvement in educational issues, wary of infringing on state autonomy and the appearance of a return to top-down New Deal politics. Sensitivities about the *Brown vs. Board of Education* case may have also contributed to a desire to further blur the separation between state and federal prerogatives. Fears about *Sputnik* overcame these hesitations, resulting in significant increases in federal support to industry and academia, for both defense and nondefense research.

During the period from 1957 to 1961, federal R&D support increased by over 100 percent. At the same time, the federal government's support of basic research more than tripled. A significant share of these new funds was designated for new scientific research equipment, low-interest college loans, and improvements in elementary and secondary science education.[2] Though the initial focus was on science and mathematics, NDEA ultimately supported almost all fields. It is interesting to note that many of today's U.S. scientists and engineers are a direct product of this push to attract students into the sciences; absent that effort, the nation would be in much poorer shape today.

Sputnik forced the nation to carefully consider its approach to science education.[3] Should we focus on the child or the discipline? On building research capacity or fostering broad scientific literacy? Who should coordinate American science education, localities or scholars? Many of these questions have yet to be decisively answered. Indeed, many observers believe that the momentum sparked by *Sputnik* has been allowed to dissipate, leading to calls for new initiatives comparable to those of fifty years ago.

Building a Pyramid of Knowledge

While the STEM education system can be thought of as a pipeline, the process of STEM learning might be thought of as a pyramid. As individuals move from childhood to adult life, they build layers of knowledge, one on top of another, that eventually add up to a basic understanding of nature and the complex systems that govern their lives. This pyramid of knowledge is one of the most distinctive features of science. For example, because a good grasp of basic mathematics is essential in all areas of science, the simple absence of a key algebra course from a student's high school curriculum could divert him or her from a scientific career. In many other disciplines sufficiently gifted students may skip any number of intermediate steps on their way to the top of the pyramid. That is hard to do in the sciences.

Much effort has been invested in defining a properly designed STEM education system, and even which skills a student should possess at the end of each precollege class level.[4] We describe these goals for elementary, secondary, undergraduate, and graduate education in very general terms later in this chapter, and analyze the nation's progress toward these goals.

At the elementary and secondary levels, students should begin to learn about space, time, and causality. They should develop an appreciation for the basic principles of such fields as astronomy, biology, chemistry, and physics, including a sense of numbers and an understanding of the scale of physical phenomena. Students should also appreciate that the physical laws that define nature's behavior at a microscopic scale also explain the behavior of systems at an astronomical scale. Based on such broad considerations, content standards have been developed by the National Research Council and widely promulgated in a publication called the *National Science Education Standards*.[5] According to the *Standards*, science curricula should cover

- Unifying concepts and processes in science
- Science as inquiry
- Physical science
- Life science
- Earth and space science
- Science and technology
- Science in personal and social perspectives
- The history and nature of science

Students' expected progress in each of these areas is further broken down by grade ranges.

Undergraduate students should study these concepts in considerably greater depth, appreciating the relationships among areas of science and the extent of current knowledge. Those who intend to pursue careers based on scientific and technical knowledge should also receive in-depth training in one or more of the basic scientific disciplines.

While college students should graduate with a general knowledge of the scientific disciplines, graduate students should familiarize themselves with the research in their field and embark on the long march toward the frontiers of their disciplines.

Science is not static. A discovery tomorrow may invalidate the dogma of decades, or even centuries. It is therefore imperative that our science education system provide continuing education to professionals in individual fields, and to society as a whole. There is a role in this process not only for formal academic programs, but for museums, magazines, television, and the World Wide Web.

The State of the STEM Education Pipeline in the United States

Many analysts believe that the STEM education pipeline is broken. In the following pages we will attempt to identify some of the major problems, and outline possible solutions.

Each stage in STEM education (e.g., K–12, undergraduate, graduate) has its own issues. Each is administered differently, and has its unique goals. Any careful analysis should treat each level separately, while also examining their interconnections.

At the Elementary and Secondary Level

K–12 STEM education is critical to ensuring a skilled American workforce and an educated citizenry. There are over 53 million school-age children in the United States, any of whom could learn to love or hate science.[6] It is essential that those who are actually interested in science not be discouraged from pursuing S&T careers by a lack of the necessary training.

To be successful at the precollege level STEM training requires motivated and trained teachers, well-designed curricula, and access to modern instructional tools. In an ideal world these ingredients would be present in every school system in the country. But in the real world there are many problems that call for a reexamination of current K–12 STEM education policies.

A diverse mix of players is involved in helping to inform and to craft STEM education and related public policies: universities and federal agencies often conduct research on new classroom techniques; state and national entities set curricular standards; states and localities generally choose the specific textbooks that will be used; and textbook publishers shape curriculum and content by designing their products to specifically appeal to meeting the needs of schools in states with well-defined standards.[7]

Elementary and secondary schools themselves take several forms, with most being public but some privately run. Many privately run elementary and secondary schools are supported by religious organizations. These schools are commonly referred to as *parochial schools*. All states also allow homeschooling. The tremendous diversity in our K–12 system poses significant challenges when it comes to crafting effective and uniform STEM educational policies. Such variations from one school district to another make it particularly difficult for the federal government to set national science, mathematics, and other S&E-related educational standards and to ensure quality in the textbooks, curricula, and teacher preparation requirements at the K–12 level.

The U.S. Constitution is not explicit on who should bear responsibility for the American educational system, and so by default oversight is left to the states and localities.[8] This means that most major decisions with regard to K–12 STEM education are made by local school boards and local and state officials. Funding for K–12 public education is also left primarily to localities, which means the quality of schools can vary greatly depending upon their geographic location and the associated economic health of the community in which a school resides.

When it comes to STEM education, students from low-income families are often doubly disadvantaged, first by the fact that their parents are not scientifically literate and second by the fact that they go to schools where the quality of STEM education is not high and available resources are often lacking. With parental pressure so often a driving force in local education policy, districts that are home to a significant proportion of scientifically minded parents are likely to pay more attention to the quality of their science education programs than other districts with less parental interest and fewer monetary resources. Solutions proposed by policymakers to address such disparities often focus around trying to provide K–12 students and their parents with more educational choices, sometimes in the form of charter schools or through school voucher programs that enable parents to send their students to schools located outside of their immediate neighborhoods. Other programs attempt to address these disparities by investing federal dollars directly into improving K–12 science and math programs in specific school districts.

In recent years, concern has arisen that American students are underperforming in science and math. This concern has been fueled by results from many national assessment tests, including the National Assessment of Educational Progress (NAEP). NAEP is the only nationally representative and continuing assessment of what America's students know in science, math, and other subjects. The NAEP governing board, the National Assessment Governing Board, is appointed by the secretary of education.[9] The NAEP assessment structure has been in place for over thirty years.

NAEP results from a 2005 assessment suggest that there is still work to be done to increase science proficiency, especially in high school: while NAEP proficiency scores had increased for fourth graders from 1996, they remained the same for eighth graders, while the achievement scores for twelfth-grade students declined by three percentage points.[10]

At the same time that concerns have grown about student proficiency in science and math, there also appears to have been declining student interest in majoring in these fields. From 1995 to 2005, the number of high school students who took the ACT college admissions test who said they were interested in majoring in engineering dropped from 7.6 percent to 4.9 percent. A similar drop—from 4.5 percent to 2.9 percent—occurred from 2000 to 2005 among ACT-tested students that expressed an interest in computer science.[11]

The underperformance of U.S. students in STEM fields at the K–12 level is often attributed in part to the fact that many teachers of STEM subjects at the K–12 level do not themselves have adequate STEM training. The NSF has estimated that on average 23 to 29 percent of public middle and high school mathematics and science teachers did not major or minor in the field they now teach.[12] This percentage is even greater in schools with significant minority enrollments, where it has been estimated that students have less than a 50 percent chance of studying with a science or math teacher who has a degree in that specific field.[13]

Many have argued that the only way to increase the number of qualified mathematics and science teachers is to pay them a higher, differential salary, especially since the inflation-adjusted salaries of public school teachers increased only slightly between 1972 and 2002.[14] But the differential approach has run into stiff opposition from teachers' unions, who argue that *all* teachers are underpaid and that any increase in salaries should be shared among the entire membership of the teaching profession, not just teachers in a single discipline.[15] This debate could be crucial as the nation works to improve student preparedness in science and mathematics at the K–12 level.

Undergraduate Level

The United States has more than four thousand degree-granting institutions of higher education, which employ approximately 1.2 million faculty and enroll roughly 17 million students.[16] At the undergraduate level, these institutions should ideally be teaching basic science literacy to students from every discipline, ranging from the humanities to engineering. At the same time, they are supposed to be training the STEM K–12 teachers who will go on to educate the next generation. Colleges and universities also have a vested interest in encouraging talented students to enter the STEM pipeline, pursuing advanced degrees and careers in S&T.

In recent decades, however, universities' undergraduate science education programs have become the targets of significant criticism. The Higher Education Research Institute at the University of California at Los Angeles has surveyed incoming freshmen about their intended major every year since 1972. In 2004, approximately one-third of white, black, Hispanic, and American Indian/Alaska Native freshmen reported they planned to major in science or engineering. Forty-three percent of Asian/Pacific Islanders intended to major in science or engineering major.[17] While these figures sound good, other studies have found that a significant number of the incoming freshmen who intend to major in a STEM field end up changing their majors to a non-STEM field by the time they graduate.[18] One 2001 study found that, of those students who had declared S&E majors as incoming freshmen at 175 colleges and universities in 1993, fewer than half had actually earned a S&E degree six years later. This same study found that women and underrepresented minorities left S&E degree programs at a higher frequency than their male and nonminority counterparts.[19] The number of four-year graduates with S&E degrees from American colleges and universities did generally increase between 1977 and 2002, but this statistic includes both citizens and noncitizens, and also masks variation across disciplines. For example, through the 1990s, the number of baccalaureates awarded in the physical sciences decreased, although it increased in the biological and agricultural sciences (see fig. 15.1).[20]

Yet another study, conducted by a special National Science Board Task Force in 1986 and chaired by one of the authors of this text—Homer Neal—found that many institutions were giving very low priority to their instructional missions; were using dysfunctional laboratory equipment; and accorded minimal recognition to outstanding teaching.[21]

The task force held a yearlong set of hearings in Washington, aimed at responding to the apparent crisis. The panel heard testimony from educational, industrial, and military leaders, federal agencies, teacher associations, the National Academies, and Congress, and raised grave concerns about whether America's graduating students were being adequately prepared in STEM fields.

The task force's report prompted action. The NSF created a new Office of Undergraduate Education, whose programs would fund undergraduate research experiences and curriculum development. A new Research Experience for Undergraduates (REU) program was created that permitted undergraduate science students to work directly with faculty on research projects during the summer, with stipends provided by the NSF. New programs were also created to provide research experiences for teachers, to upgrade instructional laboratory instrumentation, and

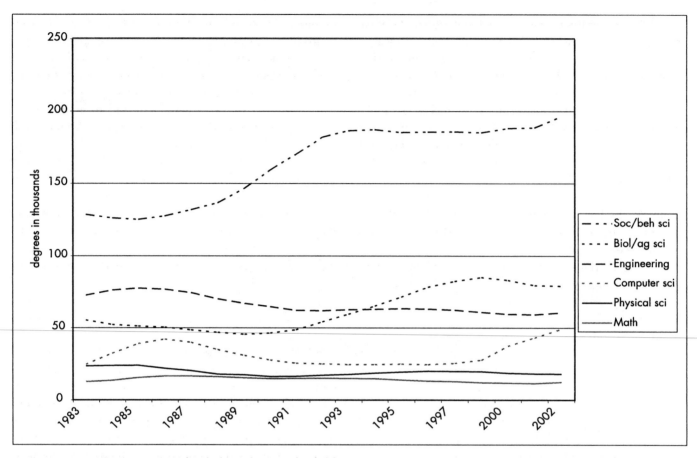

FIG. 15.1   S&E bachelor's degrees, by field: 1983–2002. Physical sciences include earth, atmospheric, and ocean sciences. Data not available for 1999. (Data from U.S. Department of Education, National Center for Education Statistics, Integrated Postsecondary Education Data System, *Completions Survey;* and National Science Foundation, Division of Science Resources Statistics, WebCASPAR database, http://webcaspar.nsf.gov; and National Science Foundation, *Science and Engineering Indicators 2006,* chap. 2, fig. 10 and app. table 2-26.)

to fund curriculum development. These and other new initiatives launched in the wake of the NSB study are currently funded at over $200 million annually.[22]

Subsequent NSB panels have called for the continuation and expansion of these undergraduate-oriented activities. One of these subsequent studies, which led to a report entitled *Shaping the Future,* was carried out by the advisory committee of the National Science Foundation's Education and Human Resources Directorate. The directorate's goal was to "consider the needs of all undergraduates attending all types of U.S. two- and four-year colleges and universities," addressing "issues of preparation of K–12 teachers in these fields, the needs of persons going into the technical work force, the preparation of majors in these areas, and the issue of science literacy for all."[23] The report's executive summary declares, "The goal—indeed, the imperative—deriving from our review is that: All students have access to supportive, excellent undergraduate educa-

tion in science, mathematics, engineering, and technology, and all students learn these subjects by direct experience with the methods and processes of inquiry."[24]

Despite its recommendations, the committee's report did not dispel criticism concerning the failure of large research universities to emphasize undergraduate education. In a 1998 report entitled *Reinventing Undergraduate Education: A Blueprint for America's Research Universities,* a national commission created under the auspices of the Carnegie Foundation for the Advancement of Teaching, commonly referred to as the Boyer Commission on Educating Undergraduates in the Research University, called for a new model of undergraduate education, suggesting that universities "take advantage of the immense resources of their graduate and research programs to strengthen the quality of undergraduate education."[25] The commission's report stressed the need for students to know how to ask questions, find answers, analyze and solve problems, conduct oral and written research, and communicate. It also urged universities to provide increased opportunities for inquiry-based and interdisciplinary learning, make better use of information technology, educate students to become teachers, and reward faculty excellence in teaching.

There is evidence of continuing cause for concern about the priority that universities assign to STEM education. Classes certainly need to be made more interesting. Perhaps some courses should be made more interdisciplinary—by combining science with management, policy, and writing—and thus more relevant to real-world problems. Too many courses are being taught by teaching assistants with poor command of English. Grading practices should be reassessed, with an emphasis on whether the strict grading curve used in many science classes is really appropriate.

Too many bright students bypass science careers because the perceived "hassle" seems so large. Focused attention needs to be directed toward retaining minorities and women in science careers, and in ensuring an adequate number of STEM majors to meet the nation's future needs.

Graduate and Postgraduate Levels

American graduate and postgraduate education has long been the envy of the world. Countless countries send their brightest students to the United States for their postbaccalaureate education, and even commit significant funds to support these students. It is instructive to examine why opportunities to study in the United States are so attractive. The answer seems to be found in the extraordinary chance students in America have to work directly with outstanding faculty on frontier research, often on a daily basis; the access they have to advanced research facilities; and the availability of research assistantships to pay for many students' education. Indeed, any comprehensive ranking of the world's universities will reveal that most of the top institutions are located in the United States.[26] Comments in what follows about the U.S. graduate and postgraduate educational programs should therefore be read in the context of a hugely successful system, subjected in recent decades to stresses that may threaten its future effectiveness.

Graduate Education

Graduate schools typically award S&E students either a master of science or doctorate degree. The number of S&E master's degrees awarded by U.S. institutions increased from 67,700 in 1983 to 99,200 in 2002, with the number peaking in 1995 and then leveling off until 2000, when it again increased. The four most popular fields among these students are engineering, social sciences, psychology, and computer sciences, with this latter field accounting for substantial growth in master's degrees awarded in recent years.[27]

The number of S&E doctorates awarded by U.S. universities and colleges similarly peaked, at about 27,300 in 1998, and then declined to 24,600 by 2002, although data collected from 2003 indicated an uptick in the number of S&E PhDs awarded (fig. 15.2).[28] Some see this profile as cause for concern. What does it suggest about the appeal of science careers, relative to other professions? Are more students opting for careers in business, law, and medicine, where the time to employment is quicker, and the entry-level salaries are higher?

Student-Teachers

Graduate students play a complex and unique role in American science programs. On the one hand, they are the students; on the other, they are teachers and researchers. This is because large research universities depend on graduate students to provide much of their undergraduate instruction and to help conduct research. Roughly speaking, the amount of research a faculty member is able to perform is directly proportional to the number of graduate students he or she employs.

Most academic departments must maintain a delicate balance between the number of undergraduate courses they offer, the number of graduate students they maintain, and the number of faculty members and research projects they support.[29] Graduate student instructors are generally funded in their first two years of study by revenues from undergraduate tuition and other local sources. Support in subsequent years more often comes from research grants, most from federal agencies such as the NSF and NIH. In 2003 alone, these two agencies funded approximately 19,300 and 24,300 graduate students, respectively, across the United States—representing most of the full-time S&E graduate students receiving federal support. Between 1983 and 2003, the percentage of graduate students supported by NIH grants increased from 23 percent to 30 percent, while those supported by the NSF increased from 20 to 24 percent.[30]

Many members of Congress do not fully understand how critical federal support is to graduate education in the sciences. One sign of their detachment from the issue can be seen in the career backgrounds of the 535 members

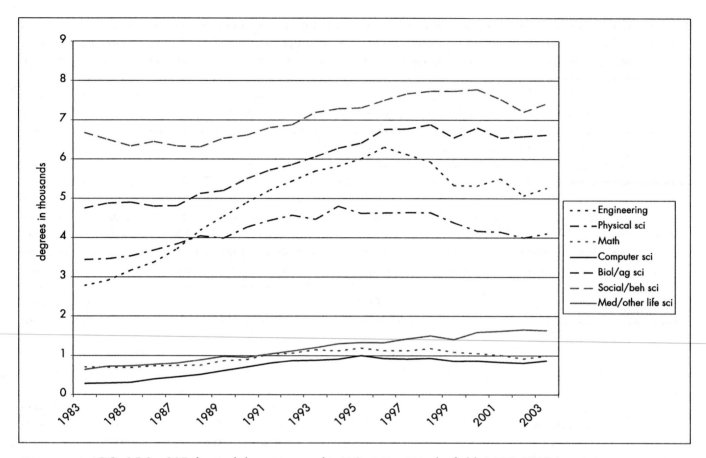

FIG. 15.2   S&E doctoral degrees earned in U.S. universities, by field, 1983–2003.
Physical sciences include earth, atmospheric, and ocean sciences. (National Science Foundation, Division of Science Resources Statistics, Survey of Earned Doctorates, WebCASPAR database, http://webcaspar.nsf.gov; and National Science Foundation, *Science and Engineering Indicators 2006*, chap. 2, fig. 16; and app. table 2-30.)

of the 109th Congress, which included only two physics PhDs, two chemistry PhDs, and a handful of others with professional experience or graduate training in the sciences. The vast majority of members of Congress graduated from business or law schools, where it is customary to finance advanced degrees with student loans.[31] This is a notable distinction between the sciences and other disciplines, and policymakers need to be made more fully aware of the impact of their budgeting decisions on the future of scientists-in-training. Furthermore, the tight coupling of graduate student support to undergraduate education has profound implications for universities' ability to carry out their educational mission: cutbacks in federal support may force departments to significantly reduce the number of graduate students they support, which in turn might mean cutting back their undergraduate enrollment.

There are many concerns about science education policy at the graduate level, including our reliance on foreign graduate students in STEM fields (approximately one-third of total doctorates awarded in 2003 went to non–U.S. citizens); the use of graduate students as employees; and the length of time to degree and degree completion rates.[32] A larger debate is also being waged over the purpose of graduate education itself. It has been argued that the American system may be too oriented toward producing PhDs who aspire to academic research careers, rather than entering the private sector. If this is the case, critics claim, then the academic system may be overproducing science PhDs. This debate will require that academics and policymakers alike assess the goals of STEM education.[33]

Postgraduate Education

No one would expect a medical student to accept his or her degree one day and undertake solo brain surgery the next. New graduates spend two further years in residency before being fully admitted to the medical profession. Similarly, new science and engineering PhDs are expected to work alongside more experienced scientists or engineers before launching their own careers. These postdoctoral apprenticeships are generally spent at institutions other than the ones where these individuals earned their PhDs. Most are supported by the federal government, although private foundations and industry do provide funding in some instances. The work itself may take place at universities, federal labs, or industry R&D facilities.

Postdocs, as they are called, are essential to the production of well-rounded scientists and engineers. In 2001, it was estimated that there were forty-three thousand postdocs, approximately thirty thousand of them in the biological, medical, or other life sciences.[34] If, for some reason, the number or quality of postdocs is compromised, fewer well-prepared researchers will be available to the workforce. That said, there is growing concern that postdocs, because there are few permanent jobs available, are not moving into the workforce to become independent researchers at the optimal rate, but are instead having to remain in the laboratories of established faculty members, where they are tapped as cheap labor, often paid at a rate equal to or less than that received by graduate students (taking into account the latter's standard tuition waivers). This is of particular concern in the life sciences, where postdocs typically continue for upwards of five years.

The plight of postdocs is a critical policy issue, especially because movement from the postdoctoral pool into permanent positions has become more and more difficult in many disciplines. Many postdocs question whether the long days of hard work, minimal pay, and minimal recognition are consistent with the goals of an internship. Some postdocs have begun organizing into advocacy groups.[35] Major entities including the American Association for the Advancement of Science and the Alfred P. Sloan Foundation have also taken an interest in the status of postdoctoral researchers, and now cosponsor the National Postdoctoral Association.[36] It is clear that universities and policymakers will have to pay greater attention to the welfare of postdocs in the years to come.

Continuing Education

Even the most successful scientists and engineers have to update their training from time to time. In medicine, for example, new surgical techniques are developed seemingly on a monthly basis. Surgeons who fail to keep up suffer professionally. And the same is true in every field from ro-

botics to elementary education. How can busy individuals keep up with the rapid pace of scientific development?

Although continuing education is not a central component in the mission of most universities—especially the large research universities—more and more academic institutions are anticipating the need to meet this challenge. Interestingly, community colleges have been at the forefront in developing continuing education programs aimed at meeting the very specialized needs of their local communities.

The need for enhanced continuing STEM education will only grow as more STEM professionals need to keep abreast of technological change. As pointed out by a 1997 National Academies report, *Preparing for the 21st Century: The Education Imperative,* government at all levels must work with postsecondary institutions to "provide coherent and high-quality training opportunities for people at various stages in their working lives."[37] Policy, for its part, will need to encourage this kind of foresight.

Scientific Outreach Programs

Universities have often been characterized as ivory towers. This image is particularly troublesome given that most of the alleged "towers" only exist thanks to public generosity. Many schools have begun trying to fight this image, working with communities to address environmental issues, offer community-based educational programs, or host public forums. Such connections have proven beneficial for both the communities and the universities themselves.

Many large university programs, labs, and research centers are also now required by their funding agencies to include outreach components in their grant proposals, literally bringing citizens into the process of project design and implementation (if only as observers). For example, had the United States decided to proceed with the construction of a supercollider—a project that would have taken ten years or more—grade schools across the country would have been able to observe the progress from prototype to full-blown design via the Internet and other educational tools. Had the project been completed, high school students would have been able to see the data it produced and even participate in its analysis. Big-science projects are arguably well suited to outreach, and could allow millions of students to directly benefit from cutting-edge scientific research, while also allowing the general public to appreciate how science is done, share in the excitement of frontier discoveries, and see for themselves how public money is being spent.

Industry also has a significant interest in outreach. For example, classroom visits from the employees of private firms can teach students about STEM career opportunities and stimulate learning about science and engineering. Tours of large industrial facilities, such as automobile simulators or virtual reality labs, can excite young minds. And summer internships can have a lasting impact on high school and undergraduate students' plans for the future.

Professional societies, national laboratories, and the federal agencies themselves have also recognized the importance of outreach. Many have designed their Web sites to attract student interest and provide teachers with teaching materials.

Museums

Contrary to the stereotype of science as being all about theory and abstraction, it is really about observation, reflection, and interpretation. When well organized, science museums allow children and adults to observe natural phenomena in context, provoking greater reflection and understanding than textbooks and articles do.

Museums also support specialized staff who can present scientific ideas in a clear and engaging way. Large museums usually employ specialists who create public displays; some facilities, such as the Smithsonian, also employ staff to conduct frontier research. This mix enables them to demonstrate scientific concepts in historical context while also conveying some of the excitement of ongoing research.

The Smithsonian Institution, San Francisco's Exploratorium, and Chicago's Shedd Aquarium, among others, have been recognized over the years for their contributions to science education. In 2005, the Association of Science-Technology Centers estimated that 74.9 million people, including 17.4 million schoolchildren, were served by over 345 such science centers and museums around the country. According to the association, a majority of these institutions offer supplementary science education programs, including school outreach (86 percent), professional development programs for teachers (78.5 percent) and science camps (68.3 percent).[38]

Any strategy to improve American science education, then, must acknowledge the role of science museums. Indeed, public funding for museums should be a top priority: when properly managed, these museums can significantly contribute to science education at all levels, from kindergarten through graduate school and beyond. The O. Orkin Insect Zoo at the Smithsonian Institution, for example, has been enormously successful in teaching the public about the insect world, and provides information unavailable to them anywhere else.[39] Special exhibits from this zoo travel across the country and offer unmatched opportunities to communities far from Washington.

The Media

American children between the ages of two and seventeen average an estimated twenty-five hours of television viewing per week. One in five watch more than thirty-five hours a week.[40] And by eighteen, the average American teenager will have spent more time in front of a television—twenty-five thousand hours—than learning in the classroom.[41] Naturally, this suggests that American children obtain a great deal of their knowledge about the world from TV: it has been estimated that 60 percent of their information comes from that source.[42]

It is clear that television has the potential to be a hugely influential medium for educating students during their formative years, and for providing science outreach to older teens and adults. Of course, TV is often treated as an escapist entertainment, making it challenging to market truly educational programming.[43] But successful programs such as *Nova* and *National Geographic Explorer* and even whole networks like the Discovery Channel and Animal Planet have demonstrated the power of the media in delivering scientific information to the general public.[44] Television dramas and police procedurals like the *CSI* series can also have a dramatic positive effect on how students, especially female students, perceive scientists and science careers.[45]

Federal agencies have long recognized TV's educational potential; indeed, many recent advances in packaging scientific information for media delivery have come about as a result of federal grants (e.g., from the NSF) provided specifically for this purpose. Targeted federal support is a wise investment of public funds, although such activities should be carefully and regularly evaluated for their effectiveness.

Magazines and movies can be similarly effective in dramatizing the awe of discovery and publicizing important breakthroughs. We need more writers capable of rendering complex scientific principles in accessible and interesting prose. Indeed, science writing should be encouraged and supported at the university level, since science writers are an important bridge between the scientific community and the public at large. And movies, too, can tantalize, offering a glimpse into the future, and motivating young people to consider a career in science. Space travel, for example, was portrayed in films many decades before it was achieved.

While the media can be a tremendous positive force in science education, the same characteristics that give it its potential for doing good also make it capable of doing real harm. Indeed, movies and television shows can and often do cause considerable damage, for instance by portraying science in a negative light or representing scientists as inherently odd, peculiar, or even deranged (e.g., the mad scientist caricature).[46]

U.S. STEM Performance: An International Comparison

The United States has historically championed new ideas, bold initiatives, and technological prowess. One would therefore expect us to have one of the most advanced systems of science education in the world. However, the evidence seriously calls this assumption into question, particularly in the area of precollege science and math performance.

Since 1995, the United States has participated in several international tests that compare our students' science and math performance to that of students from other countries. The 1995 Third International Mathematics and Science Study (TIMSS), for example, compiled the scores of fourth-, eighth-, and twelfth-grade students from forty-one countries. The test was repeated (under the name TIMSS-R) five years later, in thirty-eight countries, this time focusing on eighth-graders. The TIMSS is now called the Trends in International Mathematics and Science Study and when given in 2003 was administered to fourth and eighth-grade students.[47] Also in 2000 the Organisation for Economic Cooperation and Development (OECD) organized the Program for International Student Assessment (PISA) to assess the reading, math, and science performance of fifteen-year-olds in thirty-two countries.

In the 2003 TIMSS assessment, U.S. fourth and eighth-grade students scored above the TIMSS international averages for both the mathematics and science exams. U.S. fourth-graders outperformed their counterparts in Italy, but scored lower than fourth-graders in England, Japan, Russia, and Belgium. Eighth-graders performed below those in Japan, Netherlands, South Korea, and Belgium. U.S. students compared similarly in the science component of the assessment.[48] While U.S. students performed above the international averages, their ranking relative to other developed nations and emerging global S&T leaders raises serious questions about the adequacy of our current science education system, and suggests that the United States could learn much from studying other countries' cultural norms, valuation of science and math education, and efforts to ensure broad-based scientific literacy.

On the OECD-conducted PISA test, which presents questions based in real-world context testing science and math literacy, American students' average performance in 2003 was below the OECD average in both math and science. Twenty other countries, including Canada, France, Germany, South Korea, and Japan, had higher average scores in math, while fifteen other countries had higher scores in science than U.S. fifteen-year-olds. The scores of U.S. students did not change between 2000 (when the test was first given) and 2003; however, the scores of other countries did improve.[49]

PISA also collected data on parents' educational levels and occupations. Notably, no country showed a stronger relationship between parental occupation and student math and science scores than the United States.[50] Students whose parents had a college degree outperformed those whose parents did not have a high school diploma.

So what are we doing differently in the United States? What might explain our poor comparative showing in international science and math education? We still don't know, despite the fact that improvement in these areas has long been a presidential priority (e.g., under President Clinton's "Goals 2000: Educate America" and George W. Bush's "No Child Left Behind" initiative).[51] Instead, the United States continues to trail many other nations in student performance.

Many issues will have to be addressed before we can improve our country's international rankings and address shortcomings in American K–12 science and math education. For example, American schools have been found to assign more homework than their foreign counterparts.[52] International comparisons suggest that the large amount of homework assigned to American students may actually have a deleterious impact upon their ability to learn: the seeming overemphasis on homework may be a result of teachers trying to prep their students for the new regime of standardized tests. Such short, multiple-choice, "plug and chug"–type assignments do little, however, to help students think critically. Unless the minutiae of a particular assignment are linked to the larger picture, homework lessons become a mere listing of facts, isolated from the complex ideas they are supposed to help students understand.

There are other areas that demand our attention as well: the kind and quality of textbooks used, teacher training levels, the esteem accorded to students who are talented in math and science, and parental attitudes, among others. All affect student performance. Despite our concerns about K–12 math and science education, however, it is interesting to note that at least so far the nation has successfully met most of its needs for technological talent. If this is so, it is due in large measure to the effectiveness of the nation's higher education system, and our ability to import foreign talent. Indeed, the American system of graduate education is the envy of the world, with millions of students coming here to pursue advanced degrees and postdoctoral training. As we will see in chapter 18, however, the pressures of homeland security and increasingly restrictive immigration policies have called into question the wisdom of our reliance on foreign talent to support domestic STEM workforce needs.

Key Players in STEM Education Policy

Key players influence formal STEM education policy through a variety of means. These include leadership through persuasion, direct budget actions, and influencing those in positions to cause certain actions to occur. In what follows we review the roles of Congress and the president in influencing formal science education policy from these perspectives.

The Role of Congress

Congress can play a major role in reforming American science education, but it is not easy to focus congressional attention and interest on the issue, much less to reach agreement on what action should be taken, or which agencies should take it. Even when some kind of consensus is achieved, any number of further obstacles may make it difficult to pass and enact reform legislation. An enlightening lesson can be drawn from Congressman Vernon Ehlers's efforts to propose such legislation during the 106th Congress, and again during the 107th.

Ehlers's bill, the National Science Education Act (H.R. 4271), was intended to improve the quality of American science and math education. Among other things, it would have provided for "master teachers": outstanding instructors who could train and mentor their colleagues. The bill required "the NSF Director to award competitive grants to institutions of higher education to: (1) train master teachers who work to improve the teaching of mathematics and science from kindergarten through ninth grade in various ways, including mentoring other

teachers; and (2) assist elementary and secondary schools to design and implement master teacher programs."[53]

Congressman Ehlers had chaired a series of hearings on math and science education beginning in 1997. No serious objections were raised during the lengthy consideration process, although some participants wondered out loud whether the NSF was the best agency to oversee enforcement, or whether the Department of Education (DOEd) might be better suited for the role. H.R. 4271 was unanimously approved by the House Science Committee (36-0). Congressman Ralph Hall, ranking Democratic member of the committee, noted: "The programs established by [this bill] will address serious deficiencies in the preparation and professional development of science and math teachers. It will establish new partnerships between schools and businesses to encourage greater student interest in science and technology, and it will explore ways to employ educational technologies more effectively."[54]

But despite what appeared to be broad-based support—the bill had 110 cosponsors when it went to the floor—H.R. 4271 failed by a vote of 215-156. The primary reason for the bill's failure was a list of concerns raised by the National Education Association and other education organizations less than forty-eight hours before consideration. Many in the public education community disputed the constitutionality of a provision that allowed the "unprecedented" use of federal (i.e., NSF) funds to pay the salaries of teachers at private schools. The chairman of the House Science Committee, Congressman James Sensenbrenner, claimed that the vote against the bill was motivated by partisan politics (" the urge to play politics and to keep Congress from passing this education bill was too strong to overcome in an election year").[55]

To those accustomed to the NSF's merit-based funding competition process, the suggestion that the agency might provide funds to private as well as public schools may seem noncontroversial. The NSF makes such awards to private universities all of the time. However, the vote on Ehlers's bill was subsumed into a larger, very heated (and arguably unrelated) debate among elementary and secondary educators over the subsidization of private schools.

By the time Congressman Ehlers reintroduced his legislation in the 107th Congress, it had been modified to address these concerns. While the revised bill was not passed in its entirety, large portions, including the authorization of the training of master teachers, were ultimately incorporated into the NSF reauthorization bill signed into law by President George W. Bush in December 2002.

Congress can also influence American science education by providing special educational loans or offering loan forgiveness to students who choose to go into teaching. Or it can call the nation's attention to pressing educational issues by conducting studies or calling on other organizations to do so. Examples include the *Unlocking Our Future* study and the National Academies' *Rising Above the Gathering Storm* reports, which led to the creation of American Competitiveness Initiative.[56] Congress also has at its disposal the STEM Education Caucus. The caucus consists of member of Congress who share a desire to increase awareness of STEM education. Working closely with outside groups that also have strong interests in the issue, including the American Chemical Society, the Business Roundtable, and the National Association of Science Teachers, the caucus regularly sponsors educational briefings for members of Congress and their staff aimed at increasing legislators' understanding of important STEM policy issues. STEM education caucuses have recently been created in both the House and the Senate.[57]

The Importance of Presidential Leadership

The president also plays an influential role in science education policy. Almost every administration voices a profound commitment to improving education, and beyond the rhetoric many presidents demonstrate that commitment in practice. It may be instructive, by way of example, to consider the very different approaches taken by Presidents Clinton and George W. Bush to national educational reform.

In 1994, President Bill Clinton approved the reauthorization of the Elementary and Secondary Education Act and signed the Goals 2000: Educate America Act into law.[58] Goals 2000 established a framework within which states and local districts were expected to identify standards and measure student progress. It also provided funds that states could use to develop improvement plans, offer subgrants to school districts, and award funds for preservice and professional development.[59] Congress appropriated $105 million for the act in FY1994, and continued to fund various provisions until Goals 2000 was superceded by the No Child Left Behind Act in late 2001.[60]

Goals 2000 promised that by the year 2000 all children would start school ready to learn; the high school graduation rate would be increased to 90 percent; all students in grades 4, 8, and 12 would show competency in English, math, science, foreign languages, civics and government, economics, the arts, history, and geography; the nation's teachers would have access to high-quality professional development and training opportunities; U.S. students would rank first in the world in math and science; every adult American would be able to read and write and would have the knowledge and skills to be competitive in a knowledge- and information-based economy; U.S. schools would be drug- and violence-free; and all schools would have the capacity to stimulate parental involvement in all aspects of their children's development.[61]

The Elementary and Secondary Education Act of 1965 was amended at the same time, by the passage of the Improving America's Schools Act. Among its other provisions, the new act requested that states establish "challenging standards" against which student performance could be measured, and adopt annual "high-quality assessments" of student progress.[62]

The Clinton administration's reforms, while focused on standards and accountability, did not establish deadlines or deny funding to schools that did not make progress as expected. Rather, they provided the states with assistance for developing standards and assessment protocols, and gave them more power with regard to public education (i.e., by decreasing federal involvement).

But without deadlines or penalties, how "accountable" were schools really? In 1999, Texas governor and presidential candidate George W. Bush touted an education policy that he claimed would raise student standards and give parents more choice, while also enforcing some measure of accountability. His administration's No Child Left Behind (NCLB) Act, signed into law in January 2002, has been called the "most sweeping education reform legislation" since President Johnson's 1965 Elementary and Secondary Education Act.[63] (Technically speaking, Bush's measure is actually a reauthorization of the earlier law.) NCLB provides increased funding to schools that are failing to meet established student performance standards. The law also requires that states and local school districts set higher standards for all students, and assess student progress annually. Children in grades 3 through 8 are currently tested in math and reading, and testing in science is supposed to be added by 2007–8. High school students will also eventually be assessed in these same three areas.

The NCLB assessments are intended to make schools accountable: schools that "fail" to meet their goals two years in a row must accept outside assistance and, if still failing after two more years, are subject to "corrective action" (e.g., the replacement of specified staff). Five consecutive years of failure result in "significant penalties," which range from state takeover to the hiring of private management contractors to conversion to charter-school status.

The second key element of NCLB is parent choice: school districts are required to offer students at "failing" schools transportation to "high performing" peer institutions. NCLB also provides assistance to parents who wish to establish charter schools in their communities.[64]

No Child Left Behind takes standards and standardized testing to a new level. Schools and districts are required to make yearly progress, measured both in terms of total student population and with reference to four specific subgroups: economically disadvantaged students, racial/ethnic minority students, disabled students, and nonnative English speakers. Lack of satisfactory progress in any one of these subgroups, regardless of the success (or failure) of the student population as a whole, can lead to penalties. NCLB has been criticized for taking control away from states, for imposing unfunded mandates, and for pushing teachers to teach to the tests, at the expense of creativity and inquiry-based learning.

On the other hand, NCLB requires that teachers be "highly qualified" and certified. Teachers of core academic subjects must demonstrate competency in their fields. A high school biology teacher must be certified in biology, an American history teacher in history. This measure has created some problems for poor and rural school districts that have traditionally had difficulty attracting "high-quality" teachers, and where teachers often have to teach multiple subjects. In response to these criticisms and others, the Bush administration in 2004 relaxed the deadline for meeting NCLB's requirements.[65]

Education is one of the most important items of public interest, and most administrations attempt their own strategy to make progress in improving student achievement and teacher quality. Approaches used by the Clinton and Bush administrations were quite different, the former using persuasion to engage local cooperation, and the latter using a form of punishment of local districts that do not make progress. That neither approach achieved the success expected is one more example of the intractable nature of education. Nevertheless, efforts must continue to seek ways of improving education in the nation, because so much of the nation's future depends on its success.

Issues in the U.S. STEM Education System

American science education faces many challenges, some rooted in society's valuation of science, some in the details of how it is coordinated and funded, and some associated with perceived conflicts between science and religion. Hovering above all of these issues is the larger question of whether the nation is producing enough scientists and engineers to meet its current and future needs. We need to pay more attention to keeping students interested in science throughout their education, and creating a science- and math-literate workforce. Closer scrutiny of these issues is provided below.

Keeping Students in the "Pipeline"

As stated earlier, the long pathway from kindergarten to PhD has led some to characterize STEM education as a pipeline. Many if not most students enter the science education pipeline at an early age, but most exit somewhere along the way. Only a relative few reach the very end. Some leave to pursue nonscience interests or enter the job market. But others leave for more troubling reasons, including inadequate training. Too many students abandon their dreams of a career in science because of poor teaching, unimaginative curricula, or inadequate materials.

The journey along the pipeline is long and arduous, and any student needs encouragement to complete the trip. Even the simple knowledge that others from similar backgrounds have successfully navigated the course can be enormously sustaining. Unfortunately, because there are so few women and minorities in certain areas of science, many underrepresented students may have almost no role models to emulate.[66]

Ensuring a Balanced and Scientifically Literate Workforce

Not every student is suited to pursue an advanced degree in science, mathematics, or engineering, nor could society support them all if they did. But even most nonscience careers require a basic understanding of scientific principles. A cashier who cannot accurately count money can cost an employer hundreds or even thousands of dollars.[67] A police officer needs to understand the fundamentals of DNA in order to properly handle potential evidence; a school principal needs to know how viruses spread; a member of Congress should understand science and technology in order to legislate intelligently on issues from health care to homeland security. At a more mundane level, all citizens need a basic understanding of science and technology in order to pay bills, send e-mail, and maintain personal hygiene.[68]

Are we achieving the hoped-for balance? Is America producing the right number and mixture of highly trained scientists and engineers? Do average citizens adequately understand basic scientific principles? These are questions without easy answers, but science policy must nonetheless confront them if it is to fulfill its public service mission.

The Link between Math and Science Education

Mathematics is the language by which we codify and extend the principles of the natural world. A tourist in a foreign country can generally appreciate the country's natural setting, customs, and art without speaking the local language. But fluency is necessary in order to reap the full benefits of travel, or to contribute to local life. Mathematics is one of the basic idioms of science. Students with no grounding in math quickly lose their bearings in science and are unable to understand what is being said or done. And since science is at its core the quantitative description of nature, a lack of understanding of mathematics robs the student of the ability to understand the world. Science education should thus encourage an interest in mathematics from the earliest possible moment.

The Digital Divide

As computers have become cheaper, they have also become more readily available to the public. Even so, there is growing concern about the so-called digital divide between people who have access to computers and the Internet and those who do not. Indeed, although more people are going online every day, the gap between the technological haves and the have-nots is actually widening.[69]

This gap directly affects science education efforts. Schools are one of the main locales in which students encounter and learn about computers; but not all schools in this country are equipped with computers and Internet access. Even those that are do not necessarily have dedicated instructional computers and connectivity sufficient to maximize their value. In other words, while wealthier schools may offer a vast array of technological learning opportunities (to students who, incidentally, are also much more likely to have computer access at home), poorer schools often do not even have enough equipment to train their students in the basics. Breaking the divide down in another way, blacks and Hispanics are less likely to have ready access to computer technology than their white and Asian counterparts.

The digital divide is also evident in differences between access in urban, suburban, and rural schools. A report issued in fall of 2003 suggests that America's public schools are gradually closing the gap, however. Ninety-two percent of schools surveyed in 2002 had instructional computers and Internet access in their classrooms, compared to only 3 percent in 1994. And the student-to-instructional-computer ratio had shrunk to 4.8 to 1 in 2002—a vast improvement over the 1998 level of 12.1 to 1.[70]

But equipment is not enough. We also need to provide teachers with professional development opportunities that encourage them to use computers and the Internet in new ways. And schools can grant students access to computers outside of regular instructional time, spurring creativity and exploration. While all of this places additional demands on teachers' and administrators' time, we urgently need to reduce and eventually eliminate the disparities between students who have home access to computers and those who do not.

Attracting and Keeping Good Science and Math Teachers

Our schools lack sufficient numbers of teachers trained to teach science and math. Fewer than half of all high school students receive math instruction from teachers with an undergraduate degree in the subject; in middle school, the number is less than 15 percent. In the academic year 1999, only 63 percent of high school students received biology or life sciences instruction from a teacher who had majored in that subject, while just 41 percent received physical science instruction from teachers with a relevant major (including geosciences or engineering).[71] The U.S. Commission on National Security / 21st Century has identified the shortage of qualified math and science teachers as a critical threat to national security: "we do not now have, and will not have, with current trends, nearly enough qualified teachers in our K–12 classrooms, particularly in science and mathematics. . . . A continued shortage to the quality and quantity of teachers in science and math means that we will increasingly fail to produce sufficient numbers of high-caliber American students to advance college and post-graduate levels in these areas."[72]

Why? The shortage is explained in part by fact that students with STEM training can generally find more promising career opportunities in other professions. America's failure to value teaching as much as those other professions is having its impact. While teacher pay has recently been increasing, it remains lower than the average starting salary for college graduates in other occupations. In 2002, the salary for beginning teachers averaged approximately $30,000.[73] As mentioned previously, in many districts the lack of qualified science and math teachers is so severe that courses are taught by individuals trained in unrelated fields like physical education or English. This problem could in principle be addressed by offering higher salaries to STEM teachers, yet as previously mentioned, efforts to provide this kind of differential pay have been opposed by teachers' unions. This topic will attract growing interest as numerous national initiatives (e.g., the American Competitiveness Initiative) are calling for differential pay for teachers in critical areas.[74]

The United States must begin to address its educational deficiencies by improving teacher training, and by making teaching a more attractive career choice. Incentives must be provided to attract talented individuals, particularly those with a strong science background, to the teaching profession.

National Education Standards

Standards have long been regarded as a useful tool for insuring that all students receive a comparable education, whether in the heart of a progressive city or at the far reaches of a small rural town. Some critics, however, have suggested that poor science and math test scores might be the result of overemphasis on national and state standards and overreliance on standardized testing. While some standardization at various grade levels can be helpful, there is the possibility that overemphasis could curb creativity and classroom inquiry. This has been of particular concern in STEM fields.

President George W. Bush's No Child Left Behind initiative assigns tremendous weight to standards, directly linking school funding to student performance and student test scores. Critics of NCLB argue that schools and teachers have felt compelled to "teach to the test" in order to improve performance and avoid penalization. Moreover, they argue that punishing schools for underperforming will not help to increase their performance, but instead might make the situation worse.[75] While some standards are certainly needed, there will be an ongoing discussion of how best to measure school performance and how to improve the performance of underperforming schools without penalizing them.

The emergence of calls for national standards in higher education has provided a new twist on the old standards debate. In fall 2005, President Bush appointed the Commission on the Future of Higher Education, which was charged with investigating the quality, cost, and accountability of American higher education.[76] One of the proposals considered by the Commission was the administration of a standardized test to all college students in the country, in order to assess the relative effectiveness of their schools. The suggestion has provoked argument over whether the effectiveness of an institution of higher learning can be measured by a single exam. This is an important question, and though the commission's final report did not call for a standardized assessment, it has spawned debate about how far the government should intrude into the higher education domain when assessing academic performance.[77]

Knowing What Works

Even if there were uniform standards for students at different grade levels, and even if we had a good means of tracking student performance, we would still need to ensure that students would be able to meet these standards. The sad reality is that we understand little about which science and math educational programs work, and why.

Currently only 0.01 percent of the $300 billion federal education budget is devoted to research aimed at improving curricula and increasing teacher effectiveness in science and math.[78] If we are to truly address our education problems, we need a much better understanding of successful methods and approaches. In 2002 the U.S. Department of Education established the "What Works Clearinghouse," which collects and disseminates information on effective educational strategies and practices, including extensive databases of reported success stories.[79] The Department's Institute of Education Services collects and evaluates these data in conjunction with private educational research contractors, and makes them available on the Web. The clearinghouse is an important first step toward creating a single source for information on educational practices across the nation. Such initiatives will require greater resources in order to be successful, however, and continued effort will be needed to ensure that good ideas are contributed and used.

Conflicts between Science and Religion

Many problems with American science education are unique to the K–12 education system. This is, in part, because the way American K–12 education is run leaves the curriculum vulnerable to tensions between scientific principles and religious or moral views.

For example, science maintains that all living organisms evolved from prior organisms, beginning about 1.5 billion years ago, and that human beings in particular probably evolved from apelike hominids over many millions of years; whereas many religions posit that humans were created only tens of thousands of years ago, in much the same form as they exist today. The most recent flare-up of this controversy was sparked by an August 1999 decision by the Kansas Board of Education, which voted six to four to delete all references to evolution, natural selection, and the origins of the universe from the state's science curriculum.[80] The National Science Board (NSB), the NSF's board of directors, issued a rare public statement condemning the Kansas vote:

> The National Science Board notes with sadness and deep concern the recent action of the Kansas Board of Education to remove evolution as a topic for required teaching and testing in the state's science curriculum. Although the Kansas Board's vote allows local schools to continue teaching evolution in science classes, teaching and learning stand to suffer.
>
> Evolution is a well-documented process—and the rich scientific debate about its precise nature will continue to contribute to our knowledge base. But biology, like every science, does not exist in isolation. The Kansas action removed a key element from the body of scientific knowledge that schoolchildren need to learn and, in so doing, diminished the quality of education that they are likely to receive.
>
> At a time of already-profound concern about the quality of mathematics and science education in our Nation's schools, the Kansas action is a retreat from responsibility. A school board, whether elected or appointed, is expected to act not only in its community's best interest, but also in the national interest. America's children will someday be expected to think, vote, and participate in the global economy and local community debates—many grounded in the life sciences. An appreciation of the tensions among observation, explanation, theory, and topics will prepare them to be knowledgeable and effective citizens.
>
> Parents, educators, and policy makers should regard damaging cuts to a state curriculum with dismay. Such decisions may be "local" but, if unchallenged, they will ultimately affect the quality of life in this Nation for years to come.[81]

Other states have since followed Kansas's lead. Several now omit the word *evolution* from their science education standards. In 2001, Alabama's state board of education voted to place disclaimer stickers on textbooks, describing evolution as a controversial theory. In Georgia, state standards for middle- and secondary-school student excellence replaced the word *evolution* with the phrase *changes over time*. In another section of the same standards, the word *long* was removed from a discussion of the history of the earth in response to complaints from creationists who favored a literal reading of the Bible according to which the earth is at most several thousand years old.[82] Paradoxically, the Kansas vote provoked a political backlash: many of the board members who voted to delete references to evolution were voted out of office, and a seven-to-three vote subsequently restored the theory of evolution to state standards.[83]

Local citizens understandably have a deep interest in the content of the textbooks used for teaching their children. Since science is the study of who we are, where we came from, how the universe was created, and how it might end, conflicts between scientific ideas and religious belief are likely to characterize educational policy for some time to come.[84]

Possible Restructuring of the PhD Degree

Academia changes slowly, and suggestions about how to restructure higher education systems are not always welcome. The bachelor's, master's, and doctoral degrees have long been the foundation of traditional university education. For this reason, many eyebrows were raised when a committee of the National Academy of Sciences' COSEPUP recently questioned the value of the doctorate as it is presently structured.[85] The subject came up in part because so many PhD holders were having difficulty finding jobs in their areas of specialization, even after years of preparation.

Massive changes have occurred in the outside world, but universities retain their traditional concepts of the master's and PhD degrees. It may be time to consider alterations in the focus of these degrees, especially for students who are intending to pursue research in industry, or in allied fields outside those of their graduate study. Perhaps students intending to pursue careers in industry should bypass the PhD entirely and go directly into the job market after receiving their masters', thereby curtailing their postbaccalaureate training by many years. This topic should clearly be on the agenda of policymakers and the university community.

Minorities and Women

For many years, policymakers have struggled to explain why minorities and women are underrepresented in science- and mathematics-related professions. Research has made it clear that this underrepresentation, particularly among minorities, has its origins at the precollege level.[86] Both minorities and women tend to take fewer high school

POLICY DISCUSSION BOX 15.1

What Should Be Taught as Science in the Classroom and Who Should Decide?

Public schools in the United States are for the most part under the jurisdiction of local school boards, individuals from the community elected to serve on the board. School boards often have control over what is taught or not taught in the local classroom. For instance in October 2004, the Dover, Pennsylvania, school board elected to include in the high school biology curriculum the theory of intelligent design, adding language stating, "Students will be made aware of gaps/problems in Darwin's theory and of other theories of evolution including, but not limited to, intelligent design. Note: Origins of Life is not taught."[87] In response to this resolution, a group of parents brought a lawsuit against the Dover School Board. At the center of the case was not whether intelligent design should be taught in schools, but instead whether or not intelligent design could be taught as a scientific theory.

Intelligent design maintains that "certain features of the universe and of living things are best explained by an intelligent cause, not an undirected process such as natural selection."[88] Proponents of intelligent design argue that it is a scientific theory equal to or superior to other current scientific theories regarding evolution and the origin of life.[89] Others, including much of the scientific community, argue that intelligent design is just a modified form of creationism. They maintain that while intelligent design does not refer to God as the "designer," the designer intervened in ways that only God could have intervened. Therefore, intelligent design is a religious idea. Because science deals with the natural world, while intelligent design is about an agent independent of the natural world, intelligent design is not a science and therefore should not be taught as part of a science class.[90]

In the case of *Kitzmiller v. Dover Area School District*, a U.S. district judge agreed that intelligent design is religious in nature and therefore that teaching it in a science class violates the Constitution.[91] This leads to several questions: What should and should not be taught in science classes? Who should decide? The local school board? Parents? Teachers? Scientists? Should there be a standard definition of what "science" means in the context of the school classroom? Should this be something determined locally or at the state or national level? Should the standards for public school classrooms differ from those applied in private, religiously affiliated schools? What do you think?

---

math courses than are necessary to major in science or engineering in college. Numerous sociological studies have tried to determine the extent to which these trends are due to lack of role models, lack of encouragement and mentoring, or differences in family structure and income.[92] But the findings have been ambiguous. Any given factor may influence the success of one particular subgroup, but be irrelevant for another. We must determine which factors come into play in any given instance, and for whom, if we wish to keep the nation's research enterprise open to talented scientists of all backgrounds.

Information Technology

Since science and mathematics education proceeds along a very structured path, computers and information technology can play a more significant role in the delivery of instruction in these subjects than it might in others. An eighth-grader, for example, could take a particular course sequence on a computer. Targeted videos could be shown to illustrate experiments, after which the student would be asked to interpret the results. A series of questions could be asked to assess the student's mastery, which the student would answer via a computer program. The student would then be passed or diverted to a remedial course, as appropriate. Educators and policymakers have chased the dream of computer-aided math and science education for decades, with some of the earliest studies coming out of the University of Illinois in 1960.[93]

Educators now use computers for far more diverse purposes than simple support for traditional classroom instruction. A growing number of accredited institutions are using them to reach out to working adults, for example. These institutions offer degree credits to continuing or "nontraditional" students through night or weekend courses, heavily augmented by Internet assignments. One of the largest such nontraditional entities is the University of Phoenix, which has over seventeen thousand faculty and staff and an enrollment of over two hundred thousand adults in North America.[94] Such institutions are becoming a major force in educating the public; their role in science education, therefore, needs to be thoroughly assessed. Meanwhile, even traditional schools such as MIT are now making their course materials widely available on the Internet.

Other Web-based instructional technologies are emerging at advanced educational levels. For example, lectures can now be recorded and archived on the Web for future playback. These recordings can be viewed by anyone in the world with a computer and an Internet hookup. Such technologies hold out the promise that people everywhere could one day share access to information on complex subjects from the most knowledgeable experts in the world.

But this promise comes with a long list of questions attached. Should an advanced university course be given through this technology, rather than by a resident faculty member who may only be knowledgeable about certain subtopics of the field? If so, what does it mean to be a faculty member at that university? How should we validate the credentials of a scientist who is teaching remotely? How do we handle copyright and compensation? Should efforts be made to restrict access to information about proprietary or potentially harmful technologies? If we can resolve these questions in the years ahead, we may see dramatic changes in the delivery of science education, especially at advanced levels.

The ultimate limitation on the full deployment of computers for instruction at any level may be that we simply do not yet understand exactly how humans learn. Educators, psychologists, sociologists, and neuroscientists are now collaborating on research in this vital area. There can be no doubt that Web technologies can transform the way students access information and the way they learn, as well as how adults keep abreast of technological advances.

Structural Barriers to Addressing STEM Education

National science policy plays a vital role in determining our ability to meet America's science education goals. But the possibilities are often limited by the very structure of government, mission overlap between agencies, cultural differences between science and science education professions, and the lawmaking process itself. At any given moment policy development may, for example, require the setting of clear priorities—something that politicians normally wish to avoid. Some of the structural issues that can affect policy development are reviewed in what follows.

Priorities

The United States seems to lack a commitment to science education. Perhaps this is because many educational (particularly K–12), policy and government leaders did not themselves receive much in the way of science training. Only 8 of the 535 members of the 109th Congress could claim any professional background in science, for example.[95] Indeed, because so many of these leaders have managed to succeed without scientific training, they may be skeptical about the urgency of our educational needs.

Science education is often overshadowed by other issues. In the United States, it is socially unacceptable to be illiterate, but it is acceptable to be nearly illiterate in math and the sciences. While virtually all Americans can read, many—including some very successful and well-known people—readily admit (in some cases even proudly) that they did not do well in math and science. For policymakers, who like to champion popular issues, the lack of public interest in science is a disincentive to making our science education needs a national priority.

Cultural Differences: Scientific Community versus Education Community

Even within universities one encounters cultural barriers that impede students of education from communicating with their science and engineering peers to improve science education. Educational specialists are often critical of science and engineering faculty's teaching styles, and members of the scientific community in turn often believe that teachers don't understand how science is actually done.

These biases are often replicated among federal agency staff, deterring policymakers from addressing national needs in science education. To cite just one example, the National Science Foundation, which retains the bulk of the responsibility for science education, must work closely with the Department of Education on science policy objectives. Yet because the NSF staff generally come from a science background, while DOEd staff have usually been trained in education, they often harbor different opinions and ways of approaching the issue. As a result congressional policymakers have wasted precious time debating the relative merits of these agencies, when they could have been devising science education policies.

No matter one's perspective, coherent and effective policy requires leadership at the highest level, capable of producing consensus on a uniform approach to national goals in science education.

Agency Coordination and Jurisdiction

Consensus is useless, however, without resources and coordination among policymakers, educators, and scientists. Coordination can be elusive: federal science and education policy-making is a decentralized process, with a large number of federal agencies each responsible for its own corner of the issue. The problem is exacerbated because much of the responsibility for education policy is left to state and local governments, largely by default.

Many players must come to mutual agreement on the goals for improving science education. Each of them, from the NSF and Department of Education to teachers' unions and parent-teacher organizations, to professional scientific societies, has a stake. But each also has its own agenda and its own views on how to best achieve desired ends. While teachers' associations may emphasize the importance of higher salaries and better curricular materials, industry may focus on technical training and applied knowledge; while politicians may concentrate on standards and assessments, the scientific community may stress the value of the scientific method and hands-on scientific experience. Strong leadership is needed to broker a compromise that will enable these disparate constituencies to work in concert.

Agency Mission Overlap:
The National Science Foundation and
the Department of Education

What happens when the actors' roles overlap? The Department of Education holds jurisdiction over national educational activities, for example, while the NSF is focused on science education. This redundancy can be a source of either strength or conflict.

Many educators believe that it is counterproductive to separate one type of education from the rest. Proponents of this view voice concerns about the fracturing of the educational system and the duplication of effort. Supporters of the NSF, on the other hand, argue that science education should more properly be considered part of the conduct of science. Any hope for resolution to this debate depends on the willingness of leaders to cooperate, and on the president's ability to engineer that cooperation. But tension is never far from the surface, even in the most cooperative of administrations.

Don't both groups share a desire to improve the quality of science education? Consider the following two hypothetical viewpoints:

1. A University faculty member protests that her students have been poorly trained by a high school system replete with incompetent and poorly trained teachers, and suggests that she could do better. She harbors a longtime animosity toward the Department of Education: it once turned down her proposal for a science education program, she claims, because the agency's reviewers were unable to understand her ideas.

2. Across campus, a member of the education faculty bemoans arrogant physics professors who will not take time out of their schedules to discuss the challenges of teaching advanced scientific concepts. This faculty member has heard that undergraduates are having difficulty in introductory physics courses, but attributes the problem to the inability of poorly trained physics professors to organize and present complex concepts. The physics professor bristles at the suggestion that she should take minicourses to improve her teaching techniques.

Standoffs of this type are common, both on local campuses and in the halls of government. We have to bridge this gap between scientists and educators if science education is to succeed. Scientists need opportunities to teach science at the precollege level, and educators need opportunities to conduct actual scientific research. While we have seen significant progress toward this goal over the last few decades, we still have further to go.

The NSF's and DOEd's mission statements reveal the essential differences in what the two agencies are trying to accomplish. The NSF seems to be focused on education research, classroom content and models for effective STEM education, while the DOEd concentrates on the learning environment and on replicating and disseminating STEM education programs at the state and local levels. All of these factors are important, of course, so one might argue that the overlap in agency beliefs should be complementary and healthy. Unfortunately, however, attempts to implement joint NSF and DOEd programs have resulted in accusations of duplication of effort and confusion about the roles of the respective agencies. The story of the Math and Science Partnership (MSP) program is exemplary of this disjuncture.

The MSP was intended to unite institutions of higher education, K–12 school systems, and other partners in a campaign to help K–12 students and teachers meet higher science and math standards. To achieve this goal, two MSP programs were established: one at the NSF that would work to improve STEM education by providing competitive, peer-reviewed grants to foster the development of partnerships linking colleges and universities to elementary and secondary schools. Another at the DOEd would provide block grants to states on a formula basis. It was thought that the NSF program would focus on modeling, testing, and identification of high-quality math-science activities, while the DOEd program would disseminate these programs to the states.

Up to $450 million in funding was authorized for the MSP program through the No Child Left Behind Act, which instructed the secretary of education "to consult and coordinate with the Director of the National Science Foundation, particularly with respect to the appropriate roles for the Department and the Foundation in the conduct of summer workshops, institutes, or partnerships to improve mathematics and science teaching in elementary schools and secondary schools."[96]

The NSF's MSP program was initially better funded then the DOEd's, in part because the agencies were funded by different congressional appropriations subcommittees. As pressure grew on the Bush administration to better fund programs mandated by No Child Left Behind, however, the administration shifted funds out of the NSF program into the DOEd's MSP program. Despite congressional ob-

jections and warnings from leading higher-education associations about the need for parallel and complementary MSP programs at the NSF and the DOEd, the NSF's MSP program has now essentially been phased out. Clearly, more careful coordination will be required to ensure that STEM education roles are clearly defined in the future, and that unfortunate situations like that which beset the MSP are not repeated.

Other Areas of Potential Overlap

Of course, the NSF and the DOEd are not the only federal agencies with a finger in the science education pie. NASA has invested for years in programs and curricular materials for elementary and secondary students. In fact, NASA is the only R&D agency beside NSF to include science education in its mandate.

The Department of Energy, with its vast array of national labs, also provides significant funding for science education programs at all levels. Many of the national labs sponsor summertime programs in which selected students work alongside practicing scientists. The DOE has

POLICY DISCUSSION BOX 15.2

The Academic Competitiveness Council: How Do You Know Which Federal STEM Programs Are Effective?

In recent years, a cacophony of programs have been created by federal agencies to improve and enhance STEM education at both the K–12 and postsecondary levels. According to a 2005 GAO review of federal STEM programs, thirteen federal agencies spent approximately $2.8 billion in FY2004 to fund over two hundred such programs, with the greatest sums being spent by the NSF, NIH, and DOEd. While all of these programs were created with the best of intentions, it is unclear how effective they have actually been, and how much coordination occurs between the programs. Indeed, the GAO found that of all of the programs it looked at, only half had ever been evaluated or had an evaluation under way.[97]

In response to the GAO review, Congress established a new Academic Competitiveness Council (ACC) as part of the Deficit Reduction Act of 2005. The ACC's mission under law was to identify and evaluate the effectiveness of each STEM program, determine areas of overlap,

and recommend ways to efficiently coordinate these programs in the future. The council was chaired by the secretary of education and consisted of representatives from all of the federal agencies that maintain STEM education programs. It was required to provide a final report back to Congress. The White House was also involved in the ACT effort through the OMB, OSTP, and Domestic Policy Council. One of the greatest challenges for the ACC has been determing sound and consistent evaluation metrics, given the significant diversity among these programs and their objectives.

Both the GAO review and the creation of the ACC raised important policy questions: How do you effectively evaluate STEM programs? Should more money be devoted to education research in order to understand which programs work and which don't? Which federal agencies should take the lead in conducting such research or running STEM education programs? Is it good to have multiple federal agencies running STEM education programs? Do answers to this question differ depending on whether the programs are aimed at K–12 students, undergraduates, or graduate students?

also launched a program in which K–12 teachers spend their summers performing experiments and training with researchers at DOE-run labs. The Department of Defense laboratories are now conducting a similar program. Finally, the NIH supports numerous science education-oriented programs, hosts K–12 teachers for research projects in NIH labs, sponsors graduate training grants, and funds educational outreach (e.g., the National Institute on Drug Abuse's program called Research Education Grants in Drug Abuse and Addiction). The careful coordination and evaluation of these diverse initiatives poses a sizable challenge for national science policymakers.

Policy Challenges and Questions

There is a growing awareness among the public that a technologically skilled workforce is vital to the economic competitiveness of the nation. Yet in a 2003 survey, over half of all citizens polled agreed with the statement that "scientific research has created as many problems for society as it has solutions."[98] Many Americans understand science only through the warped lens of science fiction and disaster movies. Others associate it with nuclear weapons, chemical warfare, or the furor over human cloning. Science is too often conflated in the public mind with society's use of its discoveries. The complexity of modern science and our understanding of its salience have great implications for science policy.

In this chapter we have touched on the issues of how we can keep students interested in science throughout their education, create a workforce with a basic understanding of science and math, and more effectively link science and math education. We have noted disparities in the levels of participation among women and minorities in areas that are so critical for the nation's future. It has also been noted that the government is not optimally organized to shepherd a strong system of science education, and that public understanding of science is not as strong as it should be.

National science policy plays a vital role in our ability to meet our science education goals. In addition to the issues already discussed, science policy must also confront other questions: What standards should the federal government impose in exchange for supporting elementary and secondary science education? How should scientists voice their opinions in local debates over textbook choice? Should we have national curriculum and textbook guidelines? Is there adequate coordination among the Department of Education, the National Science Foundation, and other government agencies on science and math education issues? How should federally supported STEM education programs be evaluated? How do we provide better teacher-training opportunities at our national laboratories and universities? How can we strengthen national requirements for teacher preparation? Should undergraduate and graduate students have opportunities to teach K–12 students? How do we offer undergraduates more and better research experiences? And how do we increase opportunities for women and minorities in science?

As the nation relies more and more on its technologically literate workforce, we will witness a growing need for careful analysis of how STEM education policy can be most effective at meeting this national demand.

NOTES

1. U.S. Dept. of Education, "Overview: The Federal Role in Education," http://www.ed.gov/about/overview/fed/role.html (accessed May 26, 2007).

2. Center for Studies in Higher Education, "Federal Support for University Research: Forty Years after the National Defense Education Act (Conference: October 1, 1998)," http://cshe.berkeley.edu/events/ndeaconference1998/background.htm (accessed May 26, 2007); *National Defense Education and Innovation Initiative,* Association of American Universities, January 2006.

3. F. James Rutherford, "Sputnik and Science Education," presentation, Center for Science, Mathematics, and Engineering Education, "Reflecting on Sputnik: Linking the Past, Present, and Future of Education Reform," October 4, 1997, http://www.nas.edu/sputnik/ruther1.htm (accessed May 31, 2007).

4. National Science Board, *Science and Engineering Indicators 2006,* chap. 1.

5. *National Committee on Science Education Standards and Assessment,* National Research Council, 1996.

6. "Back to School," U.S. Census Bureau Press Release, CBO1-FF.11, August 8, 2001.

7. "School Textbooks Create National Curriculum," CNSNews.com, May 23, 2001, http://archive.newsmax.com/archives/articles/2001/5/22/161936.shtml (accessed May 26, 2007); Tamin Ansary, "The Muddle Machine," *Edutopia,* November 2004, http://www.edutopia.org/muddle-machine (accessed May 26, 2007); Chester E. Finn Jr. and Diane Ravitch, Thomas B. Fordham Institute, *The Mad, Mad World of Textbook Adoption,* September 29, 2004, http://www.edexcellence.net/institute/publication/publication.cfm?id-335 (accessed May 26, 2007).

8. While it is generally assumed that the local, state, and federal governments share the responsibility for public education, because the Constitution remains silent on the issue of education the states and localities assume the primary burden for it; the Constitution says that all duties not specifically assigned to the federal government belong to the states.

9. See National Center for Educational Statistics, "NAEP Overview," http://nces.ed.gov/nationsreportcard/about/ (accessed May 26, 2007).

10. Wendy S. Grigg, Mary A. Lauko, and Debra M. Brockway, *The Nation's Report Card: Science 2005*, NCES 2006-466 (Washington, DC: National Center for Educational Statistics, U.S. Department of Education, 2006); American Institute of Physics, "Student Performance, Assessment in Science," *FYI: The AIP Bulletin of Science Policy News*, May 31, 2006, http://www.aip.org/fyi/2006/074.html (accessed May 26, 2007).

11. ACT, *Developing the STEM Education Pipeline* (Iowa City, IA: ACT, 2006), 1.

12. A complete discussion of certification in assigned teaching fields and the college major or minor of K–12 teachers can be found in National Science Board, *Science and Engineering Indicators 2006*, 1.32–1.35.

13. Linda Darling-Hammond, "From 'Separate but Equal' to 'No Child Left Behind': The Collision of New Standards and Old Inequalities," in *Many Children Left Behind*, ed. Deborah Meir and George Wood (Boston: Beacon Press, 2004), 27.

14. National Science Board, *Science and Engineering Indicators 2006*, 1.6, 1.37–1.38; Domestic Policy Council, Office of Science and Technology Policy, "American Competitiveness Initiative."

15. Frederick M. Hess and Martin R. West, "Taking on the Teachers Unions," *Boston Globe*, March 29, 2006.

16. National Center for Educational Statistics, *Digest of Educational Statistics*, table 223 and introduction, http://nces.ed.gov/programs/digest/d05/ (accessed May 28, 2007).

17. National Science Board, *Science and Engineering Indicators 2006*, 2.11.

18. National Science Board, *Science and Engineering Indicators 2004*, 2.12–2.14.

19. Center for Institutional Data Exchange and Analysis, *1999–2000 SMET Retention Report* (Norman: University of Oklahoma, 2001).

20. National Science Board, *Science and Engineering Indicators 2004*, 2.19–2.20; National Science Board, *Science and Engineering Indicators 2006*, 2.18–2.19.

21. National Science Board, *Undergraduate Science, Mathematics and Engineering Education: Role for the National Science Foundation and Recommendations for Action by Other Sectors to Strengthen Collegiate Education and Pursue Excellence in the Next Generation of U.S. Leadership in Science and Technology*, NSB 86-100 (Washington, DC: National Science Foundation, March 1986).

22. National Science Foundation, *FY2007 NSF Budget Request to Congress* (Arlington, VA: National Science Foundation, February 6, 2006), 215.

23. National Science Foundation, *Shaping the Future*, vol. 2: *Perspectives on Undergraduate Education in Science, Mathematics, Engineering, and Technology*, NSF 96-139 (Arlington, VA: National Science Foundation, August 1998), http://www.nsf.gov/pubs/1998/nsf98128/toc.doc (accessed January 11, 2008).

24. National Science Foundation, *Shaping the Future: New Expectations for Undergraduate Education in Science, Mathematics, Engineering, and Technology*, NSF 96-139 (Arlington, VA: National Science Foundation, October 1996), http://www.nsf.gov/pubs/stis1996/nsf96139/nsf96139.txt (accessed January 11, 2008).

25. Boyer Commission, *Reinventing Undergraduate Education*. The National Commission on Educating Undergraduates in the Research University was created in 1995 under the auspices of the Carnegie Foundation for the Advancement of Teaching. It met for the first time July 27, 1995, at the headquarters of the Carnegie Foundation in Princeton, New Jersey, with Ernest L. Boyer, president of the foundation, presiding. Dr. Boyer died on December 8, 1995, and the commission was renamed in his memory.

26. See "Webometrics Ranking of World Universities," July 2006, http://www.webometrics.info/Top_100_by_Country.html (accessed May 28, 2007).

27. National Science Board, *Science and Engineering Indicators 2006*, 2.20.

28. National Science Foundation Division of Science Resources Statistics, *Science and Engineering Doctorate Awards: 2003*, NSF-05-300 (Arlington, VA: National Science Foundation, 2004), table 1.

29. Basic science departments in medical schools for the most part do not have undergraduate degree-granting programs. Functioning well for these departments thus does not rely on the number of undergraduate courses being taught.

30. National Science Board, *Science and Engineering Indicators 2006*, 2.16.

31. Mildred L. Amer, *Membership of the 109th Congress: A Profile*, Congressional Research Service, Report Number RS22007, May 31, 2005.

32. National Science Board, *Science and Engineering Indicators 2006*, 2.6.

33. Charles A. Goldman and William F. Massy, *The PhD Factory: Training and Employment of Science and Engineering Doctorates in the United States* (Bolton: Ankler, 2001).

34. National Science Board, *Science and Engineering Indicators 2004*, 2.28–29.

35. See "National Postdoctoral Association," http://www.nationalpostdoc.org/site/c.eoJMIWOBIrH/b.1461865/k.7F4C/About_The_NPA.htm (accessed May 28, 2007).

36. Ibid.

37. National Academy of Sciences, National Academy of Engineering, Institute of Medicine, *Preparing for the 21st Century: The Education Imperative* (Washington, DC: National Academy of Sciences, 1997).

38. *Science Center Highlights*, ASTC Backgrounder, December 2005, http://www.astc.org/about/backgrounders.htm (accessed May 28, 2007). See also *ASTC Sourcebook of Statistics & Analysis 2005*, No. 4-2005 (Washington, DC: ASTC, 2005).

39. Smithsonian Institution, "A Virtual Tour of O. Orkin Insect Zoo," http://www.mnh.si.edu/museum/VirtualTour/Tour/Second/InsectZoo/.

40. D. A. Gentile and D. A. Walsh, "A Normative Study of Family Media Habits," *Applied Developmental Psychology* 23 (2002): 157–78.

41. Ann Marie Barry, *Visual Intelligence: Perception, Image, and Manipulation in Visual Communication* (Albany: SUNY Press, 1997), 301. Other sources include Committee on Public Education, "American Academy of Pediatrics: Children, Adolescents, and Television," *Pediatrics* 107, no. 2 (2001): 423–26; "Taking Control . . . Guidelines for TV and Teens," Department

of Pediatrics, University of Iowa Children's Hospital, http://uihealthcare.com/topics/medicaldepartments/pediatrics/tvteens/index.html (accessed January 9, 2008).

42. Television Project, "Basic Data about Television Watching," http://www.tvp.org/Handouts%20pages/basic_data_txt.html; Screenblock.com, "Research Data," http://www.screenblock.com/data.htm (both URLs accessed May 28, 2007).

43. One only need consider the competition science education programs have: professionally created fiction, news, and entertainment shows. This competition is not merely one of audience attention; it is also for prime-time slots on the regular networks.

44. The Museum of Broadcast Communications, "Science Programs," http://www.museum.tv/archives/etv/S/htmlS/scienceprogr/scienceprogr.htm (accessed January 9, 2008).

45. Richard Jones and Arthur Bangert, "The CSI Effect," *Science Scope,* November 2006, 38.

46. See Andrew Pollack, "Scientists Seek a New Movie Role: Hero, Not Villain," *New York Times,* December 1, 1998.

47. See National Science Board, *Science and Engineering Indicators 2004,* 1.12–1.14; National Science Board, *Science and Engineering Indicators 2006,* 1.21–1.22.

48. National Science Board, *Science and Engineering Indicators 2006,* 1.20–1.23.

49. Ibid.; Organization for Economic Cooperation and Development, "Programme for International Student Assessment (PISA)," http://www.pisa.oecd.org/pages/0,2966,en_32252351_32235731_1_1_1_1_1,00.html (accessed May 28, 2007).

50. National Science Board, *Science and Engineering Indicators 2004,* 1.14–16.

51. See http://www.ed.gov/updates/PresEDPlan/index.html (accessed July 13, 2007) for details of President Clinton's plan; see note 64 for details on President George W. Bush's plan.

52. National Science Board, *Science and Engineering Indicators 2000,* chap. 5; A. Beaton, I. Mullis, M. Martin, E. Gonzalez, D. Kelly, and T. Smith, *Mathematics Achievement in the Primary School Years: IEA's Third International Mathematics and Science Study (TIMSS)* (Chestnut Hill, MA: Centre for the Study of Testing, Evaluation, and Educational Policy, Boston College, 1996).

53. H.R. 100 "National Science Education Act," summary as of July 30, 2001, http://thomas.loc.gov/cgi-bin/bdquery/z?d107:HR00100:@@@L&summ2=m& (accessed July 12, 2007).

54. "House Report 106-821: Part 1–National Science Education Act," http://thomas.loc.gov/cgi-bin/cpquery/?&sid=cp106U7ozV&refer=&r_n=hr821p1.106&db_id=106&item=&sel=TOC_126427& (accessed May 28, 2007).

55. House Committee on Science and Technology, press release, *House Democrats Torpedo Previously Bipartisan Effort to Improve Math and Science Education,* October 27, 2000, http://www.house.gov/science/106thpress/106-170.htm (accessed May 28, 2007).

56. Committee on Science, Engineering, and Public Policy, *Rising above the Gathering Storm;* House Committee on Science, *Unlocking Our Future.*

57. STEMEd Caucus Steering Committee, "Home," http://www.stemedcaucus.org/ (accessed May 28, 2007).

58. P.L. 103-227, signed into law on March 31, 1994.

59. See summary of this legislation archived at North Central Regional Educational Laboratory, "Summary of Goals 2000: Educate America Act," http://www.ncrel.org/sdrs/areas/issues/envrnmnt/stw/sw0goals.htm (accessed May 28, 2007).

60. A provision in the No Child Left Behind Act zero-funded the Goals 2000: Educate America Act.

61. See Goals 2000: Educate America Act, H.R. 1804, 103rd Cong., 2nd sess., http://www.ed.gov/legislation/GOALS2000/TheAct/index.html (accessed May 28, 2007).

62. See Improving America's Schools Act, H.R. 6, 103rd Cong., 2nd sess., http://www.ed.gov/legislation/ESEA/toc.html (accessed May 28, 2007).

63. Public Broadcast System, "Frontline: The New Rules," http://www.pbs.org/wgbh/pages/frontline/shows/schools/nochild/nclb.html (accessed May 28, 2007).

64. For additional background see Richard D. Young, "Public Education: A Primer on *No Child Left Behind* and Its Impact on South Carolina," Institute for Public Service and Policy Research, University of South Carolina, June 2004, 3–8, http://www.iopa.sc.edu/publication/NCLB%20June%202004%20FINAL.doc (accessed May 28, 2007); Andrew Rudalevige, "Research: The Politics of No Child Left Behind," Hoover Institution, 2003, adapted from Paul E. Peterson and Martin R. West, eds. *No Child Left Behind? The Politics and Practice of Accountability* (Washington, DC: Brookings Institution Press, 2003), http://www.hoover.org/publications/ednext/3346601.html (accessed May 28, 2007); "President Signs Landmark No Child Left Behind Education Bill," White House, news release, January 8, 2002, http://www.whitehouse.gov/news/releases/2002/01/20020108.html (accessed June 8, 2007).

65. National Public Radio, Michele Norris, and Anthony Brooks, "Nation: U.S. Relaxes 'No Child Left Behind' Law," March 15, 2004, http://www.npr.org/templates/story/story.php?storyId=1769154 (accessed May 28, 2007).

66. See Mike May, "Diversity: Broadening the Breadth of Science," AAAS Office of Publishing and Member Services, November 11, 2005, http://sciencecareers.sciencemag.org/career_development/previous_issues/articles/2005_11_11/diversity_broadening_the_breadth_of_science/(parent)/14014 (accessed May 28, 2007).

67. Rep. Vernon Ehlers often tells groups to which he speaks on the subject of science education of an instance when a convenience store operator decided to present his new employees with a basic math test. As a result he saved $100,000 in one year. House Subcommittee on Environment, Technology, and Standards Committee on Science, *Workforce Training in a Time of Technological Change,* 107th Cong., 2nd sess., June 24, 2002, Serial No. 107–78, 80-339PS 2002, http://commdocs.house.gov/committees/science/hsy80339.000/hsy80339_0.HTM (accessed May 28, 2007).

68. For further discussion, see National Academy of Sciences, National Academy of Engineering, and Institute of Medicine, *Preparing for the 21st Century.*

69. Sherril Steele-Carlin, "Caught in the Digital Divide," *Education World* (2000), http://www.education-world.com/a_tech/tech041.shtml (accessed May 28, 2007).

70. National Center for Education Statistics, *Internet Access in U.S. Public Schools and Classrooms: 1994–2002* (Oc-

tober 2003), http://nces.ed.gov/pubsearch/pubsinfo.asp?pubid=2004011 (accessed May 28, 2007).

71. National Science Board, *Science and Engineering Indicators 2004*, 1.28 and fig. 1.18.

72. U.S. Commission on National Security. See National Science Board, Committee on Education and Human Resources, *Task Force on National Workforce Policies for Science and Engineering*, NSB 03-69, draft for public comment, May 22, 2003, note 44.

73. National Science Board, *Science and Engineering Indicators 2006*, 1.37 and fig. 1.16.

74. Domestic Policy Council, Office of Science and Technology Policy, "American Competitiveness Initiative."

75. For further discussion see Deborah Meir and George Wood, eds., *Many Children Left Behind* (Boston: Beacon Press, 2004).

76. Karen W. Arenson, "Panel Explores Standard Tests for Colleges," *New York Times*, February 9, 2006, http://www.cae.org/content/pdf/New%20York%20Times%20Article%202.9.06.pdf (accessed May 29, 2007).

77. "Draft Report: A National Dialogue: The Secretary of Education's Commission on the Future of Higher Education," August 9, 2006, http://www.ed.gov/about/bdscomm/list/hiedfuture/reports/0809-draft.pdf (accessed May 29, 2007).

78. House Committee on Science, *Unlocking Our Future*.

79. See "What Works Clearinghouse," http://www.whatworks.ed.gov/whoweare/overview/html (accessed May 29, 2007).

80. Kansas State Department of Education Science, http://www.ksbe.state.ks.us/Welcome.html (accessed May 29, 2007). Further discussion of this case can be found in Zuleyma Tang-Matrinez, *Analysis of the Kansas Board of Education Decision on the Teaching of Evolution: Dynamics and Consequences* (Department of Biology, University of Missouri, St. Louis), http://www.umsl.edu/~mpsac/forums/forum4/t-mnotes.html (accessed May 29, 2007); Robert E. Hemenway, "The Evolution of a Controversy in Kansas Shows Why Scientist Must Defend the Search for Truth," *Chronicle of Higher Education*, October 29, 1999; *CNN.com*, "Evolution-Creation Debate Grows Louder with Kansas Controversy," March 8, 2000, http://www.cnn.com/2000/US/03/08/creationism.vs.evolution/ (accessed May 29, 2007); *CNN.com*, "Kansas School Board's Evolution Ruling Angers Science Community," August 12, 1999, http://www.cnn.com/US/9908/12/kansas.evolution.flap/index.html (accessed May 29, 2007).

81. National Science Board, *Statement on Action of the Kansas Board of Education on Evolution*, NSB 99-149 (August 20, 1999), http://www.nsf.gov/nsb/documents/1999/nsb99149/nsb99149.doc (accessed May 29, 2007).

82. Andrew Jacobs, "Georgia Takes on 'Evolution,'" *New York Times*, January 30, 2004, http://query.nytimes.com/gst/fullpage.html?sec=technology&res=9D07E2DB1138F933A05752C0A9629C8B63 (accessed May 29, 2007).

83. *CNN.com*, "Kansas Restores Evolution Standard for Science Classes," February 14, 2001, http://www.cnn.com/2001/US/02/14/kansas.evolution.02/ (accessed May 29, 2007); "Kansas Board Revives Teaching of Evolution," *Los Angeles Times*, February 15, 2001, A10, http://www.washingtonpost.com/

ac2/wp-dyn?pagename=article&node=&contentId=A7695-2001Feb15&notFound=true (accessed May 29, 2007).

84. John A. Dvorak, "Kansas Science Review Revives Evolution Debate," *Kansas City Star*, August 13, 2003.

85. Committee on Science, Engineering, and Public Policy, *Reshaping the Graduate Education of Scientists and Engineers* (Washington, DC: National Academy Press, 1995), http://www.nap.edu/readingroom/books/grad/ (accessed May 29, 2007).

86. *Land of Plenty: Diversity as America's Competitive Edge in Science, Engineering and Technology*, Report of the Congressional Commission on the Advancement of Women and Minorities in Science, Engineering and Technology Development (September 2000), http://www.nsf.gov/pubs/2000/cawmset0409/cawmset_0409.pdf (accessed May 29, 2007).

87. Joseph Maldonado, "Creation Debate Draws in Teachers," *York Daily Record, York Sunday News*, October 24, 2004.

88. Discovery Institute and Center for Science and Culture, "Questions about Intelligent Design: What Is the Theory of Intelligent Design?" http://www.discovery.org/csc/topQuestions.php#questionsAboutIntelligentDesign (accessed May 29, 2007).

89. Steven C. Meyer, "The Scientific Status of Intelligent Design: The Methodological Equivalence of Naturalistic and Non-Naturalistic Origins Theories," in *Science and Evidence for Design in the Universe: The Proceedings of the Wethersfield Institute*, vol. 9 (San Francisco: Ignatius Press, 2000).

90. American Association for the Advancement of Science, "AAAS Board Resolutionon Intelligent Design Theory," approved October 18, 2006, http://www.aaas.org/news/releases/2002/1106id2.shtml (accessed May 29, 2007). See also Alex Johnson and MSNBC, "'Intelligent Design' Faces First Big Court Test," September 23, 2005, http://www.msnbc.msn.com/id/9444600/ (accessed May 29, 2007).

91. *Kitzmiller v. Dover Area School District*, U.S. District Court for the Middle District of Pennsylvania, Case No. 04cv2688 (December 20, 2005). See also Delia Gallagher and Phil Hirschkorn, "Judge Rules against 'Intelligent Design' in Science Class," *CNN.com*, December 23, 2005, http://www.cnn.com/2005/LAW/12/20/intelligent.design/index.html (accessed May 29, 2007).

92. Ibid.

93. PLATO (Programmed Logic for Automating Teaching Operations), People, http://www.platopeople.com (accessed January 8, 2008).

94. See University of Phoenix, "About Us," http://www.phoenix.edu/about_us/ 10/21/2006 (accessed June 8, 2007).

95. Amer, *Membership of 109th Congress*, CRS Report.

96. See P.L. 107-110.

97. Report to the Chairman, Committee on Rule, House of Representatives, *Federal Science, Technology, Engineering, and Mathematics Programs and Related Trends*, GAO-06-114 (Washington, DC: U.S. Government Accountability Office, October 2005), http://www.gao.gov/htext/d06114.html (accessed May 29, 2007).

98. National Science Board, *Science and Engineering Indicators 2004*, 7.24.

Science Policy in an Era of Increased Globalization

CHAPTER 16

The Science and Engineering Workforce

What Is the Role of S&E Workforce Policy?

A strong and vibrant S&E workforce is vital to America's economic stability, as well as our quality of life, public health, and national security. Academia, industry, and the national labs are all critically dependent upon a workforce of able and interested professionals. Obviously, the federal government has a vested interest in ensuring the adequate supply of such professionals.[1]

While the strong connection between federal research investment and the quality of S&E training has long been recognized, it is not well understood. In fact, experts disagree on federal policy's effect on the supply of U.S. scientific talent. Similarly, strong disagreements rage over the impact of current deficiencies in K–12 science and math education and, to a lesser extent, in undergraduate and graduate education as well.

But how do we determine our future S&E workforce needs? If we cannot accurately predict our needs, how do we optimize our prospects in the face of uncertainty? Policymakers should develop programs that are flexible enough to respond to future needs, even if those needs cannot always be anticipated.

As in so many areas of science policy, simple solutions are elusive, the data subject to interpretation and conjecture. While concerns about the future S&E workforce have often been used to justify increasing research funding, there is still much that we do not understand about the dynamics of the funding-to-training relationship. This chapter will examine some of the U.S. S&E workforce needs and concerns and will provide data on S&E workforce trends. It will map some of the critical junctures at which sound decisions on policy are needed, and will conclude with some possible options for addressing the S&E workforce challenges discussed.

The S&E Workforce "Crisis"

It has been asserted that "the United States is facing a crisis in science and engineering talent and expertise."[2] Some worrisome trends are certainly evident, particularly in the physical sciences and engineering. The number of doctorates awarded from U.S. institutions to students in the physical sciences declined during the 1990s but has been slowly increasing in recent years. Likewise during the late 1990s and early 2000s, doctorates awarded in engineering declined.[3] The total number of undergraduate degrees issued in the physical sciences and engineering was also on the decline during the later 1990s and early 2000s, although, again, the data suggest these trends might be shifting.[4] The number enrolling in doctoral programs has generally been increasing since 1999 largely because of the increasing enrollment of foreign students. Changes in visa policy instituted after September 11, 2001, however, caused great concern that this trend might be reversed. While in the first few years after September 11, the number of S&E doctorate degree seekers on temporary visas did, in fact, sharply decline, this number has recently begun to rebound.[5]

Of course, these numbers do not necessarily indicate a crisis. As was noted in a 2004 editorial in the journal *Science*, "Time after time we have been warned of impending shortages which, with evergreen consistency, are subsequently transformed into gluts, to the dismay of those

277

most affected: the future practitioners of our disciplines. Somehow, the predictors seem to forget that calls to increase future supply should bear some relationship to the present balance between supply and demand."[6]

Why is an accurate assessment of supply and demand so hard to achieve? For one thing, the job market in the high-tech sector is tremendously volatile. During the tech boom of the late 1990s, universities could not graduate computer science majors fast enough to satisfy industry's seemingly insatiable demand. But when the bubble burst, the shortage quickly became a surplus, and a wave of layoffs quickly ensued.

So, what constitutes a crisis? Some may claim a S&E workforce crisis exists when universities are not able to fill academic research positions largely dependent upon federal funding. Others may view a S&E workforce crisis as being when there are not enough qualified workers to fulfill the needs of industry, which might be looking for a different type of expertise than universities.

Clearly, the extent of the S&E workforce "crisis" depends upon one's definitions and criteria; one person's workforce shortage is another's glut. Likewise, what constitutes a "crisis" often depends on a particular view of the relevant importance of certain national needs as well as how certain needs are weighed against others.

This makes defining scientific workforce issues particularly difficult. For instance, if one were to poll U.S. engineers on their opinion about whether there is a shortage of engineers, many might say no. However, if the same question were posed to some of the large defense contractors, such as Northrop Grumman or Boeing, the response might well be yes. Engineers often seek a specific kind of job, which many others may also be seeking. In addition, there are engineers who simply do not want to work for the defense industry because of their personal values and views. Meanwhile, the defense industry often needs individuals who can receive clearances to work on classified research. This greatly limits the pool of engineers available to them, and also means that they are often unable to hire non–U.S. citizens. These differences between workers' expectations and industry's needs thus create differing opinions and perspectives on supply and demand for engineers.

An assessment should consider that some individuals are unable to find a job for reasons unrelated to their technical or scientific qualifications. Maybe a candidate lacks communications or social skills. Job-seekers may be unwilling to relocate, or particular about whom they work for. One has to carefully filter the aggregate data on S&E employment before making statements about the balance of supply and demand.

Defining and Measuring S&E Workforce Needs

Whether a national crisis in S&E training exists may never be determined, but evidence suggests that S&E workforce issues need attention from policymakers and the scientific community. Failure to address these issues could put at risk the country's national security and its economic future.[7]

Industry and academia have become heavily dependent on foreign talent to meet their S&E workforce needs in recent decades. This dependency, however, has become problematic after September 11, and the problems are aggravated by the growing battle over national immigration policy.

When information technology (IT) and engineering employers noticed a shortage of talent in the 1990s, business and academic leaders requested that the government increase the number of H-1B visas, which are work permits for foreigners with special skills or talents.[8] The H-1B is used to address labor shortages in technical areas that are highly dependent upon business cycles, by offering employers the flexibility to look outside the country for personnel they cannot find at home. Yet in recent years the number of H-1B visas that can be issued annually has been capped, limiting industries' ability to use foreign talent to meet its workforce needs. The question of how generous or restrictive the H-IB cap is has become a major and sometimes contentious debate for policymakers.

Given the volatility of geopolitics and likelihood for change in other world regions, the United States should work to reduce its dependence on foreign S&E talent. While data suggests that many foreign students who come to the United States to study end up living and working here, there is no reason to expect that this will always be the case, particularly with high-tech and academic jobs becoming more readily available in countries such as India and China (both nations upon which the United States has relied heavily for S&E workers in the past). Moreover, even if our retention rate remains stable, the intake rate of new students from foreign countries could at any time decline, as immigration restrictions tighten in response to a terrorist attack, or as other countries build up their own R&D infrastructures.

What is the solution? Funding is one part of the answer. There is some evidence suggesting a direct correlation between the level of federal support for math, physical science, and engineering, and the number of students graduating with a degree in these disciplines, especially at the undergraduate level.[9] But how can policymakers better understand the connection? It can be argued, for example, that the doubling of the NIH budget between 1998 and 2003 has encouraged more students to enter the life sciences, and led to a significant growth in the number of life science doctorates awarded.[10] But this has led to an even greater need for NIH funding to support these newly minted scientists.[11] If funding increases the number of practicing scientists in a particular area, won't the rates at which they succeed in winning grants—often pointed to as justification for funding increases—naturally go down unless even more funding is provided? If they do, then it is reasonable to expect that unsuccessful grant applicants will become discouraged, and may abandon science. Meanwhile, the lower success rate in grant applications can itself become a justification for further increases in spending. It is hard to imagine where the cycle ends.

Interestingly, the significant increase in spending on life sciences fields has been accompanied by a growth in the average length of time for training, in particular the time spent as postdoctoral fellows.[12] This means that in general life scientists are in the training portion of the pipeline for much longer than their peers in other fields.[13] Is this good, especially since younger scientists have often proven to be more productive and innovative?[14] Are we perhaps simply allowing more senior scientists to build up major laboratories and then hire graduate students and postdocs very cheaply? Certainly, at present, given the low wages provided to postdocs, a strong case can be made that this issue should be examined closely as new funds are provided for scientific research.

Again, exactly where should the United States invest to ensure an adequate S&E workforce? The picture blurs when one considers that certain fields, like chemistry, computer science, and engineering, lead to employment opportunities in both academia and industry.[15] The recent expansion of the biomedical and pharmaceutical industries has created many new private-sector opportunities for graduates from biological and health-related disciplines. But it is by no means clear that the types of graduates being produced are the ones industry needs. Nor, for that matter, do we know that these graduates are interested in the available industry jobs, rather than traditional academic positions. Further complicating any analysis are differences between specialties: there may be an oversupply of scientists and engineers in some fields, while others desperately need more individuals with specific skills and training.

Aside from the direct impact of federal regulations and priorities, the S&E labor supply and demand is also indirectly influenced by policy from seemingly unrelated areas, particularly national defense. As the United States began to ratchet down defense spending after the Cold War, forcing defense contractors to downsize, the demand for expertise in certain engineering fields decreased significantly. Such adjustments in course are inevitable and virtually impossible to foresee, and thus will always result in unpredictable increases or decreases in the pool of required labor. The goal of intelligent policy should be to anticipate these shifts as accurately as possible, and to develop a system flexible enough to respond.

The tremendous growth in the IT and biotechnology sectors in the middle to late 1990s, mentioned earlier, exerted its influence on the S&E labor market in several ways. Desperate to meet their human resource needs, start-ups and established companies offered salaries and opportunities that the academic sector could never hope to match. Hungry for workers, and frustrated with the relatively narrow skill sets of PhD students, many biotech companies hired large numbers of workers with master's degrees, while IT firms even looked to bachelor's degree holders. Many students who might otherwise have pursued advanced degrees chose to chase a profitable career instead. As an example of one measure, the number of students enrolled in doctorate programs in S&T fields declined. Meanwhile, the perceived workforce shortage, as we have noted earlier, disappeared as the market collapsed, and many of those recently hired tech workers were left without either a job or a degree. Such fluctuations in supply and demand are common in the S&E workforce, making forecasting extremely difficult.

A Historical Perspective

Concerns about the U.S. ability to produce a sufficient supply of well-trained scientists and technical workers are not new. A whole generation of young people was propelled into scientific careers in reaction to the Russian launch of *Sputnik*, in large part through the National Defense Education Act (NDEA). The NDEA marked a whole new era in federal support of education, while *Sputnik* generated a surge in funding for research for both defense-related and non-defense-related research. This era also marked a time when major government policies were put in place aimed at training individuals in specific fields of national need.

In recent years, the federal government's attempts to predict future S&E workforce needs, however, have been viewed with some skepticism. The distrust stems, in part, from studies in the 1980s (often known as the "pipeline" studies) produced by the NSF's Division of Policy Research and Analysis (PRA).[16] The studies predicted a significant shortfall of scientists and engineers into the twenty-first century. These predictions were based largely on a 1989 paper by the PRA, which projected a significant decrease in the number of S&E degrees awarded over the next twenty-year period. The NSF and its allies in the scientific community made these findings the basis for calls to Congress and presidential administrations for large increases in the NSF budget.[17]

The pipeline studies were eventually discredited by economists and policy analysts for failing to recognize all of the external forces involved in the supply-and-demand relationship between S&E graduates and jobs. The NSF researchers had also assumed that the professorate, which had expanded significantly in the 1960s, would shrink in the 1980s and 1990s because of retirements. But many academics remained in their positions well beyond retirement age. Meanwhile, a significant number of those full-time professors who did retire were replaced with adjunct, part-time, and temporary employees.[18] The studies were also criticized by young scientists, who had expected significant career opportunities because of the predicted shortages, only to have those hopes dashed after they received their degrees and hit the job market. In short, the studies were so flawed as to be essentially useless.[19]

The House Science Committee's Oversight and Investigations Subcommittee held hearings in 1992 to scrutinize the NSF's use of the flawed pipeline studies to justify requests for increased funding. Howard Wolpe, the chair of the subcommittee, took the NSF to task. While bending the truth in Washington was certainly nothing new, Wolpe admonished, "nobody expects NSF to play that game. Everyone around here assumes that NSF's numbers are good science."[20]

What can we learn from the pipeline debacle? Clearly, such problems illustrate the difficulty of determining future S&E workforce demands. More than that, they demonstrate the danger of linking future science funding requests to workforce-related claims. Anyone interested in science policy would do well to keep these two lessons firmly in mind.

That said, the NSF, Congress, and the White House work hard to ensure an adequate S&E workforce. In 2002, the Senate appropriations subcommittee that funds the NSF and NASA expressed concern about the United States' ability to produce workers with the necessary expertise in the physical sciences and engineering, and called on the OSTP and NSTC to work with colleges and universities "to increase the number of students pursuing degrees in science and engineering."[21] During the first term of the George W. Bush administration, the NSTC convened a special subcommittee of representatives from multiple federal agencies to examine future workforce needs, which followed up on a similar effort the NSTC had undertaken during the Clinton administration.[22] PCAST has also shown a significant interest in the country's ability to meet future S&E needs.[23]

Looking past the occasional overstatements that have been made about crises, America needs to carefully assess its ability to maintain and expand the S&E workforce. If this is not done, we might wake up one day to find the country critically short of talent in areas crucial to our national security and to our economy.

Current S&E Workforce Data and Trends

In 1999, the size of the U.S. S&E workforce was somewhere between 3 million and about 10.5 million, with the sizable variance due to differences in the definition of the S&E workforce. The Bureau of Labor Statistics fixed the number at about 5.3 million, including individuals with at least a bachelor's degree, regardless of whether the degree was in a S&E area, while an independent NSF survey fixed it at just under 3.3 million by using a stricter definition for inclusion. The same NSF survey counted a little under 10.5 million individuals when it included people who had at least one S&E degree, regardless of their occupation.[24]

So how has it changed? In 2003, the estimated size of the U.S. S&E workforce had increased to between 4 million and 15 million individuals, again depending on what definition was used and who was conducting the survey. When counting individuals with at least a bachelor's degree in S&E occupations the number ranged from 4.0 million to 4.9 million. This is up significantly from 1999. The number of individuals in the workforce having at least one S&E degree in 2003 was 15.7 million, up by 5.2 million.[25]

Again, despite the difficulty of forecasting future workforce needs, some trends can be delineated, which raise interesting policy questions. For instance, while the number of S&E doctorates awarded to women and minority citizens has increased, the number of white men receiving S&E doctorates has been declining.[26] Some find this trend alarming. Does it mean that white males are moving into other fields, such as law, medicine, or business? If so, should this be a cause for concern? On the other

POLICY DISCUSSION BOX 16.1

International Workforce Comparisons: What Is an Engineer and How Many Do We Need?

In October 2005, when the National Academies released a preliminary version of its *Rising above the Gathering Storm* (RAGS) report, it reported that China had graduated 600,000 engineers and India 350,000 compared to 70,000 in the United States.[27] These numbers had been reported by *Fortune* magazine in a July 25, 2005, article after which they were repeated in press stories and by the research and business communities as a rationale for public policies to increase the number of U.S. students pursuing engineering degrees.[28]

It was not long, however, until researchers at Duke University's Pratt School of Engineering questioned the validity of these numbers on the grounds that among the 600,000 Chinese engineers cited in the RAGS report were many who had not received training comparable to U.S.-trained engineers; in fact, they had only completed training equivalent to a vocational certificate or two-year associate's degree.[29] *Wall Street Journal* numbers analyst Carl Bialik and Professor Ron Hira of the Rochester Institute of Technology, a skeptic of S&E workforce shortage claims, were quick to criticize the National Academies for inflating the number of Chinese engineers to support recommendations that would favor producing more U.S. engineers.[30] In response, the National Academies adjusted the number of Chinese engineers downward to 350,000 (in line with the estimates made by the Duke researchers) in its final report and left out India altogether.[31]

In today's increasingly global world, this example raises important questions concerning how policymakers determine the level of scientific talent needed in comparison to other nations. How do policymakers compare cohorts of U.S. science and engineering graduates to those from other nations, given the differing educational systems? Do international workforce comparisons really help policymakers gauge whether the country is ahead or behind? How would you decide the number of scientists and engineers the country needs to ensure future economic competitiveness and what levels of training for them is optimal? How would your view differ if you were the CEO of a high-tech company, president of a research university, or the director of a national laboratory?

hand, given that women make up one-half while underrepresented minorities comprise one-quarter of the total U.S. population, perhaps the data simply depict growing diversity in S&E occupations. What follows are a few more of the salient data and trends.

The S&E Workforce by Sector

Industry. The largest employer of Americans trained in S&E fields is industry. In 2003, private-sector companies employed approximately 59 percent of all people whose highest degree was in S&E, and approximately 33 percent of individuals with a doctorate in a S&E field.[32] That same year, the business and industry sector employed about 68 percent of individuals earning a bachelor's degree in a science, engineering, and health fields and approximately 58 percent of those recently earning a master's degree.[33] As of 2003, approximately half of the recent S&E bachelor's and master's degree recipients were employed in the private sector.[34]

Academia. Most members of the industrial S&E workforce are trained at American academic institutions, which are themselves a significant S&E employer. Graduate students and postdoctoral fellows are often paid to conduct research through federal research grants. In 2003, 16 percent of all S&E degree holders were employed by an educational institution: 9 percent by four-year universities and colleges and 7 percent by other educational entities. Forty-seven percent of individuals with S&E doctoral degrees were employed by educational institutions.[35] Twenty-one percent of individuals with a bachelor's in science, engineering, and health were employed by an educational institution, while this sector employed 28 percent of those with a master's degree.[36] In 2003, approximately 29 percent of all life science (biological, agricultural, and environmental) doctorates holders were employed by academia, in contrast to just over 15 percent of physical scientists, about 11 percent of engineers, and almost 9 percent of mathematicians and computer scientists.[37]

Government. Even taking into account government labs, the federal government is not a huge employer of scientists and engineers. Neither, for that matter, are state and local governments. In 2003, just under 12 percent of bachelor's degree holders working as scientists or engineers held government jobs, working in federal, state, and local governments.[38] Approximately 10 percent of S&E doctorate holders were employed by the government (almost 7 percent at federal agencies and the other 3 percent at the state or local level). In terms of field, nearly 11 percent of PhD life scientists worked for federal, state, or local government, as did 10 percent of physical scientists, about 8 percent of engineers, and just over 6 percent of mathematicians and computer scientists.[39]

Women in the S&E Workforce

A 2004 paper by Building Engineering and Science Talent (BEST) reports that white women make up close to 35 percent of the total U.S. workforce, but no more than 15 percent of the S&E workforce. (White males, in comparison, make up 70 percent of the S&E workforce yet only 40 percent of the total labor pool.)[40] Women of all ethnic and racial backgrounds constitute just over 25 percent of the S&E pool, but nearly 50 percent of the total workforce.[41] Within the general S&E category, women are better represented in some fields than in others. For example, both 1993 and 2003 figures show women constituting over half of the social science workforce (including psychology and economics) and over a third of the life sciences workforce, reaching nearly 45 percent in 2003.[42] The numbers are still low for the physical sciences, although there has been a slight increase between 1993 and 2003: 23 percent to 29 percent.[43] While the number of women entering occupations in the physical sciences has increased, in mathematics and computer sciences the number of women declined by 2 percent from 1993 to 2003.[44]

Across all disciplines, of the full-time employed doctoral scientists and engineers, about 25 percent are women.[45] Again, the majority of women PhDs are in the social sciences, and thus women more often are in social science occupations. Women also are more likely to be in nonmanagerial positions. Women with PhDs in science tend to choose careers in academia more often than men; it is not clear why this is the case.[46]

As noted in a study by the National Research Council's Committee on Women in Science and Engineering, the representation of women in science and engineering has truly changed for the better since the passage of Title IX, the Education Amendments of 1972, and the Equal Employment Opportunity Act of 1972.[47] These changes are apparent not only in the number of S&E PhDs awarded to women, but also in the number of women scientists and engineers in the workforce, and the number receiving full professorships. That said, a recent National Acad-

emies Committee on Science, Engineering, and Public Policy report notes that while women have reached gender parity in several S&E fields, such gains have not been made in leadership positions. Further, women scientists and engineers in the academy often confront barriers and biases that their male counterparts do not. While these impediments are usually unintentional, they have negative impacts on individual careers and the quality and composition of the nation's S&E workforce.[48]

While gains have been made in some areas, we have work to achieve gender parity on *all* levels. Women are still underrepresented in science and engineering; moreover, they tend to be less successful in these careers than their male counterparts.[49] They make up less than 50 percent of new PhDs, are proportionately less likely to take a full-time job in engineering or science, are less likely to hold advanced positions in industry or academia, and receive lower salaries on average (with adjustments made for age, field, and type of work).[50]

Minorities in the S&E Workforce

In 2003, underrepresented minorities (blacks, Hispanics, and American Indians/Alaska Natives) constituted only 10 percent of the S&E workforce, even though they collectively made up nearly a quarter of the U.S. population. Asians, in contrast, were 5 percent of the U.S. population but 14 percent of the S&E workforce.[51] Underrepresented minorities, similar to women, are better represented in some S&E occupations than in others, with the degree and nature of this difference varying from one group to the next. For instance, in 2003, blacks had a stronger presence than other underrepresented minorities among doctoral degree holding S&E workers in the social sciences, computer sciences and math, composing 5 percent of the social science workforce and about 5 percent of the computer and math sciences workforce, combined. American Indians/Alaska Natives were only about 1 percent of the social sciences workforce and 0.7 percent or less of the other fields, while Hispanics were represented in each of the disciplines at a level of approximately 2 to 4 percent.[52]

Figures from 2003 show that about 64 percent of black scientists and engineers, about 65 percent of Hispanic scientists and engineers, and about 64 percent of American Indian/Alaska Native scientists and engineers had completed only four years of higher education (compared to 57 percent of all scientists and engineers). Approximately 9 percent of black scientists and engineers and 8 percent of Hispanic scientists and engineers had completed their doctorate.[53]

Proportionally, blacks and American Indians/Alaska Natives tend to be even further underrepresented in the business and industry S&E sector compared to their white and Asian counterparts (62 percent of blacks and 65 percent of American Indians/Alaska Natives hold business and industry S&E jobs, versus 78 percent of whites and 83 percent of Asians; comparably 75 percent of Hispanics hold business and industry S&E positions).[54] This may be because blacks and American Indians/Alaska Natives are more concentrated in the social sciences, leading to fewer opportunities in the for-profit world. The Asian population, by contrast, is highly represented in engineering, an area rich with for-profit opportunities.[55] Black, Hispanic, and American Indian/Alaska scientists and engineers are also more likely than the members of any other group to have jobs in the government (35 percent of blacks, 22 percent of Hispanics, and 30 percent of American Indians/Alaska Natives, versus 18 percent of whites and 12 percent of Asians combined).[56]

Internationalization of the S&E Workforce

The S&E workforce is increasingly international. Whereas the greatest career opportunities used to be found in the United States, this is no longer always the case. According to the NSF, the increasing globalization of the workforce is particularly evident in two trends: First, S&E employment opportunities are more widely distributed around the globe than they were in the past, both because U.S. companies are expanding their overseas presence and because other countries are building up their own S&E sectors. Second, S&E workers are becoming more mobile, and move easily across international borders to pursue job opportunities.[57] These changes parallel the increasing internationalization of science itself, with frontier research becoming ever more dependent on international collaboration and partnerships.

The mobility of the S&E workforce is likely to grow further, as more and more U.S.-based companies move their R&D operations out of the country. It is estimated that 400,000 service jobs have been offshored since 2000, with jobs leaving the United States at a rate of 12,000 to 15,000 per month.[58] It is estimated that another 3.3 million American white-collar jobs will shift to countries such as India and China by 2015.[59]

U.S. Dependence on Foreign Talent

According to the NSF, in 2003 over one-fifth of the American S&E labor force was foreign-born.[60] Certain fields are much more dependent on foreign talent than others: over a quarter of employees and over half the doctorate holders in engineering-related fields are foreign born.[61] Similarly, in the late 1990s, the total number of foreign graduates students enrolled in U.S. universities in engineering and physical science related disciplines surpassed the total number of U.S. students enrolled in these fields (see fig. 16.1).

In addition to engineering, other areas including computer science, physics/astronomy, and chemistry are also heavily dependent on foreign workers, with these fields being more than one-quarter foreign-born. The nonagricultural biological sciences, economics, and mathematics fields follow closely, with foreign-born individuals making up almost one-fifth of these two areas. Other fields, such as agriculture science, geoscience, and social science fields (e.g., political science and psychology), are much less reliant upon foreign talent.[62] There is no doubt that much of the economic and technology competitiveness of the United States derives from the flow of talent from other countries. A National Science Board report has noted, however, that the United States may not be able to continue to rely on foreign technical manpower to meet its needs.[63]

Science policies, industrial policies, and immigration policies intersect on the matter of access to foreign talent. For example, the expiration of special H-1B visa legislation on October 1, 2003, dropped the annual ceiling on H-1B admissions from 195,000 to 65,000. This had a significant impact on the ability of industry to hire the talent it needed. Since this time, there have been numerous attempts to enact policies that would again raise the ceiling for H-1B workers. Thus far these attempts have been met with limited success.[64] More recently, attempts to lift that H-1B cap have been complicated by contentious debates over illegal immigration and the need for more comprehensive immigration reform.[65]

As we have noted earlier, the competition for S&E professionals is now global. The United States must therefore initiate policies that help the nation attract the best and the brightest from other countries, while also promoting policies that increase the crop of homegrown talent.

Federal Investment in Specific R&D Fields

We return, then, to the question of how U.S. science policy shapes the S&E workforce. As we have pointed out

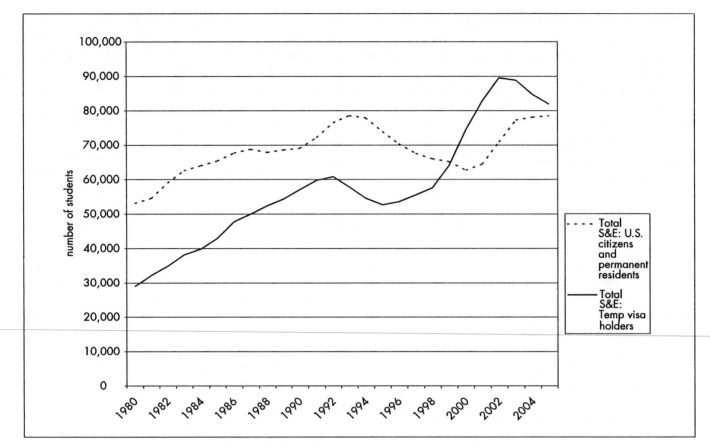

FIG. 16.1 U.S. graduate school enrollments in engineering and physical sciences: U.S. citizens compared to foreign students. Physical sciences include math and computer science. (National Science Foundation/Division of Science Resources Statistics, *Survey of Graduate Students and Postdoctorates in Science and Engineering.*)

throughout this book, a key component of science policy is the amount the federal government invests in research, particularly at the nation's universities. Thus, those who advocate increasing federal research investments often justify their position with claims that these investments stimulate interest among students in particular scientific fields. And there is some evidence that federal investments do indeed have an impact on students' decisions to pursue degrees in those fields.

One study examined the possible relationship between federal research-dollar trends and the number of bachelor's degrees awarded in math, the physical sciences, and engineering, from 1953 through 1998. The results suggest that the amount of funding appropriated for a given area correlates positively with the number of bachelor's degrees awarded in that area.[66]

Another study, conducted by the National Academies, examined federal research funding trends and enrollment

in graduate programs from 1993 through 1999.[67] The results were mixed. The level of federal funding was positively correlated with enrollment in some PhD programs (e.g., physics, chemistry, and the medical sciences), but not in others (e.g., agricultural science, atmospheric science, and civil engineering). These results support the conclusion that while funding may be one factor in students' choices, these decisions can also be influenced by personal preferences, including intellectual interests and the availability of jobs. Furthermore, while federal research funding plays a direct role in prospects for the future academic employment of PhDs, it does not have nearly as much impact on private-sector opportunities (which hold more appeal now than they did in the 1970s or 1980s).

The National Academies study also looked at the overall number of PhD graduates in relation to federal funding levels. Again, results were mixed. Notably, however, the number of foreign students earning doctorates in the se-

POLICY DISCUSSION BOX 16.2

The Debate over H-1B Visas: Does Foreign Talent Displace U.S. Citizens?

In recent years, both American companies and universities have used H-1B visas as a mechanism to hire foreign talent in technically skilled fields where qualified domestic talent, they argue, is in short supply. University officials maintain that H-1B visas help U.S. universities to attract the best talent from around the world to come to work and to conduct their research in the United States helping to make U.S. universities the best in the world. Meanwhile, industry argues that it must import talent because of shortfalls of available U.S. talent in certain technical fields. Proponents of H-1B visas maintain that industry, because it can import the best talent, is not required to move operations abroad.

Some organizations, such as the Institute of Electronics and Electrical Engineers-USA (IEEE-USA), have long lobbied against increasing the number of technically skilled foreign workers in the country on the grounds that these workers displace U.S. workers and drive down wages in high-tech fields such as computing. The IEEE-USA maintains that "dependence on foreign workers may distort the U.S. engineering labor market, adversely affecting educational opportunities for U.S. citizens and jeopardizing future economic growth."[68] They also suggest that "excessive recruiting of highly skilled foreign professionals from certain countries can result in a 'brain drain' that can threaten these countries' prospects for economic development."[69]

Does qualified foreign talent displace qualified U.S. citizens? Or are there no qualified U.S. citizens to hire in certain fields? While no one argues that we should not try to encourage more U.S. students to pursue degrees in science and engineering, could it be that our dependence on foreign talent makes high-tech careers less attractive to U.S. students? What do you think and why?[70]

---

lected subdisciplines mirrored the level of federal research dollars: when funding levels decreased, the number of non-U.S. students earning doctorates also declined.[71]

There is certainly some correlation between federal support of science and the United States' ability to produce scientific talent. The degree of correlation, however, is not at all clear. On the one hand, it is important that the scientific community not rely too much on this connection when calling for increased federal research funding. On the other hand, policymakers in Congress and the executive branch do need to be aware that federal support has a bearing on the appeal of specific fields and, therefore, on graduates' career choices. Without putting too fine a point on it, continued federal support of research is necessary to ensure that the United States can continue to produce S&E talent, especially at the graduate level.

The Risks of Failing to Ensure an Adequately Qualified S&E Workforce

As the National Science Board has pointed out, while immigration may offer industry a short-term solution to its need for S&E workers, reliance on foreign workers "is not an effective long-term strategy."[72] Perhaps the greatest risk of failing to produce an adequate domestic S&E labor pool is the threat a shortage may pose to national security.

In its 2001 report, the U.S. Commission on National Security / 21st Century, chaired by former senators Gary Hart and Warren Rudman, noted that "Second only to a weapon of mass destruction detonating in an American city, we can think of nothing more dangerous than a failure to manage properly science, technology, and education for the common good over the next quarter-century."[73]

In order to ensure America's security, the government must maintain a critical supply of talent in key scientific and engineering fields, at universities and federal laboratories. If this pool is not maintained, then the nation's future security could be jeopardized. In this sense, the future combat superiority of the U.S. military depends on the nation's ability to nurture a labor force capable of conducting the kind of high-risk, long-term research on which the military so often relies.

Any shortage of domestic S&E workers also limits America's global economic competitiveness. In 1998, 150 top decision makers from industry, labor, academia, and the public sector came together under the auspices of the Council on Competitiveness to discuss the future of U.S.

innovation. These leaders agreed that the greatest single national vulnerability was the declining U.S. talent pool in key areas of science and technology.[74] But industry's needs are often different from academia's. Some argue that universities have tended to train people to meet their needs, not those of industry and the nonacademic community. This does seem to be changing. For one thing, it has become more acceptable over the last thirty years for a PhD to work in industry, and industry is, in fact, relying more and more heavily on these advanced degree holders. In yet other cases, companies such as IBM are working with universities to create new education and research programs aimed at addressing the needs of the service industry.

Some universities have responded by creating new master's programs and other special degree programs that prepare students for careers in the private sector. However, these programs have so far failed to address the need for a new breed of academic scientists who themselves are qualified to prepare students for employment in either the academy or industry.

Overproduction versus Underproduction of S&E Talent

In the absence of good data, government policymakers need to make tough choices about whether to support programs that encourage American students to enter certain fields of science and engineering. If they decide to do so, they run the risk of overproducing S&E talent and creating a glut of skilled labor. If they do not, they risk a shortage of talent that will threaten our economic strength and national security.

Until better data become available, the nation would probably be wise to produce more S&E talent rather than less. However, common sense dictates that we not define our objectives too narrowly. Universities must offer training that serves students destined for private-sector S&E careers, as well as those who wish to work in academia. By broadening the meaning of science and engineering training, the country will reduce the chance that it will overproduce scientists and engineers.

The Impending Workforce Crisis at the National Labs

Declines in the number of U.S. students seeking S&E degrees, particularly in the physical sciences and engi-

neering, are a serious concern for government labs and research centers overseen by the Department of Defense (DOD), the Department of Energy (DOE), and NASA. These facilities are often required by law to hire U.S. citizens to carry out their research, and so are particularly vulnerable to a domestic S&E labor shortage. Unless government takes steps to address their needs, the problem may indeed reach crisis proportions.

A 2000 study of scientists and engineers at DOE laboratories found that only 26 percent were under the age of forty, compared to a national average of 40 percent in industry and other government organizations.[75] And in its own 2006 strategic plan, the DOE noted, "The average age of the workforce is increasing, and the number of skilled employees eligible for retirement suggests an impending knowledge and capability gap in the next three to seven years."[76] This graying of the DOE and national labs workforce is particularly disturbing if we recall that it is often young scientists and engineers who drive innovation. The DOE inspector general has highlighted the department's recruitment and retention problems: "The Department has been unable to recruit and retain critical scientific and technical staff in a manner sufficient to meet identified mission requirements. . . . If this trend continues, the Department could face a shortage of nearly 40 percent in these classifications within five years."[77]

Other agencies face similar S&E personnel shortages. With one-quarter of its workforce eligible for retirement within the next five years, NASA has three times as many employees over age sixty as it does under age thirty.[78] Meanwhile, one-third of the civilian scientific and technical workforce at the DOD is currently eligible to retire.[79] At least thirteen thousand DOD laboratory scientists are expected to retire within the next decade, while one-quarter of the current aerospace workforce is nearing retirement age.[80] Because NASA and DOD must often hire U.S. citizens who can receive clearance for classified research, and therefore cannot depend upon foreign talent, this pattern is even more problematic.

The Federal Pay Structure

The civil service system—the federal government pay system that has historically emphasized seniority, rather than performance measures, to determine salaries and rewards—has contributed to the government's problems with recruiting and maintaining top-notch scientific talent. In this system, it has often been the longest-tenured scientists—who may not be the most meritorious—who are the highest paid.

On its surface, this policy may seem equitable. After all, workers in many other fields receive annual cost-of-living pay raises. The problem is that some younger scientists are at least as productive, if not more so, than older scientists.[81] Moreover, scientific talent and performance may vary greatly from one scientist to the next. Indeed, Hans Mark, former director for defense research and engineering, has noted that "the presence of a few individuals of exceptional talent has been responsible for the success (and even the existence) of outstanding research and technology organizations."[82] In recent years, some agencies, such as the Department of Defense, have been working to overhaul their civil service pay systems to better recognize performance.[83]

The S&E Workforce Pipeline

Why Aren't More U.S. Students Pursuing S&E Degrees?

Setting aside any problems with K–12 science education, many very bright students do not reject the option of a S&E career until they get to college. These students may even begin college with an interest in S&E, but quickly conclude that the ratio of effort to return (in terms of starting salary) is not equal to that of a career in business or law. As Daniel S. Greenberg wrote in a 2003 article for *The Scientist*:

> Consider the economic fates of two bright college graduates, Jane and Jill, both 22. Jane excels at a top law school, and after graduation three years later, is wooed and hired by a top law firm at the going rate—$125,000 a year, with a year-end bonus of $25,000 to $50,000.
>
> Jill heads down the long trail to a PhD in physics, and after six Spartan years on graduate stipends rising to $20,000 a year, finally gets her degree. Tenure-track jobs appropriate to her rigorous training are scarce, but, more fortunate than her other classmates, she lands a good postdoc appointment—at $35,000 year, without health insurance or professional independence. Three years later, when attorney Jane is raking in $150,000 a year, plus bonuses, Jill is nail-biting over another postdoc appointment, with an unusually ample postdoc recompense of $45,000 per annum. Medicine and business management similarly trump science in earning power.[84]

If the problem does indeed come down largely to economics—young people's reluctance to invest so much time

for so few rewards—then we might ask why these disparities exist. If a society values science and technological progress, why aren't scientists compensated accordingly? The problem extends beyond academia into government itself. Scientists in the public sector, as we have seen, are tied to a federal pay structure that limits their compensation to the same levels of other career government servants with the same seniority, in spite of their individual achievements. Our society does not assign a high priority to careers in science:

> The payroll for the New York Yankee 40-man roster is 5 percent greater than that for the 1,577 scientists and engineers at the Naval Research Laboratory. This is not to suggest that government scientists be offered million-dollar contracts, but paying a competitive salary to a small number of exceptionally talented individuals would serve the nation's defense interests.[85]

Economics alone, however, do not explain young people's reluctance to become scientists. Stereotypes of what it means to be a "scientist" also pose a significant obstacle. The popular image of an individual secluded in a research laboratory for long hours, interacting only with his lab mice, has few roots in reality (except inasmuch as most jobs in any field occasionally demand long hours). Scientists do much more than work on experiments. And even when they are working in the lab, they often do so as part of a vibrant community. The scientific community needs to more convincingly convey the diverse and exciting challenges of scientific work to young people.

The Postdoc

While individuals with doctorates in other disciplines generally go directly into jobs in academia or the private sector, newly minted S&E doctorates are typically expected to continue their training as postdoctoral fellows. An estimated forty-five thousand or more such postdocs are involved in scientific research across the United States today.[86] Their difficulties, often referred to as the "plight of the postdoc," have come to wide attention in recent decades.

In principle, the postdoc process is supposed to offer the young S&E graduates an opportunity to refine their skills and work more independently, while still being well mentored. This time should be spent writing grants and manuscripts, navigating workplace politics, and developing the other skills that help young researchers succeed in the professional world.

In practice, however, postdoc positions rarely allow young investigators to establish themselves as independent researchers. While some fellows enjoy stimulating and well-supervised experiences, many others are exploited or poorly matched with their programs. In essence, rather than evenly exchanging labor for training, these young researchers find themselves working as adjunct scientists, but with substandard salaries and benefits. As COSEPUP noted in a 2000 report: "Postdocs are central to this nation's global leadership in science and engineering. It is largely they who carry out the sometimes exhilarating, sometimes tedious day-to-day work of research. It is largely they who account for the extraordinary productivity of science and engineering research in the United States. Many among them will discover fundamental new knowledge."[87] Yet a 2000 National Academies survey found that of thirty academic institutions surveyed, only 52 percent provided paid vacation time for postdocs, 45 percent provided sick leave, and only 17 percent required regular performance evaluations. Medical benefits were not guaranteed to all postdocs at any of the thirty institutions surveyed.[88] Meanwhile, the typical length of this financially, professionally, and, often, emotionally precarious time has been increasing, delaying the time when a researcher can finally achieve stable employment, and thus a stable personal and family life. Needless to say, the postdoc period has become one of the worst "leaks" in the S&E pipeline, especially for women and minorities.

Increased attention from the likes of the National Academies, the American Association for the Advancement of Science, and Sigma Xi has brought some improvements to the system, and fostered recognition of the need for further changes.[89] More and more campuses now have postdoc associations, and a National Postdoc Association was created in January 2003.[90] For the first time, an effort is being made to establish uniform expectations among all parties involved in the postdoc relationship (e.g., the fellow, the faculty principal investigator, the university). There has also been attention to ensuring that all postdocs have access to adequate health benefits. It is recognized more and more that an individual should spend no more than five years as a postdoc. One widely accepted notion is that postdocs should be able to move up on the pay scale (and further their title status) after a certain length of time in a laboratory. The goal of a postdoctoral research period is to acquire the skills to be an independent research scientist, whether in academia, industry, or government. This includes much more than laboratory skills:

writing grants, managing a lab, publishing manuscripts, and mentoring. Many universities and postdoc mentors are realizing this by attempting to incorporate these training activities into the postdoc experience.

Policy Options

How can the federal government ensure a healthy S&E workforce? We need to attract more American students to S&T careers. We need to ensure that academia properly prepares students for careers in both academia and private industry. We need to ensure a steady stream of young scientists to replace the graying S&E workforce at the national and other federal laboratories. And we need to moderate our reliance on foreign talent. We have examined many of these challenges throughout the chapter, and now turn to look at the policy options in greater detail.

Reinventing the National Defense Education Act in a New Era of Homeland Security

Recognizing the need for more homegrown talent in key S&E fields, some observers, such as M. R. C. Greenwood, former chancellor of the University of California, Santa Cruz, and associate director for science in the Clinton administration, have led the call for a contemporary version of the National Defense Education Act (NDEA).[91] The new NDEA would reduce the nation's reliance on foreign S&E talent.

In the wake of the attacks of September 11, 2001, a resolution approved by the presidents of the Association of American Universities called upon Congress and the administration to create a new graduate fellowship program that would address critical areas of national need, including in the sciences:

The Association of American Universities strongly urges the Administration and the Congress to establish a new graduate education fellowship program to address national needs in critical and understudied areas of knowledge, including but not limited to language and culture, that are important to the nation's ability to face new national and international threats and challenges. This new fellowship program should by appropriate means address national needs in a comprehensive manner that complements, and does not detract from, current federal fellowship programs. This new fellowship program should strengthen graduate study in a manner similar to that of the National Defense Education Act of 1958 and the way it invigorated graduate study in areas of national priority during the Cold War.[92]

Shortly after its creation, the Department of Homeland Security did, in fact, develop a new national Homeland Security Scholars and Fellows Program for graduate and undergraduate students interested in pursuing scientific and technological careers in relevant areas. Announcing the first rounds of awards in July 2003, DHS secretary Tom Ridge noted that "the DHS Scholarship Program will produce talented and experienced scientists and engineers that will play vital roles in securing America against terrorism." The DHS under secretary for science and technology, Charles McQueary, added: "The scientific community is in great need of new scientists. We believe that our program will not only encourage students to pursue science related degrees but will also begin to fill the gaps left by retiring scientists. This first group of scholarship and fellowship awardees will be the foundation that we can build upon in the coming years that will specifically focus on science and homeland security issues."[93] Funding cutbacks and changing priorities within the DHS have, however, put the Homeland Security Scholars and Fellows Program in jeopardy. The department suspended making new awards in FY2007, signaling an iffy future for the program.[94]

Congress has also been investigating ways to address the government's need to recruit specialized personnel for homeland security projects. One bill, the Homeland Security Federal Workforce Act (S. 589), introduced in the 108th Congress by Senator Daniel Akaka (D-HI) and passed by the Senate, would create a pilot program to repay the student loans of federal employees working in areas of critical importance to national and homeland security. The bill would also charge the Office of Personnel Management with awarding fellowships to graduate students who commit to working for the government in national security positions in the natural, physical, and computer sciences. Despite the bill's good intentions, some critics are concerned that it excludes social scientists, even though there is a great need for economists trained to assess the risk of terrorist attacks.

It is good to see careful consideration given to training American citizens in S&E areas vital to our national security. But why does it take a national crisis to move the government to action on such a critical issue? Crisis is often the engine of policy: it was, after all, the Soviet threat that drove the creation of the original National Defense Education Act. The post–September 11 focus on homeland security presents policymakers with a fresh opportunity to devise programs that address the nation's critical need for scientists and engineers.

Increasing Access and Breaking Down Barriers

Despite an apparent rise in the number of women and minorities pursuing S&E degrees, we still face major obstacles in our quest to increase the number of women and minorities in key scientific and technical fields.

The barriers women face are most often not blatant acts of discrimination. Instead, they are subject to subtler, more institutionalized biases. As one study noted back in 1975: "The principle of the triple penalty, as we have observed, asserts that women scientists are triply handicapped . . . first by having to overcome barriers to their entering science [and engineering], second by the psychic consequences of perceived discrimination—[in the form of] limited aspirations—and third by actual discrimination in the allocation of opportunities and rewards."[95] While the situation may have improved somewhat since 1975, the description is still essentially accurate.

How can we eliminate the triple penalty and increase the number of women in S&E careers? Can the system adapt so that it does not create conflicts between childbearing and the start of a career? The challenge is significant: as anthropologist Sue Rosser found in her survey of the 1997, 1998, 1999, and 2000 awardees of the NSF Professional Opportunities for Women in Research and Education program, 63 percent of the 1997 respondents cited the work-family balancing act as their biggest challenge, with that number rising to somewhere between 73 and 78 percent between 1998 and 2000.[96]

More data on this issue come from a landmark 1999 MIT report, "A Study on the Status of Women Faculty in Science at MIT." This study was initiated in 1995 by Robert J. Birgeneau, then the dean of MIT's School of Science, with interim reports in 1996, 1997, and 1998. It evolved from discussions among three tenured faculty with the dean of science about quality of-life issues for women faculty at MIT.[97] When the discussion was expanded to a survey study, the results were an eye-opener. Junior women faculty reported that they felt well supported within their department, and did not believe that gender bias would affect their careers. Most, however, added that they believed family-work conflicts might have a different impact on their careers than on those of their male colleagues. Senior women faculty, in contrast, reported feeling marginalized and excluded from significant roles within their department or unit. The study also confirmed measurable differences in space, pay, awards, and resources between men and women of equal accomplishment—and documented the persistence of these patterns over successive generations.[98]

The findings led MIT to take the analysis campus-wide and provoked other institutions to undertake similar studies. Around the same time and in parallel, the NSF created and established the ADVANCE Program. ADVANCE, which is designed to increase women's participation in the S&E workforce, encourages the representation and advancement of women in academic S&E careers.[99] It is also supposed to identify ways to improve the climate for women in U.S. academic institutions. The program gives out awards on a competitive basis, to both universities (Institutional Transformation Awards) and individuals (in the form of fellowships).

What has been learned from ADVANCE? A 2002 report by the University of Michigan's ADVANCE Program summarized the findings from its workplace climate survey and focus groups:[100]

Approximately 41 percent of women scientists and engineers reported discrimination in areas such as promotion, salary, and office space, as compared to 4 percent of their male peers.

About 20 percent of women scientists and engineers reported being the object of unwanted and uninvited sexual attention, as compared to 5 percent of the men.

Even though women scientists and engineers sat on committees at a higher rate than men did, they did not chair committees at a higher rate.

The women reported less mentoring than men. The male scientists reported an average of nearly five male mentors in their departments, while women reported an average of two.

While progress has been made, the pipeline is still leaky. A second University of Michigan ADVANCE survey, this one on workplace, also noted that female and minority S&E faculty reported less positive workplace experiences than white male faculty.[101] The report's authors recommended a list of measures: encouraging departments and units to systematically assess their workplace environment; providing department chairs with adequate support and resources for mentoring, problem solving, and conflict resolution; and holding leaders of colleges, schools,

and departments accountable for their efforts to recruit, hire, and retain women and minority faculty.

What is clear from these and other analyses are that barriers still exist for women and minorities wishing to pursue careers in science and engineering. Many of the workplace-related barriers may be overcome by conscious attempts to change individual attitudes and university culture. This, is easier said than done, of course, because as Robert Birgeneau himself noted in the 1999 MIT report:

I believe that in no case was this discrimination conscious or deliberate. Indeed, it was usually totally unconscious and unknowing. Nevertheless, the effects were and are real. Some small steps have been taken to reverse the effects of decades of discrimination, but we still have a great deal more to accomplish before true equality and equal treatment will have been achieved.[102]

But what about cultural obstacles? As we have mentioned, the MIT study suggests that many women are deterred from S&E careers by academia's tendency to create conflicts between career and family. While the workplace has recently become more accommodating to family needs—providing, for example, flexible work schedules, on-site daycare, and paternity leave—obstacles remain. How, for example, should a junior faculty member juggle her plans to have a child with the expectation that she conduct field- (and career-) advancing research at a lab halfway around the world? Until institutions overcome their insistence on fitting working women into a mold, rather than accommodating their efforts to balance personal and professional lives, the barriers will remain in place.

The current shortage of American S&E workers could be addressed by attracting more women and underrepresented minority groups to careers in the field. And, indeed, we have seen a recent proliferation of programs aimed at this goal. In too many cases, however, the government has failed to incorporate funding for program evaluation. Instead, their effectiveness is often simply asserted.

Professional Science Master's Degrees

One possible solution to the workforce problem might be to offer professional master's degrees in science, in addition to the more academic and narrowly focused doctorate. These new degrees, the S&E equivalent of an MBA or law degree, would offer training in key areas of the field (e.g., bioinformatics, nanotechnology, or information technology) that was well suited to the needs of industry, government agencies, and national and independent research labs. Shorter degree times and increased employability would make such programs attractive to many students.

The degree requirements for the PhD tend to be quite rigid. Even if a S&E major were interested in business or policy, he or she would find it difficult to fit a course from one of those areas into a program. In fact, such individuals are practically forced to pursue their degrees serially, earning a master's degree in policy, law, or business after they receive their PhD in science. The alternative would be to provide a degree option that would substantially broaden the training available in current S&E graduate programs. By offering interdisciplinary training in business, law, or public policy, as well as S&E, these new programs would create a more well-rounded and employable graduate.

The idea is already gaining popularity. The Sloan Foundation has helped thirty institutions around the country launch new professional MS (Master of Science) programs in the sciences, mathematics, and engineering. These programs have been designed jointly with industry, thereby ensuring greater professional career opportunities for their graduates. The degrees—in areas including bioinformatics, biotechnology, computational science, energy and environmental sciences, nanotechnology, medicine and health care—also prepare students for careers in consulting, banking, insurance, research management, and technology transfer.[103] Moreover, professional science master's degrees have also been endorsed by leading higher education associations such as the Association of American Universities and the Council of Graduate Schools.[104]

Alternative Scientific and Technical Careers

Of course, many S&E degree-holders are only interested in academic careers. Too often, faculty mentor individuals to pursue academic careers—in other words, to replicate themselves. For a long time, a S&E graduate student opting for a career in industry was viewed as a loss to the academic community. There is still a stigma attached to students' decision to use their training for any purpose other than research. Some professors even view such students as failures.

But there are signs that this attitude might be shifting. In order to foster a healthier relationship between academia and industry, the scientific community must fight the old prejudices. Faculty should devote themselves to training students who are versatile enough to work either in industry or in universities, as researchers, writers, administrators, or managers. Faculty, in other words, must

come to have their students' own future welfare more firmly in mind.

Better Opportunities and Better Pay

The idiosyncrasies of the federal compensation system pose a particular problem for government agencies trying to compete with industry and academia for the best scientific talent.

Interestingly, the Government Accountability Office (GAO) has outpaced all other agencies in its ability to recruit and retain talent. This is particularly surprising because most people do not think of the GAO as an attractive employer for the best and brightest. What is GAO doing right? How can other agencies learn from its success?

One of the GAO's primary advantages is its exemption from many federal civil service laws. It can pay bonuses to top performers, and can avoid the complex federal hiring requirements that bog down so many agencies.

Don't Close the Door on Foreign Talent

While the pool of domestic S&E talent must be increased, the need to transform science education, particularly at the elementary and secondary levels, makes this a long-term project. Realistically, the United States will never be able to meet its entire S&E workforce needs domestically, nor would we want to do so. Indeed, we will always have to rely on at least some talent from abroad.

Even if we could train a sufficient supply of S&E workers domestically, it would not be good for American science. There is, in fact, great value in bringing the best and brightest foreign scientists to American universities. These students and scholars often bring new ideas and knowledge from their home countries. Many will decide to stay, and even obtain citizenship. Those who return to their homes will have been exposed to our cultural values and democratic system of government, an exposure that has transformed many such individuals into international ambassadors of goodwill.

While homeland security is of great concern, of course, a letter on national workforce policies from the Association of American Universities and the National Association of State Universities and Land Grant Colleges captured the essence of the issue: "Today, higher education and science are international activities and require international participation. If the U.S. is to remain the world leader in higher education, science, and engineering and technology it cannot conduct its scientific efforts in a vacuum devoid of international participation."[105]

Policy Challenges and Questions

The federal government has a vested interest in ensuring that the nation has an adequate supply of trained scientists and engineers. But the balance between supply and demand is inherently unstable, governed by factors that are neither predictable nor easily accounted for. This makes anticipating and effectively responding to S&E workforce trends particularly difficult. Is declining American enrollment in undergraduate and graduate S&E programs a national crisis? How can we generate interest in science among our young people? What can be done to increase the success of young independent investigators? What responsibility do the government and the scientific community have for ensuring an adequate supply of scientists and engineers? How much do we depend on foreign S&E talent, and what are the risks of that dependence? Will additional federal support for science translate into increased numbers of scientists and engineers? How does industry fit into the picture, and how closely should educators tailor their curriculum to the needs of the private sector?

When examining our workforce needs, it is important to keep in mind that there are often many questions and few clear answers. That said, we believe that policymakers and scientists should be engaged in a continual dialogue aimed at addressing these questions and other issues raised in this chapter.

NOTES

1. Office of Technology Assessment, *Federally Funded Research*, 230.

2. Shirley Ann Jackson, *Envisioning a 21st Century Science and Engineering Workforce for the United States,* Report to the Government-University-Industry Roundtable (Washington, DC: National Academies Press, 2003), 5.

3. It is worth noting that the numbers can vary depending on the source. While engineering as a whole may have declined, certain subfields, for example, electrical engineering, materials engineering, chemical engineering, may have actually increased. We have relied largely on the National Science Foundation Division of Science Resource Statistics as our source of data. National Science Foundation, Division of Science Resources Statistics, *Science and Engineering Degrees: 1966–2004,* NSF-07-307 (Arlington, VA: National Science Foundation, January 2007), table 19.

4. Ibid., table 5.

5. National Science Foundation, Division of Science Resources Statistics, "First-Time S&E Graduate Enrollment of Foreign Students Rebounds in 2005," *InfoBrief,* NSF-07-312, February 2007, 1–2.

6. Donald Kennedy, Jim Austin, Kirstie Urquhart, and Crispin Taylor, "Supply without Demand," *Science* 303, no. 5661 (February 20, 2004): 1004.

7. For a discussion of current S&E workforce patterns that have negative implications for U.S. economic and national security, see Michael L. Marshall, Timothy Coffey, Fred E. Saalfeld, and Rita R. Colwell, "The Science and Engineering Workforce and National Security," *Defense Horizons,* April 2004.

8. The H-1B visa is for a limited term, as are all visas.

9. A study of this correlation was conducted and is available in an unpublished paper by Merrilea Mayo, David Bruggeman, and John Sargent, "Correlation between Federal R&D Expenditures and Bachelor's Degree Production in the Math, Physical Science, and Engineering Disciplines." A similar linkage to graduate education can be found discussed in the Committee on Trends in Federal Spending on Scientific and Engineering Research, *Trends in Federal Support of Research and Graduate Education,* ed. S. A. Merrill (Washington, DC: National Academy Press, 2001).

10. Geoff David, "Watching a Train Wreck, Part 1," *phds. org,* Science and Engineering Blog, posted December 13, 2006, http://blog.phds.org/2006/12/13/watching-a-train-wreck-part-1 (accessed June 27, 2007).

11. Brent Iverson, before the U.S. Senate Appropriations Subcommittee on Labor, Health and Human Services, Education, and Related Agencies, *National Institutes of Health 2008 Budget,* 110th Congress, 1st sess., March 19, 2007.

12. The length of time to doctoral degree has increased across most fields since the 1970s, although the increase has been slightly larger for life sciences. See Eugene Russo, "The Changing Length of PhDs," *Nature* 431, no. 7006 (September 16, 2004): 382–83.

13. Mary Beckerle, "Stepping toward Independence," *American Society for Cell Biology Newsletter,* May 2006, 2–3, http://www.ascb.org/files/0605pres.pdf (accessed December 15, 2006).

14. Ibid.; Gretchen Vogel, "A Day in the Life of a Topflight Lab," *Science* 285, no. 5433 (September 3, 1999): 1531–32; Board on Life Sciences, National Academy of Sciences, *Bridges to Independence: Fostering the Independence of New Investigators in Biomedical Research* (Washington, DC: National Academies Press, 2005), see especially pp. 2 and 18.

15. Office of Technology Assessment, *Federally Funded Research.*

16. National Science Foundation, Directorate for Scientific, Technological, and International Affairs, Division of Policy Research and Analysis, *The State of Academic Science and Engineering,* NSF 90-35, 1989.

17. An extended discussion of the NSF "pipeline" studies can be found in Daniel S. Greenberg, *Science, Money and Politics: Political Triumph and Ethical Erosion* (Chicago: University of Chicago Press; 2001), 107–28.

18. A National Academy of Sciences report based on a 1998 workshop on scientific labor forecasting pointed to some of the major flaws of the NSF study, noting that the NSF models did not anticipate a deep economic recession or the end of the Cold War. The report also notes that they failed to account for the market mechanisms that operate to bring supply and demand in balance. See National Academies, Office of Scientific and Engineering Personnel, *Forecasting Demand and Supply of Doctoral Scientists and Engineers: Report of a Workshop on Methodology* (Washington, DC: National Academy Press, 2000), 19.

19. During congressional testimony on the topic, John Andelin, then assistant director for science at the Office of Technology Assessment noted, "The long-term accuracy of such predictions isn't sufficient to guide federal policy." When asked about the value of the conclusion made by the House study, Andelin replied that "if you want press attention, yes, it is useful. If you want intelligent discourse, the answer is no." See Jeffrey Mervis, "NSF Falls Short on Shortage," *Nature* 356 (April 16, 1992): 553.

20. Ibid.

21. Senate VA, HUD, and Independent Agencies Appropriations Report, S. Rept. 107-222, July 25, 2002, 99.

22. Under George W. Bush, this committee was convened under the auspices of the National Science Technology Council (NSTC) science committee. For a report issued by the NSTC on this topic during the Clinton administration, see *Ensuring a Strong U.S. Scientific, Technical, and Engineering Workforce in the 21st Century* (Washington, DC: Office of Science and Technology Policy, April 2000), http://www.ostp.gov/html/work forcerpt.pdf (accessed June 23, 2007).

23. President's Council of Advisors on Science and Technology, *Assessing U.S. R&D Investment.*

24. National Science Board, *Science and Engineering Indicators 2006,* 3.5–3.6.

25. Ibid., 3.5–3.7.

26. Ibid., 2.22–2.23.

27. Mark Clayton, "Does the U.S. Face an Engineering Gap?" *Christian Science Monitor,* December 20, 2005, http://www.csmonitor.com/2005/1220/p01s01-ussc.html (accessed June 30, 2007).

28. Geoffrey Colvin, "America Isn't Ready (Here's What to Do about It)," *Fortune,* July 25, 2005.

29. Gary Gereffi and Vivek Wadhawa, *Framing the Engineering Outsourcing Debate: Placing the United States on a Level Playing Field with China and India* (Durham, NC: Duke University, Pratt School of Engineering, Master of Engineering Management Program, December 12, 2005).

30. Carl Bialik, "Sounding the Alarm with Fuzzy Stat," *Wall Street Journal Online,* October 27, 2005, http://online.wsj.com/article/SB113028407921479379.html (accessed June 30, 2007).

31. David Epstein, "The Disappearing Chinese Engineers," *Inside Higher Ed,* June 13, 2006, http://www.insidehighered.com/news/2006/06/13/numbers (accessed June 30, 2007). See also Gerald W. Bracey, "Heard the One About the 600,000 Chinese Engineers," *Washington Post,* May 21, 2007, B03.

32. Ibid., 3.16.

33. National Science Foundation, Directorate for Social, Behavioral, and Economic Sciences, *Recent Engineering and Computer Science Graduates Continue to Earn the Highest Salaries,* NSF 06-303 (Arlington, VA: National Science Foundation, 2006), 5, table 4.

34. National Science Foundation, *Science and Engineering Indicators 2006,* 3.22 and table 3.9.

35. This includes employment at four-year colleges and universities and other education institutions (44 percent and 3 percent respectively). This figure includes more than just tenured or tenure-track positions, for example, postdoctoral fellows, temporary employees, and those individuals holding various research and administrative positions. Examples of other educational institutions include elementary and secondary schools, medical schools, two-year colleges, and university-affiliated research organizations. See National Science Foundation, *Science and Engineering Indicators 2006,* 3.16 and fig. 3.15.

36. National Science Foundation, Directorate for Social, Behavioral, and Economic Sciences, *Recent Engineering and Computer Science Graduates,* 5, table 4.

37. National Science Foundation, Division of Science Resources Statistics, *Characteristics of Doctoral Scientists and Engineers in the United States: 2003,* NSF-06-320 (Arlington, VA: National Science Foundation, 2006), 39.

38. Ibid., 74.

39. Ibid., 26.

40. Shirley A. Jackson, President, Rensselaer Polytechnic Institute, "A Bridge to the Future: The New Engineer: Demographic Issues and BEST Practices," Engineer of 2020, National Academies Conference Center, Woods Hole, MA, September 3, 2002.

41. National Science Board, *Science and Engineering Indicators 2006,* 3.19.

42. Ibid., 3.20.

43. National Science Board, *Science and Engineering Indicators 2004,* 3.17; National Science Board, *Science and Engineering Indicators 2006,* 3.19–3.20.

44. National Science Board, *Science and Engineering Indicators 2006,* 3.20.

45. National Science Foundation, Division of Science Resources Statistics, *Characteristics of Doctoral Scientists,* 7.

46. National Science Board, *Science and Engineering Indicators 2006,* 3.20.

47. John Scott Long, ed., *From Scarcity to Visibility: Gender Differences in the Careers of Doctoral Scientists and Engineers,* a report of the National Research Council (Washington, DC: National Academy Press, 2001).

48. National Academies Committee on Science, Engineering, and Public Policy, *Beyond Bias and Barriers: Fulfilling the Potential of Women in Academic Science and Engineering* (Washington, DC: National Academies Press, 2006), 1–12.

49. Success is defined as progressing through the rank, achieving recognition for research, holding positions of leadership on panels and committees, etc.

50. Long, *From Scarcity to Visibility.*

51. National Science Board, *Science and Engineering Indicators 2006,* 3.20.

52. National Science Foundation, Division of Science Resources Statistics, *Characteristics of Doctoral Scientists,* 99.

53. National Science Board, *Science and Engineering Indicators 2006,* appendix table 3.16.

54. National Science Foundation, Division of Science Resource Statistics, *Women, Minorities, and Persons with Disabilities in Science and Engineering* (Arlington, VA: National Science Foundation), tables H-24, H-32, and H-35, http://www.nsf.gov/statistics/wmpd/pdf/race.pdf (accessed June 22, 2007).

55. National Science Board, *Science and Engineering Indicators 2002,* 3.16.

56. National Science Foundation, Division of Science Resource Statistics, *Women, Minorities,* tables H-24, H-32, and H-35.

57. National Science Board, *Science and Engineering Indicators 2002*, 3.31.

58. Sharon Otterman, "TRADE: Outsourcing Jobs," Council on Foreign Relations Web site, February 20, 2004, http://www.cfr.org/publication/7749/trade.html?breadcrumb=%2Fbios%2F11891%2Frobert_mcmahon%3Fpage%3D3 (accessed January 8, 2008).

59. John C. McCarthy, "3.3 Million U.S. Services Jobs to Go Offshore," briefing paper by Forrester Research, November 11, 2002, http://www.forrester.com/ER/Research/Brief/Excerpt/0,1317,15900,00.html (accessed June 27, 2007).

60. National Science Board, *Science and Engineering Indicators 2006*, 3.35 and table 3.19.

61. Ibid., 3.35 and table 3.20.

62. Ibid.

63. National Science Board, *Report of the National Science Board Committee on Education and Human Resources Task Force on National Workforce Policies for Science and Engineering* (Arlington, VA: National Science Foundation, 2003).

64. For a comprehensive discussion of recent legislation and policy proposals concerning nonimmigrant professional specialty (H-1B) workers, see Ruth Ellen Wasem, *Immigration: Legislative Issues on Nonimmigrant Professional Specialty (H-1B) Workers*, RL30498 (Washington, DC: Congressional Research Service, Library of Congress, May 23, 2007).

65. Jonathan Weisman and William Branigin, "Bush Continues Push for New Immigration Bill," *Washington Post*, June 15, 2007, http://www.washingtonpost.com/wp-dyn/content/article/2007/06/15/AR2007061500843.html (accessed June 23, 2007); "The Legal Visa Crunch," *Wall Street Journal*, May 30, 2007, A18.

66. This study has often been used by groups, such as the Alliance for Science and Technology Research in America (ASTRA), to support the need for increased federal funding of research (Mayo, Bruggeman, and Sargent, "Correlation"). See also Merrilea J. Mayo, "Oversupply, Undersupply: Can We Ever Get the Workforce Issues Right?" *APS News*, November 2003, 8.

67. Committee on Trends in Federal Spending on Scientific and Engineering Research, Board on Science, Technology and Economic Policy, National Research Council, *Trends in Federal Support of Research and Graduate Education* (Washington, DC: Board on Science, Technology, and Economic Policy, 2001), chap. 3, http://www.nap.edu/catalog/10162.html#toc (accessed June 27, 2007).

68. "H-1B Work Visas Enable U.S. Employers to Fill High-Tech Talent Needs," *The Institute*, July 2002, http://www.theinstitute.ieee.org/portal/site/tionline/menuitem.130a3558587d56e8fb2275875bac26c8/index.jsp?&pName=institute_level1_article&TheCat=2201&article=tionline/legacy/INST2002/jul02/fh1b.xml& (accessed June 24, 2007).

69. Institute of Electrical and Electronics Engineers, *Position on Ensuring a Strong High-Tech Workforce in the 21st Century*, February 2000, http://ieeeusa.com/volunteers/committees/cwpc/Aug05/files/9.2.Ensuring_Hi-Tech_WF_Position.pdf (accessed June 27, 2007).

70. For a discussion concerning the pros and cons of the H-1B visas, see "Does Silicon Valley Need More Visas for Foreigners?" *Wall Street Journal Online*, March 19, 2007, http://online.wsj.com/article_print/SB117388283731536825.html (accessed June 24, 2007).

71. Ibid.

72. National Science Board, *Report of the National Science Board Committee*.

73. U.S. Commission on National Security / 21st Century, *Road Map for National Security: Imperative for Change* (Washington, DC: U.S. Government Printing Office, 2001), 30.

74. Hetu and Berry, *Competing through Innovation*, 3. See also Deborah Halber and Denise Brehm, "Summit Highlights Vulnerability in U.S. Education, Research," *MIT Tech Talk*, March 18, 1998, http://web.mit.edu/newsoffice/1998/summit-0318.ht6ml (accessed July 7, 2007).

75. Sandia National Laboratory, *DOE Workforce Issue Paper* (Albuquerque, NM: Sandia National Laboratory, December 2000).

76. U.S. Department of Energy, *U.S. Department of Energy Strategic Plan*, DOE/CF-0010 (Washington, DC: U.S. Department of Energy, 2006), 23.

77. U.S. Department of Energy, Office of Inspector General, Office of Audit Services. "Recruitment and Retention of Scientific and Technical Personnel," DOE/IG-0512, July 2001, 6.

78. National Aeronautics and Space Administration, *NASA's Workforce Plan: For the Use of the NASA Flexibility Act of 2004* (Washington, DC: National Aeronautics and Space Administration, March 26, 2004), 5, http://nasapeople.nasa.gov/hclwp/Workforce-SCREEN.pdf (accessed June 27, 2007).

79. William Butz et al., *Will the Scientific and Technical Workforce Meet the Requirements of the Federal Government?* (RAND Corporation, 2004), 41.

80. U.S. Department of Defense, Research and Engineering, *National Security Workforce: Challenges and Solutions*, slide 12, http://www.dod.mil/ddre/doc/NDEA_BRIEFING.pdf (accessed June 27, 2007); Commission on the Future of the U.S. Aerospace Industry, *Final Report of the Commission on the Future of the United States Aerospace Industry* (Arlington, VA: Commission on the Future of the U.S. Aerospace Industry, November 2002), chap. 8, p. 4.

81. National Institutes of Health, Office of Extramural Research, New Investigators Program, http://grants.nih.gov/grants/new_investigators/index.htm (accessed December 15, 2006); "Youth and Scientific Innovation: the Role of Young Scientists in the Development of a New Field," *Minerva* 31, no. 1 (1993): 1–20.

82. Hans Mark and Arnold Levine, *The Management of Research Institutions: A Look at Government Laboratories*, SP-481 (Washington, DC: NASA, 1984).

83. Christopher Lee, "Rumsfeld Urges Overhaul of Pentagon Civil Service: Pay for Performance: Shift of 320,000 Jobs, Other Major Powers Sought in Legislation," *Washington Post*, April 23, 2003, A33.

84. Daniel S. Greenberg, "The Mythical Scientist Shortage," *The Scientist* 17, no. 6 (March 24, 2003): 68.

85. DeYoung, "Silence of the Labs," 5.

86. National Postdoctoral Association, "Postdoctoral Scholars Fact Sheet," http://www.nationalpostdoc.org/atf/cf/{89152E81-F2CB-430C-B151-49D071AEB33E}/Postdoc_Factsheet_2006.pdf (accessed December 9, 2006).

87. National Academy of Sciences, National Academy of Engineering, and Institute of Medicine, *Enhancing the Postdoctoral Experience.*

88. Statistics taken from the National Academy of Sciences, National Academy of Engineering, and Institute of Medicine, *Enhancing the Postdoctoral Experience,* appendix C.

89. The National Academies have been instrumental in publishing the reports bringing to light the "postdoc problem." AAAS and Sigma Xi are key collaborators with the National Postdoc Association.

90. *www.nationalpostdoc.org* (accessed June 27, 2007).

91. M. R. C. Greenwood and Donna G. Riordan, "Research Universities in the New Security Environment," *Issues in Science and Technology* (Summer 2002): Perspectives.

92. Adopted by the member universities of the Association of American Universities on October, 22, 2002.

93. Department of Homeland Security, "New Homeland Security Scholars and Fellows Program Grant Awards," press release, July 23, 2003.

94. Khodayar Akhavi, "Homeland Security Education: An Iffy Future," *Homeland Security: National Imperative or Business as Usual?* Carnegie Knight Initiative on the Future of Journalism Education, July 26, 2006, http://newsinitiative .org/story/2006/07/25/homeland_security_education_an_iffy (accessed June 23, 2007).

95. Harriet Zuckerman and Jonathan R. Cole, "Women in American Science," *Minerva* 13, no. 1 (Spring 1975):99.

96. Sue V. Rosser, *The Science Glass Ceiling* (New York: Routledge, March 2004). See also T. J. Becker, "Breaking Down Gender Barriers: New Book Looks at Roadblocks Impeding Women Scientists and Engineers," *Georgia Tech Research News,* February 16, 2004, http://gtresearchnews.gatech.edu/ newsrelease/roadblocks.htm (accessed June 23, 2007).

97. For background information see Nancy Hopkins, "MIT and Gender Bias: Following Up on Victory," *Chronicle of Higher Education,* Colloquy online, June 11, 1999, http://chronicle.com/ colloquy/99/genderbias/background.htm (accessed December 9, 2006); Robin Wilson, "An MIT Professor's Suspicion of Bias Leads to a New Movement for Academic Women," *Chronicle of Higher Education,* December 3, 1999; Deborah Halber, "Administration, Women Faculty Welcome School of Science Report," *MIT News Office,* March 31, 1999, http://web.mit.edu/news office/1999/women-0331.html (accessed December 9, 2006); "A Study on the Status of Women Faculty in Science at MIT," *MIT Faculty Newsletter,* March 1999, Special Edition, http://web.mit .edu/fnl/women/women.html (accessed December 9, 2006).

98. "Study on Women Faculty in Science at MIT."

99. The NSF's ADVANCE Program is not a result of the MIT activities. Rather, planning for the ADVANCE Program was already under way, and the launch of the program just happened to occur shortly after the 1999 MIT report.

100. NSF ADVANCE Project, Institute for Women and Gender, "Assessing the Academic Work Environment for Scientists and Engineers" (Ann Arbor: University of Michigan, September 26, 2002), http://www.umich.edu/~advproj/executivesummary .pdf (accessed June 27, 2007).

101. Ibid.

102. "Status of Women Faculty in Science at MIT," 2.

103. See http://www.sciencemasters.com (accessed June 27, 2007).

104. Association of American Universities, *National Defense Education and Innovation Initiative: Meeting America's Economic and Security Challenges in the 21st Century* (Washington, DC: Association of American Universities, January 2006); Council of Graduate Schools, *NDEA 21: A Renewed Commitment to Graduate Education* (Washington, DC: Council of Graduate Schools, November 2005), 9–10.

105. July 1, 2003 letter to Joseph A. Miller, from Nils Hasselmo, President, AAU, and Peter Magrath, President, NA-SULGC.

CHAPTER 17

Globalization and Science Policy

What Is the Role of Globalization in Science?

Modern science transcends geographical boundaries. This can only be to the good: the more widely knowledge is shared, the greater the potential for further advancement. At the same time, however, other nations' technological advancements inevitably threaten America's historical dominance in many primary scientific fields. As William Wulf aptly noted before stepping down as the president of the National Academy of Engineering, "Globalization has introduced both uncertainties and opportunities worldwide."[1]

The term *globalization* is often used interchangeably with *internationalization,* or to describe the result of long-term internationalization. Many people also use the term to refer to the supposed homogenization of global culture. Because modernization and Westernization are often associated with the introduction of new scientific and technical knowledge, science is often accused of disrupting traditional cultures. Even within the narrower framework of U.S. science policy, globalization is at best a mixed blessing: it opens up new opportunities for international scientific cooperation, but also gives other countries the means to challenge American world leadership, thus threatening our economy and national security. The term *globalization* is thus charged with multiple meanings.

These meanings often depend on the specific context in which the term is used. While serving as President Clinton's science advisor, Neal Lane defined scientific globalization as "the increasing necessity of pursuing answers to scientific questions through international collaboration and cooperation."[2] Lane thus defined the word in terms of possibilities, implications, and challenges.[3]

The recent decision by Intel to locate one of its chip-manufacturing plants in China is another example of globalization.[4] Surely this will spawn the creation of another pool of scientists and engineers who will contribute to the advancement of research and development, and create high-level jobs in an area of the world that needs them. But it has raised concerns in many quarters that a technological edge enjoyed by the United States and a few other countries will drift into the hands of what will ultimately be a juggernaut economic competitor.

This chapter looks at the opportunities and challenges for science, science policy, and the nation as a whole that are posed by the increasing internationalization of science, and of economies and cultures.

Science as an International Endeavor

Science has, of course, long been an international endeavor. One would expect as much, given that ideas and principles are not subject to borders and governments. One scientist's discovery can be understood and used by any other scientist in the world. Since the scientific process builds on earlier discoveries, no scientist or group of scientists can work in isolation for very long.

Historically, the United States has benefited from this borderlessness. The notion that electricity could be drawn from the clouds in the form of lightning was first tested in France, before Ben Franklin conducted his famous kite experiment. Indeed, Franklin made so many trips to Europe to meet with his fellow researchers that he was able to take repeated measurements of ocean temperatures in the Atlantic, thus helping to chart the Gulf Stream.[5] The French king, Louis XV, took a strong interest in Franklin and his experiments. Indeed, Franklin's science-based relationship with Louis XV is credited with helping to cement the Franco-American alliance in the American Revolution.[6]

Werner von Braun, one of the world's first and foremost rocket scientists, emigrated to the United States from Germany, bringing with him the expertise that helped build the Saturn V rocket, which carried the *Apollo 11* crew to the moon in 1969.[7] The structure of DNA was discovered through collaboration among American James Watson, the British scientists Francis Crick and Rosalind Franklin, and New Zealander Maurice Wilkins.

But international scientific cooperation is rarely easy. Political regimes change, and alliances can shift quickly. Time, distance, and language are also frequent impediments, although the transformation of English into a lingua franca (an effect of globalization) has helped break down the language barrier. Similarly, the advent of e-mail, the Internet, and the World Wide Web have largely overcome problems of distance and time. The following sections look more closely at communications technology's impact on international science.

The Internet as a Tool for Global Science

The Internet has become a relatively low-cost means of easing difficulties posed by distance and time. Detailed messages can be sent through e-mail at any time of the day or night. Large data sets can be made available on the Web. Documents, images, and audio files can be shared by attachment.

Advances in peer-to-peer (p2p) and voice over Internet protocol (VoIP) technologies have further enhanced communications over distance. The p2p technology is based on a decentralized system in which the network nodes dynamically cooperate in data traffic routing. VoIP is used for Internet-based telephone communications and is exemplified by new and emerging telecommunications services being provided by companies such as Skype and Vonage.[8]

The number of international big-science projects that are dependent upon international collaborations—and monetary contributions from multiple nations—is likely to increase in the coming years. If it does, then Internet communications and cyberinfrastructure will become even more important than they are at present. Not only will good, reliable network links be important, but standards in protocols, language, and usage rules and norms across the different scientific disciplines will become essential.

Indeed, Internet and related science policy issues are likely to take on increasing importance as more international collaborations are formed. Key issues that will need to be discussed include developing uniform Internet standards and protocols, establishing and utilizing grid computing, and ensuring transmission and Web site security. If communications are to take place among scientists seamlessly across national borders, for example, only agreed-upon standards and policies will make that possible. New policies will also be required to advance the quality of the Internet and to ensure worldwide Internet access. Moreover, how does the Internet itself change our own culture, the culture of other nations, and how different cultures interact and behave?

International Big-Science Collaborations: Benefits and Risks

It makes a great deal of sense for the United States to team with other nations on large and expensive scientific projects. Such projects often become the focal point for international scientific collaboration and for the transfer of scientific knowledge across borders.

However, more international projects means more stops and starts, and increased funding uncertainties: for example, many of the lawmakers who pushed to terminate the SSC project based their decision, in part, on a lack of support from the international partners, who were assumed to share in the cost of the project. Any large, high-visibility science project is vulnerable to funding interruptions, and changes in political and public opinion. Subjecting such a project to the whims of multiple, sometimes conflicting, political systems complicates the process exponentially. And despite the partner countries' sincere intention to cooperate, there will always be rivalries.

Still, some aspects of international projects may help overcome these challenges. For one thing, international projects may be more likely to reach completion than those funded domestically. Many formal international collaborations are also based on a treaty that guarantees support for the entire life of the project. A country may be less willing to pull out of a project if withdrawal requires the violation of a binding treaty, or otherwise suggests a lack of commitment to international cooperation.

Globalization: Good for Science

Scientists for the most part look favorably upon the internationalization of their field. They understand that more minds focused on a scientific challenge—from whatever country—will likely result in overcoming that challenge more quickly.

Most professional societies claim international memberships—even those whose names include words like *American* or *National* (e.g., the American Chemical Society). Some associations make it possible for members from developed countries to forfeit their journal subscriptions to colleagues from developing countries. Here in the United States, several federal agencies run programs that encourage collaboration with foreign scientists, including the NSF's NATO postdoctoral fellowship program, and the NIH's Fogarty International Center programs.

The globalization of science is not merely a matter of mechanics, however. It is also a function of the nature of scientific inquiry itself. In recent years, the research community has increased its attention to issues that, by their nature, require international cooperation, such as the AIDS pandemic, SARS, avian influenza, and global warming. A globally connected community of researchers has a much better chance of solving these problems quickly, because collaboration increases the total human power and resources dedicated to the issue, and improves our ability to track phenomena that span geopolitical borders.

When any one nation's policies affecting a specific area of research prove too restrictive, it may develop faster in other nations. For example, strict laws governing human embryonic stem cell (hESC) research in America, have allowed other countries, such as Sweden and Israel, to take the lead in this area. American scientists wanting to conduct stem cell research may seek partnerships with Swedish or Israeli scientists in order to facilitate their work.[9] While advancement in any field is welcomed by the scientific community, regardless of where it occurs, these advances will always be more advantageous to the host country. In the specific case of hESC research, then, the United States risks losing its technical edge to the Swedes and Israelis, even though the first stem cell lines were isolated and established here.

Globalization of science is generally regarded as healthy by the scientific community, and, to the extent acceptable to the values of the American people, having regulations that permit cooperative research opportunities for U.S. scientists will be essential for national competitiveness.

Beyond Science: Globalization's Social and Economic Effects

The globalization of scientific research is accompanied by the globalization of scientific knowledge. Fears about the nation's loss of its traditional scientific dominance can lead lawmakers to formulate policies that try to block the international dissemination of domestically produced knowledge. Such efforts were common throughout the Cold War, fueled by anxieties that the Soviets would gain a technical and military edge. In our own time, the nervousness centers more on national security and America's position in the global economy.

Many of the benefits of globalization have redounded to the United States: the combination of greater educational opportunities and high-speed Internet and telecommunications has allowed India and China to develop a large population of technical experts who can work for local companies or just as easily for American employers. Many of these scientists work much more cheaply than comparable experts in the United States. This has led to a heated debate over outsourcing, especially in the IT industry.[10] Even work that is not high-tech per se, but still relies on technology—such as financial analysis—is increasingly being outsourced.[11] This clearly challenges America's traditional identity as a global leader in innovation, research, and technology. One young Indian entrepreneur noted that, rather than complaining about outsourcing, Americans would "be better off thinking about how you can raise your bar and raise yourselves into doing something better. Americans have consistently led in innovation over the last century. Americans whining—we have never seen that before."[12]

State government officials are taking more initiative to promote economic development, and one way to do this is to attract foreign high-tech companies that want a presence in the U.S. market. States court them by offering land and tax incentives in anticipation of the high-paying jobs that will be brought to the area. Sometimes a governor and legislature will tailor legislation to a particular foreign company they are wooing. In August 2004, Michigan governor Jennifer Granholm intervened after the state's Department of Management and Budget rejected Toyota's bid to develop a new R&D facility outside of Ann Arbor, citing laws requiring that the state retain the gas and mineral rights for particular parcels of land.[13] Because Toyota's plans called for the expansion of an existing facility, Granholm asked state lawmakers to authorize the land sale and waive the gas and mineral rights re-

POLICY DISCUSSION BOX 17.1

A "Flat World": Good or Bad?

Thomas Friedman announced that "the world is flat" in his book on globalization, published in 2005. In *The World Is Flat: A Brief History of the 21st Century,* Friedman describes several stages of globalization and posits that we are currently experiencing globalization 3.0—the world has gone from small to tiny.[14] Friedman views this as a "leveling of the playing field," allowing people and countries of the developing world to take part in the exciting global economy.

This, while beneficial to some nations (e.g., India, China, South Korea), does pose challenges to the United States' continued world leadership in science and its economic superiority. The United States has always been in the forefront of S&T in many fields, giving it a distinct economic advantage over most other countries. As more nations become industrialized, however, and move into a knowledge-based economy, their pools of scientists will increase, along with their ability to compete with the United States in highly technical and industrialized fields. Perhaps ironically, America's increasing technological know-how is the very factor that has brought on this challenge.

Globalization poses broader economic, social, and environmental challenges as well. As more nations become industrialized and benefit from technological advance, their citizens' social and economic status may improve. However, these improvements may at the same time lead to increased air pollution, overcrowded cities, and depletion of natural resources. At the 2001 centennial meeting of the Association of American Universities, Zhihong Xu, president of Peking University, noted that "globalization offers no 'free lunch.' Every process has its cost. Globalization will raise new issues in economic and social development. Economic efficiency may bring greater social disparities. Because globalization restructures the economy, it creates winners and losers, even if the game is not zero-sum. Some sectors and some countries may be marginalized during this process. Efficiency versus fairness will become a major issue."[15]

Will globalization lessen the gap between the United States and other countries? Is there the potential for other nations to overtake the United States in more areas, including science and technology? Is this really a negative thing? How should the global community deal with issues of increased pollution and decreased natural resources resulting from this leveling of the playing field? Whose responsibility is it? How might science help?

---

quirements. The lawmakers passed emergency legislation clearing the way for the sale, and in March 2005, Toyota and the state of Michigan negotiated its final details.[16]

Global Science and the U.S. Higher Education System

The quality of American universities has made the United States the destination of choice for many students seeking a science or engineering education, and for science and engineering postdocs, especially from Asia.

Many other countries, especially Asian nations such as China, have been working strenuously to improve their university systems and research capacity.[17] As they progress, American universities will find it more difficult to attract the best and brightest students from around the world. Given that a significant portion of the S&E student body and workforce consists of non-U.S. citizens, a reduction in the number of foreigners at American universities may ultimately lead to a reduction in the pool of U.S.

scientific and technological talent. While data on "stay rates" suggest that noncitizens usually remain in the U.S. after completing their degrees, these data are limited.[18] As the opportunities to pursue scientific careers at home improve, stay rates will probably decline. Moreover, as overseas educational institutions increase in quality, many outstanding foreign students will likely choose not to come to study in the United States at all. This, combined with evidence indicating a mixed trend in the number of U.S. doctorates being produced, suggests that America's technological competitiveness may be in jeopardy.[19]

Globalization is also altering the nature of academic institutions themselves. Many universities are becoming truly globalized, setting up operations in multiple countries.[20] Specifically, universities are establishing campuses abroad (e.g., Monash University, an Australian University, has set up campuses in Malaysia and South Africa), creating learning centers (the British Open University has branches across Europe, as well as in several non-EU nations), and forming alliances (between Singapore and

MIT) and consortium relationships (the biomedical enterprise program run by Cambridge University and MIT). And, of course, the growing number of online programs are global by their very nature.

American academics tend to see such globalization, for the most part, as an opportunity, not a risk: a greater diversity of experiences for students, new educational opportunities for people in developing nations.[21] However, there are some risks involved. Cross-border affiliations exist at the whim of geopolitics: a change in political leadership can spell the end for the collaboration. Some analysts also worry that globalization aggravates cultural homogenization.

To address some of these concerns, the World Trade Organization (WTO) has considered regulating higher education under the General Agreement on Trade in Services (GATS).[22] The overall aim of GATS is the liberalization of trade in services and the removal of barriers to service trade. Such regulations may be beneficial to services in general, but standardization of higher education could pose unique problems, affecting subsidization, quality assurance, financial aid, and the ability to gear teaching and research to local culture and needs.[23] WTO regulation would move higher education into the global marketplace.

Global Science and the U.S. Workforce

Because nations such as India and China are increasing their technical workforce, and workers from these countries can be hired more cheaply, more and more U.S. corporations are outsourcing their technical needs, from technical support to application development. Even jobs researching technical literature in the life sciences for U.S. pharmaceutical and biotech companies, or interpreting CAT scans, can be done more cheaply overseas.[24]

State and national policymakers are naturally concerned about the exodus of white-collar jobs. In the fall of 2002, New Jersey democratic state senator Shirley Turner introduced a bill to prevent the state from outsourcing New Jersey's state Department of Human Services contract to Mumbai, India; other states have considered similar legislation.[25] In June 2003 the U.S. House of Representatives Small Business Committee held a hearing entitled "The Globalization of White Collar Jobs: Can America Lose These Jobs and Still Prosper?" and in 2004 the General Accounting Office published a study under the title *International Trade: Current Government Data Provide Limited Insight into Offshoring of Services.*[26]

The GAO report was intended to provide Congress with a detailed analysis of outsourcing's impact on the U.S. economy. The lack of data the study found was severe enough to hinder the formulation of the report itself. Surveys of imported services, for example, do not track whether these services were previously provided by U.S. citizens. The GAO report states that "reasons for the rapid growth [in offshoring] are relatively well understood. . . . On the other hand, less is known about the specific extent of offshoring. . . . Federal statistics provide some clues . . . and show . . . [that] offshoring is a small but growing trend in the US economy."[27] Another report, issued in November 2002, predicted that over three million service jobs, including many in the IT sector, would move abroad by 2015.[28]

Not everyone agrees that the future is quite so bleak. While IT-based jobs are indeed moving offshore, these analysts suggest they are for the most part less-skilled jobs in areas such as data entry, while U.S.-based jobs in fields like software engineering and database development actually increased between 1999 and 2002.[29]

As we have seen, the anecdotal evidence points to an American "brain drain." This concern is not unique to America, however, and some other industrialized countries are taking action. At the end of 1999, Canada launched a $585 million program specifically dedicated to reversing the country's loss of S&T talent. The program offers incentives designed to attract foreign and expatriate S&T specialists to Canadian research universities, and to retain Canadian researchers. As of the end of 2002, 35 percent of all new hires made under the program had come from abroad—more than doubling the 13 percent from 2000 to 2001.[30]

More alarming than outsourcing, arguably, is the decreasing number of American students pursuing graduate degrees in the sciences: after all, these are the professionals-in-training who will conduct next-generation research and produce new innovations. As of 2004, only six of the world's twenty-five most competitive IT companies were based in the United States, while fourteen were based in Asia.[31] Two of the top six pharmaceutical companies were headquartered in the United States, based on 2006 revenues.[32]

U.S. Science and Global Comparisons

One indicator of any country's global position in S&T is its rank in patent activity, scientific article output, and number of citations to those articles. Astoundingly, almost half of all U.S. patents were issued to noncitizens as recently as 2003. While this number appears to be leveling off or even declining overall, applications from several countries, including Taiwan and South Korea, have increased.[33] Further, some of the apparent leveling-off might be attributed to the unification of Europe, which has encouraged some inventors and researchers to seek patents in Europe rather than the United States.

While the United States accounted for approximately 30 percent of all publications in the world's leading science and engineering journals in 2003, as a percentage of worldwide scientific article output, this represented a significant decline from 1988. Furthermore, while the total number of articles published worldwide increased by at least one-third between 1988 and 2003, the increase appears to be largely the result of higher output from researchers in Western Europe; Japan; and the emerging S&T centers of East Asia, including China, Singapore, South Korea, and Taiwan.[34] (See fig. 17.1.)

The trends in citations are similar to those for output. In 2003, articles by Americans accounted for just over 40 percent of all citations worldwide. When adjusted to account

for total population, the United States was second only to Switzerland. However, as with article output, citations of works by U.S. authors remained level or declined between 1992 and 2003, while citations of works by authors from East Asia and Western Europe grew substantially.[35]

Meanwhile, a profound shift has been under way in U.S. R&D funding: during the 1970s, approximately two-thirds of the nation's research had been funded by the government; today, industry accounts for two-thirds of all R&D support. Industry is, as we have noted, focused on the "D" in R&D—even much of the basic research done in the private sector is highly directed.

There have been some efforts to reverse this trend, or at least counter its effects: the 21st Century Nanotechnology Research and Development Act, passed by Congress in 2003, not only put into effect a series of programs and activities endorsed by the National Nanotechnology Initiative, but also authorized over $3 billion in nanotechnology R&D funding from FY2005 through FY2008.[36]

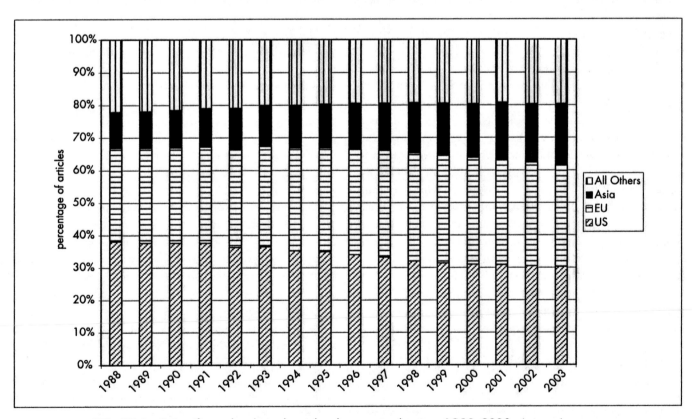

FIG. 17.1 Scientific and technical articles, by country/region, 1988–2003. Asia = Japan, China, South Korea, India, Indonesia, Malaysia, Philippines, Singapore, Taiwan, and Thailand; EU = European Union; US = United States. (Data from Thomson ISI, Science Citation Index and Social Sciences Citation Index http://www.isinet.com/products/citation/; ipIQ, Inc.; National Science Foundation, Division of Science Resources Statistics, special tabulations; National Science Foundation, Science and Engineering Indicators 2006, overview chap., fig. 17; and app. table 5-41.)

It will be important for the government to provide robust support for basic research in areas such as nanotechnology, since this is what will generate new fundamental knowledge needed to drive the economy and is research unlikely to be invested in by the private sector.

In addition to concerns about how the United States is doing in patents, articles and citations, and its investment in basic research, concern also exists that the United States might be losing ground in particular areas of science. For example, by accepting a junior role in the ITER project (see chap. 12), the United States may have pulled itself away from the leading edge in fusion research. It is difficult to imagine how an advanced nation might pass up the opportunity to take the lead in something as important as power generation. The United States chose not to even compete to host ITER. America's decision is symptomatic of a more general decrease in U.S. funding for fusion research since the 1990s, while the fusion programs in Europe and Japan—who fought with each other to host ITER, with Europe eventually winning out—have continued to grow. Meanwhile, South Korea is building an experimental fusion reactor larger than any in the United States.

The President's Council of Advisors on Science and Technology (PCAST) has expressed concern over America's declining significance in global IT manufacturing. Although the United States is still a leader in design, industry experts worry that we are losing our manufacturing advantage.[37] China is seen as a particularly strong contender for global leadership in this area: as the PCAST report specifically notes, "Because of its overwhelming population compared to other Asian competitors, China's rise as a high tech manufacturer has caused increasing concerns. China is a large emerging market and its industrial and economic policies associated with expanding this sector are likely to continue indefinitely."[38]

One of the formal comments submitted to PCAST by the NSF leadership while PCAST was researching its report, nicely summarizes the current state of affairs:

> Civilization is on the brink of a new industrial world order. The big winners in the increasingly fierce global scramble for supremacy will not be those who simply make commodities faster and cheaper than the competition. They will be those who develop talent, techniques, and tools so advanced that *there is no competition*. That means securing unquestioned superiority in nanotechnology, biotechnology, and information science and engineering. And it means upgrading and protecting the investments that have given us our present national stature and unsurpassed standard of living.[39]

Challenges of Globalization for the Scientific Community and U.S. National Science Policy

The effects of scientific globalization reach far beyond the economy. What does globalization mean for nations that have traditionally lacked the expertise to develop weapons of mass destruction? Will scientific collaboration across borders be squelched by the intense concern for national security? Current U.S. policy dictates that American journals cannot review or publish papers by scientists from nations subject to a U.S. trade embargo.[40]

As science becomes more international it is simultaneously becoming more vulnerable to international politics. Modalities for governance and management of facilities need to be carefully selected. While CERN, the high-energy physics center in Geneva, is run by the twenty member states (all European), the Gemini Observatories in Chile and Hawaii, are funded by seven countries but run by a single organization (the Association of Universities for Research in Astronomy, or AURA).

The bigger the multinational project, of course, the more complex the issues involved. How much will each partner country contribute? In which country will the facility be located? How will it be managed? What role will each participating country and its scientists play? While there are benefits to spreading the financial burden of a project among multiple national sponsors and assuring the broadest diversity of expertise, multinational agreements can also be quite inflexible, and are vulnerable to geopolitical forces.

Small science projects, of course, can also span borders. However, because these projects generally involve only a few individual researchers, national governments typically do not get involved, except to enforce some basic limits. An American scientist would not be allowed to collaborate with a scientist from any embargoed country, for example, regardless of the project's size.

A less institutionalized type of globalization has also evolved, which can have its advantages. In January 2004, French scientists called upon their government to restore funding and jobs that had been lost because of budget cutbacks. The subsequent "Save the Research!" campaign drew support from more than thirteen hundred foreign scientists, including National Academy members and Nobel laureates, all of whom signed a petition in support of their French colleagues.[41] This was a case where the scientific community behaved globally, not for research purposes per se, but in support of its members worldwide.

International Organizations That Affect Science Policy

Various international entities exert some kind of influence on America's national science policy, as well as global science policy.

NATO

The North Atlantic Treaty, signed in April 1949, sealed an alliance of nations committed to furthering peace, stability, and well-being in the North Atlantic and worldwide.[42] Today, the North Atlantic Treaty Organization (NATO) has twenty-six member nations.

NATO's Science Committee manages the Security through Science Programme. Originally established in 1958 as the NATO Science Programme, it has undergone several modifications over the years, from its beginnings when it simply trained scientists and engineers and promoted cooperation among member nations to its 2003 incarnation as the Security through Science Programme, which focuses on security-related issues and the partner countries' science priorities, including computer and networking infrastructure and water resource management.[43]

The Security through Science Programme offers fellowships and monetary awards to collaborative projects, including "Science for Peace" R&D projects, advanced research workshops, and reintegration grants. The NSF administered the NATO Postdoctoral Fellow Program, which gave recent doctoral students from NATO partner nations support to work in the United States for one year, until 2004. The program also supported visits by recent U.S. doctoral students to NATO member nations for postdoctoral work.

NATO also supports the Science, Technology, and Research Network, a Web-based set of data sources, available to members of the NATO community, that provides policymakers, scientists, and engineers with access to science and technology research information. Created in 2003, it is managed by NATO's Research and Technology Organization, which promotes cooperative research and information-sharing among the member nations, develops long-term research and technical strategies, and counsels groups and individuals on research and technology issues.[44]

The OECD

The Organization for Economic Cooperation and Development (OECD) was created to administer U.S. and Canadian aid to Europe under the Marshall Plan after World War II. Today its aim is to build and support strong economies, expand free trade, improve economic efficiency, and promote worldwide economic development. The OECD currently has thirty members, with another seventy nonmember countries subscribing to its treaties and agreements.

Because of the strong link between a nation's innovation and research capacity and its economic development and competitiveness, the OECD takes a strong interest in S&T issues. The OECD committee most directly involved in science policy is the Science and Technology Policy Committee, which is charged with "encouraging cooperation among Member countries in the field of science, technology and innovation policy, with a view to contributing to the achievement of their economic, social and scientific aims, including job creation, sustainable development and advancing the frontiers of knowledge" and with integrating "science, technology, and innovation policy with other aspects of government policy, which is of increasing importance in the emerging knowledge society."[45]

The Science and Technology Policy Committee is responsible for promoting information-sharing and discussion of science, technology, and innovation policies among its members; facilitating the identification of best practices; promoting information exchange for the purpose of building and maintaining a healthy R&D infrastructure and workforce; developing international indicators and using them to understand how science, technology, and innovation contribute to increased knowledge production, economic growth, sustainable development, and social well-being; and supporting cooperation on S&T endeavors among member and nonmember nations. Other OECD groups—such as the Committee for Information, Computer, and Communications Policy; the Chemicals Committee; and the committees and boards of the OECD Nuclear Energy Agency—also play a role in science policy-making.

The OECD publishes statistics on R&D within its member countries, covering topics from patents to workforce to industrial and university expenditures. These data are used to assess and compare member nations' S&T policies, and formulate recommendations for best practices. The assessments also help the Science and Technology Policy Committee establish benchmarking protocols and identify high-priority global science policy issues.

UNESCO

The United Nations Educational, Scientific, and Cultural Organization (UNESCO), created in November 1945, fosters peace by reinforcing educational opportunities, scientific exchanges, and cross-cultural relations. It is based on the assumption that all people benefit from science and technology.

UNESCO is made up of five thematic areas, or divisions, plus a special-focus thematic area. One of the five standard areas is a natural sciences division, a "promoter and broker of science," whose goal is to promote "creative science for the benefit of society."[46] This division includes programs on such issues as women in science, universal access to basic science and engineering knowledge, and the protection of biodiversity. Through the natural sciences division, UNESCO also supports several intergovernmental commissions—for example, the Intergovernmental Oceanographic Commission and the International Hydrological Programme. These vehicles allow member states to share scientific and technological information and cooperate to protect natural resources.

The International Council for Science

The International Council for Science (ICSU) was founded in 1931 "to promote international scientific activity in the different branches of science and its application for the benefit of humanity."[47] One of the older nongovernmental entities still in existence, the ICSU is a product of the merger of the International Association of Academies (1899–1914) and the International Research Council (1919–31).[48] Its membership includes both nations and international scientific unions or organizations. This diversity enables the ICSU to draw on a wide range of expertise and a broad representation of interests. One of its guiding principles is the Principle of the Universality of Science, which refers largely to nondiscrimination and equity across all areas of science. The ICSU has promoted this principle, importantly, by defending the right of scientists to associate freely at international scientific meetings.

The ICSU works to mobilize scientific knowledge and resources and to generally strengthen science for the benefit of the global community. Among other things, it promotes a community of scientists unrestricted by race, gender, or political viewpoint; promotes interactions among scientists across all disciplines and national boundaries; identifies issues of importance to both the scientific establishment and society at large, and, just as importantly, works to address them; serves as an independent source of authoritative advice; and stimulates conversations among governments, the scientific community, the private sector, and the global public.

The ICSU works toward these goals by creating interdisciplinary bodies and participating in joint initiatives with other organizations. It has, for example, been a partner in such past programs as the International Geophysical Year (1957–58) and the International Biological Program (1964–74). Current ICSU programs focus on issues including: global climate change, biodiversity, and the impact of humans on the environment.[49]

The InterAcademy Council

The InterAcademy Council (IAC) is a nongovernmental body created in 2000 to bring together the world's brightest scientists and engineers. The council's membership draws from every academy of science in the world, and is run by a governing board including the presidents of fifteen of these academies, from Brazil, China, France, Germany, India, Israel, Japan, Malaysia, Mexico, Russia, South Africa, Sweden, the United Kingdom, the United States, and the Third World Academy of Sciences. The IAC's mission is to provide expert advice to international entities such as the UN and the World Bank. As at the National Academies in the United States, the IAC creates panels in response to requests for data on new issues, and reports its findings and conclusions to the organization as a whole. The IAC has also initiated a panel study on national capacities for science and technology.[50]

The IAC is part of the UN Ambassadors' Symposia, which was organized at the request of UN Secretary-General Kofi Annan to update UN ambassadors on recent developments in science and technology, and their implications for society. In addition, the IAC works with the InterAcademy Panel on International Issues, a global network of science academies launched in 1993 to assist its members, particularly young science academies, in developing the tools they need to advise citizens and public officials. The idea is that networking and communicating will allow these academies to collectively "raise both their public profile among citizens and their influence among policymakers."[51]

The G8

The Group of Eight (G8) is composed of Canada, France, Germany, Italy, Japan, Russia, the United Kingdom, and the United States. The leaders of these major industrial countries attend an annual political summit where they discuss a wide range of economic, trade, security, and social issues. Several subsidiary meetings and policy research activities are also undertaken under the auspices of the G8.

The first of the political summits was held at Rambouillet, France, in November 1975. It was attended by only six of the eight current G8 nations: France, the United States, Germany, Japan, Italy and the United Kingdom (sometimes referred to as the G6). Canada first participated in the meeting a year later. Russia first began engaging with the group after the political summit in 1991 and was allowed to participate more fully in the political summit in 1997, at which time that then G7 was expanded to the G8.

Over time, the G8 has grown increasingly active on international science and science policy associated with issues such as global security, sustainable development, energy, climate change, species preservation, and world health. These science policy discussions have been promoted by meetings of the G8 science ministers and advisors.

The science ministers and advisors of the G8 countries have informally met at least twice a year. The first of these meetings was sponsored by the Carnegie Commission on Science, Technology, and Government. Since that first meeting the G8 science ministers and advisors have been referred to as the "Carnegie Group." The biannual Carnegie Group meetings provide an informal and unofficial context in which to discuss science policy and promote international communication among the G8 leaders on scientific issues.

The Impact of Science on Global Policy Issues

If global policy affects science, the reverse is also true: science has an impact on global policy. The question of how science and technology can improve global welfare has taken on great importance.

In 1999, Budapest hosted the World Conference on Science for the 21st Century, organized largely by UNESCO and the ICSU. The conference began a discussion about how to build a "social contract" for science in the twenty-first century. The participants included delegates from over two hundred organizations—national governments and institutions, educational and research establishments, the scientific community, the industrial sector, intergovernmental organizations (IGOs), and international scientific nongovernmental organizations (NGOs).[52] The participants exchanged opinions on the state of the natural sciences, their future, and their social impact. The discussions were characterized by a desire to ensure that science advanced in response to both society's expectations and the challenges of human and social progress.[53]

Rather than outlining science policies per se, the conference focused on science's power to improve global society. At the end of the conference, participants approved a World Declaration of Science and Use of Scientific Knowledge.

But what impact might the meeting and the declaration have here, within the United States? The declaration emphasizes important components of current U.S. science policy, including public funding of research; high ethical standards, including the free flow of information; and equal access to science. The document is thus likely to reinforce American policies at home, while also encouraging wider adoption abroad. The U.S. government may be asked to support some of the conference initiatives financially, and American scientists will surely be called upon to lend their expertise and support to other countries seeking solutions to global problems.

Popular skepticism toward science around the world can be generally divided into two types: true concerns and challenges, and unfounded accusations and fears. Importantly, the revised science-society contract distinguishes between these two.[54] Even more importantly, scientists and policymakers must now address the former and dispel the latter. How? Scientists can begin by actively engaging policymakers and the public. Public communication of scientific goals, methods, and values is essential. After all, the links between science and society are likely to become tighter and more numerous as global progress is made.

As Bruce Alberts, a conference participant and then president of the National Academy of Sciences noted,

one critical aspect of this close relationship between science and society is the increasing role for . . . "global citizen scientists." Our social institutions have an increasing need for individuals who can stand at the interface between new knowledge on the one hand, and major national and international societal needs on the other hand, and act as a channel to pass information in both directions between them. These individuals have responsibilities that extend both internally to the scientific community and externally to the broader society.[55]

The extent to which science will influence global issues will be determined by the willingness of the U.S. scientific community to engage in global issues in which science can be part of the solution.

In a 2003 editorial in the journal *Science,* UN Secretary General Kofi Annan echoed this sentiment, reminding the scientific community of its indispensable role in addressing current and emerging water, food security, and health crises. He also pointed to the common goals shared by the UN and the scientific community,

> There are deep similarities between the ethos of science and the project of international organization. Both are constructs of reason, as expressed, for example, in international agreements addressing global problems. Both are engaged in a struggle against forces of unreason that have, at times, used scientists and their research for destructive purposes. We share an experimental method; the United Nations, after all, is an experiment in human cooperation. And both strive to give expression to universal truths; for the United Nations, these include the dignity and worth of the human person and the understanding that even though the world is divided by many particulars, we are united as a single human community.[56]

One thing Annan's editorial makes clear is the inadequacy of approaches based on "business as usual." Science's ability to improve the world will not be fully realized until the gap between the haves and have-nots is diminished. Science and policy each have a role in that quest.

It may seem like a contradiction that the United States fights to maintain its global S&T advantage on the one hand, while on the other hand helping other nations to strengthen their research capacity. But this is not the paradox that it may at first seem. The absence of competition is not a desirable end. Each nation plays an important role in the global economy: in order to ensure that each contributes to, as well as benefits from, the collective good, all nations must have a sound science and technology capacity.

Climate Change as an Example of a Global Problem Requiring a Global Solution

Climate change knows no border.[57] It is an issue that, therefore, must be addressed internationally. There is strong scientific consensus on the trend toward global warming and its causes, both natural and man-made. Greenhouse gas emissions and changes in land use—especially deforestation—are two of the prime causes of the warming.

The scientific community has been called upon to assess the causes and implications of global warming. It also plays a role in identifying ways to reduce greenhouse gas emissions, including the development of new technologies.

It will not be easy to find a solution, and, even if we do, it will be difficult to reverse the warming that has already occurred. Rather, many in the scientific community and the general public hope at best to slow or stop the trend and to help humanity adapt to the changes that are likely to result from the warming that has already occurred.

The Kyoto Protocol, an international agreement adopted in December 1997, set binding emissions targets for developed countries at, on average, 5.2 percent below 1990 levels. While the primary responsibility for reducing greenhouse gases rests with those countries that produce the most, developing nations are also subject to limits, although these are for the most part above current emissions levels. The drafters of Kyoto did this intentionally, in order to allow for increased economic development. (Developing nations can also "sell" emissions credits to industrialized nations.) Even though the United States is one of the world's largest greenhouse gas emitters, it pulled out of the Kyoto agreement in 2001, seriously weakening the agreement's force and threatening its chances for long-term success.

Disseminating greater knowledge of the impacts of manmade climate change and building a foundation for actions that can be taken to counteract such change will require global collaboration between both scientists and governments. In 2007, the Nobel Peace Prize was awarded to former vice president Al Gore and the Intergovernmental Panel on Climate Change (IPCC) "for their efforts to build up and disseminate greater knowledge about manmade climate change, and to lay the foundations for the measures that are needed to counteract such change."[58]

The IPCC is an intergovernmental scientific body established by the World Meteorological Organization and the United Nations Environment Programme. The IPCC draws upon and helps coordinate the work of thousands of scientists and officials from over a hundred countries in an effort to provide decision makers and others with an objective and scientifically valid source of information on climate change. Hundreds of scientists from all over the world contribute to the IPCC's work which has been critical in developing an international consensus about the connection between global warming and human activities.

Policy Challenges and Questions

While globalization is good for science overall, it does pose significant new challenges for U.S. policy.

Universities, in particular, are already feeling the pressures of change. In the more security-conscious environment of the post–September 11 era, universities with an international client base of students and researchers are likely to come under increasing scrutiny from policymakers who question their commitment to America's future. We already saw signs of this shift in the early late 1980s and early 1990s, when members of Congress held hearings on a group of American universities, including MIT, that had developed programs aimed at better connecting their researchers with foreign-owned businesses. These members claimed these collaborative activities, especially with the Japanese, were "giving away America's crown jewels of technology" and represented a "circle of shame," whereby U.S.-developed knowledge, produced with federal research funding, was transferred to Japanese companies to be developed into products that were, in turn, marketed to U.S. taxpayers.[59]

Universities have also been criticized for relying on foreign students, especially graduate students. Some policymakers blame universities for this dependence, while ignoring America's inadequate K–12 educational system and the decision by many bright young Americans to pursue more lucrative careers. Just as lawmakers have in the past enacted legislation requiring companies to "buy American," some now seem to believe that U.S. universities, especially state-supported public institutions, should be in the business of educating American citizens only. As we have seen, however, the situation is not nearly so simple.

Further, U.S. industry, like universities, is also tremendously dependent on foreign talent. Indeed, it was industrial leaders who were most vocal before September 11 in supporting legislation that would make it easier for foreign nationals to remain in the United States, because staffing shortages in high-technology companies had reached a critical stage. Major international companies do have the flexibility of moving high-technology jobs to other countries when domestic conditions warrant it. This is, indeed, happening, as many U.S. firms are locating computing and design operations in India, for example, where there is a significant pool of highly trained talent willing to work at salaries only a fraction of those paid in the United States. This is a trend that policymakers who desire a strong U.S. technical workforce must carefully watch.

Policymakers must assess what globalization implies for the United States. How can the United States encourage the development of science and technology in other countries and collaborate when it is in our best interest to do so, yet at the same time ensure a robust national R&D

effort with opportunities for our own students and scientists? This is one of the major science policy challenges facing the nation in an era of increased globalization.

NOTES

1. William A. Wulf, "Changes in Innovation Ecology," *Science* 316, no. 5829 (June 1, 2007): 1253.

2. Neal Lane, remarks made at the Human Frontier Science Program, 10th Anniversary Awards Ceremony, Dwight D. Eisenhower Executive Office Building, Washington, DC, December 10, 1999, http://www.ostp.gov/html/00222_6.html (accessed June 26, 2007).

3. Ibid.

4. "Intel Chip Plant to Energize China," *Washington Times,* March 27, 2007.

5. See Lane, remarks at Human Frontier Science Program.

6. See Ira Flatow, *They All Laughed . . . From Light Bulbs to Lasers: The Fascinating Stories behind the Great Inventors That Have Changed Our Lives* (New York: HarperCollins, 1992), 1–9.

7. For additional information see Marshall Space Flight Center, "Werner von Braun," http://liftoff.msfc.nasa.gov/academy/history/VonBraun/vonBraun.html (accessed June 26, 2007).

8. "Skype vs. Vonage: The 30 Second VoIP Comparison!" http://voiphelp101.com/Skype-vs-Vonage-The-30-second-VoIP-comparison.html (accessed June 26, 2007).

9. American Association for the Advancement of Science, "AAAS Policy Brief: Stem Cell Research," http://www.aaas.org/spp/cstc/briefs/stemcells/index.shtml (accessed March 31, 2007).

10. Pete Engardio, Aaron Bernstein, and Manjeet Kripalani, "Is Your Job Next? A New Round of Globalization Is Sending Upscale Jobs Offshore," *BusinessWeek,* February 3, 2003, 8.

11. Hiawatha Bray, "Passage to India: U.S. Financial Services Firms Plan to Send More Work Abroad," *Boston Globe,* May 6, 2003, 17.

12. Thomas L. Friedman, "It's a Flat World, After All," *New York Times Magazine,* April 3, 2005.

13. Ed Garsten, "Toyota Land Bid Fails: Granholm Steps In," *Detroit News,* Autos Insider, August 31, 2004. In addition to this, DPG-York, a unit of Diversified Property Group, had put a bid on the land wanting to turn it into a housing development. DPG had actually offered over two times as much as Toyota, but because the local township was adamantly opposed to the site becoming a housing development and because Toyota's plans would benefit the state economy, state officials favored the latter. DPG sued the state, delaying the deal between Toyota and the state.

14. Friedman, *The World Is Flat.*

15. Zhihong Xu, "The Role of Universities in China's Economic Development," in *Research Universities and the Challenges of Globalization: An International Convocation,* Proceedings from the Centennial Meeting of the Association of American Universities (Washington, DC: Association of American Universities, April 22–23, 2001), 18.

16. Bryce G. Hoffman, "Toyota Upbeat on Local Land Deal," *Ann Arbor News,* March 10, 2005, http://www.mlive

.com/news/aanews/index.ssf?/base/news-12/111046923135130.xml (accessed June 26, 2007).

17. See Diana Hicks, "S&T Indicators Reveal Rapid Strengthening in Asian Research Systems," presented at the American Association for the Advancement of Science 29th Annual Forum on Science and Technology Policy, April 23–24, 2004, Washington, DC, http://www.aaas.org/spp/rd/hicks404.pdf (accessed June 26, 2007).

18. Data used are usually several years behind. For example, in G. Black and P. Stephan, "Importance of Foreign PhD Students to U.S. Science," prepared for the Science and the University Conference, Cornell Higher Education Research Institute, May 20–21, 2003, http://www.ilr.cornell.edu/cheri/conf/chericonf2003/chericonf2003_01.pdf (accessed June 26, 2007), the most recent data are from 1999. This working paper also refers to a primary source of stay rate data: the Survey of Earned Doctorates, administered by Science Resources Statistics of the National Science Foundation.

19. See Hicks, "S&T Indicators," slides 6–7.

20. "The Globalization of Higher Education," *Tomorrow's Professor,* message 637, Stanford University Center for Teaching and Learning, April 7, 2005; F. Newman, L. Couturier, and J. Scurry, *The Future of Higher Education: Rhetoric, Reality, and the Risks of the Market* (Somerset, NJ: John Wiley and Sons, 2004), chap. 2.

21. "Globalization of Higher Education."

22. Ibid.

23. WTO, *GATS—Fact and Fiction* (Geneva: World Trade Organization, 2001).

24. See Pete Engardio, Aaron Bernstein, Manjeet Kripalani, Frederik Balfour, Brian Grow, and Jay Greene, "The New Global Job Shift," *BusinessWeek,* February 3, 2003, 8.

25. State of New Jersey Senate Bill 1349.

26. Government Accountability Office, "International Trade: Current Government Data Provide Limited Insight into Offshoring of Services," GAO-04-932 (Washington, DC: U.S. Government Accountability Office, September 2004), http://www.gao.gov/new.items/d04932.pdf (accessed June 26, 2007).

27. See ibid.; Jacob F. Kirkegaard, Institute for International Economics, "Outsourcing—Stains on the White Collar?" February 2004, http://www.iie.com/publications/papers/kirkegaard0204.pdf (accessed June 26, 2007).

28. John C. McCarthy, "3.3 Million U.S. Services Jobs to Go Offshore," Forrester Research, TechStrategy Research Brief, November 11, 2002.

29. See Catherine L. Mann, "Globalization of IT Services and White Collar Jobs: The Next Wave of Productivity Growth," Institute for International Economics, International Economics Policy Briefs, PB03-11, December 2003; Kirkegaard, "Outsourcing—Stains on the White Collar?"

30. See Wayne Kondro, "New Research Chairs Mean Brain Gain for Universities," *Science* 298, no. 5600 (December 6, 2002): 1879.

31. "The Information Technology 100 Scoreboard," *BusinessWeek,* June 21, 2004, http://www.businessweek.com/pdfs/2004/0425_it100.pdf (accessed June 26, 2007).

32. See "The Top 20 Pharmaceutical Companies," *Contract Pharma,* July–August 2007, http://www.contractpharma.com/articles/2007/07/2007-top-20-pharmaceutical-companies-report (accessed January 9, 2008).

33. National Science Board, *Science and Engineering Indicators 2006* 6.28–6.30 (see also 5.48); see also Hicks, "S&T Indicators," slide 11.

34. See National Science Board, *Science and Engineering Indicators 2006,* 0.9–0.10 and 5.38–5.40; Hicks, "S&T Indicators," slide 9.

35. See National Science Board, *Science and Engineering Indicators 2006,* 5.45–5.47; National Science Board, *Science and Engineering Indicators 2004,* 5.49.

36. P.L. 108-153, http://frwebgate.access.gpo.gov/cgi-bin/getdoc.cgi?dbname=108_cong_public_laws&docid=f:publ153.108 (accessed June 26, 2007).

37. President's Council of Advisors on Science and Technology, *Sustaining the Nation's Innovation Ecosystems: Information Technology Manufacturing and Competitiveness* (Washington, DC: U.S. Government Printing Office, January 2004).

38. Ibid., 8.

39. Ibid., ii.

40. An additional feature of this policy is that correcting grammatical errors or providing referees' comments is prohibited.

41. For background, see http://recherche-en-danger.apinc.org/article.php3?id_article=222 (accessed June 26, 2007); Catherine Brahic, "French Government Concedes to Researchers," *The Scientist* 18, no. 8 (April 26, 2004): 49; Barbara Casassus, 2004, "Government Dangles Fresh Carrots to Dispirited Scientific Community," *Science* 303, no. 5665 (March 19, 2004): 1749.

42. The preamble of the treaty is as follows:

The Parties to this Treaty reaffirm their faith in the purposes and principles of the Charter of the United Nations and their desire to live in peace with all peoples and all governments. They are determined to safeguard the freedom, common heritage and civilisation of their peoples, founded on the principles of democracy, individual liberty and the rule of law. They seek to promote stability and well-being in the North Atlantic area. They are resolved to unite their efforts for collective defence and for the preservation of peace and security. They therefore agree to this North Atlantic Treaty.

For additional verbiage see http://www.nato.int/docu/basictxt/treaty.htm (accessed June 26, 2007).

43. Partner countries include members of the Euro-Atlantic Partnership Council, Russia, the Ukraine, and some countries of the Mediterranean region.

44. For additional information see http://www.nato.int/science/ (accessed June 26, 2007); http://www.rta.nato.int/Home.asp (accessed June 26, 2007); http://starnet.rta.nato.int/ (accessed June 26, 2007).

45. OECD Resolution of the Council (C(99)185/FINAL), available by searching the Online Guide to OECD Intergovernmental Activity for the Science and Technology Policy Committee, http://webnet3.oecd.org/OECDgroups/ (accessed June 26, 2007).

46. See UNESCO, "Our Mission," http://www.unesco.org/science/science_mission.shtml (accessed June 26, 2007).

47. International Council for Science, "About ICSU: A Brief History," http://www.icsu.org/5_abouticsu/INTRO_Hist_1.html (accessed June 26, 2007).

48. Ibid.

49. Ibid.

50. Interacademy Panel on International Issues, http://www.interacademies.net/ (accessed June 26, 2007).

51. Ibid.

52. Among the participants, 155 countries, 28 IGOs, and 60 NGOs were represented.

53. For background see UNESCO, "Science for the Twenty-First Century," http://www.unesco.org/science/wcs/ (accessed June 26, 2007).

54. For additional background see "Forum II: Science in Society," in Paul Hyningen-Huene, Marcel Weber, Eric Oberheim, "Science for the Twenty-First Century: A New Commitment" Background Document, version 4.0, February 2, 1999, http://www.unesco.org/science/wcs/background/21st_a.htm (accessed June 26, 2007).

55. Neal Lane, "The Scientist as Global Citizen," World Conference on Science, Budapest, Hungary, June 26, 1999, as delivered by Bruce Alberts, President, National Academy of Sciences, http://www.ostp.gov/html/998_23_2.html (accessed June 26, 2007).

56. Kofi Annan, "A Challenge to the World's Scientists," Science 299, no. 5612 (March 7, 2003): 1485.

57. For a discussion of the global characteristics of climate changes, see Eugene B. Skolnikoff, The Elusive Transformation: Science, Technology, and the Evolution of International Politics (Princeton, NJ: Princeton University Press, 1993), chap. 5.

58. The Nobel Peace Prize 2007 Press Release, Norwegian Nobel Committee, Oslo, Norway, October 12, 2007, http://nobelprize.org/nobel_prizes/peace/laureates/2007/press.html (accessed January 8, 2007).

59. Charles M. Vest, "Openness and Globalization in Higher Education: The Age of the Internet, Terrorism, and Opportunity," Center for Studies in Higher Education, Research and Occasional Paper Series, CSHE.7.06, University of California, Berkeley, June 2006, 6.

CHAPTER 18

Science and Homeland Security

How Have Things Changed for America?

The attacks of September 11, 2001, ushered in a new era of homeland security. Whereas the attacks on Pearl Harbor targeted military personnel at a little-known base thousands of miles from the mainland, the September 11 assault targeted civilians in the nation's capital and the world's principal financial center. The strikes on the Pentagon and the World Trade Center set the stage for many changes in U.S. society—changes that will exert a significant impact on science policy for decades to come.

In the latter half of the twentieth century, American defense policy was largely focused on maintaining U.S. nuclear superiority over the Soviet Union. More recently, however, this focus has shifted to deterring antagonistic nations or blocs of nations from secretly developing nuclear or other high-tech weapons, or the systems to deliver them. The surprise attacks of September 11 were all the more surprising because they used relatively low-tech weapons and nonnuclear technologies. Prior to the attacks on the World Trade Center and Pentagon, commercial aircraft were not thought of as potential weapons. The September 11 attacks, like the April 19, 1995, bombing of the Murrah Federal Building in Oklahoma City, illustrate the horrendous violence that can be committed by a determined and well-coordinated group of individuals with even a modest level of technical knowledge.

Science policy, of course, has long been intimately connected with the security of the nation. Even the 1950 act of Congress that created the National Science Foundation stated that one of the agency's primary missions was "to secure the national defense."[1] The September 11 attacks

have spawned a number of new policy questions that must be addressed.

How do we keep weapons of mass destruction out of the hands of those who wish us harm, when so many of the relevant technologies are commonly available, many of which are used by researchers in university, national, and industrial labs? Can we develop technologies that detect threats without infringing on the rights of American citizens? How do we deal with the dangerous vulnerabilities of our growing cyberinfrastructure? How do we determine the proper level of investment in security research and implementation? How can the government encourage private companies to develop and stockpile new vaccines in the absence of a current market? At what point do scientific openness and free exchange of scientific information pose a risk to national security? What role should scientists play in homeland security, and who gets to decide what it will be? Do we depend too much on foreign S&T talent, and if so, what should we do about the paucity of American citizens pursuing S&T degrees? How should universities handle the increased administrative burdens and costs of complying with new homeland security regulations? In short, what does science policy have to do with homeland security?

As many of these questions imply, efforts to protect homeland security can have unintended consequences for scientific research and advancement. To cite just one example, the free and open communication that plays such a vital role in modern, global science is threatened by restrictions meant to prevent the communication of technological information to (and among) terrorists. These well-intended restrictions may unintentionally block valuable exchanges among scientists carrying out routine research. One of the greatest challenges to policymakers and analysts is to determine when the risks of free and open communication outweigh the benefits. It is not entirely clear who should make these decisions—the scientific community or the intelligence and security community.

The challenges facing the nation in this new era are unlike any it has faced in the past, in part because of the widespread availability of very sophisticated technologies, and our increasing knowledge about the world around us. Consequently, the science policies that we develop to address this new world will need to be further-reaching, bolder, and more innovative than any the country has developed before.

This chapter offers an overview of the relevant science and technology, so that the reader can appreciate the origins of current U.S. vulnerabilities and consider why it is proving so difficult to develop policies that help us eliminate them.

Defining the Threat

Clearly, federal investment in research should be used to counter the most plausible, and also the most ominous, threats. But debates rage within the policy and scientific communities about how to make such determinations. Indeed, the needs in this area are so great that the first major university center funded by the Department of Homeland Security has focused specifically on "risk and economic analysis of terrorism events."[2]

The DHS is currently concerned with the threat of nuclear, biological, chemical, radiological, or high-explosive attacks (e.g., weapons of mass destruction), including both detection and response. But some members of Congress are also concerned about cyberattack and are urging that more federal resources be directed toward understanding and preventing cyberterrorism.[3] Increased funding has also been devoted to developing and testing a cost-effective means for protecting commercial aircraft from so-called man-portable air defense systems (MANPADS) and antiaircraft missiles. Finally, the September 11 attacks have prompted increased attention to infrastructural vulnerabilities: water and power supplies; telecommunications networks; and, of course, buildings. The National Academies has conducted studies evaluating the scientific community's role in determining and addressing such vulnerabilities.[4]

The following sections define some of the major threats to homeland security and discuss the role of science and technology in the creation of both the threats and their potential remedies. While what follows is by no means comprehensive, it illustrates the tremendous scope of the problem.

Biological Threats

Terrorism is intended to sow fear among civilians and incite chaos.[5] Few terrorist threats are more alarming than the potential use of biological agents. After all, microbes cannot be seen by the naked eye; individuals may not know when they have been infected; the pathology may not be easily recognizable to physicians; and health support systems may be overwhelmed with both the sick and the frightened. These conditions can easily cause mass hysteria. Bioterrorism captured America's attention after the October 2001 attacks, in which powdered anthrax (more precisely, *Bacillus anthracis* spores) was sent through the U.S. mail. While these attacks resulted in only twenty-two cases of infection and five deaths—compared to the over three thousand people killed in the World Trade Center and Pentagon attacks—the incidents moved bioterrorism to the top of the list of homeland security funding priorities.[6]

Biological agents are biomaterials intentionally used to harm or kill humans, animals, or plants. These agents differ from their chemical kin in that they are living organisms—microbes, viruses, rickettsiae, or fungi. They are effective weapons because relatively small doses can infect a target group: these agents are extremely virulent, can be widely dispersed, and are not susceptible to standard medical treatment. Because the period between infection and onset of symptoms is typically quite short, and the disease may initially elude diagnosis or treatment, civilians can unwittingly become carriers, dispersing the agents throughout the population. A few carriers can inadvertently expose thousands, including children and the elderly, who are typically much more susceptible to infection.[7]

Chemical Threats

Chemical weapons, unlike biological agents, do not cause disease. Rather, they are gases, liquids, or solids that have a toxic effect on humans, animals, and plants. A 1969 UN report on chemical and biological weapons includes both riot-control agents (such as Mace and pepper spray) and herbicides as chemical weapons.[8] Chemical weapons are categorized as lung agents (which irritate the eyes and throat and, in sufficient concentrations, burn lung tissue), blood gases (which, when inhaled, block oxygen transport by red blood cells), blistering agents (which burn and blister exposed tissues), and nerve agents (which impair neuronal transmission and cause uncontrolled muscle activity, usually ending in cardiac or respiratory failure).[9]

The first large-scale use of chemicals in war occurred during World War I, and their effects were devastating.[10] Some of these substances, when conveyed through aerosols or via ventilation systems or bombs, can be massively destructive. Others, including toxins (chemical substances secreted by living organisms) and genetic peptides (agents of biological origin) can severely damage the physiology of plants or animals.

Chemical weapons have not garnered as much attention as bioweapons, in the aftermath of September 11. This is not to say that they are not cause for concern. The terror that can be imposed by the use of chemical weapons is significant, as demonstrated during the 1994 and 1995 sarin nerve gas attacks in Tokyo and Matsumoto, Japan. These attacks killed twenty people, but many more were affected. And were it not for the quick response of Japanese authorities, and the ineptitude of the terrorist group behind the attacks, the results could have been much more devastating.[11] As this incident suggests, public gathering places and mass transportation systems are particularly vulnerable to chemical attacks.[12] Unfortunately, nerve agents such as sarin are probably in the possession of several terrorist groups, offering an effective weapon to motivated individuals.

Some nations, including the United States, have stockpiled *biowarfare* and chemical weapons at certain points during the twentieth century. However, military leaders have generally been wary of their unpredictable lethality, geographical distribution, and effect on friendly troops.[13] Rogue groups, on the other hand, have been attracted by biological agents' ease of cultivation, concealment, transportation, and distribution; and by their power to instill fear in the broader population. It is this threat of *bioterrorism* (as opposed to biowarfare, which primarily targets military personnel) that is at the crux of current discussions about biological weapons policy.

How is it that bioterrorism has so quickly become a threat that terrorists can now exploit? As Lewis Branscomb points out, many of our vulnerabilities today are, ironically, the result of our own scientific advances.[14] The development and subsequent streamlining of the polymerase chain reaction process has made molecular biological lab work much simpler and quicker than it was prior to 1986. Many organisms, including bacteria, have now been sequenced, and genetic engineering is a routine fact of life in legitimate research labs. Indeed, research-

ers with rather modest technical facilities can now easily turn nonvirulent strains into deadly pathogens. A technician needs little more than access to a few specimens, and to basic recombinant DNA and cell multiplication techniques, in order to produce mass quantities of deadly biological agents. In short, we must now learn to protect ourselves against the biological agents we have already learned how to create.

How did we put this cart before the horse? Why did we not consider the dangers beforehand? The answer is simple: the development of new scientific capabilities typically generates a great deal of excitement about new achievements and their potential benefits (in this case, to human and animal health). Very little attention was paid to how these technologies could be used by unscrupulous individuals to create terror and inflict great harm.

So why not simply deny terrorists access to the organisms in the first place? This is simply not possible. Many of the biological agents used in biowarfare are, after all, found naturally in the environment. Anthrax spores can be obtained from soil samples. Spoiled food contains botulism spores. And ricin can easily be rendered from common castor beans. An individual or group intent on inflicting harm on civilians, and equipped with basic scientific knowledge, will find it relatively easy to acquire the necessary agents and produce enough of them to mount a significant attack on an unsuspecting population.

Aggravating these problems is the difficulty of detecting the production and transportation of such agents. A facility that produces beer and wine, pasteurizes milk, or manufactures medicines can also produce bioweapons. We do not yet have the technology to quickly scan and detect most biological agents in transit.[15] It is clear, in other words, that any effort to strengthen U.S. homeland security must include significant attention to biological agents, and should marshal the talents of scientists at universities and our national labs. In a panel discussion on Capitol Hill sponsored by the Center for the Study of the Presidency, the late Nobel laureate Joshua Lederberg noted that America has to better equip itself to detect bioagents and rapidly treat infected civilian populations. Every hour lost in identifying the active agent may translate into the loss of millions of lives. While increased federal funding will be required, a well-defined strategic plan must be endorsed by the key actors before such funds are committed.[16]

The war on bioterrorism bears many consequences for national science policy. So far, it has led to a redirection of biological research toward the rapid identification of biological agents, the preparation of first-line responders, and the development of vaccines and inoculations for a spectrum of new diseases. This shift in priorities, particularly at the NIH's National Institutes of Allergy and Infectious Diseases (NIAID), has provoked concerns among groups advocating for research on infectious and debilitating diseases such as AIDS, SARS, West Nile virus, and asthma. Some of these groups have urged administration officials to ensure that additional funding for bioterrorism and biodefense research does not "come at the expense of research on other life-threatening illnesses."[17] Meanwhile, other critics are questioning whether we should be increasing our efforts to combat bioterrorism when we face a much greater risk of mass fatalities from a naturally occurring epidemic, of the type that could be caused by a particularly virulent strain of influenza.[18]

Nuclear and Radiological Terrorism

The nuclear bomb is one of the most widely acknowledged and frightening weapons of our time. During the Cold War, the parameters of nuclear weapons were relatively contained: only a few major countries had the technology necessary to produce such weapons, so everyone knew who had them, how many they had and of what type, and at what targets they were aimed. Adversaries shared a recognition of the threat posed by mutually assured destruction (MAD), which held that both sides would be annihilated in the aftermath of a nuclear first strike.[19] The fear of MAD created a stalemate in the midst of very dangerous circumstances, and thus, oddly, fostered a geopolitical equilibrium that persisted for several decades.

Since the end of the Cold War, however, the situation has changed dramatically. First, small terrorist groups now have opportunities to acquire or produce these weapons, meaning that we no longer know precisely who has nuclear capabilities. Second, technology has permitted the packaging of devastating nuclear weaponry into very small profiles, some the size of a briefcase, which are easy to transport and hard to detect.

It is not far-fetched to imagine that a single terrorist could transport a suitcase-sized nuclear weapon on a commercial flight into the United States, carry that device into a critical area of a major city, and then detonate it. The explosive force of such a device would be relatively modest, and might directly harm no more than a few hundred individuals, but that is beside the point. Terrorists' ability to carry out a nuclear attack on American soil would set off the nationwide panic that they seek.

While most persons believe that plutonium (the key ingredient in nuclear weaponry) is rare and difficult to obtain, hundreds of tons of this material are processed, stored, and circulated across the globe every year, in the course of civilian nuclear commerce. Since it takes less than eighteen pounds of the material to make a nuclear bomb, it is entirely feasible that a terrorist group could procure sufficient material through illicit purchase or theft to achieve a targeted objective.

There are two types of "nuclear terrorist" threats. One, like our hypothetical suitcase, involves an actual nuclear device, with its explosive impact and radioactive fallout. The other involves a conventional bomb packed with radioactive material that is dispersed over a significant area by the explosion; this is commonly referred to as a "dirty bomb," or radiation dispersal device. While terrorist groups may not have the resources to build a fully functional nuclear bomb, they could steal enough radioactive material to fashion a dirty bomb. In any event, a terrified public in the midst of a nuclear event would likely be unable to make the fine distinction.

As with biological weapons, our nuclear dilemma is an unanticipated consequence of our own technological advancements. A few decades ago, it would have been impossible to build a nuclear weapon the size of a briefcase. Now, with advances in electronics, miniaturization, and nuclear technologies, a well-funded terrorist group might be able to procure and deploy such a weapon at a time and place of its choosing. Given the reported efforts of rogue nations and groups to acquire nuclear technology, prudence suggests that this threat must be taken very seriously.

Many of the nations that already possess nuclear capability, such as Pakistan, are subject to political instabilities that could provide extremist groups with access to much more destructive weaponry. The breakup of the Soviet Union provides a slightly different example of how a country's nuclear know-how can be dispersed to rogue groups. In the wake of their country's collapse, Soviet scientists experienced a rapid fall of their own, from highly valued and relatively well-paid jobs to unemployment and poverty. Other nations eager to acquire nuclear technology were only too happy to employ these embittered or desperate scientists. While the United States mounted a major effort to offer Russian scientists alternative sources of employment during and after the collapse of the Soviet Union, some observers still worry that not enough was done to prevent the spread of potentially harmful information.[20] More advanced plans must be put in place in anticipation of such events in the future.

Bombs are not the only nuclear threats we should be concerned about. American nuclear reactors, which play a vital role in the domestic energy infrastructure, are vulnerable to conventional terrorist attacks. An attack on a large reactor could cripple delivery of power to an entire region and spread radioactivity over a vast area. If we include the fatal cancers caused by radioactive contamination, such an attack would result in several times as many fatalities as were caused by the September 11 attacks.[21]

Now that the nuclear genie is out of the bottle, containing it will be one of the greatest challenges facing the United States in the decades ahead. How might this affect science and policy? First, concerns about the nuclear threat could well encourage stricter limits on the enrollment of foreign students in relevant fields of study and research. (In the time period between the war initiatives Desert Storm and Enduring Freedom, when the United States expressed concern that Iraq was developing nuclear weapons, U.S. universities were simultaneously training many Iraqi students in nuclear physics and related disciplines.)[22] Nuclear concerns will also focus attention on the detection and tracing of nuclear materials, creating whole new areas of research at universities and national labs. The threats, as we have seen, are serious and varied; the U.S. response must be wholehearted and flexible.

Cyberterrorism

In the simplest terms, *cyberterrorism* is the creation of general fear and suffering through the abuse of computer networks.[23] Since widespread use of the Internet itself only became a reality in the early 1990s, cyberterrorism has only recently been recognized as a possible source of wide-scale disruption. Indeed, cyberterrorism is such a new concept that many people still cannot imagine how the Internet could be used to cause anything beyond the minor annoyance of a spam-filled inbox. Surely, the inconvenience of struggling to filter one's e-mail should not be equated with the loss of life from a more conventional attack. In fact, the Internet can indeed be used to disrupt human activity on a major scale—with likely casualty figures over time even exceeding those associated with the World Trade Center attacks. The United States, after all, relies on computerization for nearly everything from keeping violent prisoners locked down to powering hospitals, to securing nuclear power plants, to the mixing of baby formula in factories.[24]

The United States has come to depend on the Internet in ways most citizens do not recognize. Networked computers control airline traffic, banking transactions, and the operation of hydroelectric dams, to mention just a few examples. The pervasiveness of networks is comparable in important ways to the pervasiveness of potatoes in the diet of nineteenth-century Ireland. As we know, the fungus-induced failure of potato crops—the primary source of sustenance for much of Ireland at the time—resulted in widespread famine and the loss of millions of lives. A series of effective cyberattacks on a network-dependent society could have an even more immediate and terrifying effect than a famine, which does its deadly work over months and even years. A coordinated attack on American networks could create the impression that the entire world was coming apart in a single instant.

Computer hackers have existed for as long as there have been computers. Indeed, there is a whole subculture of individuals who try to circumvent the latest security measures and gain access to sensitive computer data and systems.[25] Some hackers regard their work as nothing more than a test of wits, while others hack for criminal purposes, stealing corporate secrets, seeking retribution against former employers, or appropriating personal information for fraudulent use. Still others hack just to demonstrate how much havoc they can wreak on society (e.g., denial-of-service attacks on large Internet service providers like AOL). Normally, such persons do not intend to cause loss of human life—just general mischief and misery. But there are also organized and highly motivated groups intent on inflicting terror and who fit our definition of cyberterrorists. Indeed, the term *cyberterrorism* has been defined by one expert as "hacking with a body-count."[26]

Science and cyberterrorism are inextricably linked by the origin of the World Wide Web, initially invented to support scientists' sharing of data. Any actions that limit the usefulness of the Web for open communication thus have an impact on scientific research. Nonetheless, new restrictions are certainly on the horizon; as the Web has become fully integrated into our society, it has created associated vulnerabilities that society cannot ignore.

What can be done? Cybersecurity research has been supported only modestly in the past, usually within the computer science field. Indeed, up until the September 11 attacks, this area was never assigned a very high priority. While that is changing now, major questions remain unanswered. How much money should be set aside for such research? How should funding be administered (e.g., should a set of national centers be established)? What network architectures should the government promote in order to ensure maximum robustness of the Internet?

We find ourselves in something of a predicament with regard to cybersecurity because we are much better at creating and extending networks than protecting them. We know more about reaching out and including people than we do about excluding them. As a consequence, the focus to date has not been on security. Even in corporate America, where losses due to intrusions already amount to hundreds of millions of dollars each year, there is a culture of underreporting incidents because the potential loss of consumer confidence would be even more costly. This laxity will certainly cease as we devote ever more attention to homeland security.

There are two principal reasons why cyberterrorism must be faced aggressively. First, we are rapidly increasing our dependence on the Internet. In the United States, Internet use has increased by about 20 percent per year since 1998: as of 2001, about two million new users were going on line each month.[27] The second reason for increasing security is that right now cyberattacks are virtually untraceable, and can be launched from anywhere in the world, at any time.

A report by the Center for Strategic and International Studies (CSIS) noted that a terror group of only thirty individuals strategically located and funded at the level of $10 million could unleash an attack on the United States with a lethality comparable to that of the September 11 attacks.[28] The report also observes that rogue countries are seeking the services of talented hackers. The militaries of many countries—including the United States—have incorporated cyberintrusion tactics into their tool sets. In January 1999, recognizing the increasing threat that a potential cyberattack posed to our increasingly computer-dependent infrastructure, President Clinton proposed $1.46 billion in funding to address these issues in the FY2000 budget.[29]

Efforts to regulate cyberspace may be necessary, but we still have only the vaguest understanding of how these changes will affect the way scientists use the Web. University researchers are being urged to secure their networks against cyberattacks, and information is being removed from government and public Web sites for fear that it might fall into the wrong hands (especially information about sensitive but unclassified research on subjects such as "critical infrastructure" and "select biological agents").[30] Meanwhile, little if any funding has been provided to universities, research labs, or industry to make these changes. Nor do most Americans fully appreciate the significance of these changes for the free and open flow of scientific information.

As public and expert demands to limit the transmission of information through cyberspace grow, nongovernmental entities like the CSIS and the Center for Security Policy will need to speak up ever more effectively about the importance of openness to scientific progress. And universities and the broader scientific community will need to articulate how excessively rigid constraints could diminish the Internet's value for general research and education purposes. Ultimately, a careful balance must be struck between homeland security and the free exchange of ideas.

Cyberterrorism is a growing threat that will occupy the attention of the executive branch and Congress for years to come. The OSTP, DOD, NSF, DOE, DOC, and other agencies will also have to be involved in the effort. Policymakers may need to accept that R&D funding on relevant subjects will have to be increased, in order to create research centers dedicated to combating this threat. Just as cyberterrorism will change the way researchers use the Web, so, too, the scientific community will have to organize a response that positively influences the government's choice of policy and regulatory efforts.

Critical Infrastructure

Citizens expect a certain level of stability in their environment. They expect transportation systems to function, telephones to work, financial institutions to process their transactions. They presume their food will be safe, energy will be available, and protection against both natural and manmade disasters will be in place. Citizens also expect that their national monuments will remain intact, their government will function, and their nuclear power plants and hydroelectric dams will operate safely.

It is thus understandable why terrorists would seek to disrupt these services. An explosion at a nuclear power plant, for example, would not only deprive citizens of electric power, but would spread fear and instability. This is why the Department of Homeland Security's mission is not only to defend the United States against catastrophic terrorism, but also to protect critical national infrastructure.

The National Strategy for the Physical Protection of Critical Infrastructure and Key Assets, issued by the White House in March 2003, describes the key objectives of this campaign, including identifying and assuring the protection of assets critical to public health and safety, governance, economic and national security, and public confidence; providing timely warnings and assuring protection against specific or imminent threat; and assuring the protection of other assets that may become terrorist targets over time, by creating a collaborative environment in which federal, state, and local governments and the private sector can coordinate their security efforts.[31]

These changes raise many issues for science policy. Unlike national security, which is primarily the responsibility of the federal government, homeland security requires a very high level of cooperation among local, state, federal, and private sectors. As citizens were reminded during the major power outage of August 2003—probably the largest in U.S. history, affecting an estimated fifty million people—much of the electrical grid for the United States is under the control of private companies.[32] It can take officials many hours, and sometimes days, to determine the source of a catastrophic problem with the power supply.

The formation of an interconnected network of university faculty, national laboratory scientists, and state and local policymakers will be critical to effectively fight against cyberattacks. The various players will have to be very clear about their individual roles and responsibilities in such a scheme. Advanced technologies will be needed for surveillance, detection, and tracking purposes. Incentives will be necessary in order to encourage private-sector R&D in areas that may not otherwise be viewed as immediately profitable. Robust communications systems will demand close attention to the development of diverse routing pathways. The nation will need enhanced capabilities for personnel surveillance, more sophisticated modeling and simulation capabilities, stronger links between the public health and health care systems, and stockpiles of medical supplies.

Critical infrastructure touches virtually every aspect of our lives, from electricity, oil, and natural gas supplies and delivery systems; to air travel, mass transit, and highways; to manufacturing and the control of toxic and hazardous wastes. The critical economic role of the U.S. Postal Service (at over $68 billion operating revenue in FY2003) as part of the national critical infrastructure is often overlooked.[33] Nuclear power plants account for almost 20 percent of the U.S. energy produced, while approximately 2 percent of the nation's electricity is generated from hydropower. Moreover, in some states as much as 70 percent of electricity is derived from hydropower.[34] The security of both of these sources must be a continuing concern.

In other words, many, many complex processes have to operate smoothly and harmoniously in order to maintain our high quality of life. Unfortunately, any one of them could be disrupted with relative ease. Because the components of this infrastructure are so intricately interconnected and their operations are based on advanced technologies, any effort to repel threats to the system will have to rely on fundamental scientific research and sound science policy.

The Role of Science and Technology in Creating the Problem and Contributing to Its Remedy

The tragedy of September 11, 2001, has focused intense scrutiny on the question of how to protect ourselves against such attacks in the future. Many of the devices and techniques that could be used against us—devices that help people see objects invisible to the naked eye; that make it possible to hear otherwise inaudible sounds; to sense body heat; and to convey the information from these observations virtually instantaneously—are the result of progress in science and engineering.

Science and technology can help protect the country against future attacks. But science and technology can also enhance terrorists' ability to achieve their goals. There are legitimate concerns about what terrorists might be able to accomplish if they gained access to cutting-edge scientific knowledge and technology. The race to make faster networks, more refined biological agents, and smaller nuclear devices can simplify the terrorist's work as effectively as it does the soldier's. Some policymakers might argue that the solution is to slow down certain types of technological development. Others strenuously argue that the benefits of such research outweigh the risks.

The Nature of the Challenge: Security and Personal Freedoms

After September 11 and the subsequent anthrax attacks, many researchers thought carefully about ways in which their work might be applied to America's new security challenges. How, for example, could scientists and engineers create effective biometric identifiers for all people entering and exiting the United States (a requirement imposed as part of the flurry of post–September 11 legislative activities)? How do you build systems that allow you to interpret these data meaningfully? How do you integrate the discrete computer systems at the FBI, CIA, INS, and Department of State to enable the efficient sharing of sensitive information? And how can this be done without infringing on Americans' cherished civil liberties?

Any defense against terrorism first requires identifying the aggressor. But this can be very difficult in an open, democratic society built on the principle of individual

POLICY DISCUSSION BOX 18.1

The Two-Edged Nature of Polio Research

In August 2002, three researchers from the State University of New York (SUNY) at Stony Brook published what some viewed as a recipe for synthetic poliovirus.[35] The researchers used the commonplace tools of molecular biology and the published genome sequence, which was available in the public domain, to synthesize infectious poliovirus from readily available materials and then published their methods in full. The researchers carefully demonstrated that this synthetic sample had the same biochemical and pathogenic characteristics as the naturally occurring virus.

Some readers may wonder why any researcher in his or her right mind would want to re-create a deadly organism that had been almost totally eradicated, at great cost, from the developed nations, and which we eventually hope to eradicate worldwide. The simple answer is that the purpose of scientific research is to test the limits of knowledge. These particular researchers had long hypothesized that viruses were "chemicals with a life cycle" and that it was feasible to synthesize them, and thus their aim was to test this hypothesis.[36] As one of the authors noted: "This work is very important to put society on alert. This is an inherent danger in biochemistry and scientific research. Society has to deal with it. . . . It won't go away if we close

our eyes."[37] This was the first time genomic data were used to create something living from chemicals. It was accomplished with tools and chemicals found in almost any molecular biology lab, and with information available to anyone with Internet access.

Some critics argued that the SUNY researchers were asking for trouble; that making genomic data openly available (standard practice in the life sciences) is akin to handing terrorists instructions. For their part, the authors believed that the security risk has been over-hyped: it would be simpler for a would-be terrorist to obtain the naturally occurring virus from a lab or human specimen, they claimed, than it would be to synthesize it. Meanwhile, they said, the furor is leading observers to overlook the potential benefits of their work, including the possibility of devising more effective vaccines.[38]

Individuals outside the security community also spoke out against the study and its publication. J. Craig Venter, founder and former chief science officer of Celera Genomics, called the work "irresponsible" and "inflammatory without scientific justification."[39]

Clearly, the de novo synthesis raises several policy issues. How accessible should genomic data and protocols be? Should the researchers' manuscript have been accepted for publication? What kind of dialogue will be needed between scientists and those concerned with security to achieve properly balanced policies?

---

freedom. In the past generally, and during the Cold War era specifically, our enemies' location and capabilities were well-known. Terrorists, on the other hand, are defined by mobility, elusiveness, and invisibility. When, as on September 11, an aggressor instigates an act of terror within our own country, we often do not know who is behind the attack until it's too late.

As U.S. citizens we have become accustomed to freedom of movement, freedom of association, freedom to acquire almost anything on the open market, freedom to sequester ourselves at home in almost complete privacy. These are exactly the types of protections terrorists take advantage of. Therein, of course, lies the central, paradoxical challenge facing the United States. How can we stop terrorists from planning and carrying out their acts without infringing on the rights of innocent citizens? How can we prevent would-be terrorists from gaining access to

scientific and technical information without impeding scientists' legitimate work?

While there is considerable debate over whether the government should be tracking the movements of individuals, most Americans seem to agree that technology should be used to detect materials of mass destruction, assess whether a terrorist attack has been launched, and assist in the response to an attack. These goals require the fullest possible use of new and existing technologies. In order to accomplish this, we need to decide, first, how to fund the necessary R&D and, second, who will carry out the required work. That is, what is the ideal balance between supporting existing technologies and investing in risky basic research that might eventually contribute to greater security? Can we rely on our national labs to carry out this responsibility, when so much talent resides in universities and industry? What, indeed, is the appropriate role of the private sector, which collectively owns most of the nation's infrastructure? And, finally, what role should university researchers be expected to play? It is urgent that we answer such questions if we are to deploy our national R&D capacity in the war on terrorism.

A Historical Perspective

It has long been known that scientific research can be of great value to the nation in wartime. Perhaps this was most evident in the development and use of the atomic bomb. There have been times in the past when the country has become suspicious of scientists' allegiance to the United States. These topics will be discussed below.

Science in Support of National Defense

The United States invested heavily in science for national security and defense purposes during World War II and the Cold War. In fact, it was science's potential to contribute to national security that spawned the current U.S. system for support of research. The example of penicillin, which helped the survival rate of troops, was enough to illustrate the value of scientific research for both military and civilian applications.

As a surrogate for military strength, science and technology also played a role in national security during and after the Cold War, serving as the engine that sustained American primacy in a world where security and military dominance were closely associated with economic superiority.

The idea of using science for security purposes is a venerable one. New knowledge enabled people to invent poison darts and gunpowder, to design swords, to build armor and fabricate cannons. Even biological agents date back to antiquity. The ancient Greeks, Romans, and Persians polluted their enemies' water supplies. During medieval times, armies catapulted diseased bodies over city walls to spread infection and force surrender.[40] Later, in a battle during the Great Northern War (1700–1721), Russian troops tossed the corpses of plague victims over the Swedish city walls of Reval. In 1763, a British general knowingly gave smallpox-laced blankets and a handkerchief to two Indian chiefs. The disease quickly spread among local tribes, likely wiping them out.[41]

Science's historical role, however, pales in comparison with what lies ahead in the era of the War on Terror. The difference is primarily due to the tremendous damage that can now be inflicted with small, easily concealed weapons. Never before has it been possible to use a portable device to kill millions of people and terrorize hundreds of millions more.

While the relationship between science policy and national security has always ebbed and flowed, current events favor a tight coordination between science policy and national defense, with significant implications for the focus of research funding, eligibility to conduct sensitive research, and the open publication of results.

Scientists Seen as a Security Risk

While the public generally appreciates science's value, it has also at times harbored concerns about its risks, particularly scientists' access to potentially harmful knowledge.

Many scientists came under increased scrutiny during the McCarthy era. It was an era when prominent people could be challenged on the basis of their beliefs and opinions, as opposed to any actual threat they might have posed to national security. Among the scientists who fell victim to such scrutiny was J. Robert Oppenheimer, who had directed the Manhattan Project under President Roosevelt. After the war, Oppenheimer, who chaired the U.S. Atomic Energy Commission, opposed the development of a more powerful hydrogen bomb, which President Truman ultimately approved. During the height of U.S. anticommunist feeling, Oppenheimer's reluctance worked against him: he was accused of communist sympathies and associating with left-wing organizations. His security clearance was ultimately revoked, effectively ending his influence on science policy.[42]

The Personnel Security Board, which reviewed Oppenheimer's case and ultimately upheld the (baseless) removal of his security clearance, noted: "There can be no tampering with the national security, which in times of peril must be absolute, and without concessions for reasons of admiration, gratitude, reward, sympathy, or charity. Any doubts whatsoever must be resolved in favor of the national security. The material and evidence presented to this Board leave reasonable doubts with respect to the individual concerned. We, therefore, do not recommend reinstatement of clearance."[43]

Edmond U. Condon, a physicist and head of the National Bureau of Standards (NBS), also fell under public scrutiny at around the same time. In 1952, the House Un-American Activities Committee (HUAC) accused Condon of being "one of the weakest links in our atomic security."[44] Committee members repeatedly questioned his security clearance, but he was exonerated every time. Even so, HUAC refused to retract its allegations. Condon eventually decided that his ability to lead the NBS had been compromised, and resigned. Even afterward, when he was serving as the head of research at Corning Glass, HUAC refused to retract its allegations. Condon's security clearance, issued for a naval project with which Corning was involved, was finally revoked.

Other U.S. scientists, such as Linus Pauling, an outspoken liberal who was awarded both the Nobel Prize in Chemistry for his work on the chemical bond structures in amino acids and a Nobel Peace Prize, were not allowed to travel abroad for fear that they might openly share their views and knowledge. Meanwhile, foreign scientists, including the British Nobel laureate Paul Dirac, also had difficulty entering the United States.[45]

Perhaps one of the most tragic cases of the McCarthy era was that of Tsien Hsue-Shen. In her book *The Thread of the Silkworm,* Iris Chang tells the story of this Chinese-born scientist who helped lay the foundation for the Jet Propulsion Laboratory (JPL). Tsien's story, writes Chang, represents "one of the most monumental blunders the United States committed during its shameful era of McCarthyism, in which the government's zeal for Communist witch-hunting destroyed the careers of some of the best scientists in the country."[46] Trained at MIT and Cal Tech, Tsien studied fluid dynamics, the buckling of structures, and engineering cybernetics; his research is said to have "made possible the early American entry into the space age."[47] After fifteen years in the United States, and just as he was about to attempt to become a U.S. citizen, he was caught up, like so many other scientists, in the anticommunist whirlwind. Accused of being a former member of the Communist Party, Tsien was arrested and deported to China, where he eventually became head of the Chinese missile program. Ironically, the U.S. government's groundless actions seem to have turned a former U.S. loyalist into the enemy. The United States lost any scientific contributions Tsien might have made, while China gained the benefit of his significant skills.

More recently, Wen Ho Lee was arrested for espionage in 1999. Lee was a physicist at Los Alamos National Lab when a story broke in the *New York Times* that the Chinese had gotten hold of U.S. nuclear information. Lee was immediately fired by secretary of energy, Bill Richardson, and was arrested several months later. He was held in confinement for nine months, even though charges against him were reduced to handling government information with an "unknown but nefarious purpose."[48] In the end, Lee was freed after pleading guilty to one felony count

of mishandling information after which the government dropped the fifty-eight other charges—all relating to national security—it had brought against him. The departments of Energy and Justice came under scrutiny for their handling of the case, but unfortunately Lee's career as a scientist was over. In such cases, security fears can lead to the ruination of scientific careers, sometimes justifiably, but just as often not.

The Scientific Community's Response to September 11

The Public's Expectations

The war on terrorism, as we have seen, poses complex challenges for the scientific community. In one corner are those who believe that scientists should set aside whatever they are doing and devote their full attention to homeland security. In another corner are those who argue that devoting too much of the nation's scientific effort to homeland security will hinder general scientific advancement. In yet a third corner is a contingent that blames scientists for the predicament the nation is in and seeks to more tightly regulate research. Policymakers and the scientific community have to define the point at which the potential risks of the pursuit of knowledge outweigh its potential benefits. This calculation is complicated by the nature of scientific research: scientists often do not know what they will learn from their work or how their findings will be used until the research is done. Government has to strike a balance between restricting scientific activities for security purposes and protecting science against unwarranted and damaging policies.

The public's support of R&D is premised on a delicate contract: an expectation that scientific research will ultimately benefit society. When the nation is threatened, the public instinctively turns to scientists for help. At the same time, however, the public becomes more fearful of the threats posed by scientific knowledge. Scientists' response in such moments can be of great importance for testing the contract's strength, and ensuring its continued viability.

Scientists' Perspective

Because of the nature of their work, scientists tend to support collaboration, the open exchange of ideas, and the right to freely publish their results. Most profess awe for the mystery of life, and avoid any actions that they believe would result in the destruction of life. To the extent that these general propositions are accurate, one might infer that secret research, or research that could result in the eradication of life, is at odds with the basic tenets of science. This tension colors any interaction between scientists and government over science's role in homeland security efforts.

There is, at present, an interesting set of exchanges under way among government officials; independent agencies such as the National Academies; scientific associations like the American Society for Microbiology and the American Association of the Advancement of Science; the publishers of scientific journals, including *Science, Nature,* and the *Journal of Bacteriology;* and university associations such as the Association of American Universities. One point of contention in this exchange has been whether the federal government should have the right to dictate to scientists and publishers which research results can be published and which cannot.

The emergent power of advances in biomedicine and other fields, in conjunction with new threats to national security, is forcing the scientific and national security communities to seek a better understanding of each other, but such negotiations are seldom easy.

The Clash of Two Cultures

In *The Two Cultures and the Scientific Revolution,* the renowned scientist-novelist Sir Charles P. Snow described the cultural split between the literary and scientific communities in Western society: "Greenwich Village talks precisely the same language as Chelsea, and both have about as much communications with MIT as though the scientists spoke nothing but Tibetan."[49] The scientific and security communities are separated by a similar divide. Scientists are, on the whole, a future-oriented group, favoring change and progress. The security community, on the other hand, uses stealth and secrecy to preserve—and protect—the status quo. Change—especially unanticipated change—is considered threatening. While we are painting both groups with a broad brush, the portraits are accurate in their basic outlines.

In the wake of the September 11 attacks, scientists have been concerned that new security restrictions would be enacted that would impede collaboration with their colleagues, or restrict access to materials needed for certain kinds of legitimate research.

Are these fears realistic? Are they justified? The sharing of information is the lifeblood of science. And yet certain types of information can be dangerous in the wrong hands. The goal should be to strike a balance that will make scientific knowledge available for legitimate use, while keeping it out of the reach of terrorists and other aggressors.

As part of this effort, we have to learn to more accurately identify those who wish us harm. We have already discussed the McCarthy era's misguided persecution of senior scientists, and similar concerns were raised in the late 1970s and early 1980s as an outgrowth of the Cold War. These alarms reached their highest pitch in January 1982, when the State Department asked American universities to deny designated foreign students access to specific courses of study and laboratories and, further, to monitor their movements—a request that many in the scientific community found "totally impractical and improper."[50] The government's requests were refused in most cases, and the control measures were eventually abandoned.

This interaction contributed to the creation of the DOD-University Forum, which was designed to improve the balance between science and security. In 1982 the NRC also created a Panel on Scientific Communication and National Security, chaired by Dale Corson, a professor of physics and former president of Cornell University. The panel concluded that (1) "security by secrecy" would ultimately weaken U.S. technological capabilities; (2) there was no practical way to restrict international scientific communication without disrupting domestic scientific communication; (3) the nation must build "high walls around narrow areas" in pursuit of "security by accomplishment"; and (4) controls should be devised only for "gray areas."[51] These discussions culminated with President Reagan's issuing of National Security Decision Directive 189, commonly known as NSDD 189, which designated classification as the primary means for controlling research with national security implications.

Classification has traditionally been the government's instrument of choice for controlling research. NSDD 189 defines "fundamental research"—the research most often conducted by universities, with their educational mission—as work whose "results . . . are ordinarily published and shared broadly within the scientific community."[52] Under NSDD 189, the dissemination or publication of fundamental research cannot be restricted. Such restrictions are reserved for classified research. In a January 2003 letter to John Marburger, George W. Bush's science advisor, the presidents of three higher education associations noted that "NSDD 189 established a clear and concise national policy framework for controlling the flow of science, technology and engineering information produced through federal funds in colleges, universities and laboratories."[53]

But is this framework intact twenty-five years later? Despite the Bush administration's assurances that the government will continue to abide by NSDD 189, federal agencies are increasingly using contract language to control the publication of results from uncontrolled and unclassified research.[54] And more and more contract provisions are also limiting the participation of foreign nationals in certain fundamental research. According to Robert Brown, former provost of MIT, such clauses create a "shade of gray between the classified and unclassified world not easily dealt with."[55]

Export Controls

A related policy area that has posed challenges for researchers is that of export controls. Export control regulations require that licenses be acquired before certain defense-related and dual-use items (*dual-use items* have predominantly commercial uses, but also have military applications) can be exported to particular foreign countries. A license is required to export technology to certain countries.

Dual-use items are regulated under the Export Administration Regulations (EAR), which receive their legal authority from the Export Administration Act of 1979. Military and defense items are regulated under the International Traffic in Arms Regulations (ITAR), which were promulgated in response to the Arms Export Control Act of 1976. The EAR is overseen and interpreted by the Department of Commerce, while for ITAR these responsibilities fall to the Department of State.

In recent years, there has been growing concerns about the impact that *deemed* export control regulations have on the ability of foreign nationals to participate in U.S. scientific research. A *deemed export* is the export or release of information about an export-controlled good, technology, or defense item to a foreign national such that information transfer itself is deemed to be an export. In such instances, an export license is required just as if the goods were being physically sent to the home country of the foreign national to which the information was provided. Because exclusions in EAR exist for education and fundamental research, universities traditionally have applied for export licenses only when they were exporting goods abroad or dealing with research not intended to be widely shared through publication. For purposes of export controls and in accordance with NSDD-189, *fundamental research* is defined as "basic and applied

research in science and engineering, the results of which ordinarily are published and shared broadly within the scientific community."[56]

In the spring of 2004, however, the inspectors general of several federal agencies, including the departments of Commerce, Defense, State, Homeland Security, and Energy, recommended the increased application of deemed export controls to foreign students and scholars involved in certain university-based research.[57] Of particular concern were recommendations made by the Department of Commerce IG that would have required that export licenses be obtained when foreign students and scholars had access to laboratory equipment that involved export-controlled technologies, even if this equipment and the associated controlled technologies were necessary to conduct fundamental research.[58]

Universities responded by arguing that the enactment of IG recommendations threatened university-based research and would stifle fundamental research critical to national and economic security. The universities also maintained that once cleared through the visa and immigration process to enter the United States, foreign visitors should be free to conduct fundamental research and use related scientific and laboratory equipment without additional background checks or a deemed export control license. After hearing significant concerns about the IG recommendations from both the research and business communities, the Commerce Department rejected the IG recommendations and appointed the Deemed Export Advisory Committee (DEAC) to make recommendations to enhance deemed export control policies.[59] The DEAC submitted its final report to the secretary of commerce in December 2007.

Many scientists, however, still have difficulties participating in international scientific collaborations and working with foreign students and scholars in research when satellites or related technology are involved. These difficulties resulted from a change in the export control law called for by the 1999 Department of Defense authorization bill, which transferred responsibility for satellite technology from the Commerce Department to the State Department. Thus, space-related research involving satellites that once were subject to the fundamental research exclusion under EAR for the first time became formally regulated by State Department under ITAR.

The impact of this change on space science has been substantial, leading top researchers to criticize export control regulations under ITAR. Discussing the impact of ITAR regulations at a May 2007 congressional hearing, Dr. Lennard Fisk, professor at the University of Michigan and chair of the National Research Council's Space Studies Board, noted that they were a "nightmare" and "probably the single biggest impediment" to international space science collaborations.[60]

New Security Challenges for the Biological Research Community

Thanks to their long history of interaction with the government, researchers in the physical sciences have developed an acute sensitivity for the security implications of their work, whereas scientists in the biological and life sciences have generally not had long experience with these issues. Thus, Executive Order 12958, issued in December 2001 and granting new classification powers to the NIH, has given rise to questions about the proper role of an agency whose research has historically been focused on the prevention of disease.[61]

Daniel J. Kelves, a Yale University historian and observer of science, technology, and society, has described the differences:

> In nuclear research, a line could be drawn between research that was and was not integral to national security. If the investigation of fissionable nuclei was critical to national defense research, research into the nuclei in most of the periodic table was not. It remained open and unclassified. In contrast, the line in biomedical research is blurred because results in almost any area of basic mo-

lecular biology may be fuel for bioterrorism. . . . A good deal of biomedical research is double purpose; it may assist bioterrorism, but it can also help defend against it and serve the needs of civilian and military health.[62]

Craig Venter, the former president of Celera Genomics, seconded this notion: "Some people argue that publishing each genome is like publishing the blueprint to the atomic bomb. But is also the blueprint for a deterrent and for a cure."[63]

In 2002, Ronald Atlas, the president of the American Society for Microbiology, spoke about the dual-application question during a hearing before the House Science Committee: "Research to make new drugs sometimes might be used to develop bioweapons. Genomic data is valuable for identifying targets for therapeutic drugs and vaccines, but such information can be viewed as potentially valuable for identifying means to increase the virulence of microbial agents to counter currently available therapies, vaccines and detection protocols."[64] This paradox is intrinsic to the relationship between the scientific and security communities. Openness is the very heartbeat of science, the means toward progress, whereas secrecy is the password of the security community, a culture in which the sharing of information jeopardizes safety. Given each community's limited understanding of the work of the other, it is amazing that the communities coexist as well as they do.

POLICY DISCUSSION BOX 18.2

Export Controls: Cold War Relic or a Necessity for National Security?

Export controls emerged during the Cold War largely out of fear that the Soviet Union might gain access to sensitive military and dual-use technologies and use them against the United States. Today, these controls protect the United States from military threats posed by terrorist groups and nations of concern.

There are many in business, academia, and security, however, who maintain that the changing nature of global threats and the loss of U.S. technological superiority in many fields mean that it is time for reform of export controls.[65] Current export control lists include technologies that are out-of-date and can easily be purchased from other countries. Meanwhile, U.S. businesses

claim that export controls hurt their ability to compete internationally, putting the United States at a disadvantage compared to nations that do not maintain the same restrictions.

With scientific and technological knowledge becoming more and more global, what regulations do you think are needed to ensure that the United States does not export technology or related information that is critical to U.S. national security and our economic well-being? What restrictions and controls are reasonable? How might regulations harm our national and economic security if they are overly restrictive or impede the free exchange of scientific information, international scientific collaborations, or the ability of international students and scholars to study and conduct their research in the United States?

Internationalization versus Isolation

As science has grown increasingly international, the need for effective knowledge transfer has intensified. Some fields, such as high-energy physics, rely heavily on international collaboration. Undertakings like the Human Genome Project and the search for the cause of SARS would not have been completed nearly as quickly without international partnerships.

The exchange of knowledge takes many different forms. It happens when researchers travel to scientific conferences. It happens when foreign students or post-doctoral fellows come to the United States, or when an American student travels abroad. It happens when papers are published in journals or made available on the Internet. The United States has long benefited from the talents of foreign scientists who have chosen to conduct their scientific research in America. Foreign students often stay in the United States after graduation, bolstering the American workforce with their much-needed technological skills. And immigrants make up a significant percentage of the total number of American scientists who have received a Nobel Prize.[66] Many of those involved in designing and building the first atomic bomb were immigrants who came here seeking asylum from fascism and war in Europe, including Enrico Fermi, Eugene Wigner, Hans Bethe, Edward Teller, and Albert Einstein.[67] The benefits of the émigrés' presence did not accrue only in the form of scientific knowledge: as Tom Ridge, former secretary of the Department of Homeland Security, noted in a speech to the presidents of leading U.S. universities, Einstein not only made great intellectual contributions but also provided the United States with urgent intelligence about Germany's quest to build an atomic bomb.[68]

It has often been said that America is a nation of immigrants. Yet September 11 and its aftermath are testing our belief in openness. Scientists have felt the pressures of this change perhaps more than other professions. Most often this additional scrutiny has been unwarranted, the result of unfounded fears about what could happen if our scientific secrets fell into the wrong hands. In a field where international collaborations are the norm, it may be natural that public anxieties focus on this possibility. What would have happened, for example, if the Nazis had obtained the secrets of our atomic weapons program?

It was fears like these that, in the wake of September 11, drove the push for faster implementation of the Student Exchange Visitor Information System (SEVIS), a database designed to track all foreign students attending U.S. colleges and universities. SEVIS was first authorized by Congress in 1996, but was never fully implemented before 2001, because of the expense and technical problems.[69] On the heels of September 11, however, Congress imposed strict deadlines for the rapid operationalization of SEVIS on campuses across the country. Educational institutions found it extremely difficult, if not impossible, to meet these deadlines, since the mandate was not accompanied by sufficient funding or support. Moreover, the SEVIS system itself was initially beset with technical glitches. At the time, some university officials described SEVIS as "a work in progress," and many complained that they were being forced to use a system that did not work.[70] Since its implementation, many of the initial bugs in SEVIS have been ironed out, and the system today is running more smoothly.

At the same time there was a push for quick implementation of SEVIS, and in an uncanny echo of the 1982 State Department episode, President Bush issued Homeland Security Presidential Directive 2:

> The Government shall implement measures to end the abuse of student visas and prohibit certain international students from receiving education and training in sensitive areas, including areas of study with direct application to the development and use of weapons of mass destruction. The Government shall also prohibit the education and training of foreign nationals who would use such training to harm the United States or its Allies.[71]

While this order initially seemed reasonable, especially in the shadow of the World Trade Center and Pentagon attacks, it was extremely difficult to enforce, providing no handle by which agencies could define which fields of study might have "application to weapons of mass destruction." Given the increasing potential for "militarization" of all scientific fields, from history to psychology to computer science, the order would seem to exclude foreign nationals from virtually every area of the physical and social sciences: a psychology student could use their knowledge to take hostages or infiltrate a security organization; a computer engineer could hack secure military servers or devise a denial-of-service attack. Consequently, implementation of the Interagency Panel on Advanced Science Security (IPASS), the monitoring system created to respond to the president's directive, never really materialized. Instead, the proposed IPASS system has been absorbed into existing mechanisms used during the process of student visas for reviewing critical technologies.[72]

New requirements have also been imposed on all student visa applicants, including more comprehensive questions about their knowledge of, or plans to study, subjects relevant to biological, chemical, or nuclear weapons development. Whereas it used to be fairly simple to obtain a U.S. student visa, this is no longer the case. In fact, if a foreign student is planning to study a subject that the government regards as technically sensitive, especially one that's included on the so called *technology alert list,* then the overseas consular affairs officer (based at U.S. embassies and consulates around the world) reviewing their visa will automatically send their visa application back to Washington for further review.

While foreign students have always fallen under some scrutiny, the events of September 11 prompted a reexamination of the government's Visa Mantis System, and major changes to the student visa policy: particularly in the kind of guidance provided to consular affairs officers by the State Department; the frequency and rigor of visa reviews was also dramatically increased.[73] At the same time, new categories of study were added to the technology alert list, including biology and urban architecture—the latter being a field that one of the September 11 hijackers had been studying. Some of these additional fields of study were later removed from the list.

Visa reviews are complicated further by the fact that the consular affairs officers are often under tremendous pressure, occupy a fairly low position in the State Department pecking order, and are poorly paid. Most have little technical knowledge of science, and in the days immediately following September 11 were simply instructed that they should automatically return any visa application to Washington if they had any doubt about whether the applicant was studying a sensitive subject. These changes resulted in long processing delays for visa applications, causing some students to miss whole semesters and leaving many university researchers wondering if their foreign graduate student assistants would ever be allowed back into the country. In a few instances, students and scholars were ultimately denied entry.

The new requirements also dictated that most all foreigners applying for a visa had to be interviewed, but no additional funding or resources were provided to hire additional staff. This resulted in long delays in the processing of student visas for a couple of years following September 11. Today, the backlogs and delays in student visa processing have, with a few exceptions, been eliminated as better systems have been implemented and better-trained staff have been added to deal with the more stringent processing requirements.

As security increased and the United States appeared less welcoming to foreign students and scholars, educators and policymakers expressed concern that talented international students were choosing not to come to the United States and instead applying to universities in other countries. According to a survey conducted by the American Council of Graduate Studies, there was a 5 percent decline in international graduate student applications from 2004 to 2005, which followed an estimated 28 percent decline in the previous year. (Reports of the 2006 numbers indicate this trend is changing; however, because of the significant decrease in the preceding years, this increase does not make up the difference.)[74] Anecdotally, many institutions also reported significant drops in their international student enrollments; at Texas A&M, for example, international student enrollment dropped by 38 percent in just one year.[75] More recent trends suggest that foreign applications are beginning to rebound, but concerns persist that the perceptions created by the problems foreign students and scholars encountered may still be having an impact on their interest in applying to U.S. institutions.

The regulations put in place following September 11, 2001, represented the efforts of a nation to guard against a threat that it had never faced before. Some provisions of these regulations may, upon reflection, not be in the best interests of the country in the longer term. In particular, any actions that impede the flow of talented students with a genuine interest in scientific studies should be carefully reviewed.

Where Are We Now?

In the intervening years since September 11 the nation has had the opportunity to assess the impact of policies and regulations put into place to better protect its national security. It has also seen some of the unintended consequences of these policies. Clearly, the step now called for is to reassess current practices in light of what is known.

Reevaluating Federal Scientific and Technological Priorities

America needs to confront the question of whether it should *reinvent* its framework for the conduct of science. There have already been calls for such a total reevaluation. Like one of her predecessors as president of the Carnegie Institution (Vannevar Bush held this position from 1939 to 1955), Maxine Singer has noted that the nation must tap the concentration of minds and talent at its universities if it hopes to identify viable, long-term solutions to the terrorist threat: "scholars, scientists, and engineers who work in our great universities, industries and research institutions can contribute much more than just novel ways of using technology. Many are trained and experienced problem solvers whose approach to difficult problems is to step 'out of the box' because that is where scientific and technical questions are most likely to yield."[76]

Others, such as President George W. Bush's science advisor, Jack Marburger, advised against fundamental change. According to Marburger the relationships that unite the federal government, universities, national labs, and industry have worked effectively, and this was not a time to create a new system for doing science: "Some have spoken of the need for a 'Manhattan Project' to satisfy the needs of homeland security. The analogy is wrongheaded. Cleverness is needed less now than a national will to use what we have."[77] In a speech before the AAAS Colloquium on Science and Technology Policy, Marburger further stated:

> As I learned more about the challenges of terrorism, I realized that the means for reducing the risk and consequences of terrorist incidents were for the most part already inherent in the scientific knowledge and technical capabilities available today. Only in a few areas would additional basic research be necessary, particularly in connection with bioterrorism. By far the greater challenge would be to define the specific tasks we wanted technology to perform, and to deploy technology effectively throughout the diffuse and pervasive systems it is designed to protect. The deep and serious problem of homeland security is not one of science, it is one of implementation.[78]

Whether or not the system itself is altered, policymakers face the challenge of determining where best to invest limited resources. How much should be devoted to developing new knowledge that could help us respond to terrorist attacks? How much should be invested in applying existing knowledge to new challenges? The balance is never easy to achieve; one notes a distinct tendency to favor the application of existing knowledge and technology. This has so far held true in the Department of Homeland Security's new S&T directorate, where relatively little has been invested in basic research.

We *should* apply preexisting knowledge. However, it would be a great mistake not to think about new types of knowledge that would help us defend our country. A better understanding of complex systems and systems engineering could help us devise systems for monitoring biological attacks or attacks on critical infrastructure. Better economic and social science models of risk assessment could assist policymakers in their decisions about where to invest resources. And basic knowledge about the nature of biological agents will be essential to developing antidotes and prophylactics against their potential use. As one university chancellor put it, we must be careful not to invest too much in "missiles and medicine" and too little in basic research that could enable us to face future challenges.[79]

Some of the areas most in need of investment are those to which little thought has been given, or where many questions remain unanswered. Biodefense, for example, warrants significantly more attention than it received prior to September 11—a fact that, judging from the initial funding provided to the NIH's new National Institutes for Allergy and Infectious Diseases (NIAID), has been recognized by both Congress and the administration. The amount of support that will be provided in other, equally vital areas, including critical infrastructure protection and cybersecurity, appears much less favorable at the time of this writing, leading some in Congress to express concern about the apparent lack of attention to these areas by executive branch agencies such as the NSF and DHS.

How important is university and college research to homeland security and national defense? If a university's primary mission is education, knowledge generation and exchange, then perhaps there are certain areas of applied science and technology in which universities ought not to engage—areas that might better be left to the government and government-supported research laboratories. Should universities engage in classified research? The search for new funding sources may lead some universities to pursue research outside of their traditional mission, in areas like homeland security and military technology. But policymakers and the academic community must take a hard look at their institutional and scientific priorities, lest this arrangement evolve haphazardly.

Balancing the Promise and the Threat of Science after September 11

As science policy expert Lewis Branscomb has noted, great harm can be done using relatively primitive tools.[80] September 11 and the anthrax attacks prove how easily common technologies can be turned into weapons of mass destruction. The September 11 attackers did not enroll in nuclear engineering or biology courses at elite universities; they studied at obscure flight schools in Florida and Arizona. Despite early reports to the contrary, only one of the terrorists was actually in the United States on a student visa; two others had applied for but not yet received theirs. The one terrorist here on a student visa was studying English—not a field that would have appeared on any federal alert list.

While risks are certainly involved in scientific advancement, it would be a mistake to halt such advancement in an attempt to prevent misuse or terrorism. Instead of trying to keep the genie in the bottle, it is better to master the genie. At the same time, scientists have to learn to better assess the potential security risks of their research and to understand that their findings may have dual use. The public and their representatives, in turn, must resist the tendency to fear what they don't understand, and avoid overreaching policies.

Making the System Work

As pointed out in previous chapters, science policy-making is an interactive process that involves balancing scientific progress with the public good. Happily, in many instances what is good for science is also good for society—but not always. The question of how to judge often comes down to opinion and personal values. The intractability of this problem is pronounced in the post–September 11 world.

But even our unusual situation is not without its precedents. Like any other time in our nation's history, the current era requires a dialogue among scientists, policymakers, security experts, and the general public. The parties involved must agree that we have certain basic goals in common, and that the difficulties lie only in deciding how best to achieve them.

Policy Challenges and Questions

How do we reach common ground? Congress and the executive branch must incorporate the views of individual citizens in their deliberations over new laws and regulations. And university scientists should use scientific and university associations and other nongovernmental organizations to share their informed opinions with policymakers. The National Academies of Science noted the importance of such dialogue in their 2002 report on the role of science and technology in countering terrorism:

The Office of Science and Technology Policy (OSTP), in collaboration with the Office of Homeland Security (OHS), and other federal security authorities, should initiate immediately a dialogue between federal and state government and research universities on the balance between protecting information vital to national security and the free and open way in which research is most efficiently and creatively accomplished. This dialogue should take place *before* enactment of major policy changes affecting universities as research and educational institutions.[81]

There are times when marriage partners, even those who love each other very much, benefit from a bit of counseling. This is one of those times. The science and policy communities would benefit from serious discussion before the new ground rules of their relationship are established.

As in any relationship, boundaries are essential: there is no room for ambiguity in any system that regulates "sensitive but unclassified information" or "sensitive areas of study." If past experience with regulations concerning the sharing of information proves anything, it is that the scientific community should not be left to determine for themselves which research projects may pose a threat to national security. These decisions ought to be made by the national security community in a comprehensive assessment conducted with scientists.

Above all else, government should embrace a "do no harm" approach toward science. At the same time, the scientific community should carefully develop criteria for determining when scientific information is so susceptible to misuse that it should not be freely disseminated. Scientists may have to learn self-restraint in this regard. If they prove unwilling to do so, policymakers with much less understanding of science will likely set limits for them. The scientific community should always keep this reality in mind.

NOTES

1. With the National Science Foundation Act of 1950 Congress established the National Science Foundation to "promote the progress of science; to advance the national health, prosperity, and welfare; to secure the national defense; and for other purposes." See http://nsf.gov/about/history/ (accessed June 30, 2007).

2. See University of Southern California, Center of Risk and Economic Analysis of Terrorism Events, http://www.usc.edu/dept/create/; "University of Southern California Chosen as First Homeland Security Center of Excellence," Department of Homeland Security, press release, November 25, 2003, http://www.dhs.gov/xnews/releases/press_release_0301.shtm (both accessed June 30, 2007).

3. *Cyber Security Enhancement Act of 2002*, House of Representative Report 107-497, 107th Congress, 2nd sess., June 11, 2002.

4. See National Academies Critical Infrastructure Roundtable, http://www7.nationalacademies.org/bice/CIRT.html (accessed March 31, 2007).

5. Michael Walzer, *Just and Unjust War: A Moral Argument with Historical Illustrations,* 3rd ed. (New York: Basic Books, 2000).

6. See Paul de Armond, "The Anthrax Letters: Five Deaths, Five Grams, Five Clues," August 16, 2002, *Albion Monitor,* http://www.albionmonitor.com/0208a/default.html (accessed June 30, 2007). For information on total spending levels for Homeland Security Research, see Kei Kozumi, "Homeland Security R&D Funding Levels off in 2006," http://www.aaas.org/spp/rd/hs06.htm (accessed June 30, 2007).

7. See Steve Bowman, "Biological Weapons," in *Biological Weapons: A Primer* (New York: Novinka Books, 2001).

8. United Nations, "Report of the Secretary General, Chemical and Bacteriological (Biological) Weapons and the Effects of Their Possible Use," Documents A/7575/, July 1, 1969.

9. See Edward M. Spiers, *Chemical and Biological Weapons: A Study of Proliferation* (New York: St. Martin's Press, 1994), chap. 1.

10. Ibid., 7.

11. World Health Organization, "Health Aspects of Chemical and Biological Weapons," working draft, 2nd ed., due for publication in December 2001, http://www.fas.org/irp/threat/cbw/BIOWEAPONS_FULL_TEXT2.pdf (accessed June 30, 2007). Organisation for the Prohibition of Chemical Weapons, *Chemical Terrorism in Japan: The Matsumoto and Tokyo Incidents* (The Hague, Netherlands: OPCW), http://www.opcw.org/resp/html/japan.html (accessed June 30, 2007).

12. Emergency Response & Research Institute, "'Terrorist Attack' in Toyko," EmergencyNet News Service, March 19, 1995, http://www.emergency.com/japanatk.htm (accessed May 31, 2005).

13. Ibid.

14. Lewis M. Branscomb, "The Changing Relationship between Science and Government Post–September 11," in *Science and Technology in a Vulnerable World: Supplement to the AAAS Science and Technology Yearbook 2003*, ed. Albert Teich, Stephen Nelson, and Stephen Lita (Washington, DC: American Association for the Advancement of Science, 2002).

15. See Terry N. Mayer, "The Biological Weapon: A Poor Nation's Weapon of Mass Destruction," in Bowman, *Biological Weapons.*

16. Dr. Joshua Lederberg, former president of Rockefeller University and a Nobel laureate, in Center for the Study of the Presidency, *Marshalling Science, Bridging the Gap: How to Win the War Against Terrorism and Build a Better Peace* (Washington, DC: Center for Study of the Presidency, http://www.thepresidency.org/pubs/cspScience.pdf) (accessed June 30, 2007), noted the need to establish an overriding policy goal, namely "a global regime of zero tolerance on the use of biological weapons, in any circumstance." Lederberg drew on his extensive experience developing the background for our national plans for marshalling resources against terrorism, arguing that the United States must keep its hands clean, conduct additional research to better treat anthrax infections, improve epidemiological modeling and surveillance, establish research priorities for the development of counterterrorism pharmaceuticals, and organize a cadre of certified physicians (with experience and judgment equal to Food and Drug Administration officials) who can "make lifesaving decisions when the public's health is endangered."

17. "Activism and Advocacy: Sign-on Letter Protesting Redirection of Funds Away from the NIAID," to Joshua Bolten, Director, Office of Management and Budget, July 11, 2003, signed by several AIDS organizations, http://aidsinfonyc.org/tag/activism/signonanthraxltr.html (accessed June 30, 2007).

18. Ronald J. Glasser, "We Are Not Immune: Influenza, SARS, and the Collapse of Public Health." *Harper's,* July 2004, http://www.harpers.org/WeAreNotImmune.html (accessed May 31, 2005).

19. MAD is a doctrine that assumes each side has the weaponry to destroy the other and that if attacked, the other side has sufficient force to retaliate. The end result would be destruction of both the attacker and the defender. The underlying assumption of this doctrine is that neither side would be so irrational as to risk its own destruction and thus would not launch a first strike. The main application of the MAD doctrine was during the Cold War.

20. National Academies Office of International Affairs, *Dual-Use Technologies and Export Control in the Post-Cold War Era* (Washington, DC: National Academies Press, 1993), 136–37; William Dunlop, "Preventing Nuclear Proliferation: The Post–Cold War Challenge," *Science and Technology Review,* September 2000, http://www.llnl.gov/str/Dunlop2.html (accessed June 30, 2007.

21. Nuclear Control Institute, "Nuclear Terrorism," http://www.nci.org/nic-nt.htm (accessed June 30, 2007).

22. See for example, Mark Clayton, "The Brains behind Iraq's Arsenal: U.S.-Educated Iraqi Scientists May Be as Crucial to Iraq's Threat as Its War Hardware," *Christian Science Monitor,* October 23, 2002, http://www.csmonitor.com/2002/1023/p01s01-wome.html (accessed June 17, 2007); Colleen Honigsberg, "Iraqi Scientists Trained on U.S. Soil: Many Top Specialists Received Degrees from American Universities," *Daily Bruin,* May 30, 2003, http://www.dailybruin.ucla.edu/news/2003/may/30/iraqi-scientists-trained-on-us/print/ (accessed June 17, 2007).

23. "Cyberterrorism is the convergence of terrorism and cyberspace. It is generally understood to mean unlawful attacks and threats of attack against computers, networks, and the information stored therein when done to intimidate or coerce a government or its people in furtherance of political or social objectives. Further, to qualify as cyberterrorism, an attack should result in violence against persons or property, or at least cause enough harm to generate fear." Testimony of Dorothy E. Denning, Professor of Computer Science, Georgetown University, before the Special Oversight Panel on Terrorism, Committee on Armed Services, U.S. House of Representatives, May 23, 2000, http://www.cs.georgetown.edu/~denning/infosec/cyberterror.html (accessed June 30, 2007). Dr. Denning is currently professor in the Department of Defense Analysis at the Naval Postgraduate School.

24. The information age has caused security to no longer be defined by armed forces standing between the aggressor and

those being defended. The weapons of information warfare can get around military establishments. Further, they can compromise the foundation of both the U.S. military and civilian infrastructures. For additional information see Center for Strategic and International Studies, Task Force on Information Warfare and Information Assurance, *Cybercrime Cyberterrorism Cyberwarfare: Averting an Electronic Waterloo* (Washington, DC: Center for Strategic and International Studies, December 15, 1998).

25. One need only type *hacker* into a search engine to find the number of Web sites supporting hacking and hackers.

26. Barry C. Collin of the Institute for Security and Intelligence is credited with coining the term *cyberterrorism* and describing it as "hacking with a body count." See Amara D. Angelica, "The New Face of War," *TechWeek,* November 2, 1998, http://gbppr.dyndns.org/spyking/cyber.htm (accessed June 30, 2007); and Barry C. Collin, "The Future of CyberTerrorism: Where the Physical and Virtual Worlds Converge," remarks at the 11th Annual International Symposium on Criminal Justice Issues, http://afgen.com/terrorism1.html (accessed June 30, 2007).

27. U.S. Department of Commerce, Economics and Statistics Administration, and National Telecommunications and Information Administration, *A Nation Online: How Americans Are Expanding Their Use of the Internet*, February 2002, http://www.ntia.doc.gov/ntiahome/dn/html/anationonline2.htm (accessed June 30, 2007).

28. Center for Strategic and International Studies, Task Force on Information Warfare and Information Assurance, *Cybercrime Cyberterrorism Cyberwarfare.*

29. President William J. Clinton's remarks, "Keeping America Secure for the 21st Century," delivered to the National Academies of Sciences, Washington, DC, January 22, 1999, http://clinton4.nara.gov/WH/New/html/19990122-7214.html (accessed June 30, 2007).

30. The government watchdog group maintained a published list of information removed from both federal and state government Web sites following September 11. The list also showed new information restriction policies that were imposed. See OMB Watch Web site, "Access to Government Information Post September 11th," originally published February 1, 2002, updated April 25, 2005, http://www.ombwatch.org/article/articleview/213/1/ (accessed June 17, 2007).

31. U.S. Department of Homeland Security, *The National Strategy for the Physical Protection of Critical Infrastructure and Key Assets* (Washington, DC: U.S. Printing Office, February 2003).

32. U.S.-Canada Power System Outage Task Force, *Final Report on the August 14, 2003 Blackout in the United States and Canada: Causes and Recommendations,* April 2004, https://reports.energy.gov/BlackoutFinal-Web.pdf (accessed December 16, 2007).

33. U.S. Postal Services, *2003 Comprehensive Statement on Postal Operations* (Washington, DC: U.S. Printing Office), http://www.usps.com/history/cs03/ (accessed June 30, 2007).

34. See Nuclear Regulatory Commission, "Nuclear Reactors," http://www.nrc.gov/reactors/power.html (accessed June 30, 2007); and Tim Culbertson, The National Hydropower As-

sociation, Testimony before the Senate Committee on Energy and Natural Resources Subcommittee on Water and Power, 110th Cong., 1st Sess, June 6, 2007; and National Hydropower Association, fact sheets, http://www.hydro.org/hydrofacts/factsheets.php (accessed July 1, 2007).

35. Cello, Paul, and Wimmer, "Chemical Synthesis."

36. Eckard Wimmer, "Synthesis of Poliovirus in the Absence of a Natural Template," presented to New York Academy of Sciences, http://www.nyas.org/ebriefreps/main.asp?int SubSectionID=553 (accessed June 30, 2007).

37. "New York Scientists Build Polio Virus; Terrorist Applications Feared," *Knight Ridder/Tribune Business News,* July 12, 2002, http://www.vaccinationnews.com/DailyNews/July2002/NewYorkScisPolio14.htm (accessed June 30, 2007).

38. While polio has been eradicated from the United States and much of the world, it is still problematic in several developing nations. See the Global Polio Eradication Initiative http://www.polioeradication.org/ (accessed June 30, 2007).

39. Andrew Pollack, "Scientists Create a Live Poliovirus," *New York Times,* July 12, 2002.

40. British Medical Association, *Biotechnology—Weapons and Humanity* (Amsterdam: Harwood Academic Publishers, 1999), chap. 2.

41. Wendy Barnaby, *The Plague Makers: The Secret World of Biological Warfare,* 3rd ed. (New York: Continuum International Publishing Group, 2002), chap. 1.

42. Vannevar Bush testified in Oppenheimer's defense at his hearing, stating that, contrary to the American system, Oppenheimer was being tried "because he held opinions," Robert Buderi, "Technological McCarthyism," *Technology Review* 106, no. 6 (2003): 8. After these accusations were made, Oppenheimer was granted an academic post as director of the Institute of Advanced Study at Princeton. During the last years of his life, "he thought and wrote much about the problems of intellectual ethics and morality."

43. In the letter submitted to K. D. Nichols, General Manager of the U.S. Atomic Energy Commission on May 27, 1954, by the Personnel Security Board regarding the findings and recommendations on the J. Robert Oppenheimer case, http://www.cicentre.com/Documents/DOC_Personnel_Security_Board_Findings_on_Oppenheimer.htm (accessed June 30, 2007).

44. Daniel J. Kevles, "Biotech's Big Chill," *Technology Review* 106, no. 6 (2003): 40.

45. Ibid.

46. Iris Chang, *Thread of the Silkworm* (New York: Basic Books, 1995).

47. Ibid.

48. See Joshua M. Marshall, "Wen Ho Lee Is Free," *Salon.com,* September 13, 2000; for additional background see Tony Clark, "Nuclear Scientist Lee Goes Home after Plea Bargain," *CNN.com,* September 13, 2000, http://www.cnn.com/2000/LAW/09/13/wenholee.free.02/ (accessed June 30, 2007); and Pierre Thomas, "FBI Director Louis Freeh Testifies on Wen Ho Lee Case," *CNN.com,* September 26, 2000, http://www.cnn.com/2000/LAW/law.and.politics/09/26/freeh.lee/ (accessed June 30, 2007).

49. Charles P. Snow, *The Two Cultures and the Scientific Revolution* (New York: Cambridge University Press, 1959).

50. Panel on Scientific Communication and National Security, National Academy of Sciences, National Academy of Engineering, and Institute of Medicine, *Scientific Communication and National Security* (Washington, DC: National Academy Press, 1982), 10, http://www.nap.edu/catalog/253.html#toc (accessed June 30, 2007).

51. The Panel on Scientific Communication and National Security report spoke to the fact that gray areas exist, yet in the final analysis, no real clarity could be provided about what those gray areas were, when exactly they existed, and how such areas should be controlled and regulated.

52. National Security Decision Directive 189, "National Policy on the Transfer of Scientific, Technical, and Engineering Information," September 21, 1985, Washington, DC.

53. Association of American Universities, Council on Government Relations, and National Association of State Universities and Land-Grant Colleges, joint letter to John H. Marburger III, Director, Office of Science and Technology Policy, January 31, 2003, http://www.aau.edu/research/Ltr1.31.03.pdf (accessed June 30, 2007).

54. Julie T. Norris, "Restrictions on Research Awards: Troublesome Clauses," a report of the joint AAU/COGR Task Force on Restrictions on Research Awards and Troublesome Research Clauses, April 2004, http://www.aau.edu/research/Rpt4.8.04.pdf (accessed June 17, 2007).

55. Sally Atwood, "Is MIT a Security Risk?" Biztech Section, *Technology Review.com*, January 1, 2001, http://www.technology review.com/Biztech/13227 (accessed January 7, 2008).

56. A comprehensive discussion of export controls and their application to university research can be found in: Council on Governmental Relations, *Export Controls and Universities: Information and Case Studies* (Washington, DC: Council on Governmental Relations, February 2004), http://www.umass.edu/research/ogca/export%20controls/Export%20Controls.pdf (accessed June 30, 2007).

57. Offices of Inspector General of the Departments of Commerce, Defense, Energy, Homeland Security, and State and the Central Intelligence Agency, *Interagency Review of Foreign National Access to Export-Controlled Technology in the United States,* D-2004-062, vol. 1 (Washington, DC, April 16, 2004).

58. U.S. Department of Commerce, Office of the Inspector General, *Deemed Export Controls May Not Stop the Transfer of Sensitive Technologies to Foreign Nationals in the U.S.*, final inspection report no. IPE-16176 (Washington, DC: U.S. Department of Commerce, March 2004).

59. Department of Commerce, Bureau of Industry and Security, "Revisions and Clarification of Deemed Export Related Regulatory Requirements," *Federal Register* 71, no. 104 (May 31, 2006): 30840–44.

60. Audrey T. Leath, "Concerns Voiced over Future of Space Science Programs," *FYI: The AIP Bulletin of Science Policy News,* May 9, 2007, http://aip.org/fyi/2007/047.html (accessed June 17, 2007).

61. The EPA and USDA were also granted classification authority under this executive order. What follows is the text: "Order of December 10, 2001, Designation under Executive Order 12958: Pursuant to the provisions of section 1.4 of Executive Order 12958 of April 17, 1995, entitled 'Classified National Security Information,' I hereby designate the Secretary of Health and Human Services to classify information originally as 'Secret.' Any delegation of this authority shall be in accordance with section 1.4(c) of Executive Order 12958. This order shall be published in the Federal Register. [signed:] George W. Bush." *Federal Register* vol. 66, no. 239 (December 12, 2001): 64345–47.

62. Kevles, "Biotech's Big Chill," 47–48.

63. Ibid., 48.

64. Ronald M. Atlas, testimony before the House of Representatives Committee on Science, "Conducting Research during the War on Terrorism: Balancing Openness and Security," October 10, 2002.

65. David R. Oliver Jr., "Current Export Policies: Trick or Treat," *Defense Horizons*, December 2001, http://www.ndu.edu/inss/DefHor/DH6/DH06.htm (accessed June 18, 2007).

66. National Academies Board on Higher Education and Workforce, Committee on Science, Engineering, and Public Policy, *Policy Implications of International Graduate Students and Postdoctoral Scholars in the United States* (Washington, DC: National Academies Press, 2004), 59–60; and Bernard Wasow, "Losing the Genius for Openness," News and Opinion Section, *Immigrationline.org*, A Century Foundation Project, February 17, 2006, http://www.immigrationline.org/commentary.asp?opedid=1217 (accessed July 3, 2007).

67. Daniel Greenburg, "The Mythical Scientist Shortage," *The Scientist* 17, no. 6 (March 24, 2003): 68. In this article, Greenburg states, "When it comes to science and technology (S&T) prowess, the United States has historically cherry-picked the world." In addition to our reliance on foreign talent to build the bomb, Greenberg also notes, "We went into space under the direction of Wernher von Braun, who pioneered rocketry in the service of Adolph Hitler."

68. Speech given by Secretary of Homeland Security Tom Ridge to the Association of American Universities, April 14, 2003, http://www.dhs.gov/xnews/speeches/speech_0104.shtm (accessed June 30, 2007).

69. The Student and Exchange Visitor Program (SEVP) is the automated process for nonimmigrant student and exchange visitor visas. It was formerly referred to as the Coordinated Interagency Partnership Regulating International Students (CIPRIS). SEVIS is the Internet-based database of the program.

70. Atwood, "MIT a Security Risk?"

71. "Homeland Security Presidential Directive-2," White House, news release, October 29, 2001, Combating Terrorism through Immigration Policies, http://www.whitehouse.gov/news/releases/2001/10/20011030-2.html (accessed June 30, 2007).

72. Interagency Panel on Advanced Science Security (IPASS) provides a specialized review of F (student), J (postdoctoral), and M (vocational) visas for students pursuing scientific study. The aim is, according to an OSTP representative, "to ensure that international students or visiting scholars do not acquire 'uniquely available' and 'sensitive' education and training at U.S. institutions and facilities that can be used against us in a terrorist attack." IPASS is composed of representatives from defense, civilian, immigration, and intelligence agencies. Whether a student's visa should be reviewed by IPASS is first established

by the Immigration and Naturalization Service using the Technology Alert List and the student's country of origin. IPASS analyzes the student's educational background, training, and work experience; country of origin; whether the field of study is uniquely available in the United States and sensitive; and whether research conducted elsewhere at the chosen school—beyond the student's major—could have national security implications.

73. The Visas Mantis System is a clearance process program that consular offices use to address concerns of U.S. intelligence and law enforcement communities regarding access to dual-use (both commercial and military) items and technologies. The program is applicable to all visa applicants—both immigrant and nonimmigrant—when the consular officer has reason to suspect a foreign national may engage in sabotage, espionage, or the unauthorized access to controlled technologies within the United States.

74. See Debra W. Stewart, "Five Trends Shaping Graduate Education; the Leadership Challenge," *CGS Communicator,* August–September 2005, 1; Council of Graduate Schools, Findings from 2006 CGS International Graduate Admissions Survey, Phase I: Applications (Washington, DC: CGS, March 2006), 3.

75. Robert M. Gates, "International Relations 101," *New York Times,* March 31, 2004, A23.

76. Maxine Singer, "Answers from Outside the Box," *Washington Post,* September 24, 2001.

77. John H. Marburger III, prepared remarks for the American Association for the Advancement of Science, *Symposium on the War on Terrorism: What Does It Mean for Science?* (Washington, DC, December 18, 2001).

78. Keynote speech given by John H. Marburger III at the 27th Annual American Association for the Advancement of Science Colloquium on Science and Technology Policy, Washington, DC, April 11, 2002.

79. M. R. C. Greenwood, "Risky Business: Research Universities in the Post–September 11th Era," in Teich, Nelson, and Lita, *Science and Technology in a Vulnerable World,* 1–20.

80. Lewis Branscomb, "Threat of Terrorism: Role of State & Land Grant Universities in Making the Nation Safer," presentation at the National Association of State Universities and Land Grant Colleges 2002 Annual Meeting November 12, 2002, Chicago, http://www.nasulgc.org/AM2002/presentations/AM2002_Branscomb.pdf (accessed June 30, 2007).

81. National Academies Institute of Medicine, *Countering Bioterrorism: The Role of Science and Technology* (Washington, DC: National Academies Press, 2002), 81.

CHAPTER 19

Grand Challenges for Science and Society

What Remains to Be Discovered?

Many people—perhaps the majority of people in the world—believe that society will always face challenges that can only be addressed through scientific research. However, others believe with equal conviction that we already know all there is to know, and that future scientific research will offer diminishing returns. John Horgan, a writer for *Scientific American,* raised this claim in his 1996 book, *The End of Science.* As he explained in an interview:

> In the future, we will be filling in details within this framework that scientists have already created with all these different theories, and there won't be any great revolutions analogous to the theory of evolution or to Einstein's Theory of General Relativity or to quantum mechanics. . . . And if you believe that science is a real process of discovery, of truths and nature, then . . . you have to accept that once we discover things, that's it. Then we have to go on to the next thing. . . . Science is a very linear process, and I think that that sort of forces you to accept that there are some limits to discovery eventually.[1]

We, the authors of this book, disagree that we will see the end of science at any time in the foreseeable future. Every new piece of knowledge reminds us how much more there is yet to be known. There are many challenges still before us, and many questions that science will have to address in our lifetimes and far beyond. Indeed, future discoveries may far outshine the most far-reaching scientific finds of human history. In addition, science is *not* a linear process. Rather it relies on continual feedback of knowledge from all stages of the process to all stages

of the process. That is, knowledge gained from basic research (one stage) can inform the questions to ask in applied research and development (another stage); likewise knowledge gained during development can inform the questions to ask in both basic and applied research. Unidirectional flow is not how science works or how science gets done.

Researchers in every scientific discipline have burning questions to which they want immediate answers. Some of these questions are unintelligible to the layperson, and their relevance to bigger issues is obscure. Then there are the "grand challenges": the big, fundamental problems whose solutions are yet unknown.[2] These grand challenges vary from one discipline to the next. In world health, for example, one of the grand challenges is to develop practical scientific and technological innovations that will address pressing health problems in the developing world and that will have a global impact.[3] Scientific grand challenges look toward frontiers that have not yet been crossed, and scientists in each field have a strong interest in collectively exploring these frontiers.

Besides scientific grand challenges, there are also national or societal grand challenges—fundamental social problems whose solutions will have a broad impact on the nation's competitiveness and citizens' well-being.[4] However, such challenges do not necessarily fit neatly within individual scientific disciplines.

Grand scientific challenges are most often defined by the scientific community, whereas grand societal challenges are generally defined by policymakers. It is easy to confuse the two, which are often closely related. In some instances, they may be nearly identical. For example, placing a man on the moon was both a scientific and societal grand challenge: scientific in breaking the bonds of gravity and delivering a human being safely to the moon and back, and societal in beating our political nemesis, the Soviet Union, to this goal. Who was more interested in the space race—scientists or policymakers?

The following sections offer an overview of some of the most significant scientific challenges, as well as societal challenges that are likely to require a major scientific contribution. Also included is a discussion of the growing importance of interdisciplinarity, which profoundly affects the shape of scientific institutions and increases the level of cooperation required from federal research agencies.

Grand Challenges for Science

Scientific challenges are everywhere. We see them when we stare up into the limitless depths of space or down into the worlds of the microscopic realm. Fascinatingly, these two apparent extremes seem to operate according to a single system of rules: much of the large-scale structure of the universe may be determined by the rules governing things as small as quarks. And what about life itself? When and how does a group of particles take on the remarkable characteristic of life? How does the brain function? How do viruses work? Any useful list of "grand challenges" must obviously be selective.

Seeing Farther (Space Exploration)

For most of the middle and late twentieth century, people thought about major scientific accomplishments in terms of our putting a man on the moon. Indeed, the "space race" stands as an example of how uncertainty about national security can be translated into support for science and engineering.

While we have now been exploring space for well over fifty years, many critical questions remain. Interest in the subject seems to be mushrooming, as improved technologies allow us to peer more deeply into the universe, searching for answers to questions that have, in one form or another, puzzled humans since the beginning of history.

The NRC Survey Committee has identified five specific challenges in the area of space science: (1) understanding the structure and dynamics of the sun; (2) understanding heliospheric structure and the interaction of the solar wind with the local interstellar medium; (3) understanding the behavior of the space environment of the earth and other bodies of the solar system; (4) understanding the basic physical principles of solar and space plasma physics; and (5) developing a near-real-time ability to predict the impact of space weather on human activities.[5] Longer-term questions also remain: Are there other stars with life-sustaining capabilities (for life as we know it)? Is the universe predominantly composed of dark matter?[6]

Getting Smaller (Nanotechnology)

The latter half of the twentieth century might be considered the age of microtechnology, where distances on the order of a millionth of a meter are probed. But the beginning of the twenty-first century is about exploring realms a thousand times smaller (i.e., the nanometer), the domain of nanotechnology.[7]

The National Nanotechnology Initiative (NNI), formalized by the 21st Century Nanotechnology Research and Development Act, passed by Congress in the fall of 2003 and signed into law by President Bush in December 2003, defines nanotechnology as (1) research and technology development at the atomic, molecular, or macromolecular levels, in the length scale of approximately 1–100 nanometers; (2) creating and using structures, devices, and systems that have novel properties and functions because of their small or intermediate size; (3) the ability to control or manipulate objects on the atomic scale.[8]

The grand challenge in the nanosciences, broadly speaking, is to understand how to use atomic and molecular clusters for several purposes, from creating specialized tools that could be dispatched into the human body to find and repair or destroy damaged blood cells, to fabricating self-repairing materials. This ambitious goal requires that nanoscientists and nanoengineers learn to characterize new nanomaterials for commercial use, identify their fundamental properties, and develop new applications in medicine, homeland security, and manufacturing, among other areas.

The physicist Richard Feynman first introduced the notion of nanoscale science and technology in 1959, asking, "Why cannot we write the entire 24 volumes of the Encyclopedia Britannica on the head of a pin?"[9] Feynman went on to suggest several related challenges, including the quest to see the world at a smaller scale (through the creation of better electron microscopes), miniaturizing computation, and rearranging atoms. Several decades after Feynman's speech, the NNI enumerated nine specific grand challenges for the field: (1) the design of nanostructured materials (led by the NSF); (2) manufacturing at the nanoscale (led by the NIST and the NSF); (3) chemical-biological-radiological-explosive detection and protection (led by the DOD); (4) nanoscale instrumentation and metrology (led by the NIST and the NSF); (5) nanoelectronics, nanophotonics, and nanomagnetics (led by the DOD and the NSF); (6) nanoscale health care, therapeutics, and diagnostics (led by the NIH); (7) efficient energy conversion and storage (led by the DOE); (8) microcraft and robotics (led by NASA); and (9) nanoscale processes for environmental improvement (led by the EPA and the NSF).[10] The aforementioned 21st Century Nanotechnology Research and Development Act formally acknowledged these "grand challenges."[11]

In order to characterize and manipulate nanoclusters of atoms and molecules, nanoscience and nanotechnology require a multidisciplinary understanding of both the chemical makeup of atoms and molecules and their physical properties. Because nanotechnologies might be used in biological systems—for example, for tissue and organ repair—knowledge of the life sciences is also essential. As the NRC noted in its 2002 review of the subject, "The development of nanoscale science and technology will require generations of interdisciplinary scientist and engineers who can learn and operate across traditional boundaries."[12]

Understanding the Origins of the Universe and the Constituents of Matter

In the past half century extraordinary progress has been made in finding and characterizing the fundamental particles of nature. Even given this triumph, a huge number of fundamental questions in physics and astronomy are awaiting resolution.

One example is that, there is no understanding of what makes up over 80 percent of the mass of the universe. The dark matter puzzle, so labeled because it is matter "not seen," is a mystery of the first order and surely represents one of the grand challenges faced by all of science. This problem will be attacked using several techniques, ranging from astronomical observations, to small-scale precision experiments that search for candidates that might account for the missing energy, all the way to exploiting the discovery potential at the new Large Hadron Collider.

In addition to looking for the source of dark energy, other daunting questions loom: What are the details of the origin of the universe? Does the Higgs boson really exist—the heretofore unseen particle that is thought to be responsible for giving all other particles mass? What symmetries exist among the physical particles? Does the

world have higher-order dimensions though we see only three-dimensional space and time? What are the smallest building blocks of matter, out of which everything else is made?[13]

These are truly fundamental questions, the answers to which must be found if we ever are to really understand the universe in which we exist. They match any definition of scientific "grand challenges."

One of the dramatic developments in recent years has been the realization of just how tightly coupled research at small scales is with cosmology, the study of the evolution of the universe itself. Indeed, scientists believe that the energy densities that had to exist a fraction of a second after the Big Bang can now be reproduced at hadron colliders, the large machines used in high-energy physics to study the properties of fundamental particles. Knowing how these basic particles behave and interact at such densities may hold the key to understanding how the universe has evolved. For this reason there will be a growing intertwining of the work of cosmologists and particle physicists in the decades ahead.

To pursue the goals embodied in these challenges will require a very significant investment in facilities and people. After decades of planning and construction the Large Hadron Collider (LHC) at CERN is to come on line in 2008. This huge multi-billion-dollar particle accelerator, housed underneath the French and Swiss countryside near Geneva, Switzerland, will open up an exciting new window for study of many of the topics we have mentioned. For example, it is hoped that within two years of its turn-on evidence will be found for the elusive Higgs boson, a particle that is essential if the so-called Standard Model of particle physics is to be confirmed. Many physicists also believe that evidence will be found for a spectacular symmetry (referred to as SUSY) between particle types in the first few years of running of the LHC. Advances on several of the other challenges already mentioned may also emerge from this facility.

Because it takes many years to plan and construct particle accelerators, the next accelerator is being planned even before the one under construction is commissioned. Even if the LHC is able to discover the Higgs boson, determining its details may require yet another facility that is specifically designed to study those properties. But that facility can only be definitively designed if one knows whether the Higgs exists and what its approximate mass is. Nevertheless, a high level of generic planning can be done in the absence of these answers, and this effort is well under way in the case of the likely successor to the LHC, the International Linear Collider or ILC. In a study conducted by the National Academies in 2006 (EPP2010), hosting the ILC was recommended as one of the highest priorities for the U.S. high-energy physics program. A decision on this project should be made in approximately 2010, with completion, in the most optimistic scenario, approximately ten to fifteen years later. Tracking this international big-science project will be a significant U.S. science policy challenge over the next decade.

Understanding Complex Systems

While in everyday language any system that includes interacting components can be considered complex, scientists have a much stricter definition of complex systems. Researchers refer to any phenomenon in the social, life, physical, or decision sciences as "complex" if it has a significant number of the following characteristics: it is comprised of "agents" or individual entities that are heterogeneous and dynamic, that change by responding to feedback, that are organized in groups or hierarchies, and that have a structure that can influence how the system evolves over time. As the agents in the system act and interact and the system evolves, system-level behaviors and properties emerge.[14] The study of complex systems by its very nature includes not just laboratory scientists but economists, political scientists, sociologists and other social scientists, as well as mathematicians and computer scientists.

Some of the grand challenges in this multidisciplinary area were set forth in one of the field's journals in 2001: How do things organize themselves? In complex systems, why is the whole greater than the sum of its parts? Relatedly, why do larger complex systems sometimes exhibit properties or behaviors that are only apparent in the larger context?[15] Other grand challenges include mapping the relationship between the connectivity and criticality of the components (or "nodes") of a system; and identifying systems that actually behave according to the models of maximum system adaptability. In other words, researchers are trying to understand the difference between "adequate" and "very best" solutions: in the world of biology, actual systems tend toward adequate solutions, yet in modeling, optimizing and finding the best solutions is a preoccupation. The fundamental question is, does nature know best?[16]

There is also a great deal to learn about the behavior of complex dynamic systems (e.g., ecosystems). Detailed knowledge of the structure and behavior of any complex system is also a grand challenge for researchers in this area.

Computing Capacity (Moore's Law)

In 1965, Gordon Moore, cofounder of Intel, observed that the number of components on a computer chip doubles every twelve to eighteen months. Moore's Law has since been rephrased several times: emphasizing, for example, that the number of transistors per integrated circuit doubles in that time, or that the law is the limiting factor on computing power.[17] It is intriguing to speculate on the drivers for this trend. Is it a fundamental measure of how fast technological discovery can proceed in this domain? Is it an optimal rate for consumers to adapt to increased computational capacity, coupled with a manageable business model in the industry?

Two of the main challenges in semiconductor research are (1) ensuring that computing processing power continues to grow by devising smaller and smaller chips without losing conducting properties; and (2) finding ways to mass-produce these chips. This means finding new materials at the nanoscale and creating more precise tools for laying circuits onto chips.[18] It is a challenge similar to those in nanotechnology, but with specific materials and systems.

Computer scientists are also using complex systems research to improve grid computing, which pools the computing or processing capacity of multiple computers, thus illustrating the key notion of complex systems: that the whole is greater than the sum of its parts.

Understanding Learning and the Brain

Neuroscience unites the study of molecular biology, biochemistry, pharmacology, neuroanatomy, physiology, and behavioral studies to study the brain and the development of the central and peripheral nervous systems, including muscle coordination and movement, communication between the brain and different organs and systems, and treatments for injuries.

The 1990s, the so-called Decade of the Brain, saw enormous progress in our understanding of the central nervous system. The sequencing of the genome, improvements in imaging technologies, and our refined appreciation of how the nervous system operates at the molecular and biochemical levels have brought us to the very cusp of a new era in brain and nervous system research.

The grand scientific challenges of neuroscience include harnessing the power of plasticity (the ability of nervous tissue to remodel itself), applying new genomic research, and mapping protein expression in the brain, the molecules that are responsible for much of the "communication" between nerves and cells in the brain. The advent of stem cell research has presented researchers with the great challenge of learning how to exploit this technology for treatment of diseased and damaged nervous tissue. Neuroscientists are also working to gain a more detailed picture of the stages through which the brain progresses as it develops and a person ages. Lastly, researchers are striving to determine exactly how the brain works, illuminating the mechanisms and underlying neural circuits that enable humans to form memories, have emotions, be creative, use language, and be attentive.[19]

Frontiers in the Social Sciences

As noted by sociologists Alvin Gouldner and S.M. Miller in 1965, "It is the historic mission of the social sciences to enable mankind to take possession of society."[20] One of the challenges of the social sciences is to analyze the interactions of institutions, people, and social structure in order to understand ways in which change might be to improve the welfare and well-being of individuals and communities.

The social sciences have historically faced enormous complexity in making predictions based on observation. Human interaction and response is inherently hard to measure, to quantify, or to translate into fundamental principles. A change in the outlook for social sciences may be, however, quite near.[21] Progress in longitudinal data collection, laboratory experimentation, advanced statistical methods, better geographic information, biosocial science advancements, and international replication are all reasons to expect a sea change in the very nature of social science research.

Longitudinal surveys, "which collect information about the same persons over many years, have given the social sciences their Hubble telescope," comment W. P. Butz and B. B. Torrey.[22] Social scientists are drawing upon and integrating data from several disciplines, including the neurosciences, to acquire a more comprehensive understanding of the economic, social, and political consequences of individual actions. The Internet provides easier access to huge samples of potential survey respondents. Moreover, sophisticated statistical analysis techniques, some of them developed outside the social sciences, are being employed to test model predictions to higher levels of accuracy. There is also an increasing emphasis in collecting data with tags on the relevant geographical and time elements, all made much easier to acquire today by the advent of "geographical information science" and inexpensive GPS systems. A bright future for the continuing evolution of social science research is on the horizon.

The preceding is little more than a sampling of science's grand challenges, which also include exploring the oceans, creating complete evolutionary trees, understanding human behavior, improving economic forecasts, and applying risk analysis models to homeland security and national defense.

Societal Grand Challenges and Science's Promise

Science is not the only domain of human existence in which we face grand challenges. Rather, there are major challenges in all areas of society. The federal government's response to social challenges has often been to create new agencies. Inasmuch as science may help us to meet these challenges, the government also often creates specific scientific offices (e.g., the DOE Office of Science) or set of institutes (e.g., the NIH) to address new problems. These mission-oriented agencies regularly draw on scientific knowledge to fulfill their overall purpose.

Transportation

Transportation is obviously an important element of life in the United States. Several federal agencies have responsibility for the nation's transportation goals and challenges, the leading agency being the Department of Transportation. In addition there are individual state departments of transportation. Both chambers of Congress maintain committees that oversee transportation and related matters.

The U.S. Department of Transportation was formed by the Department of Transportation Act of 1966, which was signed into law by President Lyndon B. Johnson. Its mission is to "serve the United States by ensuring a fast, safe, efficient, accessible and convenient transportation system that meets our vital national interests and enhances the quality of life of the American people, today and into the future."[23] While this core mission has not changed over time, the department faces new challenges after September 11, in an increasingly global environment. A 2003 strategic plan issued by Transportation secretary Norman Y. Mineta encapsulates the goal of the modern DOT as finding "safer, simpler smarter transportation solutions," sometimes extending well beyond U.S. borders.[24]

Secretary Mineta's five-year strategic objectives for the DOT were enhancing public health and safety by working toward the elimination of transportation-related deaths and injuries; advancing accessible, efficient, intermodal transportation for the movement of people and goods; facilitating a more efficient domestic and global transportation system that enables economic growth and development; promoting transportation solutions that enhance communities and protect the natural and built environment; and balancing homeland and national security transportation requirements with the mobility needs of the nation for personal travel and commerce.[25] As Mineta concluded, the transportation challenge of the twenty-first century is to interconnect our individual modes of transportation—air, highway, marine, rail, and transit—in a single, fully coordinated system, while enhancing safety and efficiency.

The Environment

Humanity must naturally be concerned with protecting and preserving the environment. In fact, many other societal challenges (e.g., in transportation and energy) include a concern for environmental well-being. There is a federal agency dedicated to environmental issues, the Environmental Protection Agency, and each chamber of Congress has either a committee or a subcommittee dedicated to environmental concerns. While there are no EPA-like units at the state level, the federal EPA maintains regional offices across the country.

The EPA, whose mission is "to protect human health and the environment," was founded by the National Environmental Policy Act of 1970, the result of increasing public demand for clean air, land, and water.[26] The act defines the following challenges for the environment: establishing a national policy that encourages productive and enjoyable harmony between humans and their environment; promoting efforts to prevent or eliminate damage to the environment and biosphere, and stimulate human health and welfare; and enriching our understanding of ecological systems and the natural resources important to the nation.[27]

While the advent of technology often brings good things, some technologies damage the environment. The goal is to achieve a delicate balance: protecting natural resources without curbing the human benefits of technology. Technological advances sometimes contribute to this balance (e.g., through the development of vehicles that do not depend on fossil fuel). In other cases, science can support the development of policies that prevent irreversible environmental damage (e.g., new means for the disposal of hazardous industrial waste).

A 2003 EPA strategic plan defined the following grand challenges: protecting and improving the air so that it is healthy to breathe; reducing risks to human health and the environment; reducing greenhouse gas intensity, that is, global warming, by enhancing partnerships with businesses and other sectors; ensuring that drinking water is safe; restoring and maintaining oceans, watersheds, and their aquatic ecosystems to protect human health, support economic and recreational activities, and provide healthy habitats for fish, plants, and wildlife; preserving and restoring the land by using innovative waste management practices and cleaning up contaminated properties to reduce risks posed by the release of harmful substances; protecting, sustaining, or restoring the health of people, communities, and ecosystems using integrated and comprehensive approaches and partnerships; improving environmental performance through compliance with environmental requirements, preventing pollution, and promoting environmental stewardship; and protecting human health and the environment by encouraging innovation and providing incentives for governments, businesses, and the public that promote environmental stewardship.[28]

Energy

Energy sources are vital to American society, providing the foundation of our economy. Outages, such as the blackout of the summer of 2003, during which many northeast and Great Lakes states lost power, highlight our dependence on energy. Not only did the lights go out, but gasoline supplies were restricted (gas pumps are electric), as was potable water (since water in some areas is supplied by electric pumping stations).[29]

The leading government agency addressing energy issues is the Department of Energy. Although its roots can be traced back to World War II, the DOE itself dates to the Department of Energy Organization Act of 1977, the product of the 1970s energy crisis and policymakers' desire to establish an agency that could create and implement a national energy plan.

The DOE's mission is "to advance the national, economic, and energy security of the United States, to promote scientific and technological innovation in support of that mission, and to ensure the environmental cleanup of the national nuclear weapons complex."[30] A DOE strategic report from September 2003 identified grand challenges including not just energy generation and distribution, but national defense (the safety, security, and reliability of the nation's nuclear stockpile), maintaining the environment (cleaning up the legacy of the Cold War), and contributing to the nation's S&T innovation.[31] With regard to energy, specifically, the challenges included improving energy security or energy independence; fostering diverse sources of reliable, affordable, and environmentally sound energy; exploring advanced technologies that make a fundamental improvement in our mix of energy options; and improving energy efficiency. Some of the technologies that will no doubt have an important role in helping to meet these national energy goals include solar power, wind power, nuclear power—both fission and fusion—and hydrogen power.

Health

The NIH, the steward of the nation's medical and behavioral research, conducts "science in pursuit of fundamental knowledge about the nature and behavior of living systems and the application of that knowledge to extend healthy life and reduce the burdens of illness and disability."[32] Its goals include fostering fundamental discoveries, innovative research strategies, and creative applications in order to improve the nation's capacity to protect and improve health; developing, maintaining, and renewing scientific human and physical resources that will assure the nation's ability to prevent disease; and expanding the knowledge base in the medical and associated sciences in order to enhance the nation's economic well-being and ensure a high return on the public's investment in research.

There are more than twenty-five different institutes and centers under the NIH umbrella. Many of them are focused on certain conditions or groups of conditions, or on organs or groups of organs: for example, the National Cancer Institute, the National Institute of Mental Health, the National Eye Institute, and the National Institute of Diabetes and Digestive and Kidney Diseases each has its own mission and goals that complement the overall mission and goals of the NIH.

The institutes aim to take research from "the bench to the bedside" (also referred to as translational research) by developing therapeutic applications. The NIH is also increasingly interested in eliminating health disparities between racial and ethnic groups and between genders and different age groups. In fact, the NIH has established new institutes or centers to address these concerns, including the National Institute on Aging, the National Institute of Child Health and Human Development, and the National Center on Minority Health and Health Disparities. Health-related initiatives have also been launched to increase public awareness of specific diseases, including the War on Cancer in the early 1970s, the campaign for AIDS/HIV awareness in the 1990s, and the recent effort to fight obesity.

The public has a keen interest in the scientific effort to improve health. The public's acceptance of health technology has recently been challenged, however, by the ethical and moral debates surrounding human embryonic stem cell research and research or therapeutic cloning. As we make further progress in the life sciences, we will face additional challenges that must balance ethics with the desire to cure and treat disease.

The protection of privacy is perhaps one of the most unexpected grand challenges in the life sciences. Our new understanding of the human genome promises great benefits, but also allows us to predict the types of diseases or conditions to which an individual may succumb. Should insurance companies or employers have access to such information?

Finally, the United States has assumed a certain obligation to meet the world's health needs, particularly in developing countries that do not have the domestic resources to fight diseases and epidemics. For example, while the United States has committed to finding an AIDS vaccine, some critics argue that we are not doing enough, particularly for the poor countries of Africa, where AIDS is rampant. Others respond that AIDS, while it affects U.S. citizens, has not reached crisis proportions here, as it has in some developing nations. Does the United States, they ask, have a responsibility to use its science and technological capabilities to benefit other countries?

The Push and Pull between "Scientific" and "Societal" Grand Challenges

The scientific and policy-making communities often battle over whether science should be conducted for its own sake, or dedicated to addressing national and societal needs.

Many in the scientific community argue that scientists should search for the answers to major questions, without being held responsible for ensuring that their findings are immediately applicable. Some even suggest that the acquisition of new knowledge, in and of itself, is the goal. Policymakers, on the other hand, are expected to justify their use of taxpayer dollars. They thus feel significant pressure to spend in ways that generate quick and identifiable results.

POLICY DISCUSSION BOX 19.1

The Social Compact: What Responsibilities Accompany Federal Research Funding?

One of the objectives of federal funding for scientific research is to address societal needs and global issues. In order for this goal to be achieved, knowledge generated from government-sponsored research and technologies resulting from that research must be available to people all over the world, including the developing world.

In an attempt to promote accessibility to health solutions, the Bill and Melinda Gates Foundation has urged that universities adopt policies that, among other things, would waive certain intellectual property rights on the sale and distribution of health products to developing countries to ensure that people in these countries have access to vaccines, drugs, diagnostic tools, and related health technologies. According to the World Health Organization, however, failure to distribute life-saving drugs and health technologies "is not the fault of patent restrictions, but the lack of funds to buy these products and inadequate infrastructure to deliver them."[33]

In March 2007 a group of universities led by Stanford outlined nine points aimed at protecting the public interest in the licensing of university technologies. Noted the group, "Universities have a social compact with society." To this end, provisions in technology-licensing agreements should address "unmet needs, such as those of neglected patient populations or geographic areas, giving particular attention to improved therapeutics, diagnostics and agricultural technologies for the developing world."[34]

What responsibility do the people and institutions who receive federal funds to conduct research have to ensure that resulting technologies are made available to the public at a reasonable price? To people in developing nations? How much should profits be a factor in determining the licensing agreements made by technology transfer officials in universities? What portion of these agreements should be aimed doing societal good? What actions may be taken to better ensure that scientific and technological advances are readily available to people in developing countries?

David Guston and Daniel Sarawitz, two social scientists at the Consortium for Science, Policy, and Outcomes, agree with policymakers. They argue that science should be conducted with an eye for its benefits to society. The research university, in particular, has a role in ensuring that social benefits are obtained from knowledge-based innovation—in terms of service, education, and research.[35] Moreover, they say that the public support of research can be rationalized by three things: advances in science are needed to generate new wealth, to solve specific societal problems, and to inform decision making.[36] The benefits from research should be equally and fairly distributed—that is, research results and benefits ought to be broadly accessible, and those conducting research with public support ought to be accountable for the resources they use, and transparent in their use of them. Guston in particular argues that science should be democratized; it should be popular, relevant, and participatory.[37]

These ideas clearly fall in line with what some lawmakers believe—that federally funded research should be "strategically" directed toward specific societal needs. While this may seem reasonable enough at first glance, many in the scientific community have expressed concerns

about the idea. Why? Scientists know from long experience that it is very difficult to conduct science with an end objective in mind: the basic researcher never really knows what problems he or she may answer.

In the early 1990s, the National Science Foundation was urged to focus on strategic uses of science. Senator Barbara Mikulski, one of the leaders of this movement, was at the time chair of the powerful Senate Appropriations Subcommittee, with jurisdiction over the NSF. Mikulski called for 60 percent of NSF research to be targeted to achieving strategic goals.[38] A commission had been established at about the same time to reexamine the NSF's mission; this commission eventually echoed the legislators' demands. As a result, the NSF's FY1993 budget was decreased from FY1992 levels, allegedly in order to send the NSF a message. Industry, which recognized the NSF's importance to science education and research, strongly disagreed with the legislators' recommendations.

Sometimes members of Congress and the public become overzealous about targeted research funds. Their definition of targeted is often much more narrow than that of the scientific community. For example, beginning in 1992, a significant amount of the defense appropriations bill was designated for breast cancer research. While patient groups and advocates were adamant that this money should be spent only on breast cancer, some members of the research community pointed out that much of what was known about breast cancer had, in fact, emerged from basic research studies that were *not* focused on breast cancer: studies of molecules like the enzyme telomerase, whose activity is tied to promoting cell division; and retinoic acid receptors, which play a role in turning genes on and off, had unexpectedly contributed to our understanding of the disease.

Government-Wide Initiatives

How can we bring the interests of science and society closer together and ensure the continued support of policymakers?

Initiatives

One way to unite science and society in common cause is to launch a major government initiative that emphasizes the potential benefits of investment in science (e.g., the National Nanotechnology Initiative) or, alternatively, increases public awareness of a societal need that such investment might help address (e.g., Nixon's war on cancer).[39]

According to Duncan Moore, former associate director of the Office of Science and Technology Policy:

As we look at the R&D portfolio, there seem to be two ways we can increase the level of funding. One way to do this is to say that science is good, so fund more of it. Scientists believe this is a compelling argument that everyone should accept, but, in fact, it is not a very compelling argument. It does not go very far outside the scientific community, or with Congress. It is good enough to get an increase of the inflation rate plus one or two percent, but at that rate, it will take 35 years to double R&D funding.

Another way of getting increased funding is to use the initiative-based argument for increasing R&D. With this method, you make such a compelling argument that a certain area is so important for some reason (you have to define the reason), that we should put huge amounts of money into this area. This is how the National Institutes of Health can get such huge budget increases.[40]

In the past, large-scale science initiatives have helped to establish government priorities for investment in science, and have provided the impetus for significant funding increases. By focusing attention on a specific societal need or scientific challenge, governmental initiatives have helped to rally public and governmental support, while also directing the attention of the scientific community toward a particular challenge.

Historically, these initiatives have tended to highlight mission-oriented research, supported by an individual federal agency. For example, the government invested huge sums in NASA's space R&D work during the 1960s. As oil prices rose and gas lines grew in the 1970s, funding was redirected toward energy R&D, much of it conducted by the Department of Energy. The 1980s were characterized by national security concerns, and increased support for research funded by the Department of Defense. The 1990s saw major increases in funding for health research, as Congress doubled the budget of the National Institutes of Health (see fig. 2.1, chapter 2).[41] The attacks of September 11, 2001, and the subsequent anthrax scare have fueled significant investment in research related to homeland security. These monies have primarily flowed to DHS and the National Institutes of Allergy and Infectious Diseases (NIAID).

In addition to funding for federal research agencies, such initiatives have also led to increased support for universities and their researchers. Early initiatives emphasized specific scientific disciplines: as a consequence, certain departments, centers, laboratories, and schools benefited, while others received little. Students follow the money: increases in federal support for specific types of university research tend to lead to increases in enrollment in related fields. For example, the strong federal investment in nuclear energy research during the late 1970s and early 1980s led to an increase in the number of students enrolling in nuclear engineering programs.[42]

More recent initiatives, such as the Information Technology Initiative and National Nanotechnology Initiative (NNI) have been more interdisciplinary, with support coming from not one but multiple federal agencies. Nor are they as mission-specific as their predecessors. While the new generation of initiatives will certainly produce practical benefits, their primary goal is to produce scientific breakthroughs. This new approach brings with it complicated issues for both government and campus officials, which will be further discussed later.

Large-Scale, Multidisciplinary, and Multiagency Science Initiatives

We have seen how large-scale initiatives can help the government direct the researchers' attention on social problems while also generating political and public support for important scientific enterprises.

Individual investigators are generally hesitant to venture out of their areas of expertise. Federal initiatives provide financial incentives for interaction among and within scientific disciplines. The NNI, for example, is encouraging discussions within individual disciplines about new questions and methods, while also encouraging collaboration among people from different fields. Finally, the NNI is also uniting scientists from universities, federal labs, and industry.

Between 1999 and 2002, several workshops were organized at the University of Michigan to encourage the formation of a cross-campus team that could respond to government requests for interdisciplinary proposals—not only the NNI, but also the Information Technology and Biocomplexity in the Environment initiatives. One of the authors of this book, who was involved in these workshops, found it interesting to watch as researchers from two different departments spoke about their research and learned that they were involved in related work.[43] Such cross-pollination can offer tremendous benefits and has, in fact, fueled some of history's most significant scientific advances.

Initiatives like the NNI also are changing how government approaches important scientific issues. One of the NNI's greatest benefits, often overlooked, is that it has prompted government agencies to think about their shared S&T objectives. The program is an excellent example of the coordinating mission envisioned when the NSTC was first established to coordinate multiagency S&T activities.

The NNI is also a valuable model for emerging interagency initiatives: in homeland security S&T, for example. The very fact that a National Nanotechnology Coordination Office exists is itself significant in this regard. Its Web site provides a single source from which the research community and the public at large can find information about what any government agency is doing with regard to nanotechnology—including conference agendas, agency solicitations, and announcements of training opportunities; the site even includes a page for children.[44]

Coordination also offers federal agencies the opportunity to calibrate their efforts with those of other agencies across the government, ensuring that they create complementary, not duplicative or competing, programs.

Finally, interdisciplinary initiatives capture the public's imagination and spark real excitement about emerging research. This, in turn, increases political support and, often, student enrollments in relevant fields.

Multidisciplinary Initiatives and the Merging of Scientific Disciplines: Implications for Institutional and Government Policy

Multidisciplinary initiatives will be the way of the future and do seem to fill the role of meeting scientific and societal grand challenges. They will not be without their challenges, though, for both the government and universities. Here we will describe some of these institutional challenges and then provide mechanisms—for both the government and universities—to use in overcoming them.

Challenges to Government

While large-scale research initiatives force agencies to work together, these efforts can be impeded by structural and cultural differences among agencies. Even the best-intentioned campaigns to promote interagency cooperation can run into significant resistance.

The decentralized nature of Congress aggravates the problem. The agencies responsible for R&D answer to a myriad of House and Senate committees. For example, oversight of the departments of Defense and Energy and the National Science Foundation is assigned to different Senate committees. Unlike the House, the Senate does not have a committee with overall responsibility for science. And even the House Science and Technology Committee does not have jurisdiction over all the key science agencies: the NIH and the Department of Defense, for example—both important players in our nation's scientific enterprise—are overseen by the House Energy and Commerce Committee and the House Armed Services Committee, respectively.

Such decentralization can make it difficult for lawmakers to agree on the goals of multiagency initiatives. Special authorization bills like the 21st Century Nanotechnology Research and Development Act, which are specifically aimed at guiding multiagency initiatives, can help.[45] But even initiative-oriented legislation can conflict with laws authorizing specific federal agencies, such as the NIH or the NSF. Congressional oversight and authorization will become more complicated with the proliferation of interagency initiatives. It will be interesting to see how legislators handle this challenge.

The diversity of appropriations subcommittees poses a similar problem, leading to situations in which the White House proposes to fund several agencies' work on a particular initiative, only to have Congress decline to fund one or more of the partner agencies. Such a logjam occurred during the first year of the Information Technology Initiative. As a part of the initiative, the George W. Bush administration proposed $70 million for a new Scientific Simulation Initiative (SSI), which involved the Department of Energy, among other agencies. Congress, however, failed to provide any funding for the DOE portion of the Information Technology Initiative.[46] Indeed, some members of Congress, failing to recognize the importance of the DOE's IT work, suggested that the NSF could do a better job of overseeing the SSI, and that there was no need for the DOE to be involved in the first place.

There is a strong tendency to oppose multiagency or interdisciplinary initiatives on the premise that they divert funds from traditional, core disciplines. But funding is not a zero-sum game: the elimination of interdisciplinary initiatives would not mean that the designated funds would automatically be redirected to individual disciplines. Indeed, interagency initiatives may well help to drive up overall funding for research.

A final challenge for the government is the need to accurately measure the effectiveness of its interagency and interdisciplinary initiatives. The Government Performance and Results Act of 1993 (GPRA) (P.L. 103-62), and, more recently, the President's Management Agenda, announced in the summer of 2001, declared the need for effective metrics to assess the effectiveness of government programs, including government-sponsored research programs. Although the OMB has developed a Program Assessment Rating Tool, also known as the PART, it will be extremely difficult to apply uniform metrics to big-science initiatives, which involve agencies with very different missions and objectives.

Challenges to Universities

Like federal agencies, universities are rife with cultural and organizational barriers to interdisciplinary collaboration. Despite growing campus interest in the subject and the proliferation of new multidisciplinary research centers, universities and their funding and reward structures are, for the most part, still organized around traditional disciplines and departments. The degree to which university researchers can respond to multidisciplinary initiatives will be determined by their ability to break out of their traditional silos.

Multiagency and multidisciplinary initiatives are not necessarily attractive to all faculty members. While large initiatives can help overcome concerns about the scope, size, cost, or complexity of new projects, they also tend to

be more structured and aimed at facilitating research for large teams of scientists. Thus these initiatives are often perceived as limiting the opportunities and independence of individual research scientists.

Universities will need to devote significant time and resources to fostering the multidisciplinary environment needed to succeed in large-scale competitions. The need for this investment might be a disincentive for universities or detract from their competitiveness.

The scope, size, and complexity of multidisciplinary projects can also deter participation. This may be particularly true of the competitions for nanotechnology centers, which will likely require that universities form significant consortia, including both other universities and private partners. Many of these competitions also require K–12 and public education components—activities that faculty members sometimes deride as irrelevant to their research. Because university researchers have little interest in meeting these requirements or lack the expertise to do so, their proposals often require significant institutional commitments. Universities with experience in collaborations and outreach programs will have a great advantage in such competitions.

Funding solicitations always carry the risk of failure. New initiatives can attract significant interest, and their award rates are often lower than those for other, more limited programs. Faculty and institutions who have invested significantly in unsuccessful proposals may become discouraged and choose not to reapply.

Large multidisciplinary initiatives will inevitably change the way research, education, and training are conducted on campus. Institutions that are successful in competing for large research awards or centers are likely to find that success changes how their people work. They must also figure out how to sustain new programs beyond the life cycle of the original award, often only three to five years. Every university that receives outside funding—and that means almost every university in the country—must have a well-formulated exit strategy which either ramps down activities in an orderly fashion or seeks new sources of support.

Overcoming the Institutional Challenges

Although universities face many challenges when responding to multidisciplinary initiatives, steps can be taken to eliminate obstacles and encourage institutional and faculty participation.

Providing governmental and institutional incentives to ensure broad-based participation. For multidisciplinary initiatives to be truly successful, they must offer adequate funding across disciplines, so that faculty members from any field who want to participate can do so. Sometimes areas or disciplines are unintentionally locked out, not because they are unimportant but because grantees do not appreciate what these individuals can offer to the program.

Special outreach efforts may be needed to explain what researchers from specific disciplines can contribute to a particular effort. Incentives can also be used to attract individuals from many disciplines. For example, the NNI includes appeals to specialists studying the ethical, legal, social, and economic dimensions of nanotechnology—not necessarily obvious constituencies when one is thinking about nanotechnology research, but clearly important ones.

Universities can also offer release time from teaching or service requirements, shared faculty positions and joint appointments, and funding for administrative support or important lecture series and symposia.

Ensuring support for the work of individual investigators and small research teams, as well as large centers. The university community tends to pay a great deal of attention to opportunities to compete for large centers. These naturally stimulate a great deal of excitement, given their scope and the amounts of money involved. It is important, however, to ensure that such initiatives also include opportunities for individual investigators and small research teams.

Providing planning and seed grant funding. Seed grants or planning grants can often help overcome campus cultural barriers by pulling people together and focusing their attention on multidisciplinary initiatives. Although such grants are often provided by the institutions themselves, government agencies should consider funding them in order to encourage institutional planning. Small grants for equipment or administrative support can also enable faculty to plan activities that make the university more competitive for grants in new and emerging fields.

Encouraging partnerships at all levels. When applying for support, universities should take into account not only their own resources, but also potential contributions from private-sector partners. An industrial partner might be able to emphasize the practical applications of a proposed research project, for example. Universities can also team up with other universities, liberal arts and community colleges, and federal labs to exploit complementary strengths.

Linking new initiatives to the academic and educational missions of universities. New government science initia-

tives should encourage not only cutting-edge research but also the creation of new courses and curricula. An influx of funding in a particular area might fuel the development of courses that can in turn disseminate the findings from the new initiative, or might encourage the creation of new degree programs that train students in the new field.

The Importance of Policy

Cultural and institutional barriers will always make it difficult for universities to engage in large, multidisciplinary initiatives. Perhaps this is a good thing: these initiatives are meant to promote investigation into new areas, and by their nature challenge traditional campus cultures and ways of conducting science.

In launching an initiative, the government is trying to refocus research agencies and the scientific community as a whole on an area where it deems the potential payoffs to be significant. The scientific community would have tremendous difficulty organizing such an effort on its own.

The importance of the government's role was perhaps best summarized by the French mathematician Pierre Louis Moreau De Maupertuis:

> There are sciences over which the will of kings has no immediate influence; it can procure advancement there only in so far as the advantages which it attaches to their study can multiply the number and the efforts of those who apply themselves to them. But there are other sciences for which their progress urgently need the power of sovereigns; they are all those which require greater expenditure than individuals can make or experiments which would not ordinarily be practicable.[47]

Policy Challenges and Questions

Some perceive an inherent conflict between the goals of the scientific community and goals of policymakers and the general public. After all, scientists are interested in creating new knowledge and understanding "how things work" simply for the sake of knowing. While they like to witness the results of their labors taken to the level of everyday use and application, many scientists are caught up in their own world of gaining knowledge and understanding. Policymakers, on the other hand, are keenly aware of the needs of their constituents and take great measures (usually) to meet those needs. Science has increasingly given birth to technologies that benefit the public; because of this, policymakers have become more and more interested in what "science can do for them." This interest in application of science has been especially true as the federal budget for nondefense R&D has grown, as it has generally since 1953. In addition, what is sometimes overlooked by policymakers is that the knowledge from basic, fundamental research lays the groundwork for downstream applications and development.

In this chapter we have outlined some of the scientific and societal challenges we believe will dominate the thoughts of scientists and policymakers. Meeting both sets of challenges is important, and the scientific community and policymakers need to work together to meet them. Are the tensions between scientific and societal goals real or perceived? Can these tensions be minimized or eliminated? If so, how? What mechanisms could be used to mesh the goals of the scientific community with the needs of society?

There are some who believe that science will end, that there will no longer be anything new to discover. This argument views science as a set of Russian nesting dolls, where one continues to open a doll to find a smaller doll until the smallest doll (which does not open) is reached. We strongly disagree with this position, viewing science as like a palm tree: On a palm tree, new fronds grow from the center of the top of the tree. They sprout out of the top as older fronds make way. These older fronds help nourish the new fronds, while the oldest fronds die, turn brown, and hang as remnants until they fall off. In science, previous findings from research feed new questions, and sometimes old theories are proven incorrect and die because of new knowledge and better instrumentation.

A palm tree remains healthy, with a continual source of fronds, if the tree is well cared for. Science in the United States, too, will remain healthy as long as it is well cared for by scientists and policymakers alike.

NOTES

1. John Horgan, interview by David Gergen, *PBS Online NewsHour,* July 26, 1996, http://www.pbs.org/newshour/gergen/july-dec96/horgan_7-26.html (accessed June 10, 2007).

2. Gilbert S. Omenn, "Grand Challenges and Great Opportunities in Science, Technology and Public Policy," *Science* 314, no. 5806 (December 15, 2006): 1696–1704; see also "Grand and National Challenges," http://archive.ncsa.uiuc.edu/Cyberia/MetaComp/GrandNat.html (accessed June 10, 2007).

3. This was the wording in the May 1, 2003, solicitation for proposals issued by the scientific board of the Global Health Initiative, which was launched earlier in 2003. See H. Varmus, R. Klausner, E. Zerhouni, T. Acharya, A. S. Daar, and P. A. Singer, "Global Health: Grand Challenges in Global Health," *Science* 302, no. 5644 (October 17, 2003): 398–99.

4. See "Grand and National Challenges," http://archive.ncsa.uiuc.edu/Cyberia/MetaComp/GrandNat.html (accessed June 10, 2007).

5. Solar and Space Physics Survey Committee, Division on Engineering and Physical Sciences, National Research Council, *The Sun to the Earth—and beyond: A Decadal Research Strategy in Solar and Space Physics* (Washington, DC: National Academies Press, 2002).

6. See Charles M. Vest, "Report of the President," Massachusetts Institute of Technology, November 1995, http://web.mit.edu/president/communications/rpt94-95.html (accessed June 10, 2007).

7. Committee for the Review of the National Nanotechnology Initiative, Division on Engineering and Physical Sciences, National Research Council, *Small Wonders, Endless Frontiers: A Review of the National Nanotechnology Initiative* (Washington, DC: National Academy Press, 2002), http://www.nano.gov/html/res/small_wonders_pdf/smallwonder.pdf (accessed June 11, 2007).

8. See National Nanotechnology Initiative, "What Is Nanotechnology?" http://www.nano.gov/html/facts/whatIsNano.html (accessed June 11, 2007).

9. Richard P. Feynman gave the talk "There's Plenty of Room at the Bottom: An Invitation to Enter a New Field of Physics" at the annual meeting of the American Physical Society in December 1959. He posed this question describing the possibilities of exploring matter at the nanoscale. The talk was first published in 1960 in the February issue of Caltech's *Engineering and Science,* http://www.zyvex.com/nanotech/feynman.html (accessed June 11, 2007).

10. See the NNI Web site, as well as Susan R. Morrissey, "Harnessing Nanotechnology," *Chemical and Engineering News,* April 29, 2004, 30–33, http://pubs.acs.org/cen/nlw/8216gov1.html (accessed June 11, 2007).

11. 21st Century Nanotechnology Research and Development Act of 2003, P.L. 108-152, 108th Cong., 1st sess. (December 3, 2003, 117 Stat. 1923).

12. Committee for the Review of the National Nanotechnology Initiative, Division on Engineering and Physical Sciences, National Research Council, *Small Wonders, Endless Frontiers,* v.

13. National Research Council, *Grand Challenges in Physics and Astronomy: Connecting Quarks with the Cosmos* (Washington, DC: National Academies Press, 2003).

14. For additional background see the definition of complex systems at the University of Michigan, Center for the Study of Complex Systems, "About the Science of Complexity," http://www.cscs.umich.edu/about/complexity.html (accessed June 11, 2007).

15. David G. Green and David Newth, "Towards a Theory of Everything?—Grand Challenges in Complexity and Informatics," *Complexity International* 8 (2001), http://journal-ci.csse.monash.edu.au/ci/vol08/green05/green05.pdf (accessed June 11, 2007).

16. Ibid., 4.

17. For additional background see Intel, s.v. "Moore's Law," http://www.intel.com/technology/mooreslaw/ (accessed June 27, 2007).

18. For additional background see James Fallows, "New Life for Moore's Law," *Atlantic Monthly,* October 2001, 44; Gene J. Koprowski, "Life after Moore's Law: Beyond Silicon," *TechNewsWorld,* July 18, 2003, http://www.technewsworld.com/story/31137.html (accessed June 11, 2007).

19. Dana Alliance for Brain Initiative, *Answering Your Questions about Brain Research* (New York: Dana Alliance, 2004); see also Robert H. Blank, "Policy Implications of Advances in Cognitive Neuroscience," in *AAAS Science and Technology Policy Yearbook 2003,* American Association for the Advancement of Science, http://www.aaas.org/spp/yearbook/2003/ch6.pdf (accessed June 11, 2007).

20. Alvin W. Gouldner and S. M. Miller, eds., *Applied Sociology: Opportunities and Problems* (New York: Free Press, 1965), vii.

21. W. P. Butz and B. B. Torrey, "Some Frontiers in Social Science," *Science* 312, no. 5782 (June 30, 2006): 1898.

22. Ibid.

23. Taken from the U.S. Department of Transportation, "Mission & History," http://www.dot.gov/mission.htm (accessed June 11, 2007).

24. For additional background see U.S. Department of Transportation, Strategic Plan 2003–8, *Safer, Simpler, Smarter Transportation Solutions* (Washington, DC: U.S. Department of Transportion, September 2003).

25. Ibid., sec. 5.

26. Taken from the Environmental Protections Agency's Web site, "About EPA"; Congress passed the bill in late 1969. President Nixon signed the bill into law on January 1, 1970, as his first official act of the new decade.

27. The National Environmental Policy Act of 1969, P.L. 91-190.

28. U.S. Environmental Protection Agency, *2003–2008 EPA Strategic Plan: Direction for the Future,* EPA-190-R-03-003 (Washington, DC: Environmental Protection Office, September 30, 2003), http://www.epa.gov/ocfo/plan/2003sp.pdf (accessed June 11, 2007).

29. "Major Power Outage Hits New York, Other Large Cities," *CNN.com,* August 14, 2003, http://www.cnn.com/2003/US/08/14/power.outage/ (accessed June 27, 2007); "Power Outage Leaves Motor City Low on Fuel," *CNN.com,* August 15, 2003, http://www.cnn.com/2003/US/08/15/blackout.detroit/index.html (accessed June 27, 2007).

30. U.S. Department of Energy, "About DOE," http://www.energy.gov/about/index.htm (accessed June 27, 2007).

31. Office of Program Analysis and Evaluation, Office of Management, Budget, and Evaluation, *The Department of Energy Strategic Plan: Protecting National, Energy, and Economic Security with Advanced Science and Technology and Ensuring Environmental Cleanup* (Washington, DC: U.S. Department of Energy, September 30, 2003).

32. See the National Institutes of Health, "About NIH, the NIH Almanac," http://www.nih.gov/about/almanac/index.html (accessed June 11, 2007).

33. Angela Rickabaugh Shears, "Curing the World: How Donors Can Help Develop Drugs Critical to World Health," *Philanthropy Magazine,* September 25, 2006, http://prt.timberlakepublishing.com/printarticle.asp?article=1420 (accessed June 5, 2007)

34. *In the Public Interest: Nine Points to Consider in Licensing University Technology,* white paper, released by Stanford and eleven other universities, March 6, 2007, 8, http://news-service.stanford.edu/news/2007/march7/gifs/whitepaper.pdf (accessed June 3, 2007).

35. David H. Guston, "Responsible Knowledge-Based Innovation," *Society,* May–June 2006, 19–21.

36. Daniel Sarewitz, Guillermo Foladori, Noela Invernizzi, and Michele S. Garfinkle, "Science Policy in Its Social Context," *Philosophy Today,* Supplement (2004): 67–83.

37. David H. Guston, "Forget Politicizing Science. Let's Democratize Science!" *Issues in Science and Technology,* Fall 2004, 25–28.

38. Susan E. Cozzens, "Quality of Life Returns from Basic Research," in *Technology, R&D, and the Economy,* ed. Bruce L. Smith and Claude E. Barfield (Washington, DC: The Brookings Institution and American Enterprise Institute, 1996), 199–200. See also Barbara Mikulski, "Science in the National Interest," *Science* 264 (April 8, 2004): 221–22.

39. Portions of this section on government science initiatives are based in part on a paper prepared by Tobin L. Smith as a part of a December 3–5, 2003, NSF workshop, Nanotechnology and Its Societal Implications. See Smith, "Institutional Impacts."

40. Duncan T. Moore, "Establishing Federal Priorities in Science and Technology," *AAAS Science and Technology Policy Yearbook, 2002,* ed. Albert H. Teich, Stephen D. Nelson, and Stephen J. Lita (Washington, DC: American Association for the Advancement of Science, 2002), 273.

41. Genevieve J. Knezo, "Federal Research and Development: Budgeting and Priority Setting Issues, 107th Congress," *CRS Issue Brief for Congress* (Washington, DC: Congressional Research Service, Library of Congress, September 24, 2001).

42. National Research Council, Committee on Nuclear Engineering Education, *U.S. Nuclear Engineering Education: Status and Prospects* (Washington, DC: National Academy Press, 1990).

43. As a federal relations officer for the University of Michigan, Tobin Smith helped to orchestrate and conduct these workshops between 1999 and 2002 for the benefit of the University of Michigan's research community.

44. This Web site is www.nano.gov (accessed June 11, 2007).

45. P.L. 108-153.

46. Michael E. Davey, "Research and Development Funding: Fiscal Year 2000," *CRS Brief for Congress* (Washington, DC: Congressional Research Service, Library of Congress, October 15, 1999).

47. P. L. M. de Maupertuis, quoted in C. C. Gaither and A. E. Cavazos-Gaither, *Scientifically Speaking: A Dictionary of Quotations* (Philadelphia: Institute of Physics Publishing, 2000), 276.

CHAPTER 20

Science, Science Policy, and the Nation's Future

What about the Future?

One can imagine how difficult it would have been for persons born in 1850 to imagine the world of today. Airplanes, x-rays, radios, the genome, the Internet, the atomic nucleus, even zippers and ice cream would all be entirely foreign. Even those with the most vivid imaginations would have been hard-pressed to foresee such novelties. In the words attributed to both Niels Bohr, Danish physicist and 1922 Nobel laureate, and Yogi Berra, baseball legend, "Prediction is very hard, especially if it is about the future."[1] Today we are not much better off: the future will always be hard to anticipate. Indeed, our excitement—and anxiety—about its prospects are a powerful force in motivating our present-day lives.

For many years, policy and science were assumed to be almost completely unrelated. Science focuses on creating new knowledge, while policy creates rules by which society functions. Science is (ideally) an unprejudiced seeking of knowledge, while policy (at least in a democratic society) seeks to do the most good for the most people, even if this means solutions are inefficient. While analysis may demonstrate that one program is more cost-efficient or likely to succeed, lawmakers may choose another program that promises immediate results for people in need of assistance. In the policy-making world, right and wrong can be subjective. Scientists often believe that policymakers lack a rational, methodical protocol, while policymakers claim that scientists are insensitive to the need for strategy.

The two groups do share some traits, however. In both arenas optimal progress is made by asking the right questions. Scientists have long grasped this fact, but policymakers have not. Because lawmakers are often driven by personal and political motives—doing what is best for the people, or searching for effective policy—their questions can sometimes seem one-sided or shortsighted.

Much depends on the government's attitude and policies toward scientific advancement. One event, like World War II or *Sputnik,* can drastically change the world, and a nation's response to such an event can shape its policy for years to come. The world we live in, in other words, has been formed in reaction to past events. Prior to World War II, there was, as we have seen, very little government support for research. As the country mobilized for war, the government recognized that scientific knowledge could help soldiers on the battlefield, and increased its research funding accordingly. Thus was forged a partnership that no one could have predicted fifty years earlier.

Many of the technological advancements of the last half-century were the result of U.S. efforts to win World War II and the Cold War. Any attempt to predict the future must take into account our ignorance of what major events await us—events that will shape our needs and hence our policies. The September 11, 2001, attacks prompted an unprecedented rethinking of our federal policies for homeland security and research related to bioterrorism. We cannot predict what events will drive major policy changes in the future. Will rising oil prices and concerns about climate change lead to increased focus on energy research? What impact will concerns about global competitiveness have? Will a global pandemic such as avian flu push us to devote scientific resources to this threat?

Policymakers sometimes respond quickly to national or international events. Undue haste, however, can lead to ill-advised policies, without thorough debate or thoughtful discussion. The USA PATRIOT Act of 2001 was written, brought to a vote in both chambers, and signed into law within a matter of weeks after September 11. The bill included provisions that might have been better written, or left out altogether. To cite just one example, its provisions on select biological agents did not specify which federal agency was responsible for enforcing the regulations. This omission initially caused great consternation among university officials, who had to ensure that their researchers were following federal law.

Policies can also be provoked by a convergence of events with a desire to pursue certain goals. For example, John F. Kennedy decided that one means of responding to the Soviet launch of *Sputnik* was to put a man on the moon, despite the tremendous costs and difficulties. In his May 25, 1961, address to Congress, President Kennedy stated: "I believe that this nation should commit itself to achieving the goal, before this decade is out, of landing a man on the Moon and returning him safely to the Earth. No single space project . . . will be more exciting, or more impressive to mankind, or more important . . . and none will be so difficult or expensive to accomplish." In 1962, after John Glenn successfully orbited the earth for the first time, Kennedy remarked: "We have a long way to go in the space race. We started late. But this is the new ocean, and I believe the United States must sail on it and be in a position second to none."[2]

The effect of this emphasis was to shift popular attention to the United States' successes in space flight and exploration. In the creation of the president's science advisor and NASA, and passage of the National Defense Education Act of 1958, the will of U.S. policymakers worked in concert with the fears stimulated by *Sputnik* to shape national science policy. Indeed, many have remarked that America needs another *Sputnik* moment to reinvigorate its support for science. Some thought that the September 11, 2001, attacks might serve this purpose, but policymakers' response to those events did not focus on science or the value of basic research.

World events influence our science and technology priorities in myriad ways. A decrease in yields, for example, may stimulate efforts to develop disease-resistant crops. Accidents, like the one at the Three Mile Island nuclear reactor, may alter attitudes toward particular areas of science or technology. Ethical and moral debates, such as those prompted by the evolution of in vitro fertilization techniques, may open up or foreclose entire areas of research.

There can be no doubt that the United States' emphasis on scientific R&D over the past fifty years has been shaped largely by perceptions of the Soviet threat. Many members of Congress and the public considered this to be sufficient justification for investment in R&D. In the late 1990s, many scientists and policy observers wondered how the dissolution of the Soviet Union would affect the public's willingness to spend tax dollars on R&D. Then came September 11, and the debate took a sharp turn. More recently, concerns regarding our national and global competitiveness, such as those outlined by the National Academies in its *Rising above the Gathering Storm* report and by Thomas Friedman in he best-selling book *The World Is Flat,* have driven policymakers to focus on policies to support science.[3] This has resulted in major new initiatives presented by President Bush and by the Democratic leaders in Congress to increase funding for science—particularly for agencies that support the physical sciences and engineering.[4] In July 2007, Congress passed and the president signed the America COMPETES Act (P.L. 110-69). The act authorized several new science programs to address these competitiveness concerns. Despite the passage of this act, and the significant fanfare that surrounded it, Congress failed to approve the funding required to implement many of these new programs in its final Omnibus Appropriations Act (P.L. 110-161).

Will emerging economic powers, including China and India, play the role formerly occupied by the USSR? How will the United States respond to the growing threat of terrorism, abroad and at home? Will the American people and their government turn to scientists to help them respond to these threats? Again, it is terribly difficult to predict the future. We can only observe that past investment in science has produced positive social and economic returns, and that future investments will likely prove similarly beneficial.

The rest of this final chapter explores some of the major advances of the past century, and makes a few predictions—some rosy, others less so—about how the situation might look fifty years from now. We close with a discussion of policy issues that will require attention in the decades ahead.

A Retrospective View of the Past Century

The scientific advances of the last century can only be characterized as phenomenal. Many were the result of a general interest in the ability of science and technology to improve our life, health, and national security.

We are now preparing to embark on a direct exploration of the planet Mars, yet at the turn of the twentieth century we knew virtually nothing about the moon. Indeed, a hundred years ago few dreamed that man could fly in a heavier-than-air device even a few feet, let alone through hundreds of thousands of miles of space. The generations of the early 1900s could only dream about where stars came from, but telescopes like the Hubble now enable us to watch the birth of stars with our own eyes. Likewise, there was a time within living memory when we had only the most primitive understanding of the mechanisms of heredity, yet now scientists have sequenced the entire human genome. The pace of progress has been breathtaking.

Our race through the twentieth century has taught us that public funding can fuel national R&D. Diseases once thought untouchable have been cured. Scientists have developed noninvasive techniques to see inside the human body. Almost no part of the earth is unreachable, and huge numbers of individuals routinely travel from one continent to another. Economies that once were regional activities have now become global. Students from a big city in southern China may study in a small university town in Kansas.

All this progress notwithstanding, we still do not always effectively allocate federal funding for science. Billions have been squandered with no result (e.g., the SynFuels Initiative, designed to develop synthetic fuels to meet the nation's growing energy needs in the 1970s, or the SSC, which was to be the next major stride in particle accelerators for basic research in the 1990s). It seems that science policy is lagging behind science in other ways, too. For instance, the environment has deteriorated measurably, in spite of technological advances that could be used to protect it. The world's petroleum reserves have been significantly depleted, with no clear vision of what will replace them when they are gone. And the debate over cloning and reproductive technology rages on without generating clear guidelines for when research should or should not—or can or cannot—be done.

Congress does not seem to be optimally organized to deal with the scientific challenges ahead. Any new scientific initiative is subject to too many competing oversight bodies, each with a different agenda. If we are to ensure continued benefits from scientific and technological research, the issue of how science policies should be created and implemented will need to be addressed.

The Technological World in 2055: An Optimistic View

Science has the potential to solve many of the world's current and future problems. It may hold the key to ending world hunger, creating cheap and renewable energy, and conquering disease. Science and technology are also important in providing education to increasing numbers of people across the globe. And while science will not be able to prevent most natural disasters for the foreseeable future, it can devise early detection and warning systems that will lessen destruction and loss of life. However, it would be naive to believe that scientists can solve all of these problems on their own. Science can only flourish in a supportive environment. Policy will influence the chances of success.

What Could Go Wrong? A Pessimistic View

While science holds tremendous promise, that promise could easily go unfulfilled. It is possible that, as science pushes moral and ethical boundaries, the public will reject it as a means of solving problems. Such a response might take root in the simplistic view that science and religion are incompatible—that one cannot accept the results of science yet believe in a Supreme Being.

The promise of science might also be impeded by public impatience for results. The demand for quick payoffs could mean less support for science for science's sake, and more emphasis on practical applications of existing knowledge. One of the challenges policymakers and scientists alike will face is proving the benefits of science and ensuring that the demand for results does not force science in unproductive directions. A balance must be struck between policies that favor conducting science with specific societal benefits and applications in mind and those that protect unfettered scientific exploration aimed at the creation of new knowledge, "science for science's sake," as it is sometimes called.

Similarly, there is always a risk that Americans may focus on their immediate needs at the expense of investments in the future. Driven by election cycles, many politicians are interested in promoting policies that clearly demonstrate results, which leads them to support short-term solutions, even if this means not sowing seeds that might bear sweeter fruit over the long term. Can any political leader prove that he or she was the reason a particular American scientist won a Nobel Prize, or that a major disease was cured? Such investments are made collectively, over time, and rarely produce rewards until long after the initial investments have been forgotten. Then, too, scientific progress is a process of accretion; even in those rare instances when it evolves in a linear (rather

POLICY DISCUSSION BOX 20.1

Science and Religion— Can You Have Both?

Science or religion? It seems that a person has to choose, especially if one is a scientist. After all, the scientific method is premised on the testing of hypotheses: experimenting, data gathering, and data analysis. Religion, on the other hand is about faith and boundless belief in some Supreme Being. Conflict between these two long-established institutions is inherent.

With opposition to evolution, human embryonic stem cell research, cloning, in vitro fertilization, and embryonic preimplantation genetic diagnosis led largely by individuals with religious beliefs, it is no wonder that when one thinks of religion and science, one thinks *conflict*. Studies have shown that the majority of scientists express doubt or disbelieve in the existence of God, and this has held over several decades.[5] Some scientists—prominent ones among them—say that science and religion cannot coexist. Evolutionary biologist Richard Dawkins, when speaking on this subject, has said, "Any belief in [religious] miracles is flat contradictory not just to the facts of science but to the spirit of science."[6] There are those

who do think that there is room for both science and religion. The widely recognized paleontologist Stephen Jay Gould believed that religion and science can coexist because they occupy "non-overlapping magisteria," or domains.[7] Others, in particular scientists who are "believers," or holders of some religious faith, take this one step further. These individuals think that not only can religion and science coexist but that there is no need for them to exist as separate worldviews, or domains. Francis Collins, the director of the National Human Genome Research Institute and a self-proclaimed "believer," says, "Science doesn't refute God. It affirms him, and gives us new respect for the beauty of creation."[8]

What do *you* think? Can science and religion coexist? What challenges does the seeming conflict between science and religion pose for science policy? Is it possible for religious leaders to support policies that allow for technologies such as preimplantation genetic diagnosis and in vitro fertilization? Of stem cell research? How do they reconcile these advances with the notion of "playing God"? How does a member of Congress reconcile scientific fact with his or her own religious beliefs and those of constituents back home?

than branched) fashion, that progress is still the result of innumerable individual contributions.

In addition, in this era of homeland and national security and the war on terror, arguments are being made that scientific progress necessarily creates increased security risks. Even though the September 11 attacks used relatively simple technology, many Americans still misguidedly believe that technology and the science behind it contributed to September 11. Others fear science's ability to help governments track and surveil individuals. The threat always looms that short-term anxieties will be the basis for bad long-term policy.

A thriving R&D enterprise depends on our fostering an environment hospitable to budding young scientists. The more pessimistic among us might point to the apparent decline in the quality of U.S. math and science education, and to decreasing student interest in science and engineering careers. The quality and size of a nation's science and engineering workforce is linked to the adequacy of its math and science education, and to the availability of appropriate opportunities for graduate students and postdoctoral fellows. Unless we invest in education at all

levels, the United States will have difficulty retaining its leadership of the global economy.

Finally, scientists themselves risk losing sight of their responsibility to conduct science with integrity. Too often they do not recall the broader context of their research, and the very reasons why they chose a scientific career in the first place.

Key Policy Issues

Our two visions of the future—optimistic and pessimistic—are not mutually exclusive. Life being neither perfect nor perfectly horrible, the most likely outcome will be some combination of the two. The extent to which the balance tips in one direction or the other depends on changes in the country as a whole: demographic, economic, social, and cultural. To realize our more optimistic view, we will need to implement a wide range of informed science policies, many of them discussed throughout this book. Sustained funding for large-scale science projects will have to be ensured, and long-term priorities for science will have to be set. We will also need to improve the channels for cooperation among R&D partners. How can universities, national labs, and industry work together to ensure innovation and sustain America's global position, not only in key fields of science but in economic terms as well? What science policies can we devise that will keep the United States on the cutting edge of technological innovation?

One way to start is by narrowing the chasm that divides policymakers and scientists, by educating policymakers about science and improving scientists' understanding of how policies are formed and implemented.

Public Support of Science

If the future is devoid of societal challenges that require scientific and technological solutions, will the public continue to support scientific research? From a scientific perspective, what is required to ensure continued public support for science?

One of the arguments made by Gunther Stent in his 1969 book, the *Coming of the Golden Age: A View of the End of Progress,* is that society will continue to support research in physics so long as it has the potential to produce tangible results, such as nuclear weapons or nuclear power. When physics reaches the point at which its value to society becomes unclear, Stent predicted, society will withdraw its support.[9]

Was his view accurate? John Horgan, the senior editor of *Scientific American* and author of *The End of Science,* remarks, "When a given field of science begins to yield diminishing returns, scientists may have less incentive to pursue their research and society may be less inclined to pay for it."[10] To many members of the public, science has become more abstract and its social value less obvious. This has led to a waning of public support for certain initiatives in science, especially as the cost of new tools and techniques grows. Recent decisions, such as the congressional vote to terminate funding for the SSC in the early 1990s, illustrate how precarious political support for science can be when the benefits of the research are not obvious. This rule was clearly at work in the failure of the SSC and the tremendous success of the Human Genome Project, whose potential public benefits were more readily apparent.

Even though the public seems to have a good grasp of the potential benefits of increased investment in biomedical and health research, many scientists are still concerned that an increasing demand for quick results could make it impossible for science to keep pace with expectations, regardless of the level of funding. When a field or line of research does not or cannot deliver timely results, the public and policymakers may decide that the funding of further research is unnecessary.

The Increasing Difficulty of Distinguishing between Basic and Applied Science

In the past, research was often viewed as linear: basic research produced new knowledge, which could in turn be applied and developed into new products, goods, and services. However, it is becoming obvious that the line is not as linear as once believed. As a researcher pursues a general line of investigation, intermediate observations may have a significant impact on the technology underlying the tools used in the investigation, thus making further advances in the basic research possible.

Ongoing attempts to make scientific results accessible not only to the scientific community but to the public at large are likely to increase the pace of innovation and scientific progress. Indeed the public will demand it. One challenge is how properly to temper the public's expectations for the readiness of new technologies. Human embryonic stem cells are being touted as the "next big thing" in the treatment of many diseases and debilitating conditions. But in reality, scientists are still far from being able to use hESC clinically in most cases. Fundamental questions need to be answered first: for example, how to control hESCs' development into different cell types of cells, and overcome the rejection response of a patient's immune system? These issues can only be resolved through continued basic research.

Another challenge is ensuring that sufficient funding is dedicated to truly basic research. If we become too wedded to the notion that research has to have a "purpose," or even a "strategic purpose," we will lose out in the end. Application and development do not just happen. Often the process begins with a fundamental discovery, whose applications may be unclear; only over time are researchers able to imagine practical applications for the new knowledge. The public and policymakers have, for the most part, come to expect good things from scientific research, and have a tendency to want to jump to the application stage, while skipping over the foundational basic research.

Many scientific institutions are organized around the distinction between basic research and applied R&D. Universities have traditionally been focused on the basic end, conducting very little applied research except in their engineering departments and schools. As the line between the two categories grows fuzzier, academics are engaging in more and more applied research, and universities are finding ways to extract revenues from these applications. They are becoming, as we have seen, more engaged in technology transfer, and thus must be more careful about insuring that there commercial activities do not compromise their traditional academic and educational work.

The Merging of Old Scientific Disciplines and the Emergence of New Ones

As science becomes more interdisciplinary, and traditional boundaries begin to crumble, whole new disciplines will likely emerge. This will place increasing pressure on universities, national and industrial labs, and the government to create programs and structures that support these new fields. Areas such as information technology, nanotechnology, bioinformatics, and complex systems are already bringing together teams that transcend traditional boundaries.

This process is likely to present challenges for both the institutions themselves and science policy. When a project combines personnel and resources from two or more existing departments, for example, who controls the project? Ad hoc answers will not be enough; major cultural changes are required all around.

As awareness of science's ability to address societal needs grows, the need for interdisciplinary research will most certainly grow along with it. Many of the research programs that are being developed to protect homeland security, for example, are already interdisciplinary. Engineers are working with risk analysts who can assess technical vulnerabilities of a proposed structure, determine the likelihood that it will be targeted, and estimate any damage that may be done to it. Interactions between the social and natural sciences are also becoming more common, thanks in large part to programs such as the National Nanotechnology Initiative and the Human Genome Project, which fund not only basic and applied research, but also exploration of the ethical, legal, and social implications of these new sciences.

The Blurring of Roles for Traditional Partners in Science Policy

The roles of those involved in science policy have traditionally been fairly well delineated. As science evolves, and the amount of time between basic discovery and practical application shrinks, these lines, too, are blurring.

Take, for example, the relationship between universities and national labs. There are serious questions about when major research facilities should be managed by national labs and when they should be built at universities. The universities' traditional management of major DOE laboratories has recently been called into question, as more and more nonprofits and private companies, such as Battelle and Lockheed Martin, compete for the management contracts. This has led to the formation of consortia uniting the public and private sectors, as in the case of the Idaho National Laboratory, which is operated by a partnership of companies, universities, and nonprofits.[11]

Government and industry have also partnered, as in the cases of programs such as the Government/Industry Co-sponsorship of University Research and the Freedom CAR. Private foundations, which provide a comparatively modest portion of the nation's total R&D budget, may nonetheless offer crucial support for types of science that are off-limits to government funds. Disease-focused foundations such as the Juvenile Diabetes Foundation and the American Stroke Foundation are keen to see progress made in hESC research, for example.

While the states are relatively new partners in science policy, they are beginning to take an active role, thus transforming the meaning of "government" funding, and introducing a new distinction—federal versus state. The resulting patchwork of policies may make some regions of the country more desirable than others for scientists, further widening the gap between the "have" and "have not" states (e.g., Massachusetts and South Dakota).

Growing Challenges concerning Intellectual Property

As boundaries shift or disappear, it will probably become more difficult to establish ownership of intellectual property. To review, the Bayh-Dole Act of 1980 declared that any IP developed with federally funded research funds belongs to the university, small business, or nonprofit that conducted the research, not the federal government, as had earlier been the case. Many research universities have formulated policies to ensure that IP developed with their resources and on their time belongs to them. Companies, however, which sponsor more research on university campuses, are interested in owning the resulting IP themselves—not merely licensing it. This of course means that universities partnering with industry have to relinquish royalty streams and potential assets. This is something universities are sometimes reluctant to do and therefore has become a source of growing tension between universities and industry.

Some industry representatives have complained that universities are greedy for licensing revenues. They say that negotiating licensing agreements is arduous because university officials allow dollar signs to obstruct the intent of Bayh-Dole, which was to get technologies resulting from federal research investments into the marketplace and made available to the public. Meanwhile, universities maintain that industry often wants exclusive rights to research results developed at universities. This can, in turn, have adverse impacts on the ability of university researchers to publish research results or to involve students in projects that are proprietary in nature.

Debate over such issues will continue, particularly given the explosion of IP in the life sciences. The University of Rochester sued the pharmaceutical company GD Searle & Co in one well-known example.[12] In April 2000 the university had been awarded a very broad patent covering the method of action of the enzyme cyclooxygenase-2, and thus the class of drugs commonly called COX-2 inhibitors.[13] Two pharmaceutical companies, Merck and Pharmacia (both of which had corporate relationships with Searle), had each by that point already developed and marketed blockbuster medications based around COX-2 inhibitors—Vioxx (which eventually went from blockbuster to millstone) and Celebrex, respectively. Rochester filed a patent infringement lawsuit against Searle, but the case was thrown out by a U.S. district judge, a move that was later upheld by the court of appeals.[14]

Universities sometimes fight hard to secure significant royalties and licensing revenues from the IP they own under Bayh-Dole. At other times, they invoke the law's "experimental exemption" clause, which exempts them from having to license a technology or pay infringement penalties as long as they're only using someone else's IP for academic research or philosophical inquiry.

In 2001, John M. J. Madey, a former Duke University professor, charged the university with patent infringement for its use of a free-electron laser created and patented by Madey. Duke had continued to use equipment based on Madey's technology even after it had terminated his employment there, without entering a licensing agreement (i.e., paying for it). Madey sued and lost in federal district court. The decision was overturned, however, by the court of appeals in 2002; in 2003, the Supreme Court denied the university's request for a review of the case.[15] Importantly, the case, known as *Madey v. Duke*, gave commercial vendors the grounds on which to make patent infringement claims for unlicensed uses of technology by academic and nonprofit researchers, challenging the experimental exemption clause many academics had become accustomed to.

Under Bayh-Dole, the federal government retains what are referred to as "march-in rights." This means that if the government believes a university is not making a good-faith effort to develop a technology resulting from federally funded research for commercial use, the government can "march in" and license that technology as it sees fit.

The government has never invoked its march-in rights, although some outside groups strongly urged the NIH to do so in June 2004 after the pharmaceutical maker Abbott Laboratories significantly increased the price of its HIV/AIDs drug, Norvir. Despite the pressure on the NIH, it ultimately determined that it would be inappropriate to use government march-in rights as a means to control drug prices. Given the high costs of some drugs, however, calls for expanded use of march-in rights as a means of cost containment are likely to continue.

Another area that has received attention is the patenting of research tools, especially in genetics and genomic sciences.[16] There have been instances in which a particular gene sequence is patented, thus limiting the use of that sequence for research purposes and for development, as well as use, of diagnostic tests. An example of this is the patenting of two breast cancer genes, BRCA1 and BRCA2, by Myriad Genetics.[17] The ownership of this IP by Myriad allows the company to charge others for use of the DNA sequence, which in turn has raised concerns about access to technology by not just scientists but also the public. Any use of the BRCA1 or BRCA2 sequence for diagnostic testing requires licensing of the DNA sequence from Myriad, contributing to medical costs.

Finally, future debates over IP will not be limited to industry and universities but are also likely to increasingly involve federal laboratories. Since the end of the Cold War, the national labs have played a growing role in America's economic growth, not least by generating technology transfers and entering into partnerships with private companies.[18] The national labs can enter into one of two general types of agreements with a firm: a cooperative research and development agreement (CRADA) or a work for others (WFO) agreement.[19] The WFO is relatively straightforward: the firm owns the IP if it is a direct sponsor of the research, while the lab owns the IP if the firm is a subcontractor on research sponsored with federal funds. CRADAs, however, require case-by-case negotiation—often arduous—over who owns the rights to new IP, the lab or the firm.

The Increasing Involvement of the Courts in Science Policy

As the distinctions between the roles of R&D performers blur, it is likely that the courts will be drawn in to settle disputes. While science depends upon the sharing of knowledge and free exchange of information, industrial competition relies on proprietary information and knowledge. The conflict emerges as universities, which have traditionally not restricted access to information, delve into applied knowledge work and enter into relationships with industry. The same is true of the national labs. Clear and forward-looking policies must be developed, in the form of both law and agency rules and regulations. But even if such clarity is achieved, the courts will still inevitably settle disagreements over IP and interpret policies more often than they have in the past.

Knowing this, the scientific community has a responsibility to educate the judiciary about science. Judges, like most legislators, come from a legal background. The court system has no inherent scientific expertise. Rather, the courts have traditionally relied on outside experts for scientific advice. This is cause for concern because the courtroom also has no capacity for peer review, and hence no way to assess the quality of expert testimony. Instead, such assessments are often made by scientifically uninformed judges and juries who decide how good the testimony "sounds."

Courts are also likely to become more involved in science policy as science pushes ethical frontiers, and laws are devised to regulate conflicts. It is important to keep in mind that, the more policies and laws are developed to regulate science, the greater the likelihood of disputes over how these laws should be interpreted.

Obtaining Better Metrics to Measure the Success of Scientific Investments

As discussed earlier, it is extremely difficult to measure the returns of specific federal R&D investments. One observer notes:

> Enormous political pressure, which is likely to increase, is being felt for better metrics and approaches to evaluating current and prospective federal R&D programs. Current methods for assessing the economic effects of federal R&D programs are weak. As a result, economists are not well positioned to respond to the political demand for evaluation, nor are they well positioned to assess the consequences of shifts in program design or reallocation of

R&D investment funds among different areas of research. One real danger is that near-term results will overshadow long-term results in such evaluation, influencing the design of program in the area.[20]

The R&D budget is part of the discretionary portion of the national budget. This has two implications: first, funding for R&D is not mandated by law, as are programs like Social Security and Medicare; and second, funding for R&D competes with popular programs from other areas of government, such as Pell grants. Overall, evaluating R&D funding programs accurately—and convincing policymakers of their worth—is a daunting challenge where further economic research is necessary and the development of new metrics essential.

The Growing Challenge of "Who Is Us"

Many countries with traditionally weak R&D capabilities are now catching up with the United States. Because science functions beyond national borders, much of the recent success of nations like India and China is due to the sharing of data and the forging of new collaborations.[21] The widespread dissemination of technology is, in many respects, a good thing. For instance, cell phones allow communications in some of the most remote places of the world; and the Internet and digital libraries enable the sharing of resources with students in developing countries, and better medical facilities in rural areas.

However, along with these global positives come some negatives. Cultures converge, and traditions and uniqueness succumb to homogenization. More immediately, nations that Americans formerly dismissed or ignored are now becoming viable economic competitors and potential military threats. Multinational companies are showing less allegiance to their home countries, increasing the likelihood that the United States will face further challenges to its international competitiveness. A global economy means that people and resources are very mobile, making it cost-efficient for companies to send jobs, including technical jobs, offshore, or even moving entire R&D operations abroad.

Many of the very largest corporations, including those that do R&D, are truly multinational. While their headquarters may be in the United States, these corporations, because of mergers and acquisitions, may maintain offices and facilities all over the globe. When a U.S. company merges with a foreign-owned firm, the new entity often keeps its facilities in both countries, not just because of cheap labor, but for financial or political reasons. Whether a company is American or not has become a matter of opinion. During the time of the existence of DaimlerChrysler Corporation, which was formed in 1998 by the merger of Daimler-Benz AG, a German company, and Chrysler Corporation, an American company, if one were to have asked an American if DaimlerChrysler was an American company, he or she likely would have answered yes. Had any German been asked the same question at the time, the answer would probably have been no. Similarly, many in the United States think of GlaxoSmithKline—formed in 2000 from the merger of Glaxo Wellcome, a British pharmaceutical company, with SmithKline Beecham, an American pharmaceutical company—as American, even though its roots are largely in the United Kingdom.[22] In other words, any corporation can have a very complex history, as the offspring of one or more mergers involving two or more nations. So, who *is* who? Should the U.S. count GlaxoSmithKline's R&D capacity as its own? And to what extent do Americans have a say in the company's affairs?

Competition is also on the rise in education. Other countries are working to build their own systems of higher education, using the U.S. system as a model. This has caused a decrease in the number of international students enrolling in American colleges and universities. An International Institute for Education survey published in November 2004 found that the number of foreign students at U.S. universities had declined by 2.4 percent in academic year 2003–4 from the prior academic year.[23] This decline is attributed to real and perceived problems in obtaining visas, along with increased educational opportunities in the homelands of many international students. Other English-speaking nations have also increased their recruitment of international students, thus making the higher education market still more competitive.[24] Ultimately, America is at risk of losing its own citizens to universities in other countries.

The question of "who is us" is complex.[25] Multinational corporations and offshoring efforts are disrupting national economies. And the globalization of higher education is increasing competition for students worldwide. Increased security measures complicate the picture further. The United States has always welcomed immigrants to its shores, but foreign students, scientists, and engineers wishing to come to the United States now run into new problems; even visiting scholars or conference attendees from abroad have had difficulties. Both real and perceived problems with visas, along with the generally poor foreign image of America's international behavior, have done little to make foreigners feel wanted. In this regard, then, the answer to "Who is us?" may be narrowing.

Balancing Homeland Security and Scientific Needs

The post–September 11 tightening of visa restrictions has raised concerns about the future of U.S. science. These concerns are aggravated by the need of policymakers and members of the security community to be assured that scientific knowledge that could be used for destructive purposes does not get into the hands of people with nefarious intentions. Scientists, especially academic scientists, have had to confront a new category of research: sensitive but unclassified. What does it mean for something to be sensitive but unclassified? Many scientists and university administrators are asking this question.

Other troubling policy moves are also afoot. As we have noted repeatedly, science is characterized by open communication and the sharing of knowledge. But since September 11, new policies restrict the publication of certain kinds of data. Journals now have to review articles to make sure they meet these requirements, and scientists are being asked to exercise caution when choosing what data to include in their manuscripts. Export control policies have also been tightened so as to limit access to particular technologies or to restrict the transfer of knowledge.

We can only lose if the United States closes its borders to foreign talent and impedes the exchange of scientific information. We must work together to balance our concerns about security and national defense with the desire to remain a worldwide leader in scientific research, ideas, and innovation.

The Growing Politicization of Science

As the environment in Washington grows ever more partisan, science is likely to get caught in the crossfire.

Traditionally, science has not been "owned" by either political party. We now see warning signs, however, that science is becoming politicized, and that decisions that should be based on sound science are instead determined by politics. As the editor of *Popular Science* noted in January 2005, "Science and politics do seem to be more and more enmeshed in this country, in ways that hamper scientific investigation and hinder scientific understanding."[26] As an example, the perception that the George W. Bush administration has politicized science has been fueled by reports that administration officials have distorted scientific facts and used political litmus tests to determine whether candidates for supposedly apolitical and scientifically based advisory committees hold "appropriate" political views.[27]

In response to what were perceived as direct attacks on science by the Bush administration, several prominent members of the scientific community launched a major effort to support John Kerry in the 2004 presidential campaign. As a part of this effort, forty-eight Nobel laureates issued a letter to the American people on June 21, 2004, calling Kerry a "clear choice for America's next president" who could "restore science to its appropriate place in government and bring it back into the White House."[28] Meanwhile, the cochairs of "Scientists and Engineers for Kerry" included several prominent scientists and engineers, including Nobel Prize winners Harold Varmus (medicine, 1989), Mario Molina (chemistry, 1995), and Burton Richter (physics, 1976).[29]

If scientists believe that their work is being politicized by the party in power, they may side with the party that is not in power. The danger here is that they may increase the degree to which one party "owns" science—in other words, they will politicize science. Scientists, while as free as other individual citizens to voice their opinions and to support specific candidates for office, as a community must resist engaging in partisan politics. It is incumbent upon scientists as a community to ensure that good science is considered when making policy decisions, whichever party may be in power.

FIG. 20.1 Science politicized.
(From Dave Coverly, Speedbump Comics.)

The Growing Tension between Science and Religion

Science and religion use different methods to answer sometimes-similar questions. The scientific method involves creating a hypothesis, testing it by gathering observable data, and then interpreting those data. It is objective, based on verifying observations and phenomena. Religion, in contrast, is subjective. In this approach, explanations are based on personal intuition, faith and belief, family tradition and experience, or on the authority of a revered prophet or text. Testing and verifying by repetition are not part of the process.

These differences have long set science and religion at odds. In the sixteenth century, Copernicus's model of the cosmos, which put the sun at the center of the universe, was rejected on the religious grounds that it contradicted biblical teachings. A century later, Galileo was charged with heresy for supporting the Copernican system. The two ways of viewing the world—the church's and Copernicus's—are inherently opposed: both make claims on the same notions—how the universe came to be, and humankind's relationship to it. How could they not be in conflict?

While plenty of people who claim some religious affiliation hold moderate beliefs—for example, embracing the Bible's story of creation while also accepting the theory of evolution—a growing number of people are taking the writings of sacred texts literally. At the same time, religious conservatives have taken a much more active role in politics, while for the general population strong religious beliefs are a characteristic they want in their political leaders. In a Pew survey conducted in August 2004, 72 percent of those polled agreed with the statement that the "president should have strong religious beliefs."[30] Another Pew survey, taken in 2003, found a widening gap in "strong religious commitment" between the two main political parties. Since 1988, Republicans have professed an increase in commitment; while, since 2000, Democrats have shown a decrease.[31] To the extent that religion is at odds with science, this divergence is likely to impact science policy, creating problems for science and scientists.

As science pushes ethical boundaries, these communities are likely to diverge. This is no more apparent than in the life sciences, where debate over human embryonic stem cell research and cloning has reached a fever pitch. Yet these arguments are mild compared to what is yet to come. In the biological sciences, for example, more and more is being learned about the origins of humankind, but these findings are at odds with the doctrine embraced by many religious fundamentalists. Behavioral studies on homosexuality and HIV/AIDS are also facing opposition from some religious groups, who view homosexuality as immoral and AIDS as divine punishment. Such groups believe that federal dollars should not be spent in support of studies of the transmission and epidemiology of HIV/AIDS.

Scientific literacy is also at stake in battles between science and religion. As we discussed earlier, the public's scientific literacy is a basic public good; conversely, scientific illiteracy enables Americans to view science as inherently antireligious. Although the general public should seek scientific literacy, members of the scientific community should also reach out, particularly to those whose belief systems contradict scientific findings. As Gilbert Omenn, then president-elect of the Association for the Advancement of Science, noted in 2004:

> We need to engage people who are mistrustful of new technologies and openly address the questions they raise. We need to engage people who do not use, understand,

teach, or even accept the scientific ways of thinking, observing, and experimenting that we value highly. We should be willing to hear and consider the criticisms and different ways of thinking.[32]

The Need for Attention to the Ethical, Legal, and Social Understanding of Scientific Advancement

The increasing prominence of intellectual property debates in science, the growing tension between science and religion, and the potential for increasing politicization of science suggest the need to examine science through an ethical, legal, and social lens. Similarly, as our understanding of science increases and the field itself becomes more complex, debates on such controversial topics as stem cell research require thoughtful analyses that go far beyond the domain of science. Beyond these controversies lie many others that can only be adequately addressed with the benefit of legal and ethical insights. What are the

POLICY DISCUSSION BOX 20.2

Peer Review versus Moral Values: Who Determines Which Research Is Worthy of Federal Support?

In fall 2003, a listing of almost 200 NIH grants compiled by the Traditional Values Coalition (TVC), a conservative religious lobbying group, was leaked by congressional staff. The TVC targeted the grants on the grounds that they were questionable scientific studies that misused taxpayer dollars. The list was composed of grants focused on gaining a better understanding of the social and behavioral aspects of issues such as HIV/AIDS transmission, homosexuality, pregnancy prevention, and drug abuse and other mental health issues. All of the grants on the list had been evaluated and awarded funding only after having undergone the rigorous NIH peer-review process, in which they were evaluated by other scientists and determined worthy of funding based upon their scientific merit.[33]

The release of this listing came on the heels of the narrow defeat of an amendment in July 2003 to defund five similar research grants already approved by the NIH. This amendment, which was defeated by only two votes, was offered on the House floor in July 2003 by Congressman Pat Toomey (R-PA) during the annual appropriations debate. Said Toomey during the debate on

the amendment, "who thinks this stuff up? And worse, who decides to actually fund these sort of things?"[34]

Members of the scientific community came to the defense of the grants on the grounds that they represented good and important science that was, in fact, valuable to understanding how, for instance, to slow the spread of HIV both in the United States and abroad. The scientific community maintained that these grants were targeted because of ideology and moral values, not because they were unsound science.[35] Speaking in defense of the studies targeted by the amendment and against congressional tinkering with the NIH process of scientific peer review, Rep. David Obey (D-WI) noted that "the day that we politicize NIH research, the day we decide which grants are going to be approved on the basis of a 10-minute debate in the House of Representatives with 434 of 435 members who do not even know what the grant is, that is the day we will ruin science research in this country."[36]

What do you think the respective roles of politicians and scientists should be in determining which science is funded and which is not? How do you think moral and religious values might affect politicians' views on which science should be funded? Do you think that peer review alone should guide how federal science funds are spent?

rules governing a researcher working with human subjects? Do research animals have rights?

The social sciences also have a role to play. Social scientists model the behavior of people, entities, or organizations in order to better understand how these interact with one another, and even to predict future outcomes. It is increasingly important that social and natural scientists work together both in the policy arena—for example, to analyze the impact of research investment on economic growth—and in basic research, as when we try to determine the interactions of biological, social, environmental, and psychological influences on human behavior.

Public support for science depends directly on whether it sees science as a shared undertaking, pursued with the intent of learning more about the universe in which we live, with a keen focus on improving the human condition, and abiding by the norms of society, including those governing the ethical standards of the time. If scientists forget this reason for public support of funding research, they do so at their own peril.

The Ongoing Problem of Attracting U.S. Students to Science

One of the main goals of national science policy should be to educate and retain more American students in the sciences and to reduce our dependence on foreign talent. In order to do this, we have to better understand how to maintain the pipeline of talented students, and we must also examine the mechanics of students' choice. Why do students choose careers in business, medicine, or law, instead of physics, chemistry or biology?

We need a surer grasp of the cultural reasons why we do not sufficiently value scientific training and literacy. We must become a society in which every citizen is expected to have at least a basic scientific understanding. Why do we find illiteracy—the inability to read—intolerable among the general population, yet find it acceptable for professionals and policymakers to be scientifically illiterate? Society has to ensure not only that people can read, but that they have a basic understanding of how their world works, and a fundamental proficiency in mathematics and key areas of science. Such lofty goals are not new, of course. The following passage is from Vannevar Bush's influential 1945 report:

We live in a world in which science lies at the very roots of community, and a mastery of scientific thinking grows more and more indispensable for the successful practice

of the arts and life. The culture of the modern age, if it is to have meaning, must be deeply imbued with scientific ways of thought. It must absorb science, without forsaking what is of value in older ways. . . .

It is a question, not of substituting the scientific culture for that which has gone before, but of reaching a wider appreciation in which the sciences in their modern development fall into their due place.[37]

The University-Government Partnership

The compact between the government and universities is critical to the health of the nation's research enterprise as a whole, and should be reinvigorated. Each of the partners we have identified in this book—the national labs, industry, state governments, and the public—at some point finds itself a part of a collaboration with the federal government and universities. The health of these extended partnerships, then, is largely dependent on the health of the government-university compact.

In recent years the government has started to treat universities more like contractors than true partners, forgetting that government once relied on universities to perform research as a service. At the same time, universities are in danger of losing sight of their mission to pursue excellence in both education and scientific research. In trying to expand university revenues and successfully compete with their peers for federal dollars, American universities must not lose interest in the attributes that have made them the envy of the world. Because this system has always recognized diversity among institutions and promoted excellence over equity, the system's built-in competition has provoked some efforts to circumvent merit-based science funding through the more parochial and politicized practice of congressional earmarking. Such efforts could compromise the excellence of publicly supported scientific research.

The government contributes to the partnership by, among other things, supporting large, interdisciplinary initiatives. In doing so, it helps to break down old disciplinary boundaries on campus, at national labs, in industry, and within the government itself. The federal government can also promote interactions among these groups and state governments, fostering a more effective sharing of resources and knowledge, and promoting economic growth and well-being.

While we do not suggest consolidating the scientific activities of the various research agencies into a central Department of Science, a great deal can certainly be done to better coordinate efforts across agencies. Such coordination is clearly needed in homeland security, for example, where—despite the recent consolidation of relevant agencies—much of the pertinent science is still being conducted outside DHS's S&T directorate.

So what office is best suited for this purpose? The OSTP, we would argue, possesses the best combination of science and policy expertise for the job. Moreover, this office is already authorized to lead interagency efforts to develop sound science policies and budgets; to work with the private sector to ensure that federal investments contribute to economic prosperity, environmental quality, and national security; and to build lasting partnerships among federal, state, and local governments and the scientific community.[38] In order to take the lead, OSTP must be given the authority and resources to do the job right. We would also suggest that in this era when science and technology are of such importance to our future national good, the President's science advisor (also the director of the OSTP) be made a member of the president's Cabinet.

The Federal Budget and Budget Process Challenges

In recent years, budget battles between the White House and Congress have grown increasingly intense. The havoc that can be caused for science programs, which require stability and long-term funding commitments, by these budget battles can be significant.

Unfortunately for science, as the amount of the federal budget devoted to mandatory spending programs such as Social Security and Medicare expands to accommodate the baby boom generation as it hits retirement age, these budget squabbles are likely only to grow more intense. Expected growth in mandatory spending means there will be fewer funds available in the budget for discretionary programs, including scientific research. Thus, a major challenge for the implementation of sound science policy will be the uncertainty surrounding the federal budget and the political volatility surrounding the budget process itself.

In this environment, while the "will" to support science may exist in both the Congress and the White House, finding a "way" to actually fund new science programs is likely to become increasingly difficult. Indeed, the budgetary environment in Washington has become so unpredictable that existing science programs—especially large and capital-intensive science projects—are likely to be continually subject to budget shortfalls and congressional

cuts, with no guarantees they will receive the funding they require from one year to the next.

The budgetary challenges facing the formulation of effective science policy were perhaps best illustrated recently by the disappointment experienced by the research community with the final funding outcomes for research at NSF, DOE, and NIST contained in the final FY2008 Omnibus Appropriations Bill. In what seemed to be a year when the White House and Congress all agreed that research agencies, including the NSF, DOE Office of Science, and NIST, deserved significant funding increases, these increases did not materialize. This left the scientific community, as well as many in the business community who had launched a strong campaign to see increases in research funding, surprised and asking the members of Congress, "What went wrong?" Just when the community finally thought that research was receiving the attention it deserved, that science might finally be the "bride" and not the "bridesmaid," disagreement between the White House and Congress over the total dollars that should be made available for overall discretionary spending resulted in essentially flat budgets for all science agencies after accounting for inflation.

Indeed, the final funding levels for research approved by Congress and the White House for FY2008 left many in the scientific community stunned given that the president and the Democratic Congress had both touted key initiatives—the president's American Competitiveness Initiative and the Democratic Innovation Agenda—aimed at boosting research funding, especially for agencies that supported the physical sciences. Indeed, in July 2007, with great fanfare the Congress passed and the president subsequently signed into law the America COMPETES Act (the full name of the bill was the America Creating Opportunities to Meaningfully Promote Excellence in Technology, Education, and Science Act), which authorized additional agency research funding and contained a number of new programs to spur U.S. innovation and enhance STEM education at all levels.

Despite all of the positive talk that surrounded the America COMPETES legislation and its unanimous and uncontested passage in both the House and Senate, decisions made in the final FY2008 Omnibus Appropriations Bill not only failed to provide additional funding, but they actually called for a termination of funds for major science projects, including the U.S. contribution to the International Thermonuclear Reactor Experiment (ITER) and the International Linear Collider (ILC). According to DOE estimates released not long after the approval of the bill, these cuts would also result in the layoff of over five hundred persons employed in scientific, technical, and administrative positions.

So, what went wrong? How was it that Congress and the president at the end of the day were unable to support the very programs that they had both suggested were a major priority? The failure here was not that they did not, in fact, believe in what they said but rather the result of a systematic failure of the budget and appropriations process itself. No longer is Congress able to pass its appropriations bills in *regular order* (i.e., separately and on time before the start of the new fiscal year), but instead they have been forced to bundle bills at the end of the year into one large omnibus, or combined, funding bill. The manner in which these comprehensive bills are developed and finalized often forgoes work previously done by the individual appropriations subcommittees and results in ill-informed and last minute funding decisions, some of which are made only by congressional staff, with little input from members of Congress themselves. These decisions can end up having a significant negative impact on science (as well as on other federally supported programs).

In the future, significant thought will need to be given to how commitments are made and kept to try to insulate science from such budgetary uncertainty and volatility. In the past, some such as Senator Pete Domenici (R-NM) have proposed moving from an annual budget and appropriations process to a biennial cycle on the grounds that it would allow for more effective long-range planning and improved congressional priority setting and oversight.[39] Reviews of such proposals are mixed, but it seems that there is some merit in trying to seek adjustments to the budget process that would ensure greater funding stability for science.[40] At the same time, the only real solution to the budgetary woes that confront the scientific community and the entire country will be meaningful entitlement reforms that try to address ballooning mandatory spending requirements. If such reforms are not undertaken, science funding is likely to continue to experience modest growth at best.

Policy Challenges and Questions: Searching for the Next *Sputnik*

Those who hope for another *Sputnik* or major transformation in science policy that will alter the conduct of science in the United States may have to wait a long time. If there is one major shift that would improve the quality of future science policy, it is most likely related, not to a specific policy, but to the people who make it.

The two cultures—scientific and nonscientific—identified by C. P. Snow still exist, and the divide between them is evident in the split between those who "do science" and those who "make policy."

In the future, scientists should more directly involve themselves in the policy-making process. Some steps have already been taken in this direction: the U.S. State Department now invites scientists as science fellows to participate in the deliberation of department's staff; fellows, many of whom are also chosen by the AAAS, also serve yearlong rotations on Capitol Hill and in other federal agencies.

But the fellows are not permanent federal employees. They will eventually finish their terms and go home, leaving behind congressional, agency, and executive branch officials and staff with the same weak backgrounds in science that they had before. Scientific literacy is also lacking among judges, members of Congress, and other high-ranking elected and appointed officials.

At a 2004 campaign event for U.S. representative and former plasma physicist Rush Holt, New Jersey governor James Edward McGreevey and former president Bill Clinton joked about the uniqueness of Holt's scientific background.[41] McGreevey quipped that when Holt was first elected to Congress, his presence "substantially raised the IQ level of the House of Representatives," while Clinton trotted out a well-worn quip about Holt actually being a "rocket scientist," and recalled that as president he used to remark that "we have to watch Rush Holt, he actually knows something. . . . That can be a dangerous thing in Washington."[42]

If scientists need to speak out more, what should they say? The first rule is to avoid jargon and simplify one's language so that it can be understood by nonspecialists. It is also important that scientists realize that what they and their colleagues find exciting may not stir great passions among the general population. The public and their representatives will want to know about the likely return on their research investment, not the details of a particular finding. In this sense, scientists need to gain more confidence when speculating on the results of their work, while realistically assessing their work's relevance to specific societal issues.

Scientists also need to learn the importance of accessibility, visibility, and accountability. For far too long, many of us in the scientific community have secluded ourselves in the laboratory, feeling no responsibility to explain how we use government funds or to ascertain whether these funds are producing valuable results. This has led to the doubly unfair accusation that "scientists with their belief in their God-given right to taxpayer dollars are little more than 'welfare queens in white coats.'"[43] This image certainly does not help scientists' efforts in policy circles.

Politicians, in turn, can learn a great deal from scientists. Most obviously, but also most importantly, they can learn more science. The English literary figure Samuel Johnson once noted that "curiosity is one of the permanent and certain characteristics of a vigorous mind," and while it is true that many politicians and policymakers have only a limited background in science, they have the same innate curiosity as any other human beings.[44] The interest is there: the material simply needs to be presented in a comprehensible way that captures their imaginations.

There is value in knowledge. Scientists understand this inherently. It is important, however, for the scientific community to remind policymakers of this value with real-world examples of what has been accomplished thanks to past investments made by the government in science. Scientists can also teach policymakers that sometimes the correct solution is driven by asking the right questions—an investigative precept that characterizes the scientific process, but is not necessarily accepted by policymakers.

Finally, increased interaction between scientists and policymakers makes good policy. Many decisions—science-related or not—would be greatly enriched by scientific input. While politicians often think about what they need to do to be reelected tomorrow, scientists tend to work with a more distant horizon in mind. Scientists must learn to appreciate their work's social relevance, then, while policymakers must learn when not to insist that research produce immediate benefits.

While policymakers have an overall responsibility to become more scientifically literate, scientists, too, must become more literate in policy. Scientists have no excuse for not knowing who their representatives are, how the budget process works, or how a bill becomes a law. In the end, the scientific community must encourage its members to take an active role in the policy-making process—a choice that, until now, has not been significantly encouraged or rewarded.

It is time to overturn the expectation that the only thing a scientist is good at is science. To retain that expectation is to ignore the critical dependence of science on government policies and public goodwill.

NOTES

1. There is not a clear agreement about who exactly said this, Bohr or Berra, and some have attributed it to other sources, including Mark Twain and Sam Goldwyn. See David Ketenbaum, "Profile: Accuracy in the Internet When It Comes to Facts," *All Things Considered* (Washington, DC: National Public Radio, transcript, April 7, 2000).

2. Sue Schuurman, "John Glenn Orbits Earth," *Weekly Wire*, February 23, 1998, http://weeklywire.com/ww/02-23-98/alibi_skeleton.html (accessed May 28, 2007).

3. Committee on Science, Engineering, and Public Policy, *Rising above the Gathering Storm*; Friedman, *The World Is Flat*.

4. Domestic Policy Council, Office of Science and Technology Policy, "American Competitiveness Initiative"; House Democrats, *The Innovation Agenda: A Commitment to Competitiveness to Keep America #1* (Washington, DC, November 2005), http://speaker.gov/pdf/IA.pdf (accessed June 7, 2007).

5. Edward J. Larson and Larry Witham, "Leading Scientists Still Reject God," *Nature* 394, no. 6691 (1998): 313.

6. David Van Biema, "God vs. Science," *Time,* November 13, 2006.

7. Stephen Jay Gould, *Rock of Ages: Science and Religion in the Fullness of Life* (New York: Ballantine, 1999), 5, 52, etc.; Douglas O. Linder, Web site *Stephen Jay Gould,* http://www.law.umkc.edu/faculty/projects/ftrials/conlaw/gouldsj.html (accessed May 28, 2007); Van Biema, "God vs. Science."

8. Francis Collins, as told to Jim Hinch, "His Beautiful World," *Guideposts,* December 2006, 44.

9. Gunther Siegmund Stent, *The Coming of the Golden Age: A View of the End of Progress* (Garden City, NY: Natural History Press, 1969).

10. John Horgan, *The End of Science* (London: Little, Brown, 1996), 11.

11. "Battelle Energy Alliance Wins Idaho National Laboratory Contract," Battelle News Release, November 9, 2004, http://www.battelle.org/news/04/11-09-04IdahoLabWin.stm (accessed May 28, 2007).

12. G. D. Searle & Company was a subsidiary of Pharmacia, which, by the time of the case reached the court of appeals, had merged with Pfizer.

13. Ted Agres, "Universities War with Big Pharma," *The Scientist* 16, no. 16 (August 19, 2002): 56–57.

14. Peg Brickley, "COX-2 Patent Case Thrown Out," *The Scientist,* news release, March 7, 2003; *University of Rochester v. GD Searle & Co., Inc., Monsanto Company, Pharmacia Corp, and Pfizer Inc.,* 03-1304 (Fed. Cir. 2004).

15. For background see "Ruling on Research Exemption Roils Universities: Finding of No Academic Privilege from Infringement May Lead to New Legislation," *National Law Journal,* December 16, 2002, http://www.foley.com/publications/pub_detail.aspx?pubid=1416 (accessed May 28, 2007); Peg Brickley, "Experimental Use Challenge," *The Scientist,* News, June 25, 2003; "Supreme Court Rejects Appeal in Experimental Use Case," American Council on Education, Department of Government Relations and Public Affairs, July 14, 2003.

16. See for example Richard Gold, Timothy A. Caulfield, and Peter N. Ray, "Gene Patents and the Standard of Care," *Canadian Medical Association Journal* 167, no. 3 (2002): 256–57; Graham Dutfield, "DNA Patenting: Implications for Public Health Research," *Bulletin of the World Health Organization* 84, no. 5 (2006): 388–92; Claire T. Driscoll, "Evolving NIH Patent and Licensing Policies and Practices for Genomic Inventions in the Post–Human Genome Project Era," presentation, DNA Patent Database, http://dnapatents.georgetown.edu/resources/ClaireDriscollNHGRINIHPresUofPenn030303.ppt#328,1, slide 1 (accessed June 2, 2007).

17. See Shubha Ghosh, "Myriad Troubles Facing Gene Patents," *Preclinica* 2, no. 5 (2004): 300; Eliott Marshall, "The Battle over BRCA1 Goes to Court: BRCA2 May Be Next" *Science* 278, no. 5345 (December 12, 1997): 1874; "Bioethics and Patent Law: The Case of Myriad," *Wipo,* August 2006, 8–9.

18. See, for example, Phillip Bond, "Partners on a Mission," presentation, Report Rollout and Panel Discussion, Rayburn House Office Building, Washington, DC, November 20, 2003, http://www.technology.gov/Speeches/p_PJB_031120.htm (accessed May 28, 2007); Task Force on Alternative Futures for the Department of Energy National Laboratories, *Alternative Futures;* Shawn Shepherd, "Labs Partner in the New Economy—Los Alamos National Laboratory; Sandia National Laboratory," *New Mexico Business Journal,* November 2000.

19. For details on the agreement types see Steven C. Wiener, ed., and Laboratory Coordinating Council, *Doing Business with the National Labs* (Industrial Technologies Program, Department of Energy, January 7, 2003).

20. Bronwyn H. Hall, "The Private and Social Returns to Research and Development," in *Technology, R&D, and the Economy,* ed. Bruce Smith and Claude Barfield (Washington, DC: Brookings Institution, 1996), 168.

21. Kathleen Walsh, *Foreign High-Tech R&D in China: Risks, Rewards and Implications for US-China Relations* (Washington, DC: Henry L. Stimson Center, 2003), 20–23.

22. SmithKline Beckman, an American company, and the Beecham Group, a British company, merged in 1989.

23. The number of international undergraduate students enrolled in U.S. higher education institutions declined by 5 percent, while the number of international graduate students enrolled in U.S. higher education institutions increased by 2.5 percent. Institute of International Education, "International Student Enrollment Declined by 2.4% in 2003/2004," press release, November 10, 2004, http://opendoors.iienetwork.org/?p=50137 (accessed June 1, 2007).

24. Ibid.

25. For further discussion, see Reich, *The Work of Nations,* chap. 25.

26. Mark Jannot, "Political Science," *Popular Science,* January 2005, 4.

27. Daniel Smith, "Political Science," *New York Times Magazine,* September 4, 2005, http://www.nytimes.com/2005/09/04/magazine/04SCIENCE.html?8hpib (accessed May 28, 2007); see also Ester Kaplan, *With God on Their Side: George W. Bush and the Christian Right* (New York: New Press, 2005), chap. 4.

28. "An Open Letter to the American People," June 21, 2004, http://www.gwu.edu/~action/2004/kerry/kerrynobel062104.html (accessed June 1, 2007).

29. Geoff Brumfiel and Emma Marris, "Nobel Laureates Spearhead Effort to Put Kerry in the White House," *Nature* 430, no. 7000 (2004): 595.

30. Pew Forum on Religion and Public Life, *GOP the Religion-Friendly Party* (Washington, DC: Pew Research Center for the People and the Press, August 24, 2004), 2.

31. Pew Research Center for the People and the Press, *Evenly Divided and Increasingly Polarized: 2004 Political Landscape* (Washington, DC: Pew Research Center for the People and the Press, November 5, 2003), 3, 79.

32. "Omenn Sees 'Grand Challenges and Great Opportunities'," *Science* 306, no. 5705 (December 24, 2004): 2204–5.

33. Jocelyn Kaiser, "NIH Roiled by Inquiries over Grants Hit List," *Science* 302, no. 5646 (October 31, 2003): 758.

34. Stacye Bruckbauer, "Politics and Health: Recent Debate in Congress Questions Some Government Grants," *Journal of the National Cancer Institute* 96, no. 3 (February 4, 2004): 169–70.

35. Alan I. Leshner, "Don't Let Ideology Trump Science," *Science* 302, no. 5650 (November 28, 2003): 1479.

36. Bruckbauer, "Politics and Health," 169–70.

37. Bush, *Science—the Endless Frontier,* 151, 152.

38. Office of Science and Technology Policy, "About OSTP, What We Do," http://www.ostp.gov/html/_whatwedo.html (accessed May 28, 2007).

39. "Biennial Budget Urged," *Issues in Science and Technology,* Spring 1997, http://www.issues.org/13.3/hill.htm (accessed January 20, 2007); American Association for the Advancement of Science, "Domenici Proposes Biennial Budget Legislation," *Science and Technology in Congress,* March 1997, http://www.aaas.org/spp/cstc/pne/pubs/stc/bulletin/articles/3-97/bienbudg.htm (accessed January 20, 2007).

40. Congressional Budget Office, *Biennial Budgeting* (Washington, DC: U.S. Congress, February 1988); Stan Collender, "Budget Battles: The Reason for Biennial Budgets," *Government Executive.Com,* March 8, 2000, http://www.govexec.com/dailyfed/0300/030800bb.htm (accessed January 20, 2008).

41. In his remarks at the 2005 William D. Carey Lecture made at the 30th AAAS Science and Technology Policy Forum, Washington, DC, Loews L'Enfant Plaza Hotel, April 2005, Representative Holt joked about the bumper stickers his constituents use, "My Representative is a Rocket Scientist." He also half-jokingly made the point that he is not a rocket scientist. He is a physicist. However, because many in the public perceive science, any science, to be incomprehensible, all scientists are placed in a distinct separate category to which any and all science-y questions should be directed. An example Representative Holt used was the anthrax attacks of 2001. His colleagues insisted on turning to him for a better understanding of anthrax as disease, its causes and dissemination, and so on. Holt, who is not a microbiologist, knew no more about these questions than did other members of Congress. The difference was that he was not hesitant to do research to find out the answers, whereas his colleagues had the attitude "It's science—I won't understand it anyway."

42. Lilo H. Stainton, "Ex-President Stumps for Kerry, Holt, His Own Book: Clinton Visits Garden State," *Home News Tribune,* August 5, 2004, 1.

43. Begley, "Gridlock in the Labs," 44.

44. "The Quotations Page," s.v. "Samuel Johnson," Quotation 28785, http://www.quotationspage.com/quote/28785.html (accessed June 1, 2007).

Index

Italicized page numbers indicate figures and tables.

ing twice" for R&D, 148; peer review vs. moral values, 355; philosophical differences, 20; polio virus synthesis, 317; science and religion, 348; scientific infrastructure funding, 224; scientist- vs. society-driven decisions, 10; self- vs. governmental regulation of research ethics, 234; S&E workforce, compared internationally, 281; S&E workforce visas, 285; social compact, 338; stem cell research, 160; STEM program effectiveness, 269; student tuition as funding for research, 101

policymakers: executive branch as, 27–39; iron triangle of, 53–54, 70n13; judicial branch as, 25–26, 42–44; legislative branch as, 39–42; motivations of, 25, 40, 228; overview of, 25–27; public interests and, 175; on R&D tax credit, 145–46; response to events, 345–46; scientific vs. political choices of, 167; scientific vs. societal challenges for, 331–32, 337–39; scientists' dialogue with, 69, 326; scientists vs. society as, 10; STEM education policy role of, 260–62; university-government partnership concerns of, 113; web of, 54

policy-making: congressional committees' role in, 56–57, 58–59; focus of, 345; key decisions in, 54–55; mechanisms for, 55–67, 56; metaphors for, 25, 47n1, 53; models of, 53–54; overview of, 25–27, 52; web of interests in, 54

Policy on Global Affairs Division (NRC), 44

Policy Research and Analysis (PRA) Division (NSF), 279–80, 292n18

policy support organizations, 46. See also ad hoc advocacy groups and coalitions; advisory groups and reports

polio research, 196n55, 317

political scientists: policy-making view of, 53–54

politics: data falsification case and, 231–32; eVoting initiatives in, 139; of local votes on evolution, 265; of media coverage of science, 171, 172; of regulations and rules, 67, 68; of science, 353–54; of science education legislation, 260

Popular Science (periodical), 173, 353

pork and pork barreling. See earmarks and earmarking

Porter, John Edward, 69, 82

postdoctoral researchers (postdocs): NATO program for, 298; pros and cons of, 287–88; research role of, 98–99, 281; "Responsible Conduct of Research" course for, 244; STEM residencies for, 256–57; time spent as, 279; U.S. programs' prestige and, 255

PRA (Policy Research and Analysis) Division (NSF), 279–80, 292n18

Pratt School of Engineering (Duke University), 281

PRCs (Primate Research Centers), 222, 226

Preparing for the 21st Century (1997), 257

Prescription Drug User Fee Act (PDUFA, 1992), 139

president: agency-creation powers of, 18; appointments by, 55–56; directives and executive orders of, 62–63; as NSTC chair, 30; persuasive power of, 48n11; science advisor's relationship with, 28–29, 30, 48n11; science policy role of, 27–28, 55; signing and veto authority of, 55, 60, 66; statement of administration policy of, 60; State of the Union address, 64; STEM education policy role of, 261–62. See also executive branch; executive orders; presidential directives; specific presidents

presidential directives: function of, 56, 62–63; on national security, 320–21; on student visas, 323; on university-government partnership, 110

President's Commission for the Study of Ethical Problems in Medicine and Biomedical and Behavioral Research, 240

President's Council of Advisors on Science and Technology (PCAST): establishment of, 62; expansion of, 48–49n23; reports of, 63; science policy role of, 29–30; S&E workforce initiative of, 280; specific recommendations of: global IT manufacturing, 302; R&D tax credit, 146; research funding, 79–80, 82; technology transfer, 107

President's Council on Bioethics, 242

President's Information Technology Advisory Committee, 48–49n23

President's Management Agenda (PMA), 31, 77, 78, 341

President's Science Advisory Committee (PSAC), 28, 29

President's Scientific Research Board (PSRB), 20–21

Preventing and Defending against Clandestine Nuclear Attack (DSB, 2004), 187

Primate Research Centers (PRCs), 222, 226

primeval soup model, 53, 69n9

print media: public confidence in, 170; science coverage of, 170–74; STEM education efforts of, 258–59

privatization, of federal laboratories' function, 130

Procter and Gamble, 157

professional journals: articles in, by country, 300–301, 301; bibliometric uses of, 74, 87n15, 150n4, 300–301, 301; data falsification case in, 231–32; founding of first scientific, 230, 244n9; function of, 207; gift-authors in, 232; peer review of, 230; plagiarism in, 230–31; post-9/11 discourse in, 320; publication requirements of, 245n23; publication restrictions on, 302, 353

Professional Opportunities for Women in Research and Education program (NSF), 289

professional organizations: auto industry, 138; information technology, 139; international members of, 298; outreach activities of, 176, 258; pharmaceutical research, 139; post-9/11 discourse in, 320; science policy role of, 45; science writing, 171; scientific literacy concerns, 165; semiconductor industry, 143–44

Program Assessment Rating Tool (PART), 31, 33, 78–79, 341

program associate directors (PADs), 31, 64

Program for International Student Assessment (PISA), 259

Progressive Policy Institute, 46, 155

Project BioShield, 136

Project HINDSIGHT, 74, 196n42

proteome (proteins in human cells), mapping of, 204

PSAC (President's Science Advisory Committee), 28, 29

pseudoscience, defined, 168–69

PSRB (President's Scientific Research Board), 20–21

public: digital divide in, 263; expectations of, 315; haves and have-nots in, 161–62, 263; industry relations with, 147; media's impact on, 170–74; national security vs. personal freedoms of, 310–11, 315, 316–18; as "paying twice" for R&D, 148; policymakers' relationship with, 175; post-9/11 neglect of science by, 4; R&D support of, 319; science policy role of, 55; in science policy web, 54; scientific community and, 170, 170, 174–75, 176; scientific literacy of, 165–66, 167, 250, 262, 355; sources of information for, 171, 173; S&T support from, 169–70; threats to gathering places of, 312; university relations with, 97; university technology transfer and, 107

—attitudes and beliefs: animals in research, 237–38; conflict of interest, 236; environment, 167–68, 168; ethics, 228; pseudoscience, 168–69; religion, 354; science, 166–67, 270, 349, 354–56

public good concept, 73, 86n8